WCDMA (UMTS) DEPLOYMENT HANDBOOK

WCDMA (UMTS) DEPLOYMENT HANDBOOK

Planning and Optimization Aspects

Editors

Christophe Chevallier
Christopher Brunner
Andrea Garavaglia
Kevin P. Murray
Kenneth R. Baker

All of QUALCOMM Incorporated
California, USA

John Wiley & Sons, Ltd

Published in 2006 by John Wiley & Sons Ltd, The Atrium, Southern Gate, Chichester, West Sussex PO19 8SQ, England

Telephone (+44) 1243 779777

Email (for orders and customer service enquiries): cs-books@wiley.co.uk
Visit our Home Page on www.wiley.com

Reprinted September 2006

Other Wiley Editorial Offices

John Wiley & Sons Inc., 111 River Street, Hoboken, NJ 07030, USA

Jossey-Bass, 989 Market Street, San Francisco, CA 94103-1741, USA

Wiley-VCH Verlag GmbH, Boschstr. 12, D-69469 Weinheim, Germany

John Wiley & Sons Australia Ltd, 42 McDougall Street, Milton, Queensland 4064, Australia

John Wiley & Sons (Asia) Pte Ltd, 2 Clementi Loop #02-01, Jin Xing Distripark, Singapore 129809

John Wiley & Sons Canada Ltd, 6045 Freemont Blvd, Mississauga, ONT, L5R 4J3, Canada

Wiley also publishes its books in a variety of electronic formats. Some content that appears in print may not be available in electronic books.

British Library Cataloguing in Publication Data

A catalogue record for this book is available from the British Library

ISBN-13: 978-0-470-03326-5 (HB)
ISBN-10: 0-470-03326-6 (HB)

Typeset in 10/12pt Times by Laserwords Private Limited, Chennai, India.

This book is printed on acid-free paper responsibly manufactured from sustainable forestry in which at least two trees are planted for each one used for paper production.

Contents

List of Contributors

The technical content of this book was developed by:

Kenneth R. Baker
Christopher Brunner
Patrick Chan
Christophe Chevallier
Andrea Forte
Andrea Garavaglia
Kevin P. Murray
Sunil Patil

The content was edited by:

Diana Martin
Lynn L. Merrill
Donald Puscher

Figures were created by the content developers and by:

Max Campanella.

All from QUALCOMM Incorporated.

Foreword

Mobile wireless communications has already dramatically affected our lives, and will continue to do so as usage, services, and coverage rapidly expand with the adoption of the Third Generation of cellular wireless, and the transition to Internet-protocol-based networks.

First Generation cellular networks used analog FM and circuit switching. Despite low voice capacity, uneven quality, limited roaming, and bulky, expensive handsets with limited battery life, the rapid increase of voice subscribers necessitated the adoption of digital transmission technology.

Second Generation (2G) networks, including TDMA-based GSM, PDC, IS-54, and CDMA-based IS-95, allowed rapid expansion of voice subscribers and the introduction of some data services, including short message service (SMS). These Second Generation digital technologies featured advanced coding and modulation, offering greater voice capacity and quality, and supporting digital control channels. The result? More robust and secure signals, smaller and lower-power handsets, enhanced roaming, and a rapid expansion of subscribers worldwide. Even with the limited data capabilities of 2G technology, it became clear that a next generation of cellular networks should focus on even greater capacity, high speed data, and increased reliance on packet switching.

Third Generation (3G) wireless, encompassing three forms of CDMA–CDMA2000®, including 1X and EV-DO; WCDMA, also called UMTS and 3GSM; and most recently TD-SCDMA–has been introduced by many operators and is rapidly gaining subscribers. Both plug-in cards and integral modems are supporting broadband mobile communications directly to laptops. An abundance of powerful handsets are now reaching the market, which support a wide variety of services including music, streamed and stored video, multiplayer games, multiparty instant messaging, and location-based services.

Such growth in usage and applications poses great challenges for the network operators, test equipment vendors, infrastructure manufacturers, and the technical staff that plan, deploy, and operate these networks. This book focuses on the knowledge needed to effectively deploy Wideband CDMA (WCDMA) networks, much of which has been either publicly unavailable or widely scattered across various journals and other sources. In gathering and distilling this knowledge in a readable and coherent form, the authors have achieved their goal of further speeding the deployment and optimization of WCDMA networks.

Third Generation cellular networks will enhance our lives in many ways, rapidly reaching every part of the world and supporting education, business, entertainment, health, and government. The demand for knowledgeable practitioners will continue to grow. This book should provide welcome assistance.

We have come a long way. I look forward to the excitement of further rapid change.

Irwin Mark Jacobs
Chairman of the Board
QUALCOMM Incorporated

Preface

In our day-to-day activities, as part of the Engineering Services Group of QUALCOMM®, we consult with network operators throughout the world. In working with them, we have realized that operators repeatedly face the same four challenges: improving RF optimization, properly tuning system parameters, increasing the reliability of inter-system transitions, and providing better indoor coverage. These issues, among others, cannot be resolved simply by studying the communication standard; consequently, they have not been widely addressed in the literature.

In this book, using the experience we have gained from performing many network assessments, we focus on the day-to-day tasks and real world choices that confront operators. We have chosen to minimize paraphrasing of the standard. This is not to say that we disregard the ample documentation written by the Third Generation Partnership Project (3GPP), also known as the standard. We do refer to the standard throughout this book but rather than present its concepts in a dry manner, we introduce only the sections that readers can use to deepen their knowledge on specific topics. We selected these topics to help network planners and optimization engineers make a better transition from GSM to WCDMA while understanding how to perform the required tasks.

This volume attempts to provide as many answers as possible to the complex questions that planners or engineers encounter in their daily activities. As we were writing, we had to make difficult choices about what to include. Without these choices, of course, we would still be writing. Here are the basic questions that we tried to answer in each chapter:

- **Introduction to UMTS networks.** What nodes are necessary in a WCDMA network? What are their basic functions? What is WCDMA anyway? What differentiates WCDMA from other technologies, such as GSM? What are the key terms and concepts of the technology?
- **RF planning and optimization.** What is a typical Link Budget for the different services offered in WCDMA? Is the Downlink or the Uplink limiting? What are the main factors that determine the coverage? How can the coverage of a WCDMA network be qualified?
- **Capacity planning and optimization.** What is the capacity of a WCDMA cell? How does soft handover affect the capacity of a WCDMA network? How do the different services affect the overall capacity? How can the capacity of the network be maximized? Will microcells affect the capacity of the network?
- **Initial parameter settings.** What are the most important parameters to focus on? What is a good starting point for each parameter? How can you verify the values that are broadcast, and where?
- **Service optimization.** How should the optimization process be started? What are the basic procedures that will affect all services? What should you look for to resolve

typical failures? What differs from one service to another? Do any parameters apply only to particular services?

- **Inter-system planning and optimization.** Why rely on other systems? When should you start looking at inter-system issues? What parameters are involved in inter-system changes? What are good starting points for their respective settings?
- **HSDPA.** What is HSDPA? What advantages does it offer compared to a WCDMA (Release 99) network? How does it differ? How and where should HSDPA be deployed? What parameters are available in HSDPA? How do these parameters affect the coverage and capacity of the entire network?
- **Indoor coverage.** Why is indoor coverage different? When should indoor coverage be provided? How can it be achieved and optimized?

By the time you have read this book, you will no doubt be ready to ask several more questions. Hopefully, with the aid of this book, you will have the skills to find the answers you need.

Acknowledgments

Writing a book of this scope is a major undertaking, one that perhaps cannot be fully appreciated by those who have not wrestled with it. It cannot be achieved without a great deal of help. Adequately acknowledging all the people who have assisted us is difficult, even more so when an entire team has contributed to the effort, even if indirectly. With that in mind, we would like to acknowledge the entire Engineering Services Group of QUALCOMM Incorporated. In particular, we would like to acknowledge the initiators of this project: Dan Agre, Steve Anderson, Richard Costa, and Thomas Erickson. We would also like to thank those who have contributed by reviewing the material, within QUALCOMM and beyond: Jay Dills, Mauricio Guerra, Tony Guy, Pat Japenga, Ben Miller, Mukesh Mittal, Peter Rauber, Mustafa Saglam, Salil Sawhney, and Ralf Weber. Finally, we apologize to all of those we may have neglected to mention here.

Acronyms

1xEV-DO	Code Division Multiple Access technology compliant with revision 0 of the IS-856 standard, Evolution-Data Optimized of CDMA2000 1X
2-D	Two-Dimensional
2G	Second Generation
3-D	Three-Dimensional
3G	Third Generation
3GPP	Third Generation Partnership Project
ACK	ACKnowledge
ACLR	Adjacent Channel Leakage Ratio
AGC	Automatic Gain Control
AICH	Acquisition Indicator Channel
AM	Acknowledged Mode
AM	Amplitude Modulation
AMC	Adaptive Modulation and Coding
AMR	Adaptive Multirate
ANSI	American National Standards Institute
AS	Access Stratum
AS	Active Set
ASET	Active Set
ASN.1	Abstract Syntax Notation One
ASU	Active Set Update
ASUC	Active Set Update Complete
ATM	Asynchronous Transfer Mode
AuC	Authentication Center
AUTN	Authentication Token
AWGN	Additive White Gaussian Noise
BCCH	Broadcast Control Channel
BCH	Broadcast Channel
BDA	Bidirectional Amplifier
BHCA	Busy Hour Call Attempts
BLE	Block Error
BLER	Block Error Rate
BMC	Broadcast/Multicast Control
BPL	Building Penetration Loss
BS	Base Station
BSC	Base Station Controller
BSIC	Base Station Identification Code
BSS	Base Station sub-System
BTS	Base Transceiver Station

CC	Call Control
CCCH	Common Control Channel
CCTrCh	Coded Composite Transport Channel
CDMA	Code Division Multiple Access
CDMA2000 1X	Code Division Multiple Access technology compliant with revision 0 or later of IS2000 standard
CE	Channel Element
CELL_DCH	Basic Connected state following a successful call origination, or termination
CFN	Connection Frame Number
CIO	Cell Individual Offset
cm	centimeters
CM	Connection Management
CM	Compressed Mode
CN	Core Network
CPCH	Common Packet Channel
CPICH	Common Pilot Channel
CPICH_E_c/N_o	Pilot channel quality energy per chip over total received power spectral density
CPICH_RSCP	Receive signal code power of the Pilot channel
CQI	Channel Quality Indicator
CRC	Cyclic Redundancy Check
CS	Circuit Switched
CTCH	Common Traffic Channel
DAS	Distributed Antenna System
dB	Decibel
dBc	Decibels below carrier power
dBi	Decibels Isotropic
dBm	Decibel referenced to 1 milliwatt
DCCH	Dedicated Control Channel
DCH	Dedicated Channel
DCR	Dropped Call Rate
DCS1800	Digital Cellular Standard for 1800 MHz band
DL	Downlink
DPCCH	Dedicated Physical Control Channel
DPCH	Dedicated Physical Channel
DPDCH	Dedicated Physical Data Channel
DRAC	Dynamic Resource Allocation Control
DRX	Discontinuous Reception
DSCH	Downlink Shared Channel
DTCH	Dedicated Traffic Channel
DTX	Discontinuous Transmission
E1	European (CEPT) standard data rate of 2.048 Mbps
E_b/N_t	Energy per bit over the effective noise power spectral density
E_c/I_{or}	Energy per bit over the total transmit power spectral density
E_c/N_o	Energy per chip over total received power spectral density

EDGE	Enhanced Data rates for GSM Evolution
EFR	Enhanced Full Rate
EIR	Equipment Identity Register
EIRP	Effective Isotropically Radiated Power
EMR	Electromagnetic Radiation
ERP	Effective Radiated Power
ETSI	European Telecommunications Standards Institute
FACH	Forward Access Channel
FAF	Floor Attenuation Factor
FDD	Frequency Division Duplex
FDMA	Frequency Division Multiple Access
FEC	Forward Error Correction
FPLMTS	Future Public Land Mobile Telecommunication Systems
FTP	File Transfer Protocol
G	Geometry
Gbyte	Gigabyte
GERAN	GSM/EDGE Radio Access Network
GGSN	GPRS Gateway Support Node
GHz	GigaHertz
GIS	Geographic Information System
GMM	GPRS Mobility Management
GMSC	Gateway Mobile Switching Center
GoS	Grade of Service
GPRS	General Packet Radio Service
GPRS-CN	General Packet Radio Service, Core Network
GPS	Global Positioning System
GSM	Global System for Mobiles
GSM900	Global System for Mobile communication operating in the 900 MHz band
HARQ	Hybrid Automatic Repeat Request
HCS	Hierarchical Cell Structure
HLR	Home Location Register
HO	Handover
HORF	Handover Reduction Factor
HPA	High Power Amplifier
HPSK	Hybrid Phase Shift Keying
HSDPA	High Speed Downlink Packet Access
HS-DPCCH	High Speed Dedicated Physical Control Channel
HS-DSCH	High Speed Downlink Shared Channel
HS-SCCH	High Speed Shared Control Channel
HS-PDSCH	High Speed Physical Downlink Shared Channel
HTTP	HyperText Transfer Protocol
Hz	Hertz
IAF	IntrA-Frequency
IC	Integrated Circuit
IE	Information Elements

IEEE	Institute of Electrical and Electronic Engineers
IEF	IntEr-Frequency
IMT-2000	International Mobile Telecommunications-2000
$I_{oc}/\hat{I}_{or}$	Ratio of other-cell interference to same-cell received power density
I_{oc}/N_o	Ratio of other-cell interference total received power spectral density
IP	Internet Protocol
IR	Incremental Redundancy
IRAT	Inter-Radio Access Technology
IS-95	Code Division Multiple Access technology compliant with Release 0 or later of the TIA-IS-95 standard
ISCR	Inter-System Cell Reselection
ISHO	Inter-System Handover
ISO	International Standards Organization
ISDN	Integrated Services Digital Network
ITU	International Telecommunication Union
Iub	Interface between RNC and Node B
K	Kelvin
k	Boltzmann constant (1.38×10^{-23} Joules/Kelvin)
kbps	Kilobits Per Second
kHz	kiloHertz
km/hr	Kilometers per Hour
KPI	Key Performance Indicator
L3	Layer 3
LA	Location Area
LAN	Local Area Network
LAU	Location Area Update
LNA	Low-Noise Amplifier
LNF	Lognormal Fading
LOS	Line of Sight
MAC	Medium Access Control
MAPL	Maximum Allowable Path Loss
Mbps	Megabits per second
MB	Megabyte
MCM	Measurement Control Message
Mcps	Megachips per second
mErl	milli-Erlangs
MHz	MegaHertz
MIB	Master Information Block
MM	Mobility Management
MMS	Multimedia Messaging Service
MO	Mobile Originated
MOS	Mean Opinion Score
MoU	Minutes of Use
MPEG	Moving Picture Experts Group
MRM	Measurement Report Message
ms	Millisecond

MS	Mobile Station
MSC	Mobile Switching Center
MT	Mobile Terminated
mW	milliWatts
NA	Not Applicable
NAK	Negative Acknowledgement
NAS	Non-Access Stratum
NBAP	Node B Application Part
NF	Noise Figure, or Noise Factor
NLOS	Non-Line of Sight
ns	Nanosecond
NSS	Network and Switching sub-System
O&M	Operation and Maintenance
OA&M	Operations, Administration, and Maintenance
OBS	Obstructed (opposite of Line of Sight)
OMC	Operation and Maintenance Center
OOS	Out of Service
OVSF	Orthogonal Variable Spreading Factor
PA	Power Amplifier
PAMS	Perceptual Analysis Measurement System
PAR	Peak to Average Ratio
PC	Personal Computer
PCCH	Paging Control Channel
PCCPCH	Primary Common Control Physical Channel
PCH	Paging Channel
P-CPICH	Primary Common Pilot Channel
PDA	Personal Digital Assistant
PCU	Packet Control Unit
PDC	Personal Digital Cellular
PDCP	Packet Data Convergence Protocol
PDSCH	Physical Downlink Shared Channel
PDP	Packet Data Protocol
PDU	Protocol Data Unit
PESQ	Perceptual Evaluation Speech Quality
PHY	Physical
PI	Page Indicator
PICH	Paging Indicator Channel
PLMN	Public Land Mobile Network
PO	Power Offset
PRACH	Physical Random Access Channel
PS	Packet Switched
PSC	Primary Scrambling Code
P-SCH	Primary Synchronization Channel
PSNR	Peak Signal-to-Noise Ratio
PSQM	Perceptual Speech Quality Measurement
PSTN	Public Switched Telephone Network

QAM	Quadrature Amplitude Modulation
QCIF	Quarter Common Intermediate Format
QoS	Quality of Service
QPSK	Quadrature Phase Shift Keying
RA	Routing Area
RAB	Radio Access Bearer
RACH	Random Access Channel
RANAP	Radio Access Network Application Part
RAU	Routing Area Update
RB	Radio Bearer
RF	Radio Frequency
RLA	Received Signal Level Averaged
RLC	Radio Link Control
RNC	Radio Network Controller
RNS	Radio Network Subsystems
ROT or RoT	Rise Over Thermal
RRC	Radio Resource Control
RSCP	Received Signal Code Power
RSSI	Received Signal Strength Indicator
RTT	Round Trip Time
RV	Redundancy Version
Rx	Receive
SCCPCH	Secondary Common Control Physical Channel
SCH	Synchronization Channel
sec	Second
SF	Spreading Factor
Sf_HORF	Softer Handover Reduction Factor
SGSN	Serving GPRS Support Node
SIB	System Information Block
SID	Silence Descriptor
SIM	Subscriber Identity Module
SINR	Signal-to-Interference-and-Noise Ratio
SIR	Signal-to-Interference Ratio
SM	Session Management
SNR	Signal-to-Noise Ratio
SPER	Sub-Packet Error Rate
SQCIF	Sub-Quarter Common Intermediate Format
SRB	Signal Radio Bearer
SRES	Signed Authentication Response
SSC	Secondary Scrambling Code
S-SCH	Secondary Synchronization Channel
T1	Trunk Level 1, Digital transmission line, data rate of 1.544 Mbps
TB	Transport Block
TBS	Transport Block Size
TCP/IP	Transmission Control Protocol/Internet Protocol
TDD	Time Division Duplex

TDMA	Time Division Multiple Access
TFCI	Transport Format Combination Indicator
TFCS	Transport Format Combination Set
TFRC	Transport Format Resource Combination
TGCFN	Transmission Gap Connection Frame Number
TGD	Transmission Gap Distance
TGL	Transmission Gap Length
TGP	Transmission Gap Patterns
TGPL	Transmission Gap Pattern Length
TGPRC	Transmission Gap Pattern Repetition Count
TGPS	Transmission Gap Pattern Sequence
TGPSI	Transmission Gap Pattern Sequence Identifier
TGSN	Transmission Gap Slot Number
TM	Transparent Mode
TMA	Tower Mount Amplifier
TPC	Transmit Power Control
TRX	Transceiver
TSP	Transmit Status Prohibit
TSN	Transmission Sequence Number
TTI	Transmission Time Interval
TTT	Time-to-Trigger
TV	Television
Tx	Transmit
UARFCN	UTRA Absolute Radio Frequency Channel Number
UE	User Equipment
UL	Uplink
UM	Unacknowledged Mode
UMTS	Universal Mobile Telecommunications Systems
URA	UTRAN Registration Area
UTM	Universal Transverse Mercator
UTRA	Universal Terrestrial Radio Access
UTRAN	Universal Terrestrial Radio Access Network
UV	UltraViolet
VoIP	Voice over Internet Protocol
VLR	Visitor Location Register
VT	Video-Telephony
W	Watts
WAF	Wall Attenuation Factor
WCDMA	Wideband Code Division Multiple Access
WGS	World Geodetic System
WLAN	Wireless Local Area Network
WLL	Wireless Local Loop
YUV	Video format where luminance (Y) and chrominance (U and V) are a weighted function of R(ed) G(reen) B(lue) signal

1

Introduction to UMTS Networks

Patrick Chan, Andrea Garavaglia and Christophe Chevallier

Since their inception, mobile communications have become sophisticated and ubiquitous. However, as the popularity of mobile communications surged in the 1990s, Second Generation (2G) mobile cellular systems such as IS-95 and Global System for Mobile (GSM) were unable to meet the growing demand for more network capacity. At the same time, thanks to the Internet boom, users demanded better and faster data communications, which 2G technologies could not support.

Third Generation (3G) mobile systems have evolved and new services have been defined: mobile Internet browsing, e-mail, high-speed data transfer, video telephony, multimedia, video-on-demand, and audio-streaming. These data services had different Quality of Service (QoS) requirements and traffic characteristics in terms of burstiness and required bandwidth. More importantly, the projected traffic for these types of data services was expected to surpass voice traffic soon, marking a transition from the *voice paradigm* to the *data paradigm*. Existing cellular technology urgently needed a redesign to maximize the spectrum efficiency for the mixed traffic of both voice and data services. Another challenge was to provide global roaming and interoperability of different mobile communications across diverse mobile environments.

Toward these ends, the International Telecommunication Union (ITU), the European Telecommunications Standards Institute (ETSI), and other standardization organizations collaborated on the development of the Future Public Land Mobile Telecommunication Systems (FPLMTS). The project was later renamed International Mobile Telecommunications-2000 (IMT-2000). The goal of the project was to achieve convergence of the disparate competing technologies by encouraging collaborative work on one globally compatible system for wireless communications.

Set to operate at a 2 GHz carrier frequency band, the new 3G mobile cellular communication system needed to be backward-compatible with the 2G systems while improving system capacity and supporting both voice and data services. The system was expected

WCDMA (UMTS) Deployment Handbook: Planning and Optimization Aspects QUALCOMM Incorporated

to support both circuit switched (CS) and packet switched (PS) data services. For the PS domain, the supported data rates were specified for the various mobile environments:

- Indoor or stationary – 2 Mbps
- Urban outdoor and pedestrian – 384 kbps
- Wide area vehicular – 144 kbps

Of the various original proposals, the two that gained significant traction were based on Code Division Multiple Access (CDMA): CDMA2000 1X and Universal Mobile Telecommunication System (UMTS).

- CDMA2000 1X was built as an extension to cdmaOne (IS-95), with enhancements to achieve high data speed and support various 3G services. CDMA2000 1X further evolved to support even higher data rates with a data optimized version: CDMA2000 1xEV-DO [1].
- UMTS was based on the existing GSM communication core network (CN) but opted for a totally new radio access technology in the form of a wideband version of CDMA (Wideband CDMA: WCDMA). The Wideband Code Division Multiple Access (WCDMA) proposal offered two different modes of operation: Frequency Division Duplex (FDD), where Uplink (UL) and Downlink (DL) traffic are carried by different radio channels; and Time Division Duplex (TDD), where the same radio channel is used for UL and DL traffic but at different times. Evolution to support higher data rates was achieved with the recent introduction of High-Speed Downlink Packet Access (HSDPA) [2].

The goal of this book is to address the deployment aspects of the FDD version of the UMTS IMT-2000 proposal – namely WCDMA network planning and optimization. While it is accepted that deploying a WCDMA network requires a thorough knowledge of the standard, this book leaves that to other existing works such as Refs [3] and [4], and concentrates instead on the key aspects necessary to successfully deploy and operate a WCDMA network in a real-world scenario. For newcomers to this technology, however, this chapter describes the basic network topology and underlying concepts associated with the technology.

1.1 UMTS Network Topology

When deploying a WCDMA network, most operators already have an existing 2G network. WCDMA was intended as a technology to evolve GSM network toward 3G services. Paralleling that evolution, this chapter first discusses GSM networks, then highlights the changes that are necessary to migrate to Release 99 of the WCDMA specification. The discussion then moves on to Release 5 of the specification and the network changes needed to support HSDPA.

1.1.1 GSM Network Architecture

Figure 1.1 illustrates a GSM reference network [5], showing both the nodes and the interfaces to support operation in the CS and PS domains.

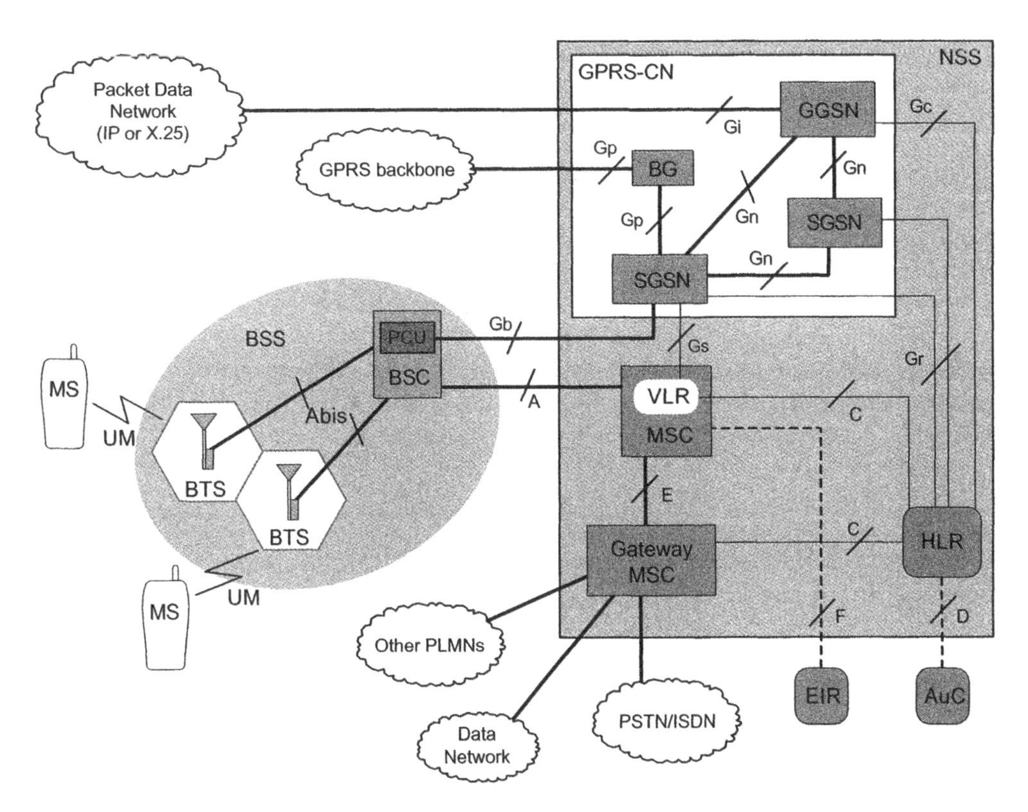

Figure 1.1 GSM reference network

In this reference network, three sub-networks [6] can be defined:

- **Base Station Sub-System (BSS) or GSM/Edge Radio Access Network (GERAN).** This sub-system is mainly composed of the Base Transceiver Station (BTS) and Base Station Controller (BSC), which together control the GSM radio interface – either from an individual link point of view for the BTS, or overall links, including the transfers between links (aka handovers), for the BSC. Although the interface connecting both nodes was intended to be standard, in real-world implementations the BTS–BSC links are closed to competition, particularly in terms of Operation and Maintenance (O&M). When data functionality was added to GSM with the deployment of General Packet Radio Service (GPRS), an additional node was added to the interface between the GPRS-CN and the radio interface, that is the Packet Control Unit (PCU). Interfaces toward the Network and Switching Sub-System (NSS) are limited to A for the CS domain and signaling, and Gb for the PS domain traffic. For simplicity, Figure 1.1 does not show the GSM/Edge Radio Access Network (GERAN). GERAN, a term that was introduced with UMTS, is the sum of all Base Station Sub-System (BSS) within the GSM Public Land Mobile Network (PLMN).
- **Network and Switching Sub-System (NSS).** This sub-system mainly consists of the Mobile Switching Center (MSC) that routes calls to and from the mobile. For management purposes, additional nodes are added to the MSC, either internally or externally.

Their main purpose is to keep track of the subscription data, along with associated rights and privileges, in the Home Location Register (HLR), or to keep track of the subscribers' mobility in the HLR and Visitor Location Register (VLR). Two other nodes manage security issues: the Equipment Identity Register (EIR) verifies the status of the mobile phone (i.e., the hardware), while the Authentication Center (AuC) manages the security associated with the Subscriber Identity Module (SIM). The last node listed in Figure 1.1 is the Gateway-MSC (GMSC). For all practical purposes, the MSC and GMSC are differentiated only by the presence of interfaces to other networks, the Public Switched Telephone Network (PSTN) in the GMSC case. Typically, the MSC and the GMSC are integrated. The interfaces listed (E, F, C, D) are not detailed here but mostly enable the communication between the different nodes as shown.

- **General Packet Radio Service, Core Network (GPRS-CN).** Within the NSS, two specific nodes are introduced for the GPRS operation: the Serving GPRS Support Node (SGSN) and the Gateway GPRS Support Node (GGSN). In the PS domain, the SGSN is comparable to the MSC used in the CS domain. Similarly, in the PS domain, the GGSN is comparable to the GMSC used in the CS domain. These nodes rely on existing BSS or NSS nodes, particularly the VLR and HLR, to manage mobility and subscriptions – hence the interfaces to the Gs and Gr interfaces (to the VLR and HLR respectively). Figure 1.1 also shows the Border Gateway (BG) that supports interconnection between different GPRS networks to permit roaming, and the PCU to manage and route GPRS traffic to the BSS.

1.1.2 UMTS Overlay, Release 99

As mentioned earlier, UMTS is based on the GSM reference network and thus shares most nodes of the NSS and General Packet Radio Service, Core Network (GPRS-CN) sub-systems. The BSS or GERAN is maintained in the UMTS reference network as a complement to the new Universal Terrestrial Radio Access Network (UTRAN), which is composed of multiple Radio Network Systems (RNS) as illustrated in Figure 1.2.

Compared to the GSM reference network, the only difference is the introduction of the Radio Network Controller (RNC) and Node Bs within the newly formed RNS. Essentially, these two nodes perform tasks equivalent to the BSC and BTS, respectively, in the GSM architecture. The main difference is that the interface Iu-PS to the PS-CN is now fully integrated within the RNC.

With the addition of these new nodes, a number of new interfaces are defined: Iub is equivalent to the Abis, Iu-CS is equivalent to the A, and Iu-PS is equivalent to the Gb. In addition, the Iur interface (not shown in the figure) is created to support soft handover (HO) between RNCs connecting multiple RNCs within the same UTRAN.

From a practical standpoint, the common nodes between GSM and UMTS would actually be duplicated, with the original nodes supporting the 2G traffic and the added nodes supporting the 3G traffic.

1.1.3 UMTS Network Architecture beyond Release 99

The initial deployments of WCDMA networks comply with Release 99 of the standard [7]. This standard, or family of standards, began to evolve even before being fully implemented, to address the limitations of the initial specifications as well as to include

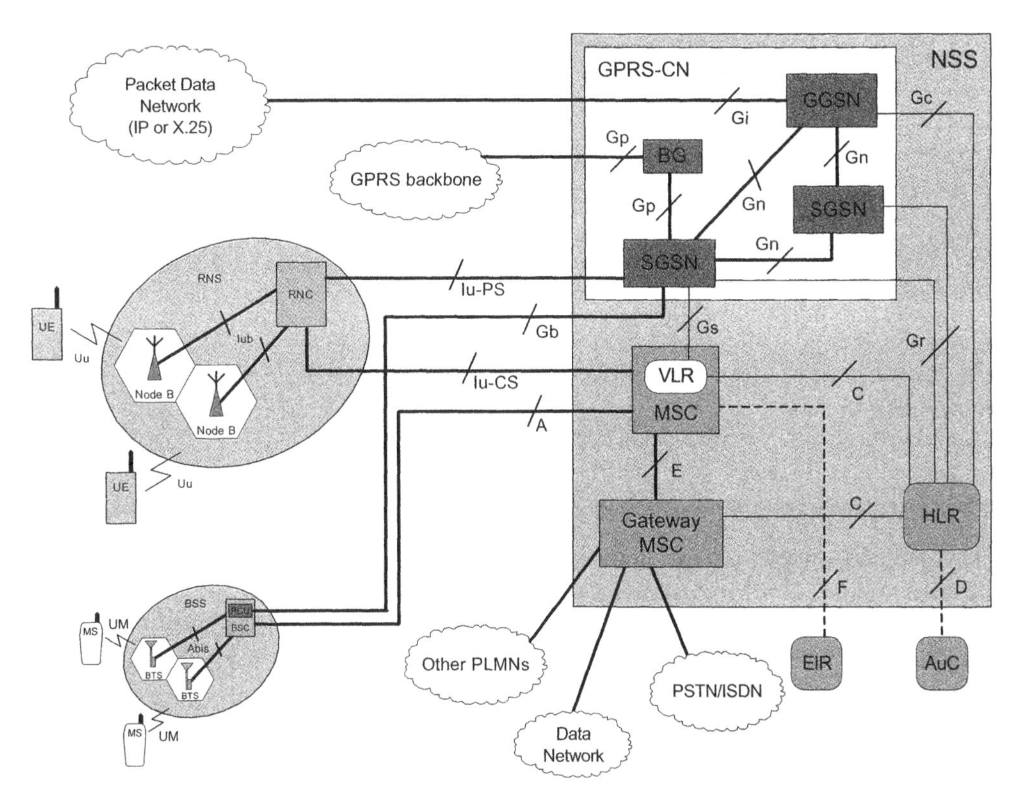

Figure 1.2 UMTS reference network

technical advancements. At a higher level, migrating from Release 99 to Releases 4, 5, and then 6 does not change the structure of the network. However, the details do differ: for example, the transport for the interfaces changes from Asynchronous Transfer Mode (ATM) in Release 99 to all Internet Protocol (IP) in Release 5. In addition, the layering changes in Release 5, to support HSDPA and Node B scheduling (see Section 1.2.2).

1.2 WCDMA Concepts

Figure 1.3 summarizes the physical aspects of the WCDMA air interface, where the flow of information at 3.84 Mega chips per second (Mcps) can be divided into 10 ms radio frames, each further divided into 15 slots of 2560 chips. Here the notion of chips is introduced instead of the more typical bits. Chips are the basic information units in WCDMA. Bits from the different channels are coded by representing each bit by a variable number of chips. What each chip represents depends on the channel.

This section discusses the most fundamental concepts used in WCDMA: channelization and scrambling, channel coding, power control, and handover. The section then defines how the different channels are managed (layers and signaling), and finally defines the channels at the different layers: logical, transport, and physical.

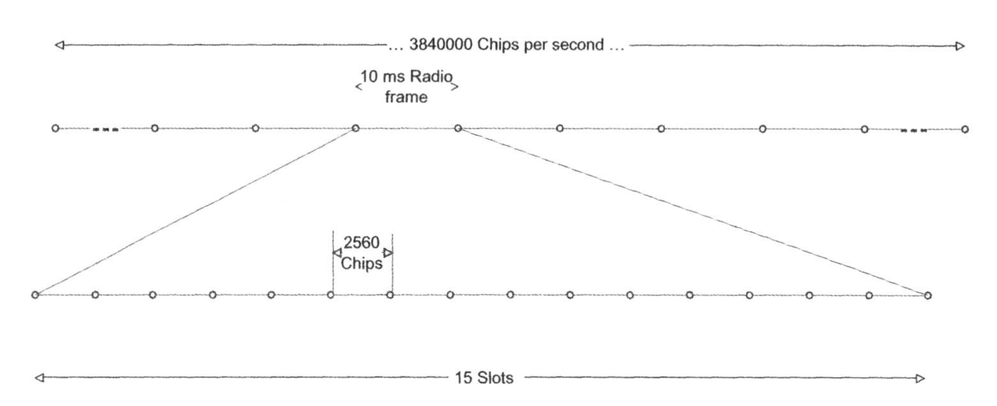

Figure 1.3 WCDMA air interface architecture

1.2.1 WCDMA Physical Layer Procedures

In the selection process for 3G standards, air interface efficiency – which translates to capacity – was one of the main criteria used to assess the different proposals. By that time, significant worldwide deployments of 2G CDMA-based networks had proven the technology's ability to deliver systems with high spectral efficiency. The concepts described in the following sections are vital in any CDMA technology.

1.2.1.1 Power Control

In CDMA technology, power control is critical. It ensures that just enough power is used to close the links, either DL, from the Base Station to the mobile device, or UL, from the mobile to the Base Station. Of the two links, the UL is probably more critical. The UL ensures that all instances of user equipment (UE) are detected at the same power by the cell; thus each UE contributes equally to the overall interference and no single UE will overpower and consequently desensitize the receiver. Without power control, a single UE transmitting at full power close to the Base Station would be the only one detected. All the others would be drowned out by the strong signal of the close user who creates a disproportionate amount of interference.

On the DL, power control serves a slightly different purpose, because the Node B's power must be shared among common channels and the dedicated channels for all active users. On the DL, all channels are orthogonal to each other (with the exception of the Synchronization Channel); thus the signal, or power, from any channel is not seen as interference. Ideally, the other channels do not affect the sensitivity. However, power control is still required to ensure that a given channel is using only the power that it needs. This increases the power available for other users, effectively increasing the capacity of the system.

Conceptually, two steps are required for power control:

- Estimate the minimum acceptable quality.
- Ensure that minimum power is used to maintain this quality.

Outer loop power control handles the first step; inner loop handles the second. Ideally, the outer loop should monitor the Block Error Rate (BLER) of any established channel and compare it to the selected target. If they differ, the quality target, estimated in terms of Signal-to-Interference Ratio (SIR), is adjusted. The closed loop power control can then compare, on a slot-by-slot basis, the measured and target SIR, and send power-up or power-down commands. Power control processes run independently in the UL and DL, each signaling to the other the required adjustment by means of Transmit Power Control (TPC) bits: the DL carries the TPC bits indicating the UL quality, while the UL carries the TPC bits indicating the DL quality.

On the basis of the frame and slot structure (10 ms radio frames consisting of 15 slots each), we can deduce that the TPC bits are sent at 1500 Hz, which is the rate of the inner loop. The outer loop, on the other hand, is not as strictly controlled by the standard and is thus implementation-dependent: neither its rate nor the step sizes are signaled to the other end. Moreover, although the purpose of the closed loop is to ensure that the BLER target is met, the implementation may be based on other measurements such as SIR, or passing or failing the Cyclic Redundancy Check (CRC).

1.2.1.2 Soft Handover

Soft handover refers to the process that allows a connection to be served simultaneously by several cells, adding and dropping them as needed. This feature is possible in WCDMA because all cells use the same frequency and are separated only by codes: a single receiver can detect the different cells solely by processing, with a single Radio Frequency (RF) chain. The need for soft handover in a WCDMA system is intertwined with the power control feature. Supporting soft handover ensures that a UE at the boundary among several cells uses the minimum transmit power on either link. On the UL, it is necessary to avoid overpowering the other UEs connected to the cell. On the DL, it is not as critical, but is a good practice because it maximizes capacity and increases link reliability. Once soft handover is enabled in a system, meaning that a UE must monitor and use the best possible link, additional benefits can ensue:

- On the DL, the UE can combine the different received signals to increase the reliability of demodulation. By combining the signals from different links, the effective SIR increases, which reduces the transmit power even when compared to the power required over the best link only. This is termed *soft combining gain*. In addition, the fact that the UE can be connected to multiple servers at once increases link reliability and thus provides a diversity gain, typically called *macro-diversity gain*.
- On the UL, if macro-diversity gain is observed, the same is not always true for the soft combining gain. If the cells in soft handover do not belong to the same Node B, it is not possible to combine the signals before they are demodulated. Instead, all the demodulated frames are sent to the RNC, which decides which one to use. This process still provides a gain compared to a single link, since it increases the probability of having at least one link without error. This is the selection gain, also a macro-diversity gain.

As we have seen, soft handover offers advantages: it increases the reliability of transmission and reduces the power requirement for each link used. Unfortunately, soft handover has drawbacks, too. Since information must be sent over multiple links, that repetition decreases the efficiency of resource utilization. As subsequent chapters (mainly Chapter 4) will show, balancing handover gains with resource utilization is a delicate process, controlled by multiple parameters. Clearly, the optimal balance is achieved only when the links that contribute significantly to the transmission quality are included in the Active Set.

1.2.1.3 Spreading, Scrambling, and Channelization

Soft handover is possible in a WCDMA system because all the cells of the Node Bs transmit using the same frequency. This universal frequency reuse – or 1 to 1 frequency reuse in Time Division Multiple Access (TDMA)/Frequency Division Multiple Access (FDMA) terminology – requires several codes to differentiate between cells and users. These codes must be introduced on both the UL and the DL, since the constraints on each link are different.

On the DL, the first requirement is to differentiate among different cells. In the TDMA/FDMA world, this is achieved by using a different frequency for each cell. In the WCDMA world, cells are discriminated by using Primary Scrambling Codes (PSCs). To understand how they work, imagine a coded message. When viewed, the coded message is perceived only as random letters – or *noise* in radio terminology. Only a reader using the proper ciphering key – or PSC in WCDMA terminology – can make words out of the random letters. These words can be further assembled into sentences, either on a single subject or on different topics. The topics can be compared to different channels for which proper rules must be defined: these rules would correspond to the different channeling codes that allow the decoded words to be assembled into sentences. Just as only a limited set of rules makes up a language to ensure that everybody understands it, only a limited number of channelization codes are used to simplify the implementation.

To extend the analogy, several words in a sentence typically express a single idea; this is the same principle as spreading, where several chips represent one bit. Just as in a language, where losing a single word does not prevent comprehending the idea, losing the exact value of a chip does not compromise the demodulation of the corresponding bit. In the WCDMA world, spreading, or conveying a single bit over multiple chips, is done at the channel level, before the PSC ciphers the entire message. Multiplying a signal with a PSC does not achieve spreading, it only randomizes the signal, as illustrated in Figure 1.4(a).

Within the cell, the different channels are separated by their own set of rules, the Orthogonal Variable Spreading Factor (OVSF). The OVSF handles the signal spreading, as illustrated in Figure 1.4(b). OVSF has two main characteristics: an orthogonality property, and the fact that the orthogonality is conserved between OVSFs of variable lengths.

- The OVSF orthogonality property ensures that different users of the same cell do not interfere with each other. If a signal coded with a given OVSF is decoded with a different OVSF, the resulting signal gives an equal number of 1s (-1) and 0s ($+1$). The result is an average null signal.
- The variable aspect of OVSF supports different data rates from the same code tree: low data rates can be coded with long OVSFs, while high data rates are coded with

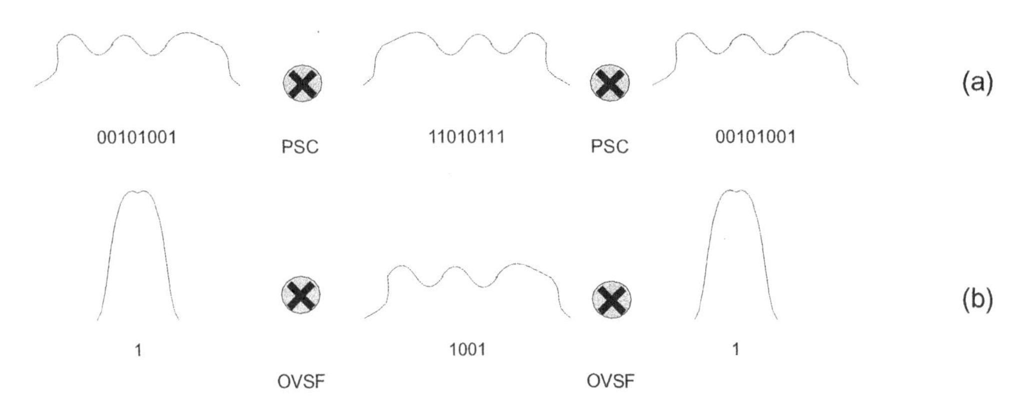

Figure 1.4 PSC and OVSF

short OVSFs. The length of the OVSF refers to the number of chips for a single input bit: a bit coded with OVSF length 256 would be represented by 256 chips, while a bit coded with OVSF length 4 would be represented by four chips. Using a long OVSF has the advantage of adding redundancy to the transmitted information. The impact of this redundancy is seen in the spreading gain, that is, the ratio of user bits to transmitted chips.

In combination, and with only a limited number of codes, PSC and OVSF can distinguish between cells and users. Without the PSC, the receiver cannot reconstruct the words sent by the different cells. Once the words are reconstructed, the same set of rules OVSF can be reused to understand (demodulate) the messages.

1.2.1.4 Channel Coding

The Physical Layer procedures described in Section 1.2.1.1 through Section 1.2.1.3 are required for efficient implementation of WCDMA. In addition to these mandatory procedures, channel coding further protects against transmission errors caused by repeating information multiple times, and spreading the retransmissions over time.

For channel coding, either convolutional or turbo coders can be used. Convolutional coders primarily apply to delay-sensitive information, since the resulting delay is relatively short and affected by the code rate and constraint length. Turbo coders, on the other hand, must consider a block of data before outputting the block. For turbo coding to be efficient, the block should contain a large amount of data, usually more than 320 symbols, thus causing significant delay in the coding and decoding processes.

1.2.2 UMTS Signaling Concepts

To understand signaling – or more generally the exchange of data – in WCDMA, it is important to understand layering and its relationship to the various nodes, for both the control plane (signaling) and the user plane (user data). As will be demonstrated in the following sections, WCDMA offers a highly structured protocol stack with clear delineation between the functions of each entity.

1.2.2.1 Layering Concepts

From an overall network point of view, the first distinction that can be made is between the radio access functions (Access Stratum, or AS) and the CN functions (Non-Access Stratum, or NAS). For WCDMA and GSM the Non-Access Stratum (NAS) is similar, so we will not discuss it further. The second distinction is between the control or signaling plane (control data) and the user plane (user data). For the control plane, all layers terminate at the operators' controlled nodes; for the user plane, the top layer ensures the user-to-user connection. Figure 1.5 illustrates these concepts as they apply to the CS domain.

WCDMA-specific processing occurs at the Access Stratum (AS) on the three lower layers, which are similar in both planes:

- **Radio Link Control (RLC).** Sets up the delivery mechanism ensuring that the data sent is received at the distant end.
- **Medium Access Control (MAC).** Permits multiple information flows to be sent over a single physical channel.
- **Physical Layer (Layer 1).** Transmits the combined information flow over the WCDMA air interface (Uu).

For each layer, different channels are defined and mapped onto one another: logical channels are associated with Radio Link Control (RLC), transport channels with Medium Access Control (MAC), and physical channels with Physical Layers.

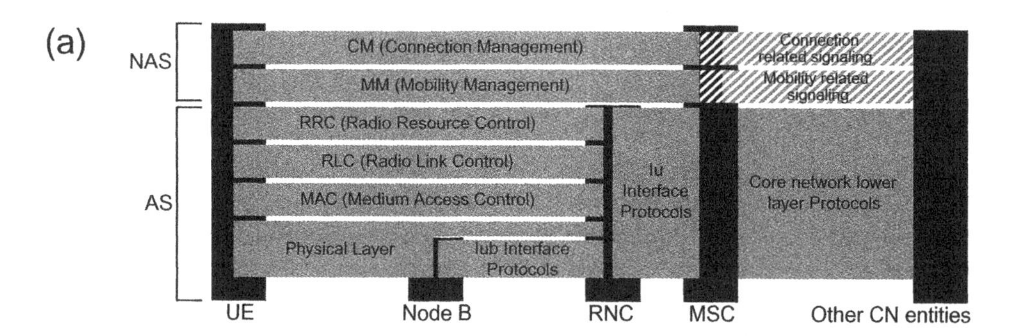

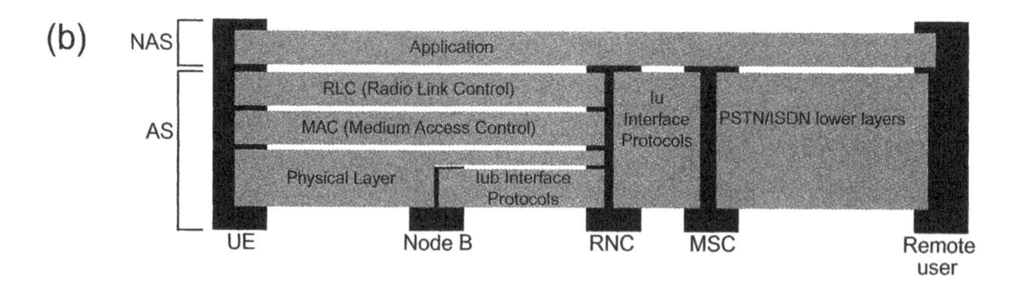

Figure 1.5 Control plane (a) and user plane (b) layering in the CS domain

In Figure 1.5, note that the AS Physical Layer is not entirely contained within the Node B. Outer loop power control and frame selection occur at the RNC, even though they are Physical Layer processes. This shows that in the UTRAN, the RNC is a critical node. In addition to some of the Physical Layer processing, the RNC is responsible for link supervision (RLC) and any multiplexing/assembly (MAC) for the channels.

At the RLC level, the link supervision is done in one of three modes: Transparent Mode (TM), Acknowledged Mode (AM), or Unacknowledged Mode (UM). The appropriate mode depends on the time constraints and error tolerance of the information:

- **TM.** Typically used for the user payload for speech services. For speech, delivery of vocoder packets at a constant rate is more important than error-free transmission; the vocoder can conceal errors if they are below a few percent, typically 1 or 2%. In TM, RLC does not verify the packets only passes them to the higher layers. In the worst case, retransmission can be achieved at the application layer when the user requests "can you repeat please."
- **AM.** Used when information must be sent error-free. Each packet is given a sequence number to be individually acknowledged and delivered in sequence to higher layers. A typical application is e-mail messages, where the content is more important than any latency in delivery.
- **UM.** Used for applications that must receive packets in order, but are neither delay- nor error-constrained. Examples are media streaming and some types of signaling. UM is appropriate for packets that may be processed at the RLC layer, but for which no retransmission is requested if errors are detected. For UM packets, processing is usually limited to reordering, ciphering, or segmentation/concatenation.

In the PS domain the structure is similar for the control plane with the exception of the terminating nodes, as shown in Figure 1.6.

The PS user plane (Figure 1.6(b)) shows more pronounced differences from the CS user plane, with an added layer in both the AS and NAS. The Packet Data Convergence Protocol (PDCP) on the AS is mainly used for header compression, to transfer TCP/IP packets more efficiently over-the-air interface. The Packet Data Protocol (PDP) on the NAS creates and manages the associated variables for the packet data sessions. For example, when an IP session is required, the IP addresses that identify the UE for a session are assigned at that layer.

As systems evolve and incorporate HSDPA, the PS domain layering will change, as shown in Figure 1.7. To speed up Physical Layer processing, the entire Layer 1 terminates at the Node B. The drawback is that the MAC layer must extend to the Node B, with the introduction of a specific MAC entity dedicated to high speed data; the MAC-hs.

Note that Figure 1.7 does not show the control plane; this is because HSDPA supports the user plane only. The control plane is maintained in the PS domain, Release 99 architecture.

In addition to the three basic layers used in all domains (packet/PS or circuit/CS) and all planes (user data/user plane or signaling data/control plane), a layer is defined in the control plane that defines the messages exchanged between the RNC and the UE: the Radio Resource Control (RRC). The RRC defines the messages exchanged between the RNC and the UE, which initiate connection set-up, tear down, or reconfiguration.

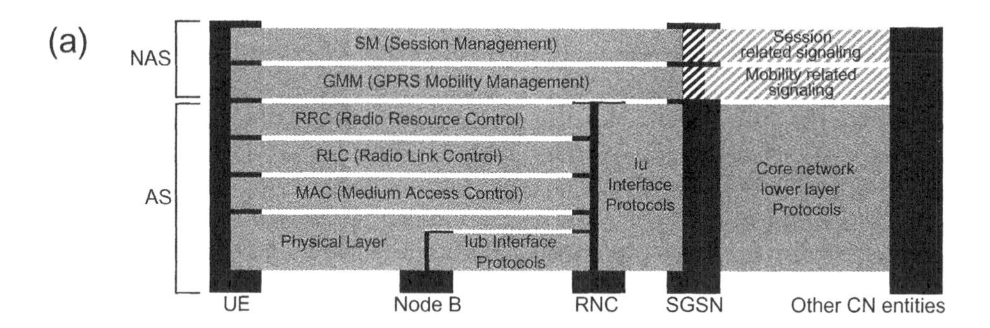

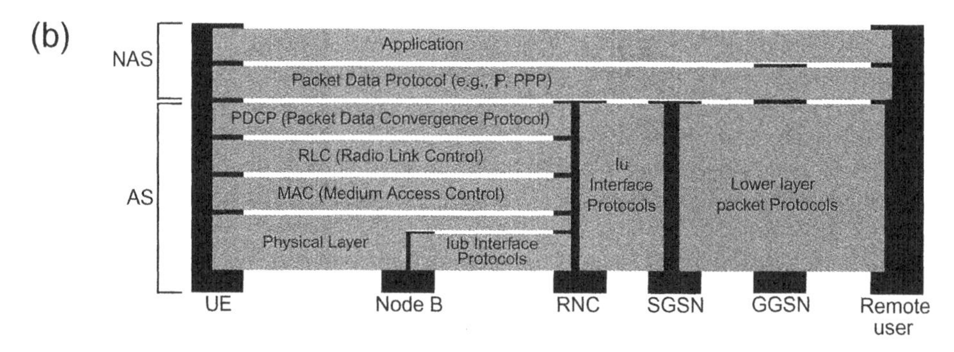

Figure 1.6 Control plane (a) and user plane (b) layering in the PS domain

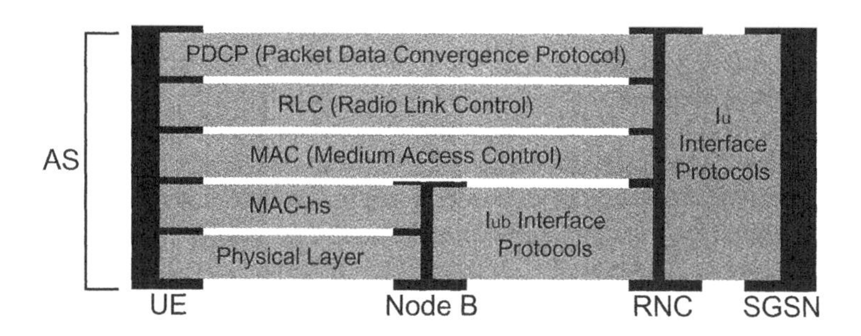

Figure 1.7 User plane layering in the PS domain, HSDPA architecture

All the call flow examples in later chapters (Chapters 5 and 6) of this book are based on RRC messaging [8], which is exchanged between the UE and the RNC over the Signaling Radio Bearers (SRB).

Prior to any message exchanges, the SRBs must be established during the RRC connection setup procedure. Depending on vendor implementation, three or four SRBs are established and mapped onto a single Dedicated Channel (DCH) by MAC.

1.2.3 Physical, Logical, and Transport Channels

Section 1.2.2 introduced the concept of mapping logical channels onto physical channels. Figure 1.8 shows the different channels for Release 99 and HSDPA operation, along with

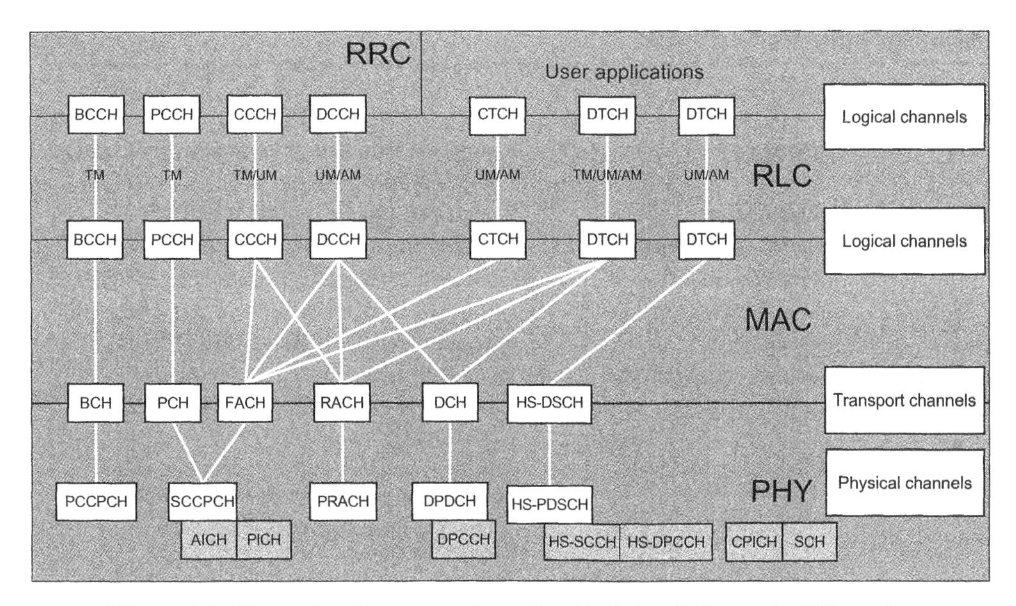

Figure 1.8 Physical and transport channels with their relation to the ISO model

how they map between different layers. At the Physical Layer, some of the channels, for example Synchronization Channel (SCH) and Common Pilot Channel (CPICH), are not mapped onto any transport channels. This is because these channels only support Physical Layer procedures; no actual data from higher layers is transmitted over them.

During optimization, it is important to understand which processes are associated with each type of channel. Physical channels are associated with all the coding and closed loop power control processes. Transport channels are associated with some of the critical channel measurements, such as BLER or SIR targets, since these values are set per transport channel.

Other physical channels are used for physical procedures or scheduling but do not directly map onto transport channels; however, they do carry information related to these physical procedures. Channels in this group include the Acquisition Indicator Channel (AICH), Paging Indicator Channel (PICH), Dedicated Physical Control Channel (DPCCH), High-Speed Shared Control Channel (HS-SCCH), and High-Speed Dedicated Physical Control Channel (HS-DPCCH). For example, the DPCCH is a channel that does not carry any user or signaling information but contains information to help the receiver decode the information carried by the Dedicated Physical Data Channel (DPDCH). Complete function details for all of these channels can be found in Ref [9] and [3].

Table 1.1 lists the channels shown in Figure 1.8, along with their main uses.

In addition to showing the logical, transport, and physical channels, Figure 1.8 also shows the mappings between them, along with the RLC mode typically used for these channels. Instead of describing all possible mappings, the following discussion analyzes one example, where signaling and user speech payload are transmitted over a single DL physical channel. Figure 1.9 illustrates the example.

In this example, a single physical channel carries seven logical channels: four for signaling and three for voice [10]. The information, coding, and transport block sizes are

Table 1.1 List of WCDMA channels

	Channel name	Description
BCCH	Broadcast Control Channel	*Logical channel* that sends System Information Block (SIB)
BCH	Broadcast Channel	*Transport channel* carrying the BCCH
PCCPCH	Primary Common Control Physical Channel	*Physical channel* carrying the BCH
PCCH	Paging Control Channel	*Logical channel* carrying the pages to the UE
PCH	Paging Channel	*Transport channel* carrying the PCCH
CCCH	Common Control Channel	*Logical channel* carrying the common signaling, e.g., RRC Connection Setup message
FACH	Forward Access Channel	*Transport channel* carrying common and dedicated control channel as well as user payload in certain connected states (Cell_FACH)
SCCPCH	Secondary Common Control Physical Channel	*Physical channel* carrying the PCH and FACH channels
AICH	Acquisition Indicator Channel	*Physical channel* used by the cell to ACK the reception of RACH preambles
PICH	Paging Indicator Channel	*Physical channel* used by the cell to inform a group of UEs that a page message can be addressed to them
DCCH	Dedicated Control Channel	*Logical channel* used to carry dedicated Layer 3 (RRC) signaling to the UE
RACH	Random Access Channel	*Transport channel* used by the UE to carry signaling or user payload
PRACH	Physical RACH	*Physical channel* used to carry the RACH
CTCH	Common Traffic Channel	*Logical channel* used to carry common payload, e.g., broadcast or multicast services
DTCH	Dedicated Traffic Channel	*Logical channel* used to carry user payload
DCH	Dedicated Channel	*Transport channel* used to carry dedicated signaling (DCCH) or payload (DTCH)
DPDCH	Dedicated Physical Data Channel	*Physical channel* used to carry the DCH
DPCCH	Dedicated Physical Control Channel	*Physical channel* used for carrying information related to physical layer operation, e.g., dedicated pilot or power control bits
HS-DSCH	High-Speed Downlink Shared Channel	*Transport channel* used for carrying user payload. Unlike the DCH, only user payload is carried over the HS-DSCH; no signaling (DCCH) is carried by HS-DSCH
HS-PDSCH	High-Speed Physical Downlink Shared Channel	*Physical channel* used for carrying the HS-DSCH

Table 1.1 (*continued*)

	Channel name	Description
HS-SCCH	High-Speed Shared Control Channel	*Physical channel* used for carrying HS specific control information, e.g., modulation, Transport Block Size (TBS), or HARQ related information
HS-DPCCH	High-Speed Dedicated Physical Control Channel	*Physical Uplink channel* used by the UE to carry Channel Quality Indicator (CQI) and Acknowledgment information
CPICH	Common Pilot Channel	*Physical channel* used for cell identification and channel estimation
SCH	Synchronization Channel	*Physical channel* used by the UE to detect the presence of WCDMA carrier (Primary SCH: P-SCH) and synchronize with radio frame boundary (Secondary SCH: S-SCH)

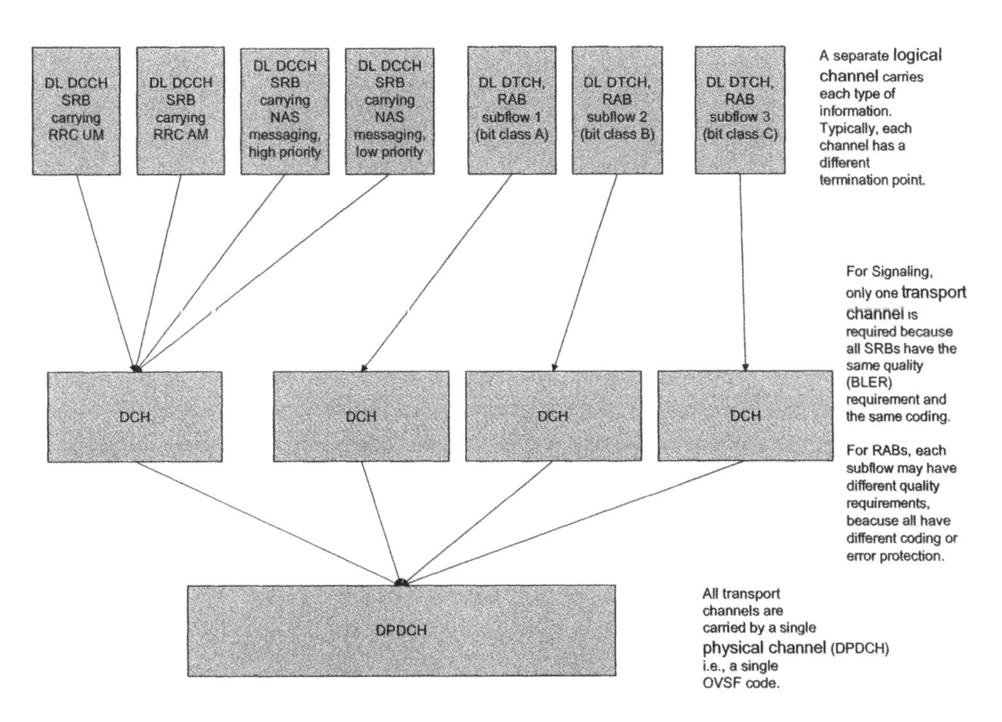

Figure 1.9 Mapping of logical to transport to physical channels for speech and signaling

all uniquely defined [10] and summarized in Table 1.2. The different SRBs and Radio Access Bearers (RABs) are each associated with an RLC entity. RLC also defines the size of the packet exchanged with the higher layer's Packet Data Unit (PDU). Headers can be added to the information if any multiplexing will be performed at the MAC level, or to verify the delivery of the packet. For speech, the user payload does not include

Table 1.2 Main channel characteristics for Speech + SRB example

Higher layer		RAB subflow #1	RAB subflow #2	RAB subflow #3	SRB #1 RRC	SRB #2 RRC	SRB #3 NAS	SRB #4 NAS
RLC	Logical channel type	DTCH			DCCH			
	RLC mode	TM	TM	TM	UM	AM	AM	AM
	Payload sizes, bit	39, 81	103	60	136	128	128	128
	Max data rate, bps	12200			3400	3200	3200	3200
	TrD PDU header, bit	0			8	16	16	16
MAC	MAC header, bit	0			4	4	4	4
	MAC multiplexing	N/A			Four logical channel multiplexing			
Layer 1	TrCH type	DCH	DCH	DCH	DCH			
	TB sizes, bit	39, 81	103	60	148			
	TFS TF0, bits	0 × 81	0 × 103	0 × 60	0 × 148			
	TF1, bits	1 × 39	1 × 103	1 × 60	1 × 148			
	TF2, bits	1 × 81	N/A	N/A	N/A			
	TTI, ms	20	20	20	40			
	Coding type	CC 1/3	CC 1/3	CC 1/2	CC 1/3			
	CRC, bit	12	N/A	N/A	16			
	Max number of bits/TTI after channel coding	303	333	136	516			
	RM attribute	180–220	170–210	215–256	155–230			
TFCS	TFCS size	6						
	TFCS	(RAB subflow #1, RAB subflow #2, RAB subflow #3, DCCH) = (TF0, TF0, TF0, TF0), (TF1, TF0, TF0, TF0), (TF2, TF1, TF1, TF0), (TF0, TF0, TF0, TF1), (TF1, TF0, TF0, TF01), (TF2, TF1, TF1, TF1)						
DPCH Downlink	Spreading Factor	128						
	Format	8 [9]						

Bits/slot	DPDCH Bits/slot		DPCCH Bits/slot			Transmitted slots per radio frame N_{Tr}
	Ndata1	Ndata2	Ntpc	NTFCI	Npilot	
40	6	28	2	0	4	15

such headers because TM is used and no MAC multiplexing is performed, although it is necessary for the signaling aspects.

At the Transport Channel level, the payload plus any necessary headers are put into transport blocks for which both a size and a Transmission Time Interval (TTI) are defined. For all the voice subflows, the TTI is 20 ms, or two radio frames, consistent with the frequency at which the vocoder generates packets. Alternatively for the SRBs, the payload is spread over four radio frames, with a TTI of 40 ms. For higher-speed data rate RAB, the basic payload size is at the most 320 bits (in Release 99), while the TTI is usually 10 to 20 ms and multiple blocks (RLC PDU) are transferred during each TTI.

The number of channels in this example and the fact that their multiplexing does not follow a predetermined pattern leads us to the concept of the Transport Format Combination Set (TFCS). The TFCS determines how the blocks corresponding to the different DCH channels are combined on the Physical Channel. In this example, not all combinations are allowed; only six are permitted, as per the standard. During the setup of the radio bearer (i.e., the Physical Channel) the possible TFCSs are signaled to the UE as a table. Within each radio frame, the Transport Format Combination Indicator (TFCI) signals an index pointing to that table. From this, the UE can reconstruct the DCH channels received. In the example (as compared to Table 1.2), slot format 8 does not reserve any bits for TFCI. In this case, the UE guesses which format was used, using a process called *blind detection* [11].

1.3 WCDMA Network Deployment Options

GSM networks boast an advanced network architecture, where macrocells, microcells, and indoor cells all interact. This flexibility is enhanced by the many available GSM network products, from a macro BTS handling a dozen transceiver modules (TRXs) per sector, to a pico BTS handling only a few TRX over a single sector.

Similarly, since WCDMA was designed to interact closely with the deployed GSM network, it will follow the GSM trend in terms of ubiquity and deployment options. From practical and economic points of view, WCDMA deployments are not initially justified outside the high-traffic areas, typically the city centers. This section discusses the main WCDMA deployment options and presents their relative advantages and shortcomings. All of the options assume that at least two carriers are available for deployment. If this is not the case, then deploying multiple layers – micro, macro, or indoor – greatly affects capacity, as Chapter 3 illustrates.

1.3.1 1 : 1 Overlay with GSM, Macro Network

A 1 : 1 overlay of WCDMA onto a GSM network has so far been one of the most popular deployment options, although not necessarily the best one. The main advantage that explains its popularity is that this approach largely simplifies the site acquisition process; the only acquisition needed is an additional antenna position within the existing structure. This option is sometimes further simplified, from a site acquisition point of view, by replacing the existing antenna with a multiband or wideband antenna. In this case, the design and optimization options for the WCDMA network are quite limited, thus leading to suboptimal performance.

This situation applies to any site reuse between GSM and WCDMA. From a network-planning standpoint, technical differences in the air interface between the two networks make it difficult to share sites. Key differences include coverage, mainly due to WCDMAs higher frequency band, and capacity, mainly because of WCDMAs improved spectral efficiency. Furthermore, the universal frequency reuse in WCDMA makes it difficult to deploy Hierarchical Cell Structure (HCS) in WCDMA networks, whereas it is widely used in GSM.

In GSM, using HCS is beneficial because large cells (with tall antennas) can provide coverage, while small cells (with low or medium height antennas) can provide capacity. With the 1 : 1 frequency reuse of WCDMA, deploying HCS would allow a UE to reselect the most appropriate layer; however, in Connected Mode, the UE would be in constant handover between the layers, or, if parameters are set to prevent handover, the resulting intercell interference would decrease the capacity advantage. Chapter 3 explores this in greater detail.

As a result, a WCDMA overlay onto a GSM network is usually not 1 : 1, but would exclude the tallest sites of the network as well as the microcells, at least initially. Also, initially the overlay is not made over the entire GSM coverage area but only where the capacity requirements are the highest. Two issues usually determine this choice: economics and coverage. These may be linked to some extent. From a coverage perspective, the initial WCDMA deployment occurs at 2100 MHz (1900 MHz in North America) while GSM is widely deployed at 900 MHz (850 MHz in North America). This gives GSM a 10 to 15 dB Link Budget advantage in terms of RF propagation. This can translate to a site count for WCDMA, for coverage only, of four to seven times the GSM site count. This offsets any economic justification for deploying WCDMA in rural areas, where coverage requirements are high but capacity needs are low. Because of this limitation, planners may rely on GSM coverage outside of urban centers. Unfortunately, the mechanism for inter-system transition, even if simple in principle, requires careful planning and optimization. To facilitate that, Chapter 6 discusses inter-system transitions in detail.

1.3.2 1 : 1 Overlay with GSM, Macro, Micro, and In-Building

The preceding discussion about applying a WCDMA overlay to a macro network could easily apply to all layers, if micro and pico Node Bs are available. In the early years of WCDMA growth, only macro Node Bs were available but that situation is rapidly changing.

To introduce a micro or indoor layer, it is necessary to have multiple carriers available. In this case, the macro layer operates on one carrier, while the micro and indoor layers operate on a separate one. This better isolates the layers, providing significant advantages in terms of resource utilization. The drawback, of course, is that managing mobility between layers – via cell reselection or handover – becomes more complex (see Chapter 6).

Not only is mobility management more complex but a 1 : 1 overlay with GSM may also use the available capacity inefficiently. In GSM, the spectrum – and hence the capacity – can be allocated in 200 KHz increments. But in WCDMA, increments are fixed at 5 MHz, which for most operators represents 33 to 50% of the available spectrum. For this reason, operators might decide to deploy the micro or indoor layers on the same carrier. However, they must then consider the spatial isolation between layers, which could affect capacity. Chapter 3 explores this issue.

1.3.3 WCDMA-Specific Network Plan

Another option for overcoming the limitations found in GSM overlay deployments is to create a network plan specifically for WCDMA. Here, one does not rely on GSM site locations but only on the expected WCDMA traffic and coverage requirements. As long as coverage and capacity issues (see Chapters 2 and 3, respectively) and indoor issues (see Chapter 8) are properly handled, a WCDMA-specific deployment results in a network that is easier to optimize. Unfortunately, because of the ever-greater obstacles to site acquisition, this option is only partially achievable. One possible solution is to start with a WCDMA-specific plan, then select the sites from an existing GSM site portfolio when they fulfill the coverage and capacity objectives for the WCDMA network.

1.4 The Effects of Vendor Implementation

In any WCDMA deployment scenario, vendor-specific implementation plays an important role, especially during network optimization. In spite of its 28 volumes containing hundreds of specifications, the Release 99 standard [12] does not cover every detail. As a result, vendors have a great deal of freedom to implement these details differently, to differentiate themselves from competitors.

Vendor implementation affects several areas of network deployments:

- **Node availability.** As mentioned earlier, micro, pico, or other flexible coverage solutions are only beginning to become available in the market. Eventually, all vendors will offer multiple Node Bs for different applications, but today the choice is limited. Vendor-specific implementations of RNC scalability are important for consideration. For example, an RNC with limited backhaul connectivity should have ample Iur connectivity and capacity to ensure that soft handovers can be supported across the RNCs.
- **Hardware architecture.** This is another area that will continue to evolve in the next few years. From a deployment point of view, the architecture itself is not critical but network planners should evaluate the performance associated with the architecture, as well as its expandability. The expandability should be considered as a two-dimensional space, where features and supported nodes are the axes. The importance of feature expandability is already an issue with High-Speed DL Packet Access (HSDPA; see Chapter 7), and High-Speed UL Packet Access, which is being finalized and deployed while Release 99 networks are still being rolled out. This continuous rollout – together with the standardization of WCDMA in different frequency bands, and node availability – also emphasizes the importance of expandability.
- **Performance.** This is only guaranteed by the standard to a limited extent. The standard mainly defines RF performance; however, there are other performance aspects to be considered. From a user's point of view, processing speed performance is at least as critical as RF performance. Even in lightly loaded networks, the processing speed for signaling can affect user performance. An example is how Measurement Report Messages (MRMs) are processed: an architecture that queues the MRMs instead of processing them in parallel is likely to retain obsolete members in the Active Set. When setting up parameters (see Chapter 4), optimization engineers must weigh such performance issues and their trade-offs.
- **Parameter settings.** This also is limited to a large extent by the architecture performance. To simplify initial implementation, most parameters can be set only at the RNC level, even if cell-level or area-level setting might eventually be required. But eventually, as in other maturing systems, such as GSM or CDMA2000 1X, the whole range

of standard parameters will be set at the Node B or cell level. This affects the workload of optimization engineers, since the number of parameters must then be multiplied by the number of possible permutations. Understanding the parameters in detail is so important that this book devotes an entire chapter to that subject; see Chapter 4.

References

[1] C.S0024-A. CDMA2000 high rate packet data air interface specification. 3GPP2; 2004.
[2] 25.858. Physical layer aspects of UTRA high speed downlink packet access. 3GPP; 2002.
[3] Richardson A. *WCDMA Design Handbook*. Cambridge University Press; 2005.
[4] Tanner R, Woodard J. *WCDMA Requirements and Practical Design*. Wiley & Sons; 2004.
[5] Mischra AR. *Fundamentals of Cellular Network Planning and Optimization, 2G/2.5G/3G... Evolution to 4G*. Wiley; 2004.
[6] Mouly M, Pautet M-B. *The GSM System for Mobile Communications*. Cell & Sys; 1992.
[7] R1999 Specification. ftp://ftp.3gpp.org/specs/2000-06/R1999/. 3GPP; 2000.
[8] 25.311. Radio Resource Control (RRC) protocol specification. 3GPP; 2004.
[9] 25.211. Physical channels and mapping of transport channels onto physical channels (FDD). 3GPP; 2002.
[10] 34.108. Common test environments for User Equipment (UE) conformance testing. 3GPP; 2002.
[11] 25.212. Multiplexing and channel coding (FDD). 3GPP; 2002.
[12] 3GPP specification. http://www.3gpp.org/specs/specs.htm.

2

RF Planning and Optimization

Christophe Chevallier

2.1 Introduction

For Wideband Code Division Multiple Access (WCDMA), as with any cellular technology, the first step in ensuring good Quality of Service (QoS) is to have strong, reliable coverage. Unlike in Frequency Division Multiple Access (FDMA)/Time Division Multiple Access (TDMA) planning, WCDMA coverage and interference planning should be done at the same time because improvements in coverage may create interference and any attempt to control interference may cause coverage holes. Since the frequency reuse number in WCDMA is 1, WCDMA network planning becomes challenging when finding the balance between coverage and interference.

This chapter begins with an overview of the entire deployment process, from initial planning to commercial operation. This is followed by step-by-step details on coverage planning and optimization, including sample Uplink and Downlink Link Budgets to provide reliable coverage. The following sections describe network planning and modeling tools that can be used to achieve a balance between coverage and interference by theoretical analysis and simulations. The final sections show how real field measurement data can be used to support and improve the overall process.

2.2 Overview of the Network Deployment Process

The network life cycle can be represented in three steps, as shown in Figure 2.1. First, the network is planned. Then the network is deployed, at which time initial optimization can begin. When acceptable QoS is achieved, the network can enter commercial operation, at which point the continuous optimization process begins, which ensures that evolving performance and capacity needs are met.

2.2.1 Network Planning

Network planning consists of two steps: initial planning and detailed planning, as shown in Figure 2.2.

WCDMA (UMTS) Deployment Handbook: Planning and Optimization Aspects QUALCOMM Incorporated

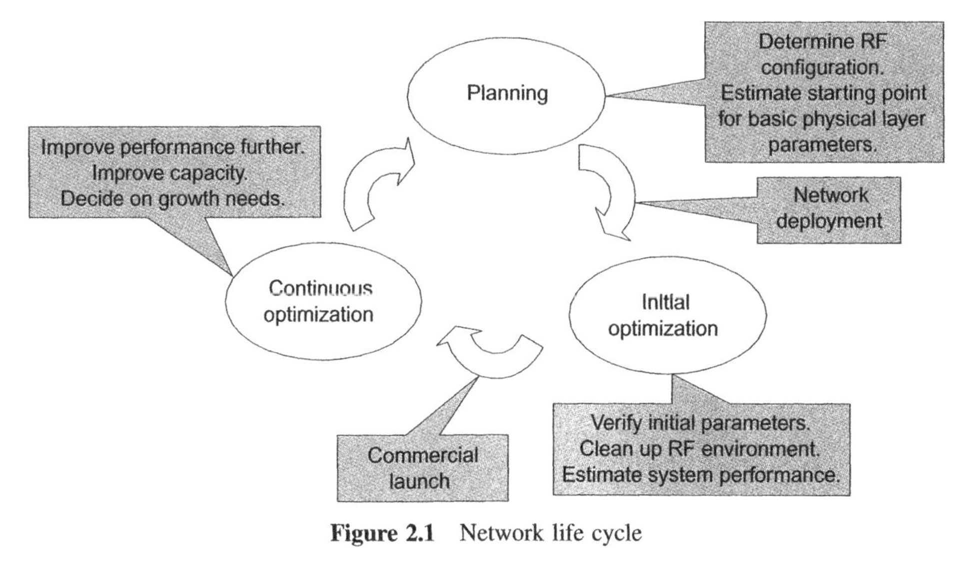

Figure 2.1 Network life cycle

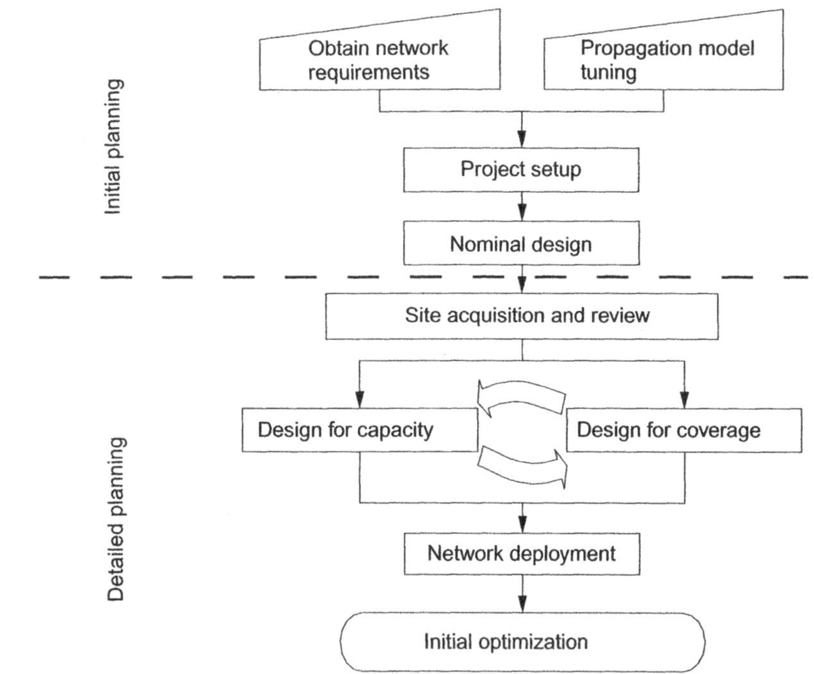

Figure 2.2 Network planning process overview

During the initial planning – or nominal design – phase, the number of sites needed to cover the target area is estimated to ensure that the requirements for both coverage and capacity are met. The coverage requirements can be verified with Link Budgets and Radio Frequency (RF) models. Section 2.3 describes Link Budgets. This book does not discuss RF models because their accuracy depends greatly on the morphology of the site. RF models are best considered on a case-by-case basis, or adapted from similar models developed at a similar frequency – typically 1800 MHz (Europe) or 1900 MHz (US). Chapter 3 discusses extensively capacity requirements, as well as coverage and capacity trade-offs.

During detailed network planning, site placement is estimated to ensure that coverage and capacity requirements are met. This is done with the help of network planning tools, which require extensive inputs to provide accurate results. Using these tools, coverage can be estimated both in terms of signal level and interference. The main advantage of these tools is that they can use Monte Carlo or similar simulations to estimate the interaction between coverage and capacity. Ultimately, the network planning stage governs where and how nodes are deployed in the network.

Once the network is deployed, its behavior should be tested against requirements and expectations. A gradual process is recommended. It is begun by verifying that the conditions required to start optimization have been set. First the RF conditions, then the low-level services (voice), and then higher-level services (such as video telephony and packet switched [PS] data) are measured and adjusted. This approach simplifies the optimization process by ensuring that only a limited set of parameters is considered during each step.

2.2.2 *Initial Optimization*

The optimization process, illustrated in Figure 2.3 starts with the preoptimization tasks aimed at verifying that all RF configuration and system parameters are set according to plan, and that all features are working properly under ideal RF conditions.

RF optimization primarily evaluates the RF configuration – that is, the antenna type, azimuth, tilt, or height – to ensure that the measured signal level (Receive Signal Code Power [RSCP]) and interference (E_c/N_o) meet the design target. During this process, in addition to RF configuration, a limited number of parameters can be tuned: Primary Scrambling Code (PSC) assignments and Monitored Cell Lists. Although Common Pilot Channel (CPICH) power allocation parameters can be allocated on a per-cell basis, this is not recommended because of their impact on link balancing; consequently, CPICH power allocation is not considered as a settable parameter during network optimization.

Parameter tuning should start with service optimization. At that stage, reselection, access, and handover parameters are of primary importance. These parameters are initially tuned for voice service. No further access and handover parameter tuning is necessary because the access and handover processes are similar for all types of service. For reselection, further tuning may be necessary for PS data, especially for connected modes (CELL_FACH, notably).

For video telephony or PS data service, additional parameters should be evaluated: target Block Error Rate (BLER), power assignment and power control, admission and congestion control, rate and type switching, and Radio Link Control (RLC) parameters.

Preoptimization tasks are a critical part of initial optimization and should be performed before sending a data collection team. The preoptimization tasks listed below ensure a smooth transition to all the subsequent optimization steps by verifying that all nodes and features are available.

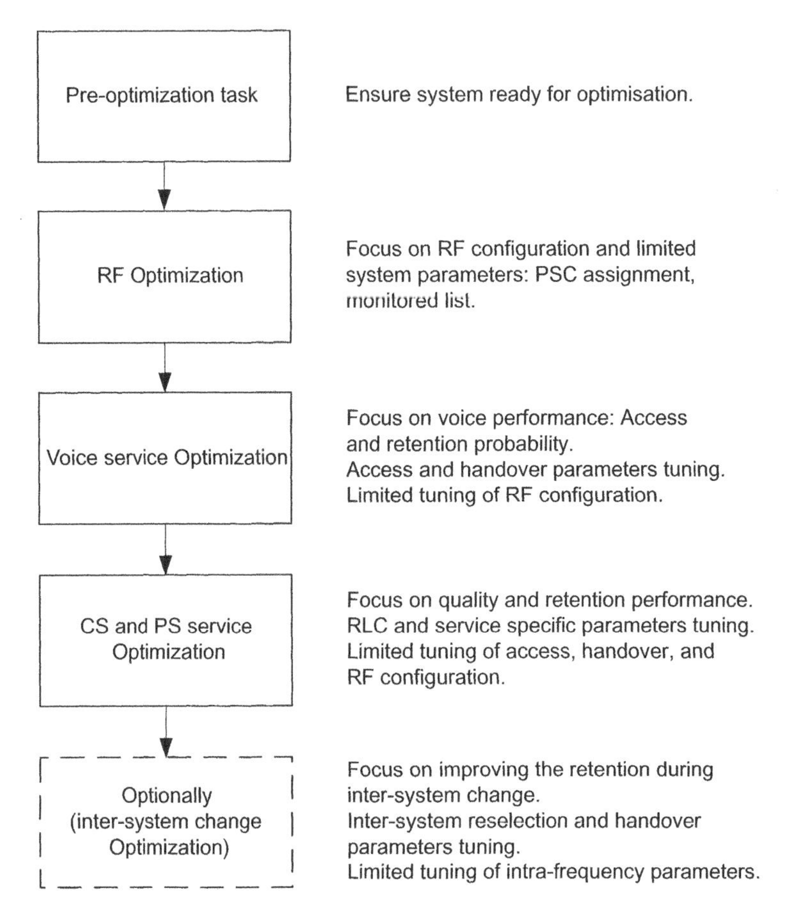

Figure 2.3 Overview of optimization process

- **Basic feature verification.** Both call access (origination and termination) and call retention performance tests should be done in controlled, stationary RF conditions. *Controlled RF conditions* means a stable, strong received signal with limited interference. A target received signal of −70 dBm ± 10 dBm will fulfill part of the conditions and is within the user equipment (UE) operating range of −107 dBm to −25 dBm [1]. Interference should be limited by the number of cells the UE can detect in the test location as well as by the low load under which each cell is operating. No maximum number of cells required for this verification can be defined, but a minimum of two cells belonging to different Node Bs is sufficient to reliably verify all the features, including cell reselection and soft handover testing. This testing can be done under static conditions because the goal is to ensure that call access and call retention are reliable; RF is not the main focus at this stage. Any issues discovered in this test will be in the realm of interoperability or commissioning, rather than optimization. With this in mind, only a few nodes need to be tested (typically a test bed). In addition to call access and retention, basic feature verification may include intra-system cell reselection and, depending on the implementation, inter-RNC handover.

- **Site verification.** Site verification is similar to basic feature verification except that only a limited number of tests are performed, and these tests are done for each Node B. Typically, a good RF environment is chosen for this test. However, it is best to run the verification independently for each cell of each site. Since this is a functional test and not a performance test, only a few calls in each cell are required.
- **Site inspection.** Site inspection ensures that all the sites in the tested area have been built as designed in the network plan. The inspection focuses on all parameters that affect RF propagation. This includes the antenna system (feeders, jumpers, Tower Mount Amplifiers [TMAs], and so on), the exact position of the site, and the parameter settings. For an impartial assessment, it is highly recommended that a trusted third party, other than the installation and commissioning team, be responsible for site inspection.

Section 2.8 and Chapter 8 discuss the RF and service optimization processes.

2.2.3 Continuous Optimization

Once the desired QoS for each service has been achieved and the network is commercially deployed, the process of continuous optimization begins.

The early stage of continuous optimization verifies that performance meets expectations after the first commercial users are introduced. Differences between the initial optimization and the continuous optimization are likely to be observed because commercial traffic patterns rarely match test traffic patterns. Different tools are needed to monitor commercial traffic. Because it is impractical to log every user, a statistical view based on network counters should be employed. As the commercial traffic grows, network capacity must be adjusted. This may be addressed by adding sites, carriers, or technology, all of which require further planning and optimization.

2.3 Link Budgets

Link Budgets estimate acceptable signal power levels by calculating the Maximum Allowable Path Loss (MAPL). During this analysis, several assumptions should be made and documented. These assumptions concentrate on a few scenarios that represent the deployment area. A drawback of Link Budgets is that they present an oversimplified view of the network. For the best possible accuracy, the assumptions should be selected as carefully as possible.

One key assumption is homogeneity of the parameters across all cells. In a deployed network, interference varies for each user and each cell because of different UE manufacturers, imperfect power control, different channel conditions, different RF configurations, and different bit rates.

To complement the early phase of dimensioning, typically done with a spreadsheet, complete network simulations are necessary. These would better represent the performance of the network, assuming that all the simulations are run using accurate parameter settings. However, simulations are time consuming and impractical when multiple scenarios are required, as it is the case during initial planning.

Link Budgets are further limited by their inability to estimate coverage and capacity trade-offs. Although Link Budgets include a loading term (for Uplink or Downlink), this term usually does not represent a true distribution of traffic over time, service, and location. These factors affect loading and are important for understanding the coverage and capacity trade-off.

In contrast to how it is commonly done in FDMA/TDMA, network planning for WCDMA examines the Uplink path before the Downlink path, from both the CPICH and the different Radio Bearers. The Uplink is examined first because the assumptions are simpler and more widely accepted than they are for the Downlink. This is due to several factors, notably the following:

- **Single reception point (Uplink) versus multiple reception points (Downlink).** This factor causes the conditions at the receiver to be identical for all users. On the Uplink, perfect power control can be assumed. This means that all users receive the same power and create the same amount of interference for other users, even if the channel conditions are different for each path. On the Downlink, each user is in different RF conditions, in terms of channel conditions, interference, and handover conditions.
- **Selection (Uplink) versus combining (Downlink) handover gain.** For each link, the soft/softer handover has a positive impact. On the Uplink, the fact that the best link can be selected at any given time leads to a gain that can be considered a reduction of Log-normal Fading (LNF). This can be estimated analytically using widely accepted formulas [2] and as presented in Section 2.3.1.18. On the Downlink, handover has multiple impacts, all of which reduce the traffic channel power required to overcome interference.
- **Independence (Uplink) versus interaction (Downlink) of users.** On the Uplink, the maximum transmit power is limited by the amount of power available to each user, as determined by the capability of each UE. The required transmit power is affected by other users in proportion to the interference they create. On the Downlink, interference from other users in the same cell is reduced, because of the orthogonality of the channels. However, the amount of traffic channel power available depends on all users, because a single power amplifier is used.

2.3.1 Uplink Link Budgets

Table 2.1 shows a sample Uplink Link Budget for Adaptive Multi-Rate (AMR) voice. The subsequent sections discuss each item in the Link Budget and address its effect on speech AMR, CS64 bearer (typically used to support video telephony), and PS data bearer services.

The references and formulas in Table 2.1 can be used in a spreadsheet program, such as Microsoft Excel. By convention, lowercase references are used for logarithmic (dB or dBm) values, while uppercase references are used for linear values. Gain is represented as a positive value (+), while loss is represented as negative value (−). As an example the Rise over Thermal (RoT) is entered as a negative number, meaning a loss or a reduction in the MAPL. The following sections explain each entry in Table 2.1.

2.3.1.1 Maximum Transmit Power

Maximum transmit power is dictated by the UE class. Table 2.2 shows the maximum transmit power for each UE class and the associated tolerances [3]. A voice-centric (i.e., handheld) UE is usually class 3 or 4. A data-centric (i.e., data card) UE is usually class 3. If the network can mix UE classes, the Link Budget should be drafted for the highest class

Table 2.1 Link Budget example for Speech AMR Uplink

Reference	Description	Values	Units	Formula
a	Maximum transmit power	21.0	dBm	Input
b	Cable and connector losses	0.0	dB	Input
c	Transmit antenna gain	0.0	dBi	Input
d	ERP	21.0	dBm	$= a + b + c$
e	Thermal noise density		dBm/Hz	$E = kT$
				k: Boltzmann constant (1.38×10^{-23})
				T: temperature in K (typically 290 K)
		−174.0		$e = 10 \times \log(E \times 1000)$
f	Information full rate		dB-Hz	$f = 10 \times \log(F)$
		40.9		F input
g	Thermal noise floor	−133.1	dBm	$= e + f$
h	Receiver noise figure	5.0	dB	Input
i	Load (fraction of pole capacity)	0.5		Input
j	Rise over Thermal	−3.0	dB	$= 10 \times \log(1 - i)$
k	Required E_b/N_t	7.2	dB	Input
l	Sensitivity	−117.9	dBm	$= e + f + h - j + \mathrm{k}$
m	Receive antenna gain	17.0	dBi	Input
n	Cable, connector, combiner losses	−3.0	dB	Input
o	Rx attenuation and gain	14.0	dB	$= m + n$
p1	Cell edge confidence	0.9	%	Input
p2	Standard deviation	8.0	dB	Input
p	Log-normal fading	−10.3	dB	Per equation 2.3
q	Handover gain	4.1	dB	Per equation 2.4
				$\times 1.6 - (8 - \mathrm{p2})/10$
r	Diversity gain	0.0	dB	Input
s	Car penetration losses	0.0	dB	Input
t	Building penetration losses	−20.0	dB	Input
u	Body loss	−3.0	dB	Input
v	Propagation components	−29.2	dB	$= p + q + r + s + t + u$
w	Maximum Allowable Path Loss	123.7	dB	$= d - l + o + v$

Table 2.2 Maximum transmit power and associated tolerance for each UE class

UE class	Maximum Tx power [dBm]	Tolerance [dB]
1	33	+1/−3
2	27	+1/−3
3	24	+1/−3
4	21	+2/−2

(lowest maximum transmit power). For a more conservative estimate, the Link Budget can use the lowest acceptable transmit power for the class, considering the tolerances, such as 19 dBm for UE class 4.

2.3.1.2 Cable and Connector Loss

Because a handheld UE typically does not use an external antenna, this item is null. For data cards, external antennas are commonly available. If this option is widely used in the network, the loss associated with the cable should be counted. In addition, the Link Budget also must account for the antenna gain (typically 3 to 6 dBi).

2.3.1.3 Transmit Antenna Gain

Because a handheld UE typically does not use an external antenna, the gain is set to null (0 dBi).

2.3.1.4 Effective Radiated Power

Effective Radiated Power (ERP) is calculated for all gains and losses in the transmitter chain.

2.3.1.5 Information Full Rate

Information Full Rate is the bearer data rate. On the Uplink, the main data rates are 12.2 kbps for speech, and 64 kbps for PS or Circuit Switched (CS) data services. Release 99 of the standard [4] supports higher data rates of up to 384 kbps for PS (limited to 64 kbps for CS), but these rates are not commonly implemented.

2.3.1.6 Thermal Noise Floor

The thermal noise floor in this Link Budget is calculated using bit rate, not chip rate, to ensure consistency with the sensitivity calculation. Thermal noise floor affects the sensitivity.

2.3.1.7 Receiver Noise Figure

The Node B noise figure should be consistent with the site configuration. A value of 4 to 5 would assume that no TMAs are used. If TMAs are used, the cascaded amplifier noise factor in the linear domain is calculated using Equation 2.1.

$$NF = NF_1 + \frac{NF_2 - 1}{G_1} + \frac{NF_3 - 1}{G_1 \times G_2} + \frac{NF_4 - 1}{G_1 \times G_2 \times G_3} \tag{2.1}$$

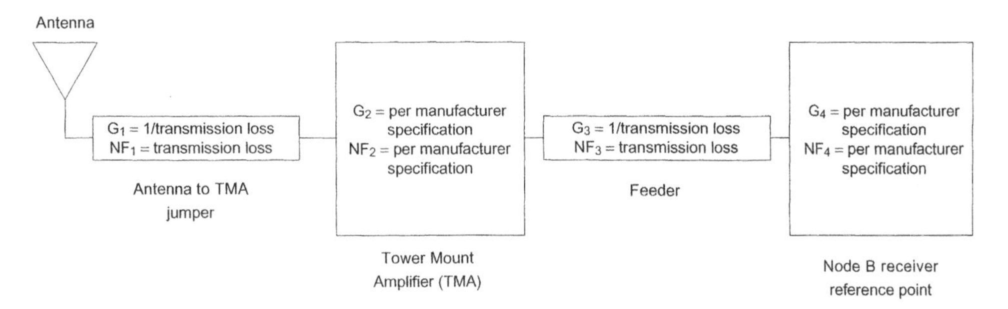

Figure 2.4 Receive chain considered for noise figure calculation

This equation is limited to four stages, shown in Figure 2.4, which would normally be sufficient for a receive path. Stage 1 is the jumper from the antenna to the TMA. Stage 2 is the TMA. Stage 3 is the feeder. Stage 4 is the Node B receiver. For each stage, NF_n refers to the noise factor and G_n refers to the gain. For passive (e.g., feeder) components, the gain would be lower than 1; to reflect a loss, the noise factor should be calculated as the inverse of the gain.

2.3.1.8 Load

In the Link Budget, load is the only input representing traffic. When estimating the trade-off between coverage and capacity, the coverage and capacity load estimations are set to the same value. Loading of 50% is typical for symmetric traffic with links of similar capacity. However, this assumption does not apply if data services are in the traffic mix. With data services, Uplink loading is expected to be about 35 to 40%. This is further discussed in Chapter 3.

2.3.1.9 Rise over Thermal

RoT is also known as *Interference Margin* or *Load Margin*. The RoT is estimated from the loading value by means of Equation 2.2.

$$\mathrm{RoT} = 10 \times \log_{10}(1 - load) \tag{2.2}$$

The result of this equation will be negative to reflect a loss in the Link Budget.

2.3.1.10 Required E_b/N_t

The required E_b/N_t, energy per bit-over-total interference (N_t) ratio is influenced by four factors: the coding of the bearer, channel conditions, the target BLER, and the quality of the receiver.

Error-resistant coding helps reduce the E_b/N_t requirement. For instance, bearers that support turbo coding (mainly used for interactive, or PS data, applications) typically have a lower E_b/N_t requirement compared to bearers supporting only convolutional coding (mainly used for conversational, or CS, applications).

Similarly, if an application can sustain a higher probability of errors (BLER), the E_b/N_t requirement is reduced.

Channel conditions, namely multipath effects and UE speed, also affect the required E_b/N_t. More unresolvable multipaths require higher E_b/N_t because Rake receivers and power controls become less efficient. Thus, higher energy per bit is required to recover the same amount of energy. Time diversity provided by the coding is not sufficient at UE speeds between a few kilometers and under 30 kilometers per hour.

The required E_b/N_t can be estimated from link-level simulations, or from manufacturer specifications. For Link Budget purposes, an important assumption is the diversity under which the E_b/N_t numbers were derived. Typically, they correspond to two-pole diversity, with sufficient decorrelation between the paths. This assumption can be verified for antenna separation of over 10 wavelengths using polarization diversity in an urban environment. Table 2.3 shows the minimum performance requirements [1].

The requirements in Table 2.3 are defined for the test conditions (cases) that correspond to the delays and the relative power for the multipaths shown in Table 2.4.

Table 2.3 Example of E_b/N_t requirements for different bearers

Bearer	Target BLER [%]	Minimum requirement [dB]				
		Static (AWGN)	Case 1	Case 2	Case 3	Case 4
Speech AMR 12.2 kbps	1	5.1	11.9	9.0	7.2	10.2
CS64 kbps	1	1.7	9.2	6.4	3.8	6.8
PS64 kbps	10	1.5	6.2	4.3	3.4	6.4
PS128 kbps	10	0.8[a]	5.4[a]	3.7[a]	2.8[a]	5.8[a]
PS384 kbps	10	0.9	5.8	4.1	3.2	6.2

[a] Value corresponding to 144 kbps rather than 128 kbps.

Table 2.4 Multipath profiles [1] used in the minimum performance determination

Case, per 25.101	Case 1, speed 3 km/hr		Case 2, speed 3 km/hr		Case 3, speed 120 km/hr		Case 4, speed 3 km/hr	
	Relative delay [ns]	Relative mean power [dB]	Relative delay [ns]	Relative mean power [dB]	Relative delay [ns]	Relative mean power [dB]	Relative delay [ns]	Relative mean power [dB]
First detected path	0	0	0	0	0	0	0	0
Second detected path	976	−10	976	0	260	−3	976	0
Third detected path	NA	NA	20000	0	521	−6	NA	NA
Fourth detected path	NA	NA	NA	NA	781	−9	NA	NA

2.3.1.11 Sensitivity

Sensitivity is calculated on the basis of concepts presented in Figure 2.5. The minimum signal level received (at the chip level) for successful demodulation is based on how much bit energy is received and whether the bit energy is sufficiently above the noise and interference generated from thermal noise and other users. Thermal noise is calculated at a bit rate rather than a chip rate. This is equivalent to calculating the thermal noise at the chip rate, then spreading it over the bandwidth with a ratio of W/R_b, where W is the spreading bandwidth in chips per second (3.84 Megachips per second, Mcps for WCDMA) and R_b is the bearer bit rate (12.2 kbps for speech AMR, and 64 kbps for either CS or PS data).

2.3.1.12 Receive Antenna Gain

By convention, antenna gains are given in dBi, referenced to an isotopic element. The antenna gain used in the sample Link Budget allows for a beamwidth of 65 degrees and still is less than 2 meters tall.

2.3.1.13 Cable, Connector, and Combiner Losses

This entry accounts for the losses associated with all the components between the antenna and the Node B reference point. The values are similar on both the transmit (Downlink) and the receive (Uplink) sides. These losses should be set according to the expected deployment configuration, including whether the antenna will be shared with other tech-

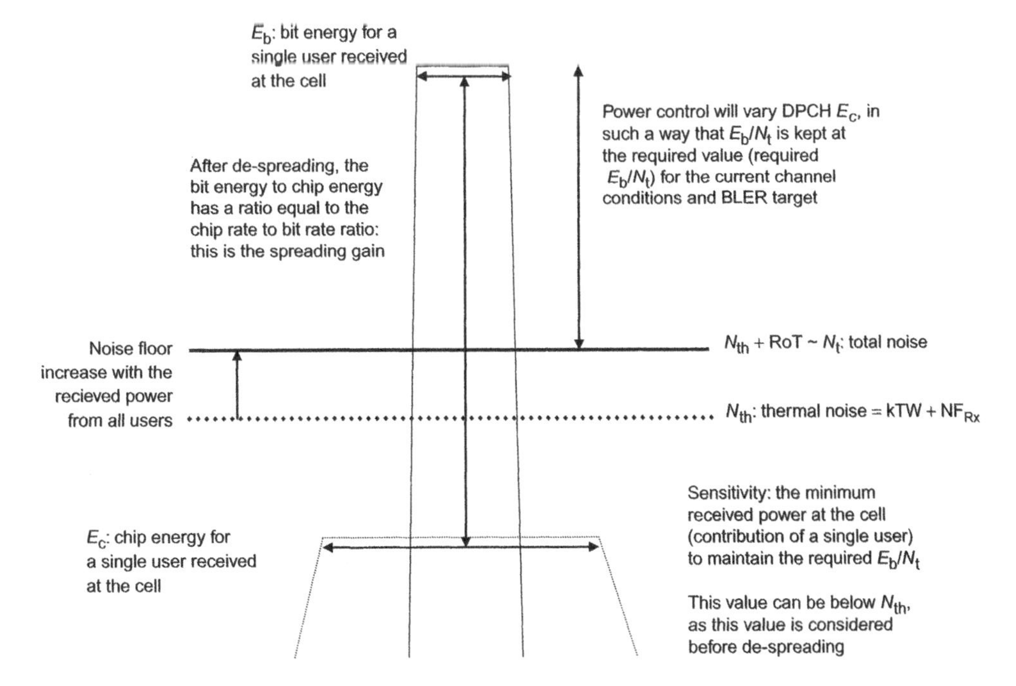

Figure 2.5 Uplink receiver sensitivity concepts

nologies (Global System for Mobiles [GSM] notably), whether existing feeders will be reused, or whether tower-mounted amplifiers will be used.

At the IMT-2000 frequency band (2.1 GHz), the feeder is usually the main source of loss.

2.3.1.14 Receive Attenuation and Gain

The receive attenuation gain, or loss, is calculated by adding all the gain and loss in the receive path.

2.3.1.15 Cell Edge Confidence

The cell edge confidence, with a standard deviation, is used to calculate the LNF margin. The cell edge confidence represents the probability that coverage will be available at the periphery of the cell, on the basis of a statistical distribution of the path loss. Instead of the cell edge *confidence*, the cell area *probability* can be used to estimate the network coverage QoS.

The cell area and cell edge can be linked. The graph shown in Figure 2.6 [2] represents the sensitivity to the path loss exponent, r, and the standard deviation, σ. For the typical value of 90% cell edge confidence, the cell area confidence is 97% over a wide range of r and σ. The cell area confidence should be set in accordance to the expected QoS. For example, for 90% cell edge confidence and 97% cell area confidence, the maximum target for call success should be 97%. Expecting better QoS than cell area confidence would be equivalent to expecting service in areas where coverage is not available.

2.3.1.16 Standard Deviation

Standard deviation represents the dispersion of the path loss or received power measured over the coverage area. The morphology of the area – manmade structures and natural obstacles – disperses the signal propagation by altering the line of sight and causing diffraction, refraction, and reflection. The standard deviation shows limited correlation with frequency, but this can vary greatly with morphology [5–7]. In cities, the standard deviation can be estimated as 4 to 12 dB. Table 2.5 shows standard deviations by land use density.

2.3.1.17 Log-normal Fading

Considering that the received signal at a given location can be represented by a Gaussian distribution, the LNF can be estimated as a complementary cumulative probability distribution function. For easy implementation in a spreadsheet, this can be determined from the cell edge probability (ε) and the standard deviation in the area (σ), as shown in Equation 2.3.

$$\mathrm{LNF} = -\mathrm{NORMINV}(\varepsilon, 0, \sigma) = \sigma \times \mathrm{Q}(\varepsilon)$$

where

$$Q(\varepsilon) \cong \int_{\varepsilon}^{\infty} \frac{1}{\sqrt{2 \times \pi}} \times e^{-\lambda^2/2} d(\lambda) \tag{2.3}$$

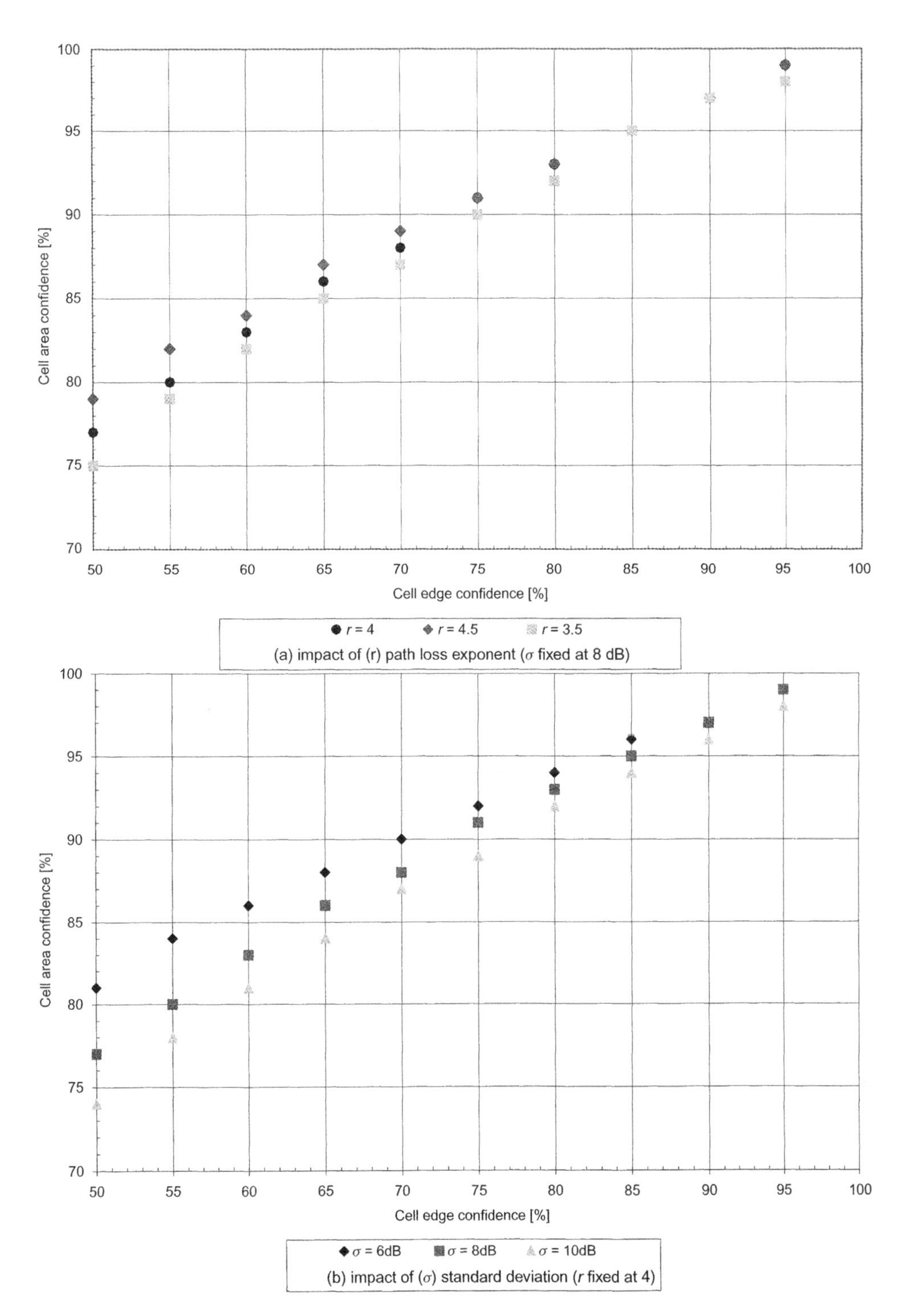

Figure 2.6 Cell edge confidence versus cell area probability, as a function of path loss exponent (a) and standard deviation (b)

Table 2.5 Standard deviation per land use category

Land use category	Practical range for standard deviation [dB]
Suburban	4–8
Urban	8–10
Dense urban	10–12

2.3.1.18 Handover Gain

On the Uplink, the handover gain can be seen as a reduction in the LNF. This assumption is valid if we consider that the paths to the different cells serving a call are independent, or have a limited correlation. Path independence for cells from different sites is obvious, because the geographical locations of the receiver lead to independent path obstructions. For cells of the same site, as in the softer handover case, paths could have greater correlation, resulting in a reduced statistical handover gain. The Link Budget does not take this into account, because it is partly compensated for by possible soft combining. Assuming that both paths are equal, the combining gain would be 3 dB for both directions. Equation 2.4 estimates the handover gain for 50% correlation.

$$Gain_{HO} = \sqrt{NORMINV(\varepsilon, 0, \sigma) - NORMINV(1 - \sqrt{1-\varepsilon}, 0, \sigma)} \times \left(1.6 - \frac{8-\sigma}{10}\right) \tag{2.4}$$

2.3.1.19 Diversity Gain

Diversity gain is set to 0 dB in this Link Budget because it was included in the E_b/N_t requirement. Including two-pole diversity in the E_b/N_t requirement is a standard practice. Two-pole diversity can be achieved either with two spatially separated and vertically polarized antennas or with a single cross-polarized antenna. However, if sufficient signal decorrelation cannot be achieved, then vertically polarized antennas must be separated by a minimum of 10 wavelengths to achieve reasonable decorrelation. Separating the antennas even more will further improve decorrelation. For cross-polarized antennas in a macrocellular network, the decorrelation achieved in dense suburban or higher density environments yields a diversity gain comparable to space diversity (10 λ). For morphologies with few manmade obstructions, such as rural environments, the cross-polarization gain can be reduced by up to 2 dB, compared to urban and dense urban environments [8].

Table 2.6 shows how the various adjustments affect antenna diversity gain.

2.3.1.20 Car Penetration Losses

Car penetration losses depend on car type and construction, as well as local regulations. Historically, for network design, car penetration losses were set between 3 dB and 6 dB. For newer car materials, such as heat-efficient glass, this value can be increased up to 10 dB.

Table 2.6 Diversity gain adjustment for different morphologies and diversity techniques

Morphology	Space diversity			Polarization diversity
	20 λ or over	10–15 λ	5 λ	
Rural				−2
Suburban	Reference	−0.5	−1	−1
Urban				0

Many countries have regulations that ban cell phone use while driving unless a hands-free kit has been installed. The location of this car kit equipment causes great variation in car penetration loss.

Taking all these factors into account, the conservative value for car penetration loss is 6 to 10 dB.

2.3.1.21 Building Penetration Losses

Building Penetration Loss (BPL) is discussed in Chapter 8. In our sample Link Budget, BPL is a single number, although in reality the value depends on the area of expected coverage. Values below 20 dB are usually sufficient to cover the ground floor area immediately inside the outermost walls of the building. Values up to 45 dB would be required to cover 95% of the ground floor space. This amount of coverage may require other deployment scenarios, such as microcells or an indoor solution, to overcome the interference created by the external building walls.

2.3.1.22 Body Loss

Body loss is affected by the evolution of handsets and how people use them. With hands-free kits, the UE can be located anywhere on the user, not necessarily close to the head. This could cause large variations in body losses, but no definitive characterization has been done.

For video-telephony applications, body loss can be ignored or reduced as compared to voice applications, because users will be holding the UEs at arm's length away from their bodies.

For PS data usage, the body loss depends on both the UE and the application. For mini-browser applications, the UE is held in the hand so the user can navigate the built-in browser. Body loss is assumed to be similar to that in video-telephony applications. In contrast, for mobile office applications on a UE with a Personal Computer (PC)-card, equipment loss is a greater factor than body loss. The loss created by this equipment depends on the type of antenna (fixed integrated, swivel integrated, or external) and the computer to which the card is connected.

The 3 dB value in the Link Budget can be considered conservative. Measurements can reveal attenuation from 2 to 5 dB for a UE held at head-level, depending on the UE antenna design and its direction relative to the main server.

2.3.1.23 Propagation Components

In a Link Budget, the penetration losses and user-related losses are similar: both represent additional attenuation that affects signal and interference equally. Thus, penetration losses and user-related losses can simply be added together.

2.3.1.24 Maximum Allowable Path Loss

The MAPL can be calculated from the Effective Radiated Power (ERP), the Receive sensitivity ($Rxsens$), the received attenuation and gain ($Rx_{gain+loss}$), and the propagation components ($Propagation_{gain+loss}$), using Equation 2.5:

$$MAPL = ERP - Rxsens + Rx_{gain+loss} + Propagation_{gain+loss} \tag{2.5}$$

2.3.2 Downlink Link Budget for CPICH

Downlink coverage varies according to the mechanism by which a UE accesses a Universal Mobile Telecommunications Systems (UMTS) network. Figure 2.7 presents a simplified diagram of the access procedure, from turning on UE power until a bearer is established and the UE is in Idle Mode.

This diagram shows three distinct coverage states: system acquisition, access probability, and maintaining a dedicated channel (DCH).

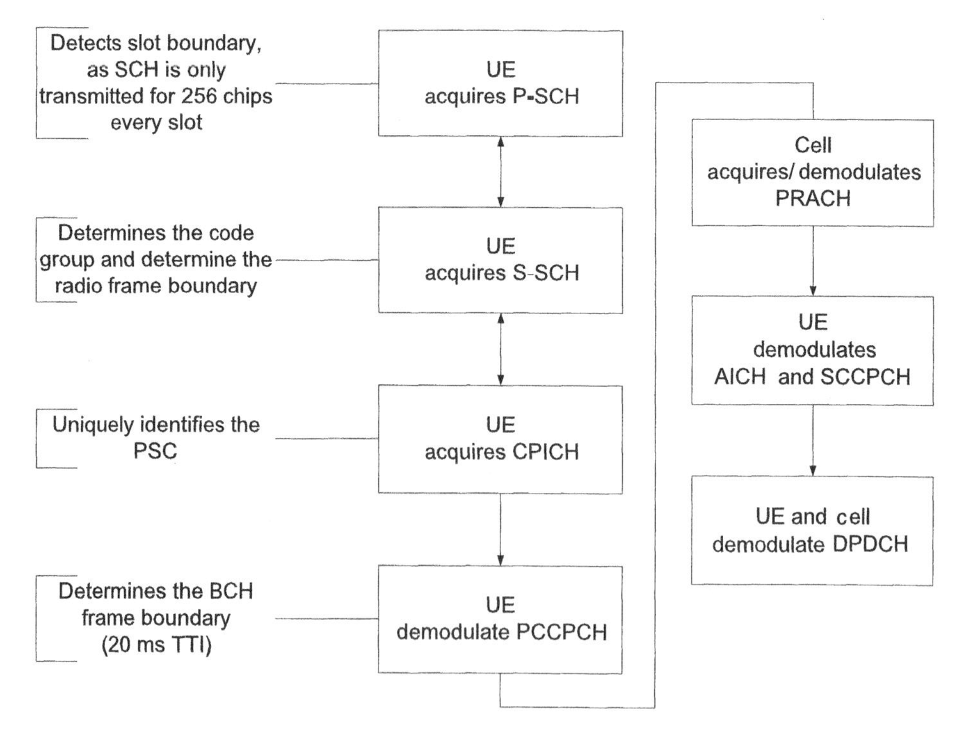

Figure 2.7 Simplified flow diagram from UE power on to Idle Mode

- **System acquisition.** During system acquisition, coverage can be defined as the probability that the system will be acquired when the UE is powered on. Four channels must be acquired or demodulated in the following order: Primary Synchronization Channel (P-SCH), Secondary Synchronization Channel (S-SCH), CPICH, and Primary Common Control Physical Channel (PCCPCH). The acquisition or demodulation of any of these channels depends on their assigned power, the channel conditions, and the performance of the UE. A Link Budget can be written for each of these. Since none of these channels is power-controlled and the entire process relies only on Downlink channels, this is a good definition of coverage for a Downlink Link Budget representing the maximum extent of cell coverage.
- **System access.** During system access, system coverage can be defined as the probability of successful call termination or call origination. Defining a Link Budget in this case is not practical, because both Uplink and Downlink channels affect the overall results.
- **Connected state.** Connected state corresponds to the Link Budgets defined in Section 2.3.1 for Uplink and Section 2.3.3 for Downlink, with the log-normal margin accounting for the random nature of the process. In the Connected Mode, the CPICH should be monitored at all times, because handovers are based on CPICH measurements.

Because CPICH is required for system acquisition and must be monitored at all times afterwards, it would be natural to use this channel to define the maximum possible coverage. This idea is further justified when we consider Idle Mode processes. CPICH-measured values are compared to broadcasted thresholds to evaluate the need for reselection or to declare out-of-service area occasions, indicating that the UE is outside the coverage area.

Having established that a Link Budget based on the CPICH would represent an absolute Downlink coverage, Table 2.7 presents a CPICH-based Downlink Link Budget example. When evaluating the coverage of WCDMA services, the coverage for all the common channels is assessed to ensure proper acquisition, reselection, and access. This assessment can be made using the CPICH Link Budget for channels that are not demodulated (Synchronization channel mainly), and the Downlink service Link Budget for channels that are demodulated (Acquisition Indicator Channel [AICH], Paging Indicator Channel [PICH], PCCPCH, and Secondary Common Control Physical Channel [SCCPCH]). The target E_b/N_t for any of the demodulated channels should be known and combining gain should not be considered. Using soft handover (macrodiversity) for any of the channels used in Idle or Access state assumes that reselection works perfectly and the UE can instantaneously reselect to the stronger server. If the reselection process is delayed, notably when using long Treselection (see Chapter 4), this assumption is not valid and handover should be set to 0 dB.

The most important parameters in this Link Budget are the minimum RSCP and the target E_c/N_o; we recommend that they be set to Qrxlevmin and Qqualmin, respectively. (Chapter 4 explains the reason for this.) This Link Budget assumes that the values correspond to the edge of coverage definition of a fully loaded Code Division Multiple Access (CDMA) system. Once the minimum RSCP is known, the MAPL can be calculated and used to compare Uplink and Downlink coverages.

2.3.3 Downlink Link Budget for Various Services (Connected Mode)

Table 2.8 shows a Link Budget example for the Downlink that illustrates the effects of coverage and interference. Interference is estimated by comparing the achieved E_b/N_t to

Table 2.7 Absolute downlink link budget, based on CPICH

Reference	Parameter	Log values	Units	Formula (uppercase represents linear units; lowercase represents log values)
a	CPICH transmit power	33.0	dBm	Input
b	CPICH E_c/I_{or}	−10.0	dB	$= a - f$
c	Cable, connector, combiner losses	−3.0	dB	Input
d	Transmit antenna gain	17.0	dBi	Input
e	CPICH ERP	47.0	dBm	$= a + c + d$
f	HPA max transmit power	43.0	dBm	Input
g	Maximum ERP	57.0	dBm	$= f + c + d$
h	Target load (power)	0.0	dB	Input
i	ERP at target load	57.0	dBm	$= g + h$
j	Thermal noise density	−174	dBm/Hz	See Table 2.1
k	Information full rate	65.8	dB-Hz	$10 \times \log(3840000)$
l	Receiver noise figure	7.0	dB	Input
m	Thermal noise	−101.2	dBm/Hz	$= j + k + l$
n	$I_{oc}/\hat{I}_{or}$	3	dB	Input
o	E_c/N_o	−14.8	dB	$= \dfrac{E/AE}{M + \dfrac{N \times i}{AE}}$
p	Minimum E_c/N_o	−16.0	dB	Input
q	RSCP from target cell	−111.0	dBm	$= e - ae + ad$
r	Minimum target RSCP	−111.0	dBm	Input
s	Receive antenna gain	0.0	dBi	Input
t	Cable, connector, combiner losses	0.0	dB	Input
u	Rx attenuation and gain	0.0	dB	$= s + t$
v	Cell edge probability	0.9	%	Input
w	Standard deviation	8.0	dB	Input
x	Log-normal fading	−10.3	dB	See Table 2.1
y	Handover gain	4.1	dB	See Table 2.1
z	Diversity gain	0.0	dB	Input
aa	Car penetration losses	0.0	dB	Input
ab	Building penetration losses	−20.0	dB	Input
ac	Body loss	−3.0	dB	Input
ad	Propagation components	−29.2	dB	$= x + y + z + aa + ab + ac$
ae	Maximum Allowable Path Loss	128.8	dB	$= e - r + aa + u$
af	Target RSCP counting propagation components	−81.8	dBm	$= q - ad$

Table 2.8 Downlink Link Budget example based on Downlink-Dedicated Physical Channel (DL-DPCH)

Reference	Description	Values	Units	Formula
a	Maximum DPCH Power	29.3	dBm	Input
b	Minimum DPCH E_c/I_{or}	−13.7	dB	$b = f - a$
c	Cable, connector, combiner losses	−3.0	dB	Input
d	Transmit antenna gain	17.0	dBi	Input
e	Maximum DPCH ERP	43.3	dBm	$e = a + c + d$
f	HPA maximum transmit power	43.0	dBm	Input
g	Maximum total ERP	57.0	dBm	$g = f + c + d$
h	Thermal noise density	−174.0	dBm/Hz	See Table 2.1
i	Information full rate	40.9	dB-Hz	See Table 2.1
j	Receiver noise figure	7.0	dB	Input
k	$I_{oc}/\hat{I}_{or}$ (1/geometry)	3.0	dB	Input
l	Beta (combining gain)	3	dB	Input
n	Achieved E_b/N_t	8.9	dB	Per [10] and equation 2.7
o	Required E_b/N_t	8.9	dB	Input
p	Sensitivity, discounting interference, and propagation components	−117.2	dBm	$p = h + i + j + o$
q	Maximum Allowable Path Loss, without interference	160.5	dB	$q - e - p + w$
r	Received power from target cell	−103.5	dBm	$r = g - q$
s	Other-cell interference power density, I_{oc}	−163.4	dBm/Hz	$s = r + l + k - 10 \times \log(3840000)$
t	Sensitivity	−112.0	dBm	$= 10 \times \log(10^{((h+j)/10)} + 10^{(s/10)}) + i + o$
u	Receive antenna gain	0.0	dBi	Input
v	Cable, connector, combiner losses	0.0	dB	Input
w	Rx gain and losses	0.0	dB	$w = u + v$
x	Log-normal fading	−10.3	dB	See Table 2.1
y	Handover gain	4.1	dB	See Table 2.1
z	Diversity gain	0.0	dB	Input
aa	Car penetration losses	0.0	dB	Input
ab	Building penetration losses	−20.0	dB	Input
ac	Body loss	−3.0	dB	Input
ad	Propagation gain and losses	−29.2	dB	$ad = x + y + z + aa + ab + ac$
ae	Maximum Allowable Path Loss	126.1	dB	$ae = e - t + w + ad$

the required E_b/N_t. Coverage is estimated from the MAPL, which directly depends on the assigned Dedicated Physical Channel (DPCH) transmit power.

Many entries in Table 2.8 are the same as those described for the Uplink Link Budget and are described in Section 2.3.1. The following sections explain the entries that are different 1 or added for the Downlink.

2.3.3.1 Maximum DPCH Power

Unlike for the Uplink, in which the maximum transmit power is standardized for each UE class, this value is configurable by the operator for the Downlink. The value can be set as an absolute power or as a value relative to the CPICH power. Depending on the vendor implementation, this value could be the same for all Radio Bearers or settable according to the type of Radio Bearer: voice, CS64, PS data. Setting a value for each Radio Bearer is more adaptable. The power assignment can change to match the desired coverage at the target BLER, or to limit the coverage for a given Radio Bearer to favor capacity. The setting in this sample Link Budget was chosen to meet the E_b/N_t requirement, given all the other assumptions described below. This resulting value is not necessarily the recommended maximum DPCH power – the recommendation should consider extreme cases and capacity, as described in Chapter 4.

2.3.3.2 Minimum DPCH E_c/I_{or}

This value is not directly used in our Link Budget example. We mention it nonetheless because it is a common output of link-level simulators. In such simulators, required transmit powers are not usually specified in absolute values, but as fractions of the maximum transmit power. This emphasizes the fact that Downlink transmit power is set mainly to compensate for interference, rather than to compensate for path loss. If a link-level simulator is used to determine the DPCH transmit power, the Link Budget could be modified to accept the minimum E_c/I_{or} as an input. Then the maximum DPCH transmit power could be calculated from the maximum High Power Amplifier (HPA) power and the minimum E_c/I_{or}.

For example, Figure 2.8 shows the required E_c/I_{or} for several geometries and channels, based on a link-level simulator.

2.3.3.3 HPA Maximum Transmit Power

The vendor determines the maximum transmit power at which the HPA is driven. This value is used only to estimate the Downlink interference.

2.3.3.4 Receiver Noise Figure

The Receiver Noise Figure value should reflect the minimum performance of the UE deployed in the network. Typical values range from 6 to 9 dB.

2.3.3.5 Other- to Same-Cell Interference $I_{oc}/\hat{I}_{or}$

The Link Budget example uses the ratio of other-cell interference to same-cell contribution ($I_{oc}/\hat{I}_{or}$) rather than the better-known term *geometry*, G, which denotes its inverse, as shown in Equation 2.6 [9].

$$I_{oc}/\hat{I}_{or} = G^{-1} \quad (2.6)$$

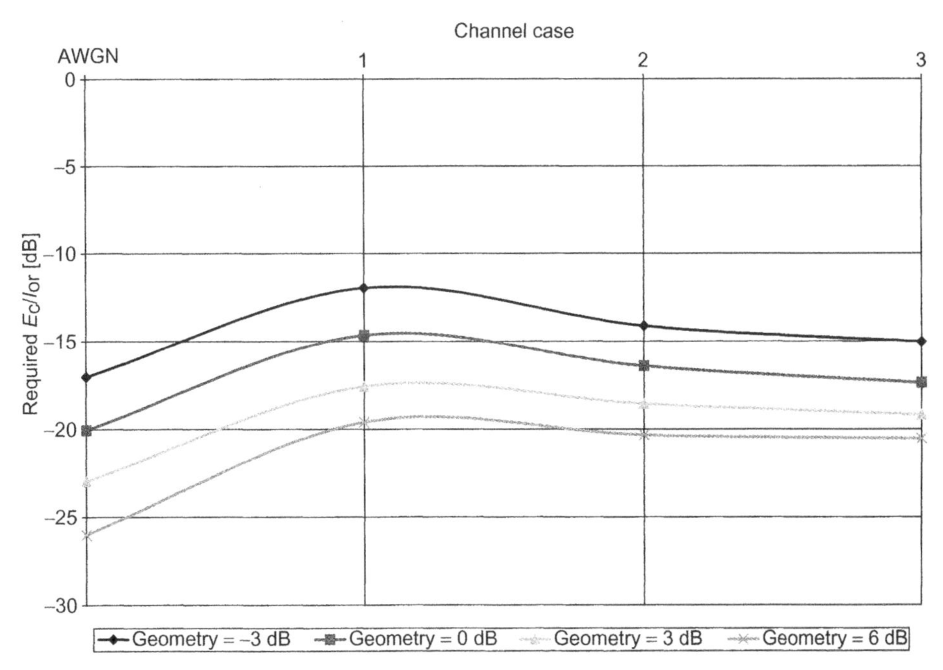

Figure 2.8 Simulated minimum E_c/I_{or} required for various channel conditions, based on a link-level simulator

In this ratio, I_{oc} (other-cell interference) represents the total received interference spectral density for cells not in soft handover. This can be determined from the effective transmit power for all the cells not in soft handover, and the path loss from these cells to the location of interest. This definition shows that I_{oc} depends on how well the coverage of the cells not in handover is contained. The coverage overlap between cells should be limited as much as possible, to minimize other-cell interference. Alternatively, I_{oc} can be controlled by including all the cells detected in the area of interest in the Active Set, but this reduces efficiency. $\hat{I}_{or}$ represents the received power spectral density of the Downlink of all cells that are in soft handover, that is, that contribute to the link. Both terms consider the total transmit power of the cell at a given time. Hence, the ratio depends on the current load situation.

Figure 2.9 illustrates how $I_{oc}/\hat{I}_{or}$ evolves within a cell. It presents the geometry, G, estimated through Monte Carlo simulation, under a full and evenly loaded hexagonal grid network. In this type of network, the most significant improvement is achieved when the Active Set size is increased from 1 to 2 (about 3 dB). When the Active Set size is further increased to 3, the geometry at the edge improves by about 1 dB. Increasing the Active Set size from 3 to 4, not presented in the figure, yields only marginal improvements, typically less than 1 dB. This observation implies that an Active Set size of 3 should be sufficient, assuming handovers are performed without delays and the overlap between cells is well contained.

Geometry cannot be computed easily from field measurements. From the same Monte Carlo simulation, Figure 2.10 compares the geometry to the combined E_c/N_o. Table 2.9 shows the average geometry and the 5th and 10th percentile tail of the geometry distribution for the AS = 1 through AS = 4. The mean geometry can be used in a Link Budget to

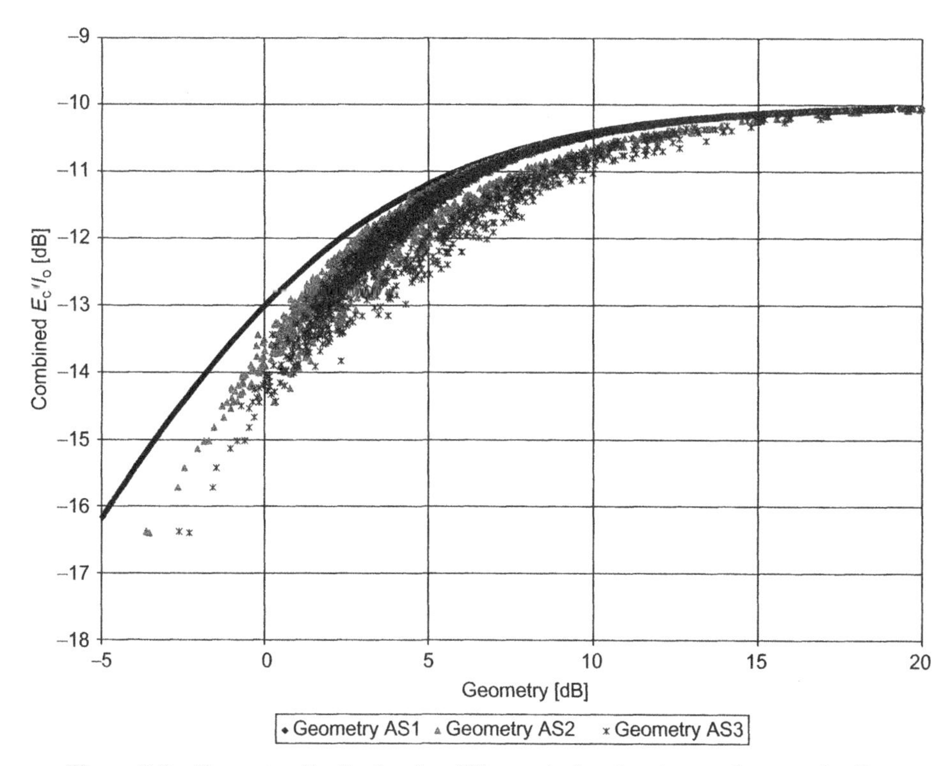

Figure 2.9 Geometry distribution for different Active Set sizes under even loading

estimate coverage, while the 5th or 10th percentile can be used to determine the maximum DPCH required. The percentile used should match the expected probability of service, also known as the *cell area confidence*, to ensure that the Downlink will not drop because of insufficient DPCH power.

2.3.3.6 Beta (Combining Gain)

On the Downlink, the combining gain represents the possibility that the multiple DPCHs in the Active Set can be combined. This reduces the required transmit power from each leg in soft handover. Using the maximum ratio combining equation [10] and assuming that the paths to the different serving cells and the loading are equal, the combining gain can be calculated as 2 (3 dB) for two servers and 3 (4.8 dB) for three servers.

2.3.3.7 Achieved E_b/N_t

Achieved E_b/N_t (DCH E_b/N_t) can be estimated using Equation 2.7, given the following:

- The portion of the total power used for the DCH (DPCH E_c/I_{or})
- The chip and information bit rate
- Combining gain (β)
- The relative weight of the multipath components (a_i)
- The geometry ($I_{oc}/\hat{I}_{or}$)

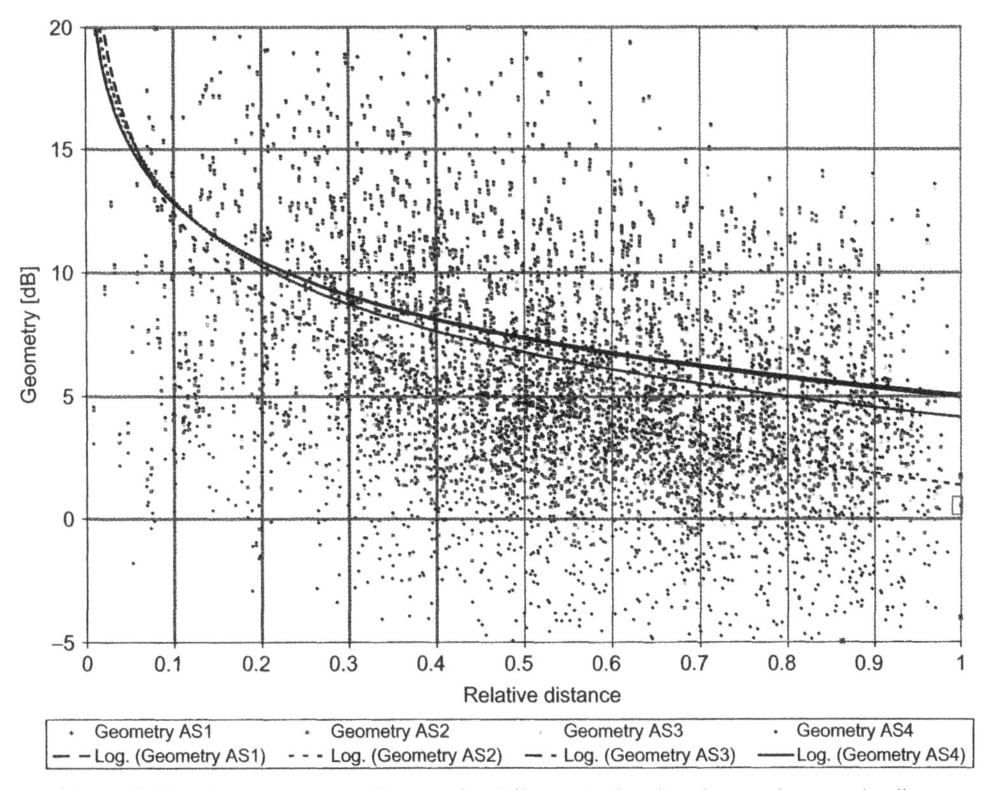

Figure 2.10 Geometry versus distance for different Active Set sizes under even loading

Table 2.9 Geometry for different handover parameter settings

Conditions	Mean [dB]	10th percentile [dB]	5th percentile [dB]
AS size = 4, AS threshold = 3	7.6	2.5	1.8
AS size = 3, AS threshold = 3	7.5	2.4	1.7
AS size = 2, AS threshold = 3	7.0	1.6	0.9
AS size = 1	4.9	−1.9	−3

This formula is an extension of the DCH_E_b/N_t presented in [10]. For any combining gain, it is valid only for servers of equal strength.

The formula in Equation 2.7 is estimated in the linear domain and is not practical for spreadsheet calculations unless equal-power multipaths are assumed. The Link Budget example in Table 2.8 considers two paths of equal strength. To show the effect of multipaths on the achieved E_b/N_t, Figure 2.11 estimates the achieved E_b/N_t for most common 3GPP channel cases [1].

Figure 2.11 shows that the channel with two equal paths is appropriate for initial network dimensioning, because it represents the average between case 2 (slow mobility) and case 3 or 4 (high mobility). This graph also illustrates that the system will be less efficient when multipaths of similar strength are present. With power control, the achieved

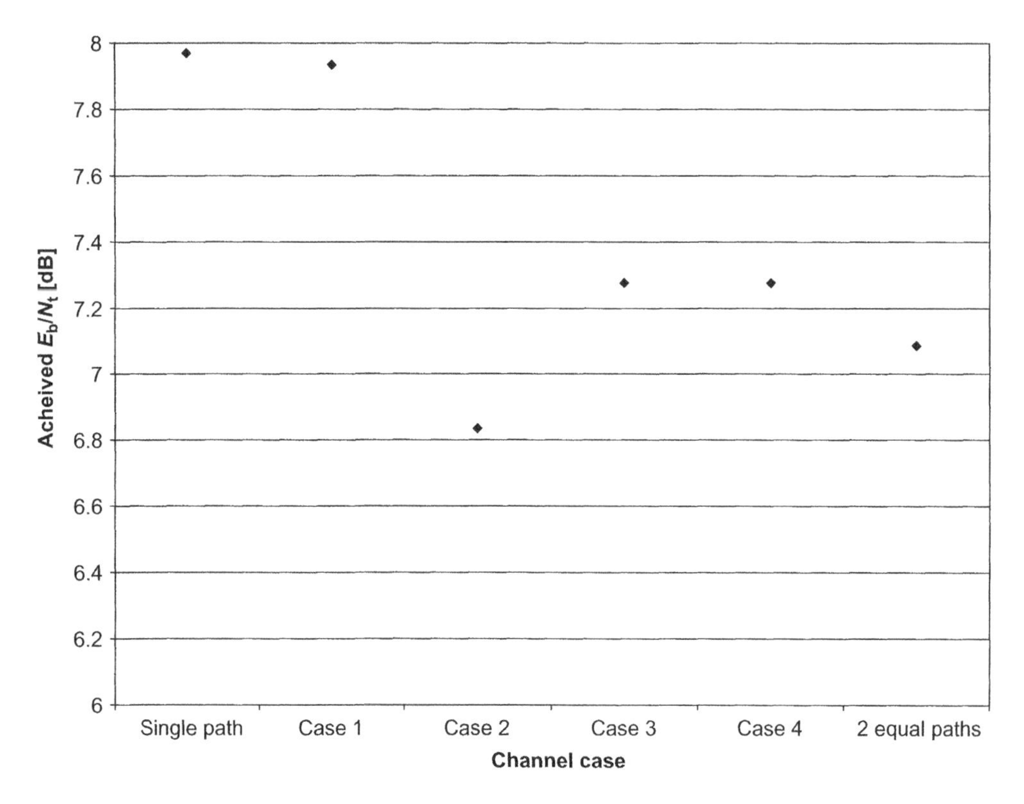

Figure 2.11 Achieved E_b/N_t for different channel conditions

E_b/N_t should converge to the required E_b/N_t. Equation 2.7 can be used to estimate the required DPCH (DPCH E_c/I_{or}) to maintain the required E_b/N_t for a given channel condition.

$$DCH\frac{E_b}{N_t} = \frac{DPCH\_E_c}{I_{or}} \times \frac{Chip\_rate}{Information\_data\_rate} \times \sum_{i=1}^{L} \frac{\beta \times a_i^2}{\frac{I_{oc}}{\hat{I}_{or}} + (\beta - 1) + (1 - a_i^2)} \tag{2.7}$$

2.3.3.8 Required E_b/N_t

On both the Downlink and the Uplink, the same variables affect the target E_b/N_t for a given service. For the Downlink Link Budget, the receiver is a UE, not the Node B cell. The difference in implementation and power control changes the typical range for Required E_b/N_t compared to the Uplink values presented in Table 2.3. This difference is apparent when we consider how the standard defines the minimum performance for the UE [3] compared to the Node B cell [1]. For the Downlink, and thus for the UE, the minimum performances are based on minimum DPCH_E_c/I_{or} rather than on E_b/N_t.

Table 2.10 Minimum Downlink performance in E_c/I_{or} and E_b/N_t

Bearer	Target BLER [%]	Minimum requirement DPCH_E_c/I_{or} [dB][a] Static (AWGN) DCH_$E_c/\hat{I}_{or}$	Static (AWGN) $I_{oc}/\hat{I}_{or}$	Case 3 DCH_E_c/I_{or}	Case 3 $I_{oc}/\hat{I}_{or}$	Estimated required E_b/N_t [dB] [b] Static (AWGN)	Case 3
Speech AMR 12.2 kbps	1	−16.6	1	−11.8	3	5.6	8.93
CS64 kbps	1	−12.8	1	−7.4	3	2.5	4.6
PS64 kbps	10	−13.1	1	−8.1	3	2.3	3.9
PS128 kbps	10	−9.9	1	−9	−3	2.2	4.0
PS384 kbps	10	−5.6	1	−5.9	−6	2.1	4.4

[a] From [3]
[b] Calculated from [3] and Equation 2.7

Equation 2.7 can be used to convert between both values, as shown in Table 2.10. For simplicity, this table presents only the Additive White Gaussian Noise (AWGN) and case 3.

2.3.3.9 Sensitivity Calculation

The sensitivity calculation for the Downlink is similar to the Uplink calculation, but the Downlink interference is created mainly from the other cells. Interference from other users is mitigated by their orthogonality.

Figure 2.12 presents a conceptual view of the Downlink sensitivity. On the basis of this view, we need to estimate other-cell interference, I_{oc}. Other-cell interference is calculated in two steps. Discounting any interference, the sensitivity calculation is similar to the calculation used for Uplink. The sensitivity is used to estimate the receive power from the main server at the cell edge. With this value, the other-cell power density (interference) can be calculated using the geometry. Receiver sensitivity can then be calculated by comparing the total noise and interference (N_t) with the target E_b/N_t.

For simplicity, the Link Budget does not include the effects of nonorthogonal signals in the same cell on the Downlink. Instead, Figure 2.13 shows the impact of the orthogonality factor on the total interference density. The baseline is I_{oc}, as included in the Link Budget. For a nondispersive channel, the orthogonality would be close to 1, without ever reaching 1, because the synchronization channels (P-SCH and S-SCH) are nonorthogonal to the other channels [11]. The orthogonality factor (α = 1- orthogonality) is generally used to estimate same-cell interference.

2.3.3.10 Handover Gain

In addition to the combining gain used for the achieved E_b/N_t estimation, the macro diversity (handover) offers the same reduction in LNF margin as seen in the Uplink.

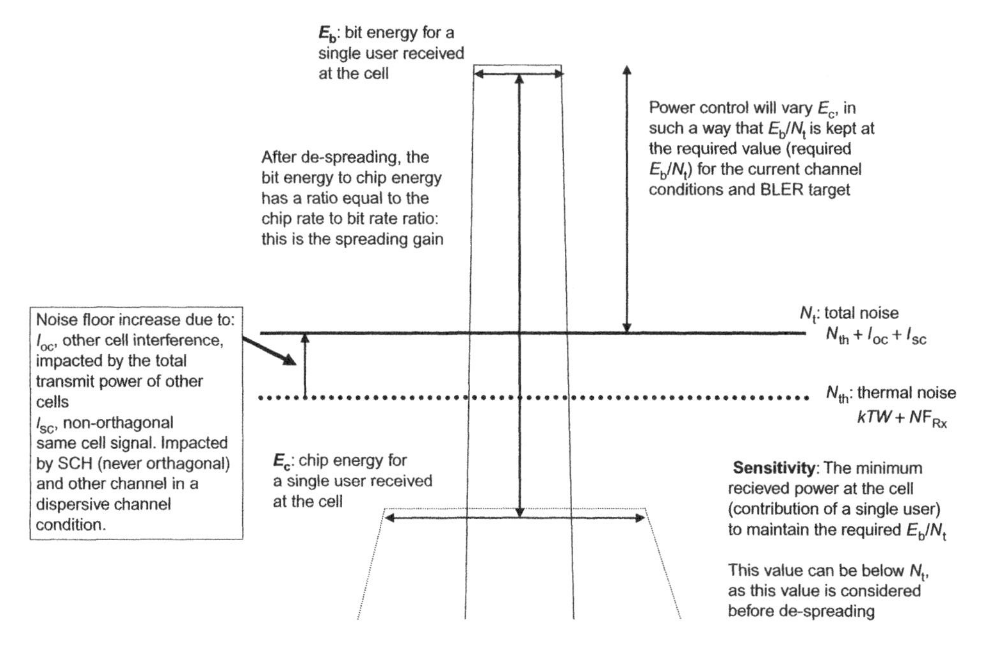

Figure 2.12 Downlink sensitivity: conceptual view

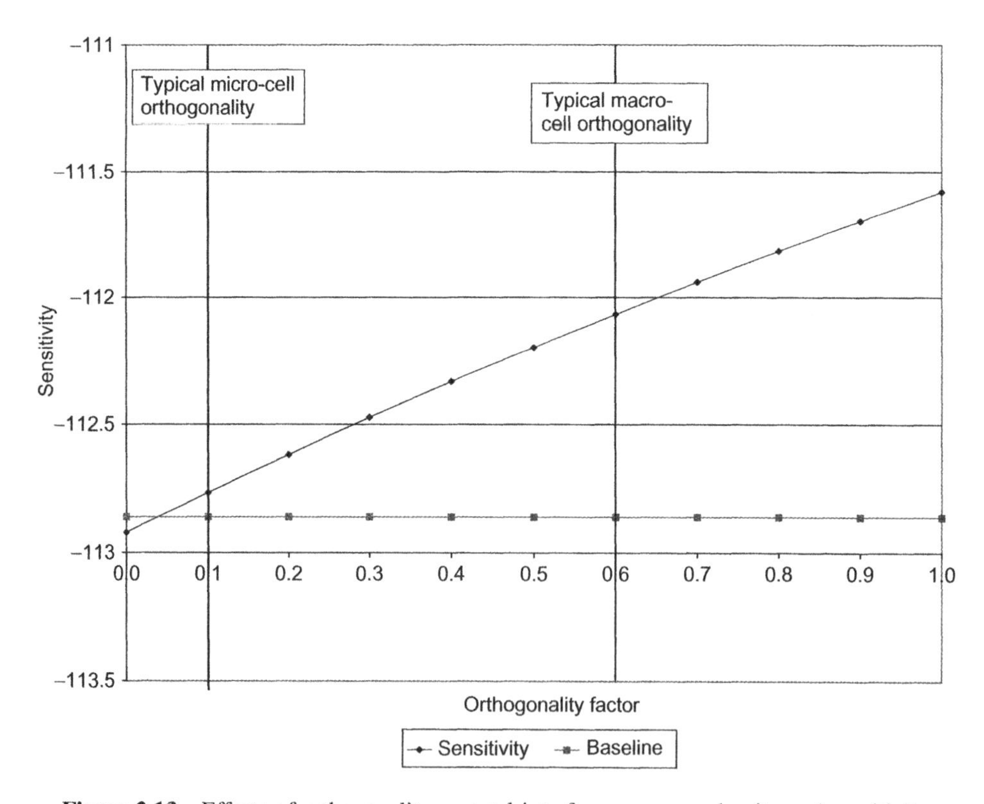

Figure 2.13 Effects of orthogonality on total interference power density and sensitivity

2.3.3.11 Diversity Gain

For the Downlink, the E_b/N_t requirement does not consider diversity. Receive diversity is not yet widely used for the Downlink because of the added cost and complexity it would impose on the UE. When UE receive diversity becomes available, that gain should not be used in the initial dimensioning until this feature is commonly available. By that time, transmit diversity should also be available, so both transmit and receive diversity could be considered together.

2.3.4 Uplink and Downlink and Service Comparison

So far we have defined Uplink coverage for voice service, CPICH coverage, and Downlink coverage for voice services. Comparing the MAPL for CPICH ($L_{MAPL,CPICH}$) and for voice service ($L_{MAPL,DL}$), the CPICH coverage exceeds service by about 3 dB. This unbalance is typical and allows the CPICH of a given cell to be monitored accurately well before it becomes the best cell. More importantly, the Uplink and Downlink MAPL ($L_{MAPL,UL}$ and $L_{MAPL,DL}$) should be compared.

Comparing $L_{MAPL,UL}$ calculated from Table 2.1 and $L_{MAPL,DL}$ calculated from Table 2.8, the Uplink is limited by about 3 dB. This result is consistent with the concept that Uplink is the limiting link in a WCDMA system or, more generally, in a cellular system. This result depends on several assumptions, some optimistic, mainly on the Downlink. One of the assumptions relates to geometry, which can be limited in a real network, compared to the ideal network used for the Link Budget calculation.

From the Link Budget comparison, the Uplink and Downlink can be balanced by decreasing the $L_{MAPL,DL}$. The CPICH power is set to 30 dBm to balance the CPICH and voice service Downlink MAPL. This supports the minimum RSCP, but the achieved E_c/N_o could be lower than the minimum target previously defined. In that case, the maximum HPA power is reduced to 40 dBm to maintain a constant maximum E_c/I_{or} for the CPICH. Alternatively, the target E_c/N_o is reduced, and proper UE operation is verified. Considering the minimum performance requirement [12] for the UE, measurement accuracy is defined for the CPICH E_c/N_o level as down to -20 dB. Decreasing the target to that level will not hinder UE operation. It is important to ensure that common and DCHs can still be demodulated correctly, meaning they are unchanged in terms of absolute power.

In a commercial implementation, any change in CPICH power should be accompanied with increasing the ratio between CPICH and other common channels of the same value. If the CPICH power is reduced from 33 to 30 dBm, the ratio between CPICH and PCCPCH should be increased by 3 dB.

Before going to the extreme of adjusting all the common control channels to balance the coverage, we need to review the initial assumptions to determine if the Uplink or the CPICH coverage is limiting in all cases.

For the Uplink, class 3 UE is considered, but class 4 is currently more common. This difference increases the $L_{MAPL,UL}$ by 3 dB, which balances the links for UE class 4 voice services.

The Uplink Link Budget is calculated without TMA. If TMAs were used, the gain would be seen first on the Receiver Noise Figure, then on the cancellation of the receive feeder losses. This would cause the Uplink MAPL to increase by 4 to 5 dB, and the

Downlink will become limiting. In this case, the CPICH power should not be increased to match the Uplink coverage because this will adversely affect the capacity. This is explained in Chapter 3, which discusses capacity. When coverage is unbalanced in favor of the Uplink, power control ensures that the UE transmit power is limited to the required value, to save UE battery life.

In summary, the CPICH Link Budget can be used for an initial pass at link balancing. This link balancing matches the CPICH to the Uplink coverage for the least limiting service. A full analysis should be done to ensure that the link balancing remains valid for all the considered services.

In an Uplink Link Budget, the following parameters primarily determine the difference between voice and other services:

- **Information full rate.** This variable increases up to 64 kbps for both CS and PS data. Other rates, 128 or even 384 kbps, are available for PS data, but are rarely implemented. This increase in information rate is the main factor in the Link Budget, because of the reduced spreading gain. Compared to voice, for which the spreading gain is 25 dB, PS data at 64 kbps has a spreading gain of less than 18 dB. Because of the higher coding scheme available, this 7 dB difference is only partially compensated for by the reduction of required target E_b/N_t.
- **Required E_b/N_t.** Table 2.3 shows sample E_b/N_t requirements for the main bearers. The higher coding available for data service can reduce the required E_b/N_t by 3.8 dB on an average, for all channel conditions. Each case should be considered individually, but this average shows that the decrease in spreading gain cannot be offset by the turbo coding capabilities available for high-speed service. During dimensioning, E_b/N_t can be affected not only by the type of service but also by the mobility and speed of the users.

Table 2.11 compares Uplink MAPL for three types of services (AMR voice, CS64, and PS64) and two channel conditions (AWGN and case 3). For static channel conditions, all services have a similar footprint, with a slight advantage for voice as compared to the other services. Case 3, in contrast, has greater footprint differences between the services, while retaining an advantage for voice. This illustrates the importance of fully understanding how each service will be used, to avoid over- or under-designing the network.

On the Downlink, a service-dependent Link Budget (as defined in Section 2.3.3) and the absolute coverage defined in Section 2.3.2 should be considered. For link balancing, the required DPCH power to meet the path loss corresponding to the CPICH coverage

Table 2.11 MAPL for Uplink, for different services and channel conditions

Service [dB]	AWGN	Case 3
AMR voice	125.8	123.7
CS64	125.0	119.9
PS64	125.2	120.3

should be set for each service. Similar to the Uplink, several Downlink factors affect the required DPCH power for each service.

Table 2.12 is based on the Downlink Link Budget presented in Section 2.3.3. Given the CPICH absolute path loss, the table shows the required DPCH for each service to achieve this coverage. In the static case (AWGN channel), all services could be achieved over the service area, but the required power for high-speed services would be expensive. For case 3, data rates should be restricted to 128 kbps because the required power for 384 kbps is higher than the -3 dB limit on E_c/I_{or} that is typically observed on commercial equipment. The required power listed in this table does not consider capacity limitations; consequently, the resulting values are not recommendations for power settings. Furthermore, the results are only for two scenarios, while a recommendation should account for all typical cases.

On the basis of the assumptions discussed so far, the Uplink is limiting for all services. Unless the coverage of the Downlink is reduced using a less powerful HPA, coverage balancing would require Uplink changes such as TMA or higher order diversity.

On the Downlink, service areas can be equalized by increasing the DPCH power. Service in this context means a type of traffic rather than a data rate. For PS data in particular, service means the availability of a PS data bearer at any data rate. A high data rate might not be available over the entire area, depending on geometry and channel conditions. On the Uplink, this power balancing is not possible because of the limited UE power. Therefore, the network plan should be designed to accommodate the limiting service, typically CS64.

Before comparing Uplink and Downlink coverage, both links should be balanced, with similar path loss and interference. Downlink path loss can be estimated from measured RSCP, and interference can be estimated from the E_c/N_o. On the Uplink, the only available measurement at the UE is transmit power. Power control affects this value by either path loss or interference. Ideally, to determine whether the links are balanced, a planner should compare the UE transmit power to both the RSCP and E_c/N_o to ensure that path loss and interference are similar on both links. A more practical method of verifying the Downlink interference is to compare the best server RSCP to E_c/N_o. In a well-planned network, this curve should match that shown in Figure 2.14. In this figure, the upper limit corresponds to a single cell with low load (CPICH $E_c/I_{or} = -3$ dB) and the lower limit corresponds to three equal cells, fully loaded (CPICH $E_c/I_{or} = -10$ dB).

Table 2.12 Downlink Link Budget comparison for different services and channel conditions

Service	AWGN	Case 3
CPICH path loss [dB]	128.8	
Required DPCH voice [dBm]	28.9	31.6
Required DPCH CS64 [dBm]	30.7	32.8
Required DPCH PS64 [dBm]	30.5	32.1
Required DPCH PS128 [dBm]	33.4	35.2
Required DPCH PS384 [dBm]	38.1	Not possible (over 40 dBm)

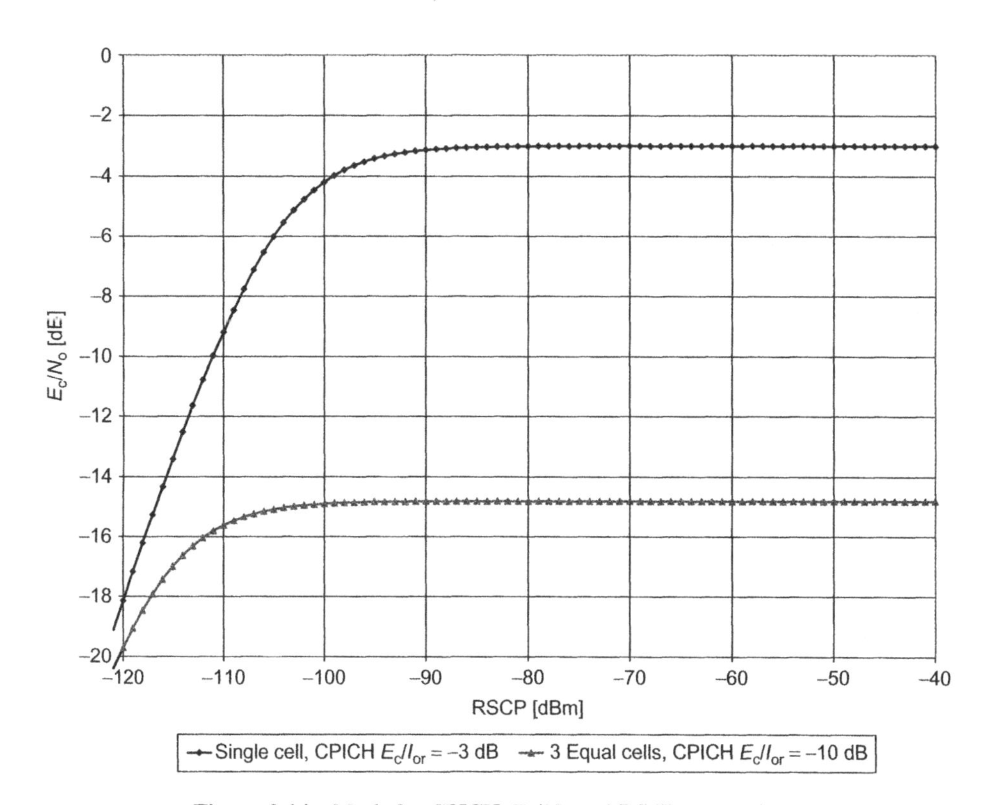

Figure 2.14 Mask for CPICH E_c/N_o and RSCP comparison

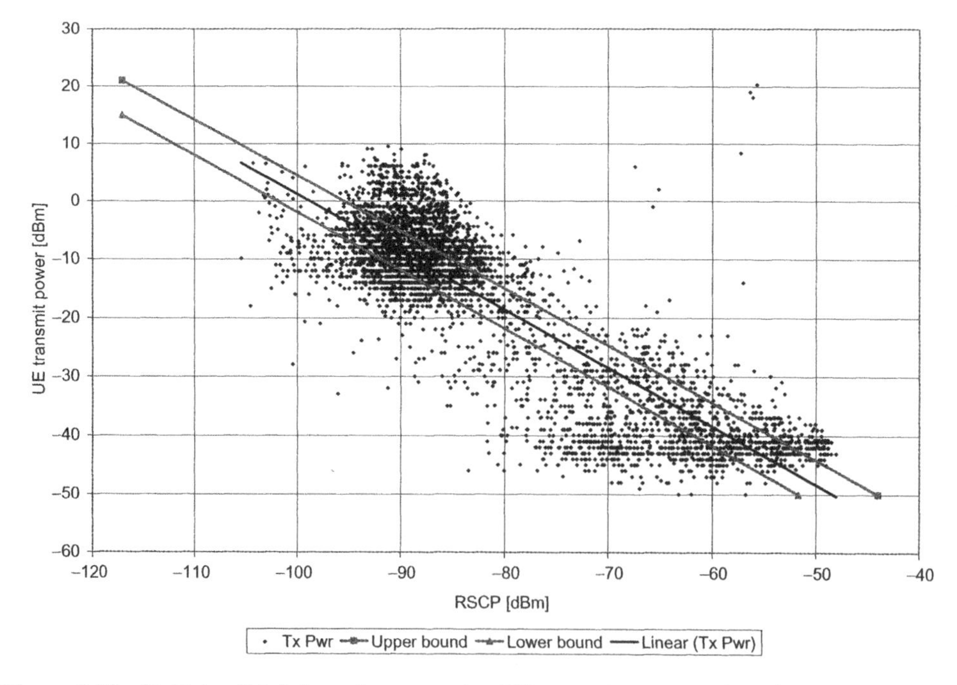

Figure 2.15 Verifying link balance by comparing UE transmit power and received CPICH RSCP

After verifying the Downlink interference, the UE transmit power and CPICH RSCP are compared to estimate the presence of Uplink interference. The performance requirements for the UE [3] specify its minimum and maximum transmit power, as well as the maximum and minimum CPICH RSCP level. These values represent the upper limit for link balance verification. The lower limit must consider the maximum load expected in the network, typically 75% on the Uplink, as represented in Figure 2.15. In this example for voice, the trend of UE transmit power versus CPICH RSCP is within the boundaries, indicating that the links are balanced. For services other than voice, the slope of the curve should be modified to follow the change in spreading gain. For a given path loss, as estimated by the CPICH RSCP, the UE transmit power should be higher for a low spreading gain service (PS data for example) than it is for voice.

2.4 Network Planning Tools

A network planning tool estimates coverage on the basis of a set of assumptions and inputs. The assumptions are similar to the entries used in the sample Link Budget. Other inputs include maps representing the terrain, clutter, traffic, and similar information. From this, a path loss can be calculated between each transmitter and each location, where location is defined as a square of a given size (Bin), from a few meters to hundreds of meters. From the path loss information, several other variables can be calculated. The most commonly used variables are received signal power, interference (or E_c/N_o), and required transmit power.

For WCDMA, calculating only location-based Link Budgets is not sufficient. As seen in the Link Budget section, network loading affects the final results. WCDMA network planning tools must also include a traffic distribution module. The traffic distribution module, often referred to as a Monte Carlo simulation, randomly distributes users according to modifiable rules. These rules distribute traffic according to known traffic density maps and known traffic profiles.

Table 2.13 summarizes WCDMA network planning tool options.

2.4.1 Network Planning Tool Input

Network planning tools require several inputs. Each is discussed in the following sections, in the same order they would be applied when setting up a project.

2.4.1.1 GIS Data: Projection and Coordinates System

Maps are two-dimensional representations of the earth surface, which is a complex shape commonly idealized as an ellipsoid. Projection, the process of converting a three-dimensional shape into two dimensions, introduces some errors. Because of earth's irregularities, the extent of these errors varies from location to location. Historically, localized projections have minimized errors at the expense of larger errors at distant locations. Table 2.14 shows sample local projections and the areas where they apply. This table includes two worldwide projections: transverse Mercator and Universal Transverse Mercator (UTM). These projections divide the earth into 60 zones, minimizing the error in each zone. To completely define a coordinate system, the ellipsoid and datum are required. Local ellipsoids give a localized idealization of the earth surface, while datum gives the

Table 2.13 Network planning tool selection criteria

Criteria	Priority	Comment
Compatibility with GIS data	Medium to high	Compatibility with existing GIS data is important mainly for inter-system planning
Compatibility with site data	Low to medium	Ability to export from an existing network planning tool and import into the WCDMA network tool reduces the need for both tools to be fully compatible
Compatibility with traffic data	Medium to high	Compatibility with 2G traffic data is mitigated by the potential difference with 2G traffic (PS data mainly)
Compatibility with RF Model	Low (for 900 MHz model)	RF model tuned at 900 MHz cannot be directly reused
	Medium (for 1800 MHz model)	RF model tuned at 1800 MHz can be reused initially, if properly adjusted for the frequency difference
Coverage predictions	High	At least RSCP coverage should be available. Path loss prediction availability is a good complement
WCDMA predictions	High	Specific WCDMA predictions should be available for accurate service area prediction. Main plots in this category will include CPICH E_c/N_o, Uplink and Downlink achieved E_b/N_t, and handover-related predictions
Traffic modeling	High	PS traffic modeling is the main differentiator between network planning tools
Monte Carlo simulations	Mandatory	Monte Carlo simulation is the only way to accurately predict capacity

Table 2.14 Earth projections and applicable areas

Projection	Area where applicable	Comments
ED 50	Western Europe	Further divided into zones
Lambert Conformal Conis	North America	Mercator projection also widely used
Transverse Mercator	Worldwide	Further divided into zones and locally associated with datum and ellipsoid
UTM	Worldwide	Further divided into zones and locally associated with datum and ellipsoid

precise measurement of a point of reference. With the ubiquity of Global Positioning System (GPS), the World Geodetic System (WGS) 84 datum and its associated ellipsoid are commonly used. The coordinate system should match the one used to create the map, support any local reporting requirements, and be compatible with measurement devices.

2.4.1.2 GIS Data: Terrain Data

Site-to-site distance is an important consideration during network planning. For the best prediction, the Geographic Information System (GIS) data resolution should be at least 10 times higher than the minimum site-to-site distance. A maximum resolution of 50 meters is acceptable. The more typical 30-meter resolution is suitable for general predictions. If microcells are included, the resolution is increased to a maximum of 10 meters. At that resolution, building data is usually available. It is important to understand what is included in the terrain data. Ideally, only the terrain should be available, with building height data on a separate layer. In the RF prediction, the terrain elevation mostly affects diffraction and the effective antenna height.

2.4.1.3 GIS Data: Clutter

After terrain elevation, the second most important piece of data about a geographical area is surface utilization. The classification of how the surface is used is known as *land use categories*, or *clutter categories*. These categories are not standardized among map providers, and can have very different meanings among countries. For example, *suburban morphology* would consist of low density one- to two-story buildings in the United States, but would consist of low-to-medium density high-rise buildings in other countries. The discussion of resolution, as it applies to terrain data in Section 2.4.1.2, also applies to clutter information. In addition, the number of clutter classes must be defined. The need for a high number of classes depends on how the clutter information will be used – for RF prediction or traffic distribution. For RF prediction, the RF model might influence the required number of clutter classes.

2.4.1.4 GIS Data: Vector

Vector maps are commonly used and present linear features such as rivers, roads, and administrative boundaries. These items are generally grouped under the name of vector data. The vector layer has two main purposes: verifying site positions relative to the street layout, and determining traffic distribution. For verifying site position, only the location accuracy of the vector is important. For traffic distribution, accuracy is less critical but different vector categories, such as those that correspond to different street widths, should be available. This allows traffic to be weighted differently.

2.4.1.5 GIS Data: Traffic

A traffic map can provide subscriber density or traffic density. Either density can be used if the call model is defined properly. For subscriber density, the call model should reflect the expected traffic per subscriber. For traffic density, the call model should reflect a traffic of 1 Erlang.

2.4.1.6 GIS Data: Area

Several user areas are defined during a network planning project. The following are usually necessary:

- **Target area.** A target area defines the extent of the network area to be planned. It may extend over a city, a region, or a country. This reference area is used to calculate whether the simulation results meet the target.
- **Clusters.** The target area could be divided into clusters to divide the work into manageable entities. Clusters should be drafted so that they are isolated from each other. Alternatively, clusters may include sites in similar countiguous morphologies.
- **RNC/LA/RA boundaries.** Boundaries are used to balance traffic between areas and should be set in low traffic area to minimize ping-pong effect. The Radio Network Controller (RNC), Location Area (LA), and Routing Area (RA) boundaries might not be defined initially. Unless dictated by existing constraints, boundaries are best drafted after some RF planning has been done.

2.4.1.7 RF Models

In the early phase of detailed design, accurate RF models tuned for an individual cell may not be available. RF models based on drive tests of a limited number of sites are more likely to be available. Because of the limited number of sites driven, only a few models, typically two to five, will be available. These models can be accurate with a mean error close to 0 dB and a standard deviation ranging from 8 dB for a good model to 10 or higher dB for a limited accuracy model. Despite the average accuracy of such generic models, local errors can still be detected.

2.4.1.8 Equipment: Cell-Related Parameters

Table 2.15 lists the main parameters that affect network planning simulations. The parameters set during simulations should be consistent with actual parameter settings. Some of these parameters may also be tuned, or optimized, during the design process. However, it is preferable to tune these parameters to improve system performance after the RF configuration has been optimized. This ensures that parameters are not set to mask RF issues, but are set to provide the best level of service given the best possible RF configuration. During network design, any parameter optimization will be limited because of the static nature of the simulation, and because a network planning tool cannot fully model network behavior.

2.4.1.9 Equipment: Antenna and Antenna Near Field

This category includes antennas, TMA, and feeders. Some network planning tools might include additional equipment in this category, such as Remote Electrical Tilt, but these usually have little or no effect on the simulations. The antenna should include the main

Table 2.15 Cell parameters that affect network planning simulations

Parameters	Setting	Comment
Frequency carrier	Consistent with spectrum allocation	–
Receiver noise figure	Consistent with manufacturer specification	–
PSC	–	Can be automatically assigned by the tool
Maximum power	Consistent with manufacturer specification	–
CPICH power	33 dBm typical	Can be optimized during design to balance Uplink and Downlink coverage and capacity
P and S-SCH power	Consistent (±3 dB) with CPICH power	Setting should consider the tool implementation; i.e., if the time multiplexing with PCCPCH is taken into account
PCCPCH	Consistent with UE performance characteristics (Idle and Access)	Could be optimized during design to ensure satisfactory UE performance in Idle and Access state
SCCPCH	Consistent with UE performance characteristics (Idle and Access) and used SF	Could be optimized during design to ensure satisfactory UE performance in Idle and Access state
AICH	Consistent with signaled setting, −5 to −7 dB compared to CPICH typical	Effective power of this channel varies with load, because no power is transmitted if no information is sent
PICH	Consistent with signaled setting and number of PI per frame, −6 to −7 dB compared to CPICH for 18 PI per frame	Effective power of this channel varies with load, because no power is transmitted if no information is sent
Neighbor list	–	Should be optimized during design, typically automatically generated by the tool

antenna used for the project. It is typically 65 degrees, 16 to 18-dBi gain, with a few alternatives for specific cases. Specific cases include wide and narrow beamwidth, high and low gain, and different sized antennas to accommodate special site acquisition requirements. For antennas that have continuously variable tilt, the pattern should be available in single degree increments because network planning tools can accurately estimate only mechanical tilt effects, not electrical tilt effects. The antenna pattern should be available using some type of null fill to avoid unrealistic shadowing in close proximity to the site. Other pieces of equipment or their effects can be effectively configured, mainly in terms of noise figure and attenuation, can be set at the cell level.

2.4.2 Coverage Considerations during Network Planning

The main approach to estimate coverage with a network planning tool is to evaluate the RSCP of the received CPICH power at the UE. The first step of coverage analysis does not consider load and service, but only CPICH transmit power and path loss. It can be scaled up or down to represent different coverage levels, corresponding to factors such as building penetration losses. Figure 2.16 illustrates this concept. The simplicity of this method, however, is also its main drawback: Uplink and Downlink services cannot be estimated, nor can the impact of load on the coverage.

Coverage prediction starts with placing sites to provide continuous coverage, on the basis of the CPICH Link Budget. In this process, clutter, terrain, and achievable antenna height determine whether a regular grid design can be maintained. Most likely, the result will be an irregular network.

After continuous coverage in terms of RSCP is achieved, the interference level must be verified and optimized. The interference level can be determined from the best server CPICH E_c/N_0. Although the RSCP calculation includes a margin, including such a margin for E_c/N_0 yields a pessimistic view of the network. This can be explained by using Equation 2.13 to estimate the E_c/N_0. Figure 2.17 illustrates this conceptually.

If only one of the servers is affected by fading (10 dB in the figure), that server drops from the best server list and, at the same time, the E_c/N_0 of the best server improves, as less interference is observed. At the other extreme, if all links are affected by 10 dB

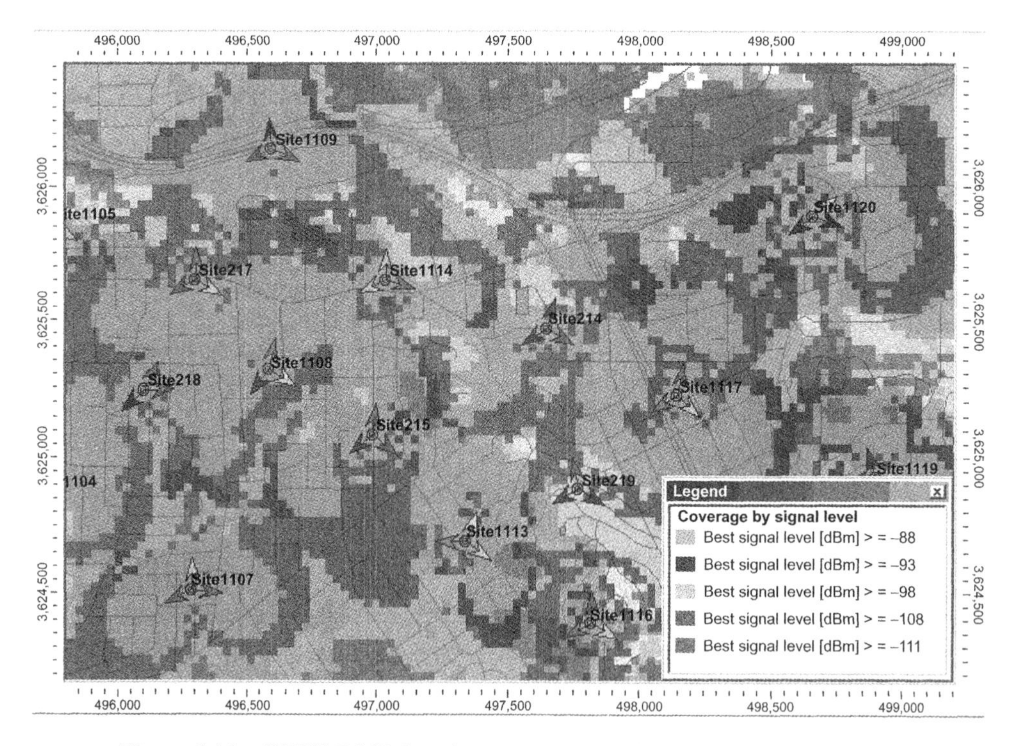

Figure 2.16 CPICH RSCP-based coverage, presenting different levels of BPL

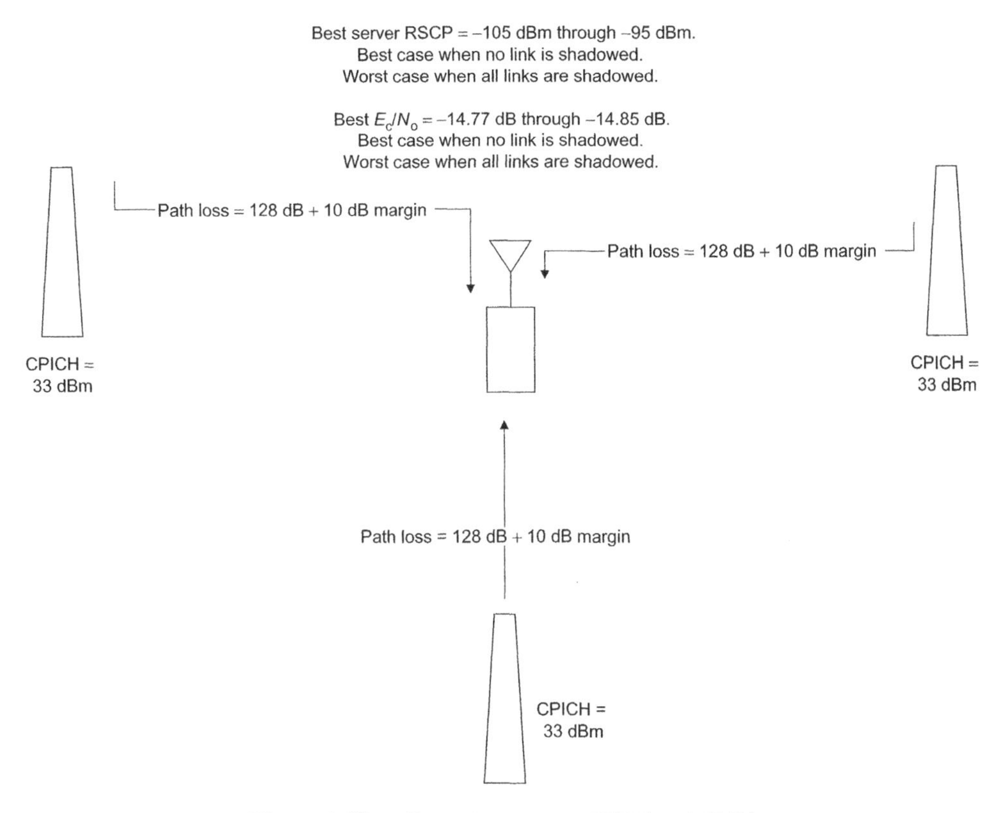

Figure 2.17 Effect of margin on RSCP and E_c/N_o

fading, the relative interference between servers is preserved, and the E_c/N_o remains approximately constant, even if the RSCP has dropped. If all path losses increase, the thermal noise becomes more important and would lead to degradation of E_c/N_o.

Our example Link Budget arbitrarily defined the minimum RSCP as -111 dBm. This value is based on several assumptions, the first one being the criteria for a good design. In a regular grid pattern with perfectly contained cell coverage, the worst case would be the detection of three equal cells. Cells should be spaced so that signals from any of the three detected cells are equal to the thermal noise, as shown in Equation 2.8, where P_j represents the power received from a given cell, N_f represents the noise figure, and kTW represents thermal noise:

$$\left[\frac{E_c}{I_o}\right]_j = \left[\frac{E_c}{I_{or}}\right]_j \times \frac{P_j}{kTW + N_f + 3 \times P_j} \tag{2.8}$$

Assuming a fully loaded system where CPICH is set at 10% of the total power ($E_c/I_{or} = -10$ dB) and $Pj \sim kTW + Nf$, Equation 2.9 can be solved as follows:

$$\left[\frac{E_c}{N_o}\right]_j = -10 + 10 \times \text{Log}\left(\frac{1}{1+3}\right) = -16 \text{ dB} \tag{2.9}$$

To estimate the RSCP, we need to estimate the value of Received Signal Strength Indicator (RSSI), which can be assumed to be equal to I_o, as shown in Equation 2.10:

$$RSSI = kTW \times Nf + 3 \times Pj = kTw \times Nf(1 + 3 \times Pj/kTW \times Nf) \qquad (2.10)$$

Knowing the thermal noise and the noise figure (7 dB for example), and assuming that P_j is equivalent to the thermal noise, Equation 2.11 can be solved as follows:

$$RSSI = -108.13 + Nf + 10 \times \log(1 + 3) = -95.1 \text{ dBm} \qquad (2.11)$$

With E_c/N_o and RSSI, the RSCP can be calculated as in Equation 2.12:

$$RSCP = E_c/N_o + RSSI = -16 - 95 = -111 \text{ dBm} \qquad (2.12)$$

This value cannot be compared directly with the minimum DPCH_RSCP value of −117 dBm defined in the standard [3]. The standard considers absolute sensitivity in the absence of other-cell interference, a highly unrealistic case for field deployment.

Once the minimum RSCP is defined, it should be derated for different building penetration losses and LNF. If the last two values can be included in the prediction, based on clutter type, the coverage verification can be done for all clutter types at once.

In areas where the coverage is deficient (lower than the expected value), coverage can be improved in a limited number of ways. The main techniques are discussed below, based on the example in Figure 2.18 and assuming that the RF model has been tuned. Each technique has unique advantages and different implementation costs.

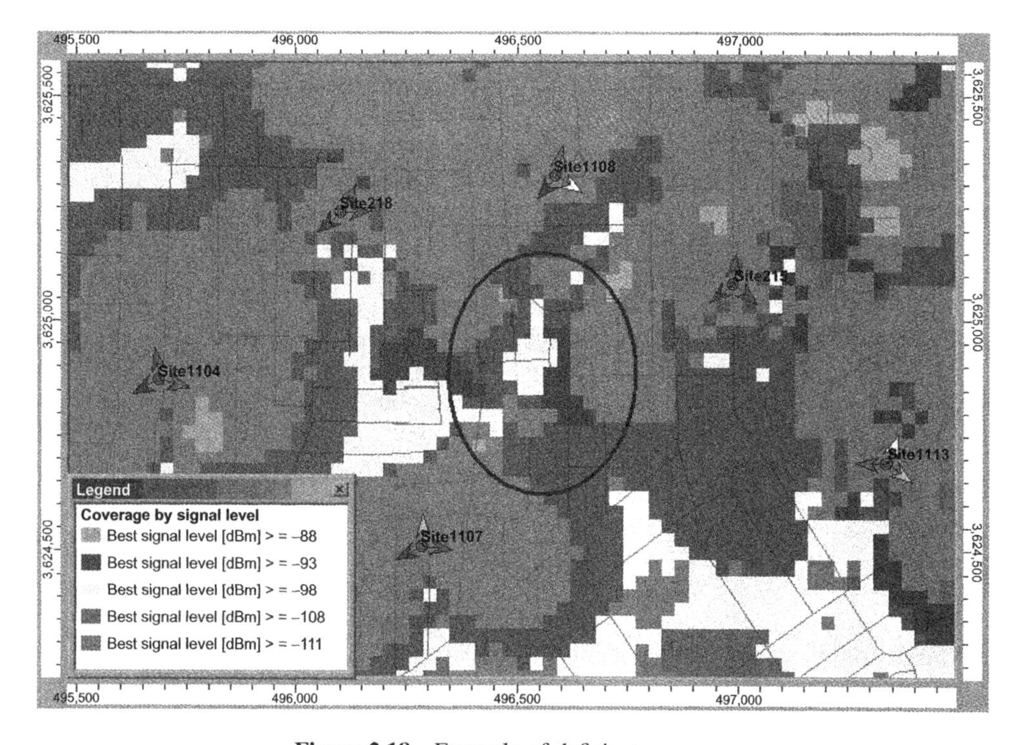

Figure 2.18 Example of deficient coverage

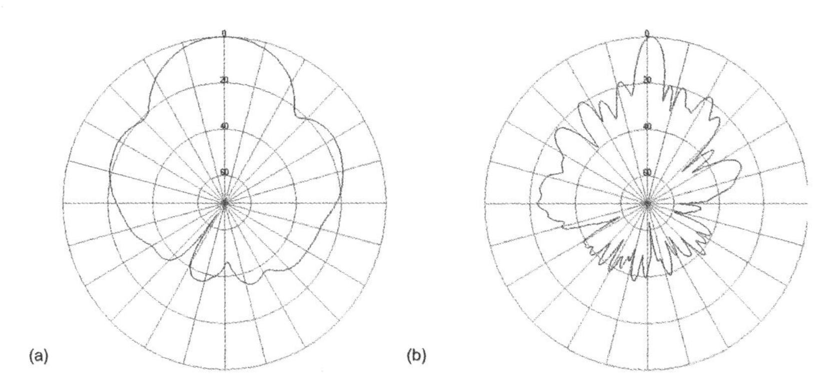

Figure 2.19 Antenna pattern examples: (a) horizontal pattern and (b) vertical pattern

- **Reorienting sectors.** Reorienting can be used only when the coverage deficiency is located outside the boresight of the antenna. The possible correction depends on the antenna pattern. For the antenna pattern in Figure 2.19, reorienting the Site1108_3 antenna would yield a maximum improvement of about 20 dB.
- **Changing antenna tilt.** To improve coverage, tilting the antenna is less effective than reorienting it. This can be explained by looking at the vertical pattern shown in Figure 2.19. In this example, gaining 20 dB would be possible only in the immediate vicinity of the sites and in the boresight, where a tilt greater than 10 degrees would compensate for the antenna null. Antenna tilt is used more often to reduce the interference caused by a site rather than to increase its coverage.
- **Increasing antenna gain.** Antenna gain can improve coverage only over a given area. The gain of an antenna is directly related to its number of elements, and thus to its physical dimensions, or size. A high gain, large antenna changes the beamwidth. Since the antenna concentrates the energy only in one direction (but does not amplify it), any increase in gain in a given direction reduces the beamwidth: horizontal, vertical, or both. As a result, any improvement at a given location might produce degradation in all the areas off the boresight.
- **Increasing antenna height.** In a COST-Hata-type model, increasing the antenna height decreases the path loss exponent, as seen in Figure 2.20. The path loss exponent is reduced because higher antennas increase the line-of-sight area and tend toward free space loss. Raising the antenna height increases the coverage of the cell. This increased coverage increases interference to distant cells. Changing antenna height is recommended only when it reduces the standard deviation of the antenna height (height above average terrain) distribution. It is easier to control interference during planning when all the antennas are approximately of the same height.
- **Moving the site.** Moving the site is a drastic change, and it can be difficult to acquire a new site. Moving the site closer to the source of traffic could improve coverage and capacity in the corresponding area, but such reconfiguration is expensive.
- **Increasing power.** Of all the coverage enhancement methods, increasing CPICH power is the least preferred. Increasing CPICH power, while keeping the maximum HPA at the same level, effectively reduces the capacity of the system. Coverage is also

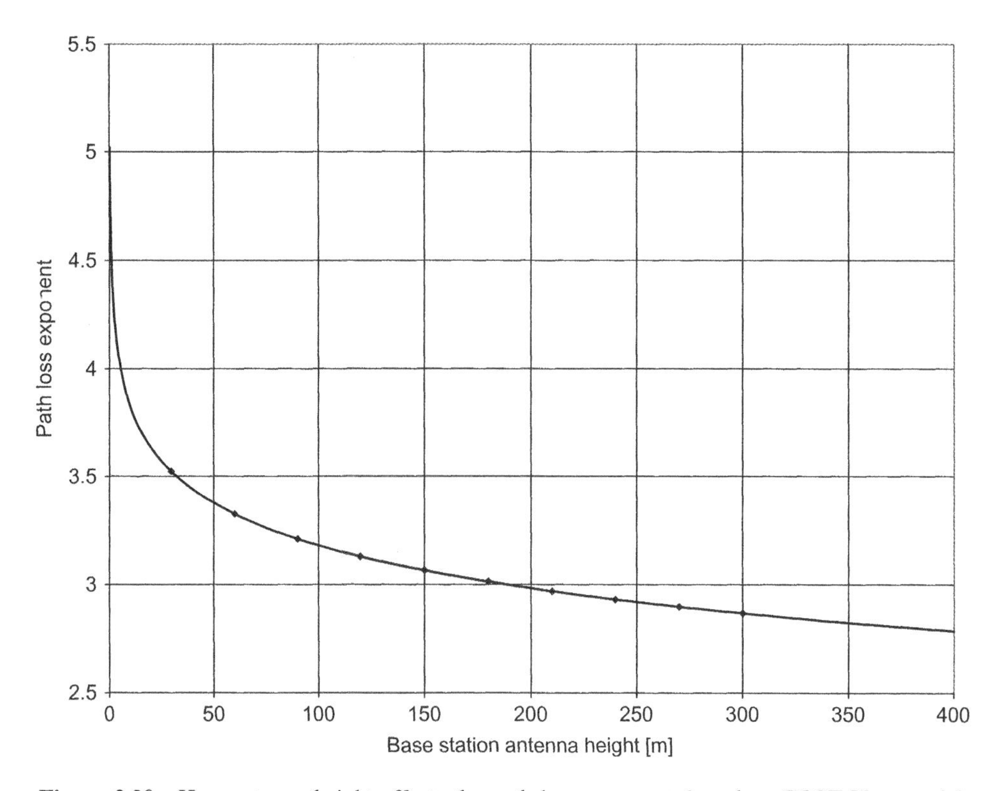

Figure 2.20 How antenna height affects the path loss exponent, based on COST-Hata model

compromised because increasing one channel, or increasing total HPA power, creates imbalances in the other areas.

- **Ignoring the coverage deficiency.** If the cost to overcome the deficiency is not balanced by future gains, or if the coverage assumptions are known to be inaccurate, coverage deficiency can be ignored. In this case, local knowledge of traffic patterns and subscriber expectations is invaluable.

2.5 Interference Considerations during Network Planning

Designing a network based on RSCP is only the first step. The network plan must also account for interference. A level of interference that will ensure proper network performance without having to run Monte Carlo simulations for every minor change of RF configuration (antenna tilt, azimuth, height, etc.) is used.

This can be achieved by using CPICH E_c/N_o, assuming that the design thresholds are consistent with loading. Unless a Monte Carlo simulation is run, only the unloaded condition, in which the common control channels are present, can be accurately known. Equation 2.7 can be used to determine the unloaded CPICH E_c/N_o. To do this, the CPICH E_c/I_{or} must be determined first. Table 2.16 shows the average powers for all common control channels. This averaging is necessary during network planning because networkplanning tools do not offer dynamic simulations, and some of the channels are only intermittently active.

Table 2.16 CPICH E_c/I_{or} estimation based on Common Control Channel only

Channel	Typical power assignment dBm/W	Duty cycle	Average power dBm/W	Comment
CPICH	33/2	100	33/2	–
SCH	30/2	10	20/0.1	Typical power setting range 28–33 dBm
PCCPCH	30/2	90	29.5/0.9	Typical power setting range 30–33 dBm
SCCPCH	30/1	100	30/1	Typical setting for SF 128
AICH	27/0.5	–	–	AICH active only in the presence of traffic
PICH	27/0.5	–	–	PICH active only in the presence of traffic
Total	NA	NA	36/4	Typical unloaded CPICH $E_c/I_{or} = -3$ dB

With the estimated CPICH E_c/I_{or} in unloaded condition being −3 dB, the target E_c/N_o for an un-loaded condition is −9 dB according to Equation 2.7. When load increases, up to a maximum CPICH E_c/I_{or} of −10 dB, E_c/I_o degrades correspondingly; therefore, the target E_c/N_o in a loaded condition should be set to −16 dB.

2.6 Topology Planning

Our previous discussion on network planning assumed that all cells were from a single layer. This section considers the planning of microcells within the macronetwork.

When planning picocells, microcells, and macrocells, use of proper equipment characteristics (transmit power, number of channel elements, number of RF carriers, etc.) and proper RF propagation model is ensured.

The following points are reviewed to ensure that all cells integrate into the network:

- **Coverage.** When microcells are used to fill coverage holes that cannot be addressed by macrocells, both overall coverage and individual cell coverage are considered. For individual cell coverage, the overlap between cells is checked. Adding a microcell could introduce further cell fragmentation and handover. In a high mobility traffic area, the UE perceives fragmentation as interference when reselection or handover does not have time to complete.
- **Interference.** Overlap between cells affects the level of interference (CPICH E_c/N_o). In the immediate vicinity of the microcell, E_c/N_o is expected to improve, especially when the microcell is placed in a high traffic area. The presence of a dominant server decreases the total transmit power ($\hat{I}_{or}$) at the surrounding sites, improving E_c/N_o over the associated coverage area. At the periphery of the microcell, further lack of dominance may be detected if micro- and macrocell coverage is not coordinated properly. Because of the static nature of the simulation, a network planning tool would not detect interference caused by reselection or handover delays and failure.

- **Handover area.** The limited ability of network planning tools to detect interference can be compensated for by handover plots, or (even more relevant) a plot showing the number of servers within the handover threshold. This plot could show the areas where servers are strong, but cannot be included in the Active Set; in other words, areas of Pilot pollution.

While TDMA/FDMA network planning tools can perform some level of Hierarchical Cell Structure (HCS) planning, this feature is not supported in WCDMA network planning tools. This is largely due to the one-to-one frequency reuse, which would require dynamic simulations of UEs moving at various speeds over defined trajectories. Monte Carlo simulations cannot be considered dynamic, because each UE is distributed to a fixed location.

2.7 Parameter Settings and Optimization during Network Planning

In addition to determining the RF configuration, which is the main goal of network planning, a limited number of parameters can be estimated. These are only estimations; network planning tools usually do not support dynamic simulations, which provide more accurate parameter settings. The parameters that can be estimated with a network planning tool, and their limitations, are listed below. These parameters can be estimated only after the RF configuration has been optimized.

- **Maximum DPCH setting.** During network planning, the maximum DPCH power can be estimated by observing the reason for failure during a Monte Carlo simulation. Failure caused by insufficient achieved Downlink E_b/N_t indicates that the Downlink DPCH power is set too low. Setting this value in a network planning tool may produce an inaccurate estimate of the required E_b/N_t. The network planning tool usually does not estimate E_b/N_t target values; instead, they are set manually on the basis of an expected channel condition. The parameter setting affects both coverage and capacity. An optimal value for this parameter would result in a similar probability of failure due to either "insufficient Downlink E_b/N_t" or "cell power limitation". The target E_b/N_t for different channel conditions should be taken from the minimum performance requirement, as done in Section 2.3, or determined from a link-level simulator.
- **Active set threshold (or event 1a reporting range).** This parameter ensures that the strongest cells at any given location are included in the Active Set, up to the maximum Active Set size. The accuracy of the threshold setting is limited in static simulations, because fast-rising servers usually cannot be considered in a network planning tool. For a given maximum DPCH power, the optimal setting would be the lowest possible Active Set threshold that does not affect the call success rate.
- **Maximum active set size.** This parameter ensures that all the overlapping sectors in a prediction can be admitted to the Active Set. During network planning, this parameter setting depends mainly on the Active Set threshold. The resulting value of the Active Set size does not consider the dynamic changes typically observed in a network, and is likely to be underestimated. Once the Active Set threshold is set, Active Set size can be determined by observing the Monte Carlo simulation results.

The optimal value would be the lowest possible value that does not affect the call success rate.
- **PSC setting and neighbor relationship.** These parameters are standard in most WCDMA network planning tools and can be considered as RF configuration parameters.

2.8 RF Optimization

No matter how extensive the network planning is, RF optimization must be the first step of network optimization. There are many reasons for this. During network design, coverage and capacity were optimized on the basis of network planning tools. Estimating how the network will actually perform is limited by the accuracy of both the RF model and the input data. The terrain model is accurate to about 10 to 20 meters. The building model, when used at all, considers the outline of the building but not the building structure and construction materials. Long-term statistical RF models are accurate overall, but short-term statistics usually fluctuate around the mean, by several dBs. The list could be extended and more inaccuracies could be uncovered for each model. After all the careful planning, the actual deployment adds its share of inaccuracy. For example, feeders could be longer or shorter, causing more or fewer losses due to actual routing, or antenna orientation may be based on magnetic north, rather than true north.

RF optimization can be seen as the first, and perhaps only, opportunity to observe the system in its entirety, under known conditions. During this exercise, the effects of various inaccuracies can be observed and the actual settings known. Corrections can be introduced to align coverage and capacity expectations with the network plan.

RF optimization for a WCDMA system should verify the coverage (RSCP) and minimize interference (E_c/N_o) over the intended coverage area. Interference takes priority over coverage because interference affects both the usable area and, in the long term, the capacity of the system.

In WCDMA, coverage can be seen as the area where the CPICH is higher than the specified quality and signal strength thresholds. In Idle Mode, the quality threshold is known as *Qqualmin*, or E_c/N_o. The signal strength threshold is known as *Qrxlevmin*, or RSCP. Each of these is defined in Chapter 4, along with other related parameters. The signal strength RSCP is easier to understand and estimate because it depends only on the power allocated to the CPICH, and the path loss between the cell and the UE. Increasing coverage based on CPICH RSCP requires only increasing the power or reducing the path loss. Given finite HPA power, increasing the CPICH power reduces capacity and is not a good long-term solution. On the other hand, reducing the path loss can be achieved only by reducing the site-to-site distance, possibly affecting quality, or E_c/N_o. This is where controlling interference comes into play.

When site-to-site distances decrease, the risk of overlap increases. Unless the coverage of cells is properly controlled, there is a risk of reduced E_c/N_o, which creates a ripple effect. In Idle Mode, the UE must acquire and measure the CPICH and read all the System Information Blocks (SIBs). Lower E_c/N_o affects the demodulation of the PCCPCH, unless its power is sufficient to maintain the required E_b/N_t. This is also noticeable when accessing the system. When only a single server is used, low E_c/N_o directly correlates with low geometry, which increases the transmit power needed to meet the E_b/N_t requirements of PCCPCH, SCCPCH, PICH, and AICH. It is possible to increase the power assignment for these control channels, but this would reduce the power available for DPCH and, therefore, reduce capacity.

In Connected Mode, handover can partly compensate for low RF performance. With handover, the geometry and the required E_c/I_{or} are not directly related to the CPICH E_c/N_o of the individual server, but to the combined CPICH E_c/N_o of all the servers in the Active Set. The required E_b/N_t of the DPCH can be met by combining the energy from multiple links.

Although handover is one of the main advantages of WCDMA in terms of code channels or from a power perspective, handover also can affect capacity. The effect of handover on code channels is most visible for low Downlink spreading factors (Orthogonal Variable Spreading Factor [OVSF] codes). This is shown in Figure 2.21, which plots the capacity per cell, for PS384 bearers, in Erlangs, against the handover overhead factor. The handover overhead factor, also called *Handover Reduction Factor* (HORF), is the average number of connections serving a call. It can be used to gauge the efficiency of the network, when all other metrics are equal. A network with a lower HORF can carry more traffic. Capacity decreases sharply as the HORF increases because of the properties of Erlang theory, illustrating the connection between handover and capacity. As the example in Figure 2.21 shows, an HORF higher than 1.5 significantly reduces capacity.

Channel rate switching can compensate for the code tree limitation or for capacity limitations due to DPCH power. This is most effective for PS data applications. Channel rate switching can downgrade a bearer to a lower data rate when the DPCH power

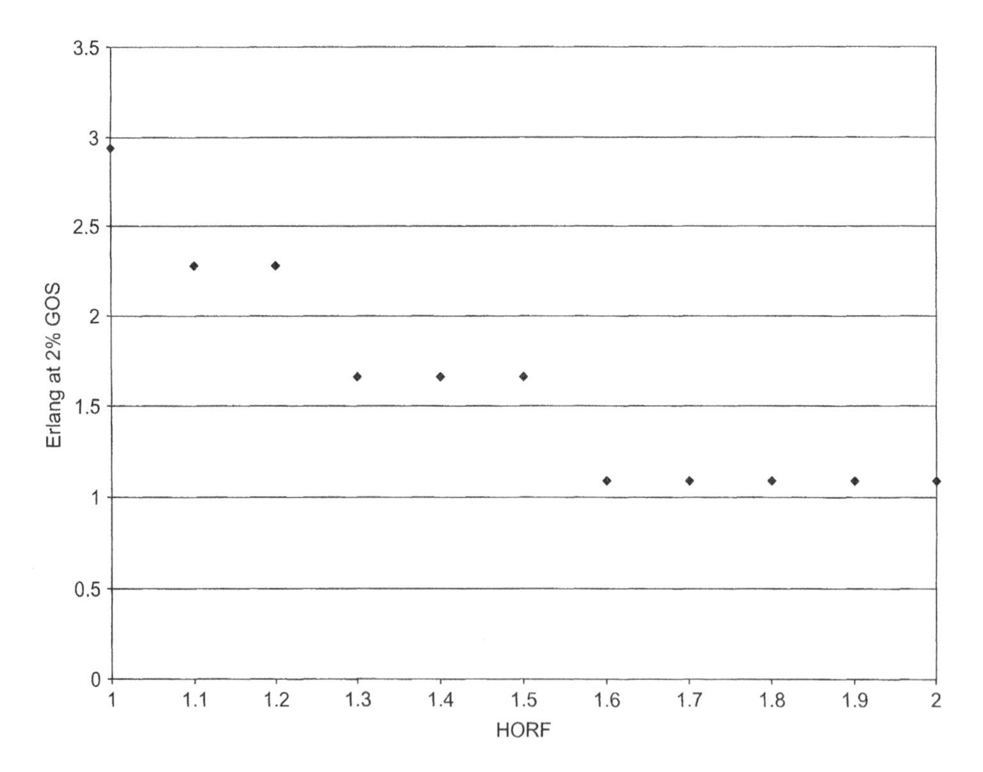

Figure 2.21 How HORF affects cell capacity for SF = 8

increases beyond a specified threshold. With channel switching, the throughput responds to changes in geometry, as shown in Figure 2.22. In this figure, geometry varies from −6 to 9 dB. For all the bearers (PS64, PS128, and PS384), the maximum DPCH is limited to −7 dB E_c/I_{or}. Rate switching should be favored in areas of degraded geometry (i.e., cell edge), where best throughput is achieved with PS64. In this example, switching from 384 to 128 kbps should occur around a geometry of 1 dB. Switching from 128 to 64 kbps should occur around a geometry of −4 dB. In an actual network, rate switching would not be triggered by geometry, but by reaching a DPCH power or reported E_c/N_o threshold.

The gain provided by using handover to compensate for poor E_c/N_o is offset by the total consumed power per connection [4]. The total consumed power is the sum of the DPCH power for all the handover links serving a given connection, expressed in terms of E_c/I_{or}. This value is compared to the expectation of the average active cell size E{N}. Figure 2.23 illustrates the handover gain per link by the monotonic decrease of per-link DPCH power, represented as DPCH_E_c/I_{or}. As the mean Active Set size increases, the DPCH required per-link decreases. The total consumed power, or per-Active Set power, initially decreases, corresponding to the area of maximum handover gain, but then the per-Active Set power increases. In that region, the handover gain is not optimal, ultimately wasting Downlink resources. This observed sweet spot is a function of the quality of

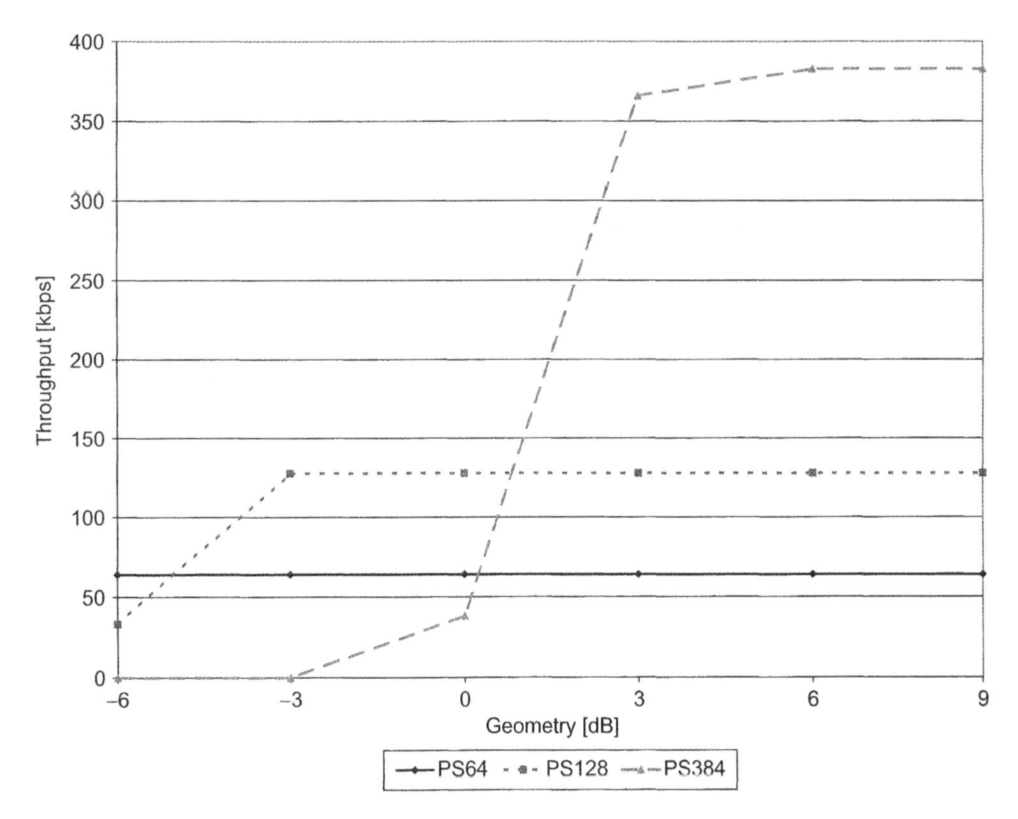

Figure 2.22 How geometry affects user throughput for various bearers, with E_c/I_{or} limited to −7 dB

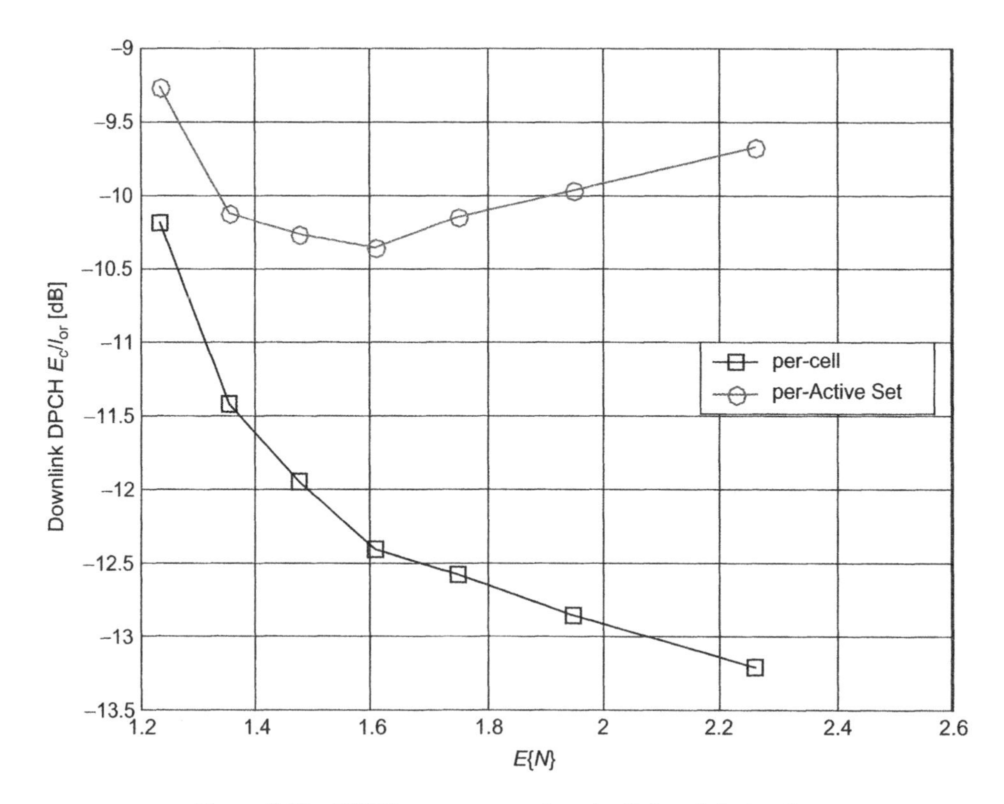

Figure 2.23 DPCH power comparison by link and Active Set

the RF optimization. At equal parameter settings, the best RF optimization results in the lowest average Active Set size.

These examples illustrate the importance of optimizing the RF configuration when the CPICH E_c/N_0 is maximized at the given load. This ensures that the remaining optimization will be as efficient as possible.

2.8.1 Quantitative Optimization

In the early phases of overlaying an existing network with a new technology, the new network must compete with established networks that may have reached a very high level of service after years of continuous optimization. Achieving the same QoS upon commercial implementation of the new network is neither practical nor economical. For this reason, a limited set of practical optimization objectives should be applied to attain the design objectives and provide acceptable QoS to the end users. At each stage of the optimization process, objectives can be set through corresponding Key Performance Indicators (KPIs).

KPIs should be defined for each step of the optimization process, starting with a particular preoptimization task and successively increasing in scope with each subsequent step. Later sections discuss this in detail. Table 2.17 and Table 2.18 list the KPIs for preoptimization and RF optimization.

Table 2.17 Preoptimization KPIs

Task	KPI	KPI target
Network planning review	CPICH E_c/N_o	> −9 dB over 95% of the area
	CPICH RSCP	> design target over 100% of the area, considering appropriate margins
Single site inspection	Site installation and commissioning	All sites in the tested area installed and commissioned
Site verification	Site operation	All sites in the tested area processing calls
Basic feature verification	Call access performance in controlled RF environment	>99% for all services
	Call retention performance in controlled RF environment	>99% (or <1% call dropped) for all services

Table 2.18 RF optimization KPIs

Task	KPI	KPI target example
RF optimization	Measured RSCP	> −88 dBm (for 20 dB BPL) over 97% of area
	Measured E_c/N_o	> −9 dB over 95% of area
	Cell overlay	<3 cells within r1a over 95% of area
	Qualitative distribution	No cell overshoot Minimal cell fragmentation Minimal change in best server Minimal differences between prediction and measurements

The RF KPI examples in Table 2.18 should be reviewed for each optimization project, according to the network planning. The RSCP threshold can be calculated from the minimum acceptable RSCP, then derated according to any margin or loss in the Link Budget, as Table 2.19 shows. The link margin (especially LNF) is not directly included when measurements are performed; it is indirectly factored as the probability of measuring a value below the threshold. Link margin should be consistent with the area coverage probability, assuming that the standard deviation corresponds to the value used for planning.

The threshold used for E_c/N_o and the cell overlay should be consistent with the value used during network planning. An important parameter for the E_c/N_o calculation is actual loading at the time each cell is tested. This can be calculated using Equation 2.13, which estimates the E_c/N_o of a given server, j, from one of these two options:

- The total transmit power ($\hat{I}_{or}$) and path loss (L), *or*
- The RSCP and the loading (E_c/I_{or}) of each server

Table 2.19 Calculation of RSCP optimization threshold

Line item	Value [dB] or [dBm]	Comment
Minimum RSCP	−111	Corresponds to Qrxlevmin
Body loss	3	Per design, for worst service
Link margin	0	Not considered for the measurement threshold; used only for coverage probability
Building penetration loss	20	Per design
Antenna gain	−3	Typical for omni mag mount antenna
Antenna cable loss	3	Corresponds to 4 m of RG-58 cable
Calculated RSCP optimization threshold	−88	

Although their values are equivalent, the second option is more convenient because it does not require knowing the path loss, only the RSCP. The E_c/I_{or} used in this equation should be measured or read from the network element, but when no users are present, it can be estimated from the sum of the control channel powers.

$$\left[\frac{E_c}{N_o}\right]_j = \frac{RSCP_j}{kTB + Nf + \sum_{i=1}^{n}(I_{or})_i \times L_i} = \frac{RSCP_j}{N_{th} + \sum_{i=1}^{n} RSCP_i \times (E_c/I_{or})_i} \quad (2.13)$$

Although the number of servers is reflected in the E_c/N_o calculation, it is a good idea to count the number of servers within the reporting range independently. The number of servers can be used to estimate the overhead factor, which will ultimately affect system capacity. Although the E_c/N_o and number of server KPIs are important during optimization, they do not completely portray the quality of the network. Qualitative criteria can complement these metrics, as discussed in the next section.

2.8.2 Qualitative Optimization

The qualitative metrics discussed in this section influence system performance in the long term, but might not be immediately obvious from looking at E_c/N_o, RSCP, or the number of servers within a few dB.

- **Cell overshoot (boomer).** Cell overshoot is usually generated by boomer cells, namely cells that are taller than the average terrain height. Figure 2.24 shows a cell overshoot of Site1080_1 from either the best server plot or the RSCP plot. From the best server plot (Figure 2.24(a)), cell overshoot can be detected as the best server at locations that extend beyond the first tier of the originating cell. From the RSCP plot (Figure 2.24(b)), a cell overshoot can be detected in which strong RSCP occurs beyond the first or even second tier of the originating cell Site1080_1. The best server plot will not show cell overshoot as reliably as the individual RSCP plot, because a server could extend outside its intended coverage area yet not be the strongest server. Cell overshoot causes several issues in a network. In particular, overshooting cells have large handover areas that exhaust resources. Once resources are exhausted, call origination in this area fails and

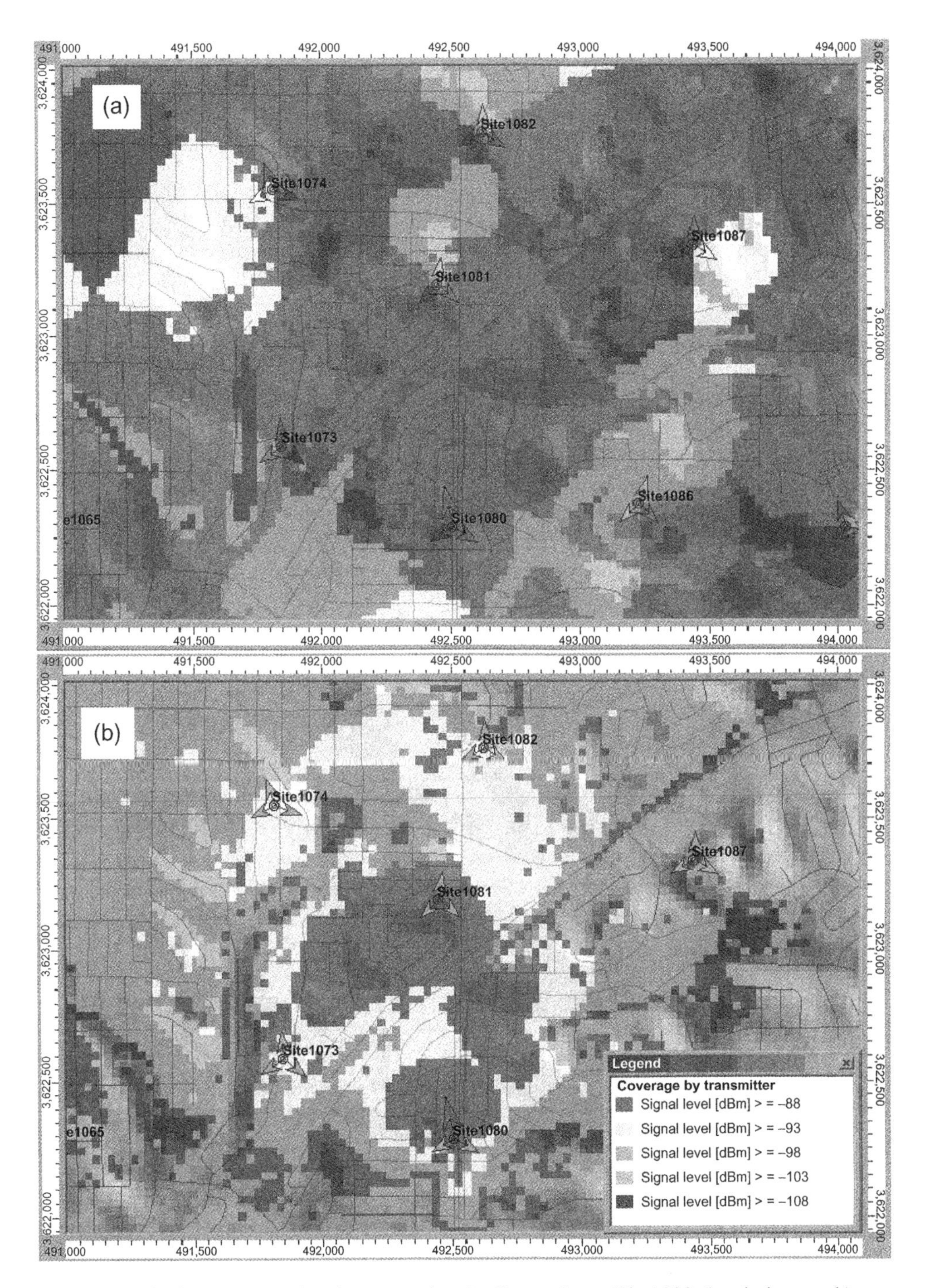

Figure 2.24 Network planning example of cell overshoot (Site1080_1 pointing north)

resource requests for handover are denied. When handovers are denied, cell overshoot could affect the quality of the connection. When handover is not granted, a call experiences lower geometry due to a strong cell that is not in the Active Set. When the geometry degrades further and additional power cannot be allocated to the DPCH, the quality (BLER) degrades and eventually the call drops.

- **Cell fragmentation.** Cell fragmentation can be defined as the noncontinuous area where a cell is detected as the best server. Cell fragmentation can be a direct consequence of cell overshoot, but may also be detected in areas of weak cell coverage, such as obstructed areas or areas with antenna of different heights. Handover is affected by cell fragmentation. Specifically, fragmentation usually increases Active Set size. The increase in Active Set size, or HORF, varies according to the handover parameter settings. This affects reselection and, possibly, call origination performance. The effect on cell reselection depends on the Treselection parameter and user mobility. Figure 2.25 illustrates CPICH E_c/N_0 measurements over time, measured by a Pilot scanner for various observed cells. This example shows that at speeds higher than 50 km/hr, and a distance of 600 meters, a Treselection setting above 1 sec prevents the UE from reselecting to cell 1007_2; it remains on cell 1008_2. The E_c/N_0 of cell 1008_2 could fall below −20 dB, at which point call origination is likely to fail unless all the control channel powers are set to overcome such channel conditions.

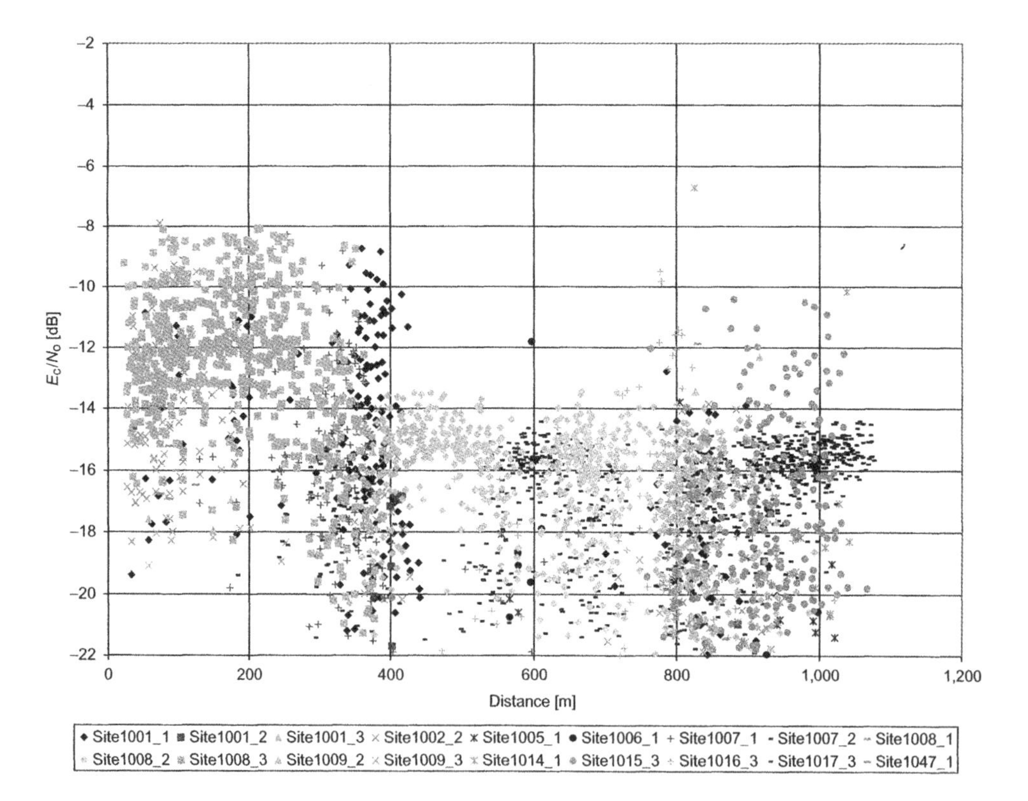

Figure 2.25 Cell fragmentation example, as observed with a Pilot scanner

- **Frequent change in best server.** Unlike cell fragmentation, in which a UE selects a given server multiple times, frequent change in best server is characterized by selecting (or attempting to select) different servers. This can be caused by a high density of cells, or by the absence of a dominant server due to imperfect optimization. High cell density should not cause problems if coverage for each cell is well contained, and if the reselection parameters and handover parameters (to a lesser extent) are set to accommodate a rapid change of best server. For reselection, this means minimizing Treselection while accommodating the mobility of users in the area. Frequent change in best server due to the absence of dominant servers is more critical because this reduces E_c/N_0. Low E_c/N_0 affects call access and retention.
- **Differences between prediction and measurement.** While minor differences between predicted and actual measured behavior are expected, any major differences in server performance or expected values (RSCP or E_c/N_0) should be analyzed and resolved. Differences in detected servers usually indicate installation issues, swapped cables, obstruction in the antenna near the field causing reflection, or pattern distortions.

2.8.3 Idle Mode Optimization

The RF optimization discussion in Section 2.8 assumed that only a Pilot scanner could be used. After the early stages of RF optimization are complete, it is helpful to employ a UE in Idle Mode to estimate UE reselection performance.

Chapter 4 details the trade-off between reselection parameters and performance. Table 2.20 shows the main performance criteria for Idle Mode.

Table 2.20 Idle Mode performance criteria

Performance criteria	Target	Comment
Reselection rate	Variable, depends on mobility	During reselection, the UE consumes more power; thus, a low reselection rate is desirable Reselection should still be dynamic enough to ensure that the UE camps on the best possible cell
Camp cell E_c/N_0	−9 dB over 95% of area	Metrics similar to the E_c/N_0 KPI defined for RF optimization. Only difference is that this metric is measured by the UE for the cell it camps on
Tail of E_c/N_0 distribution	> −16 dB over 97% of the area. Percentage should be consistent with coverage area probability	Minimize poor coverage
Number of Out-of-Service (OOS) conditions	0	An OOS condition indicates that a UE lost service for an extended period of time (>12 sec); this would affect call delivery performance

All criteria except number of Out-of-Service (OOS) conditions can be estimated from Layer 1 measurements, available from a test UE. Measure and compare performance over multiple tests using different parameter configurations to determine the optimal setting. The first step is to ensure that the UE detects no OOS condition. This condition cannot be extracted from Radio Resource Control (RRC) messages. It can be detected only by a test UE and logging software that can output low-level messages. The standard [13] defines an OOS condition as the inability of the UE to acquire a serving cell when it wakes up from a Discontinuous Reception (DRX) cycle.

After all the OOS conditions are eliminated from the system, the other metrics should be considered together because they are interdependent. For example, slow reselection may cause a UE to camp on a cell that is not the best one, thus affecting the measured E_c/N_0.

References

[1] 25.104. UTRA (BS) FDD: Radio transmission and reception. 3GPP; 2003.
[2] Jakes WC. *Microwave Mobile Communication*. Wiley & Sons; 1974.
[3] 25.101. User Equipment (UE) radio transmission and reception (FDD). 3GPP; 2004.
[4] 34.108. Common test environments for User Equipment (UE), conformance testing. 3GPP; 2003.
[5] Ata OW, Garg H, *Path Loss Correlation between PCS and MMDs/ISM Bands in Suburban Morphology – An Empirical Model*. IEEE; 2004.
[6] Bertoni HL. *Radio Propagation for Modern Wireless Systems*. Prentice Hall PTR; 2000.
[7] Mogensen PE, Eggers P, Jensen C, Andersen JB. Urban area radio propagation measurements at 955 and 1845 MHz for small and micro cells. *Global Telecommunications Conference*; 1991 Dec 2–5; vol. 2; p 1297–1302. Phoenix Arizona Digital Object Identifier 10.1109/GLOCOM.1991.188579.
[8] Laiho-Steffens JK, Lempiainen JJA. *Impact of the Mobile Antenna Inclination on the Polarization Diversity gain in a DCS1800 Network*. IEEE; 1997.
[9] 21.905. Vocabulary for 3GPP specifications. 3GPP; 2001.
[10] 25.942. RF system scenarios. 3GPP; 2002.
[11] 25.211. Physical channels and mapping of transport channels onto physical channels (FDD). 3GPP; 2002.
[12] 34.121. Terminal conformance specification, radio transmission and reception (FDD). 3GPP; 2002.
[13] 25.133. Requirements for support of radio resource management (FDD). 3GPP; 2004.

3

Capacity Planning and Optimization

Christophe Chevallier

3.1 Basic UMTS Traffic Engineering

Traffic engineering has been in existence almost as long as telecommunication itself. In fact, the unit used to measure traffic is named after the first engineer to summarize traffic in an equation, A.K. Erlang, who originally published his work in 1909.

The Erlang theory can be used to estimate the probability of blocking when a given amount of traffic is carried over a limited number of resources. Equation 3.1 shows the Erlang B formula.

$$B = \Pr(Blocking) = \frac{A^N}{N! \times \sum_{i=0}^{N} \frac{A^i}{i!}} \tag{3.1}$$

In Equation 3.1, the blocking probability B can be calculated from the traffic intensity, or offered traffic A, and the number of servers N.

The probability of blocking can also be expressed in relation to Grade of Service (GoS), as in Equation 3.2:

$$B = 1 - GoS \tag{3.2}$$

By extension, Erlang theory can calculate any one of the following when the other two are known: blocking probability, number of resources, and traffic. This calculation is normally done during the initial design, to determine the number of resources required to carry the assumed traffic while maintaining the expected GoS.

In a Wideband Code Division Multiple Access (WCDMA) network, the Erlang formula can be used almost directly in the circuit switched (CS) domain, because resources are dedicated to a user for the duration of a call. The main difference between WCDMA and Frequency Division Multiple Access (FDMA)/Time Division Multiple Access (TDMA)

WCDMA (UMTS) Deployment Handbook: Planning and Optimization Aspects QUALCOMM Incorporated

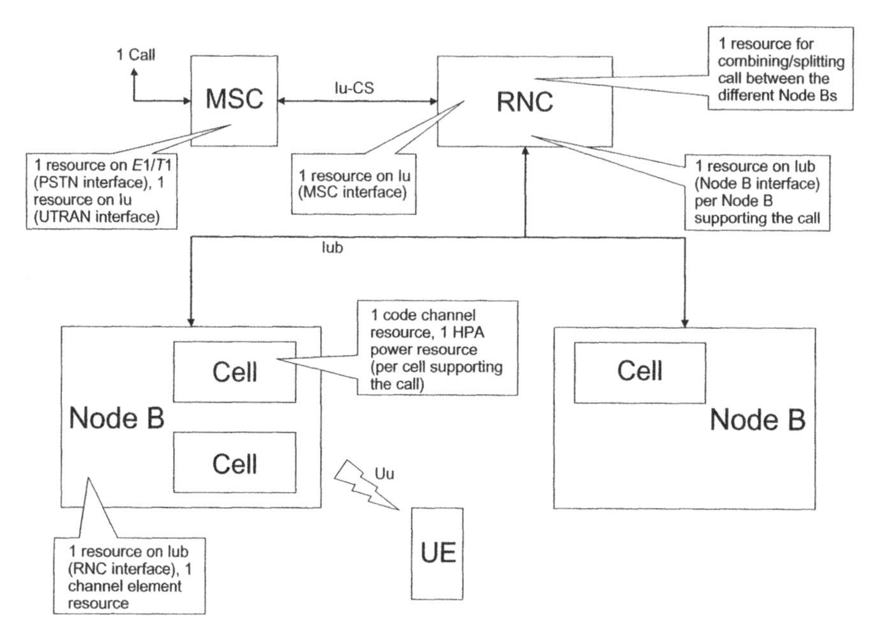

Figure 3.1 Resource definition in a WCDMA system

technology is the multiple definitions of resources, as illustrated in Figure 3.1. As the figure indicates, resources should be defined and planned for each component of the WCDMA network.

- **At the Mobile Switching Center (MSC).** Resources are assigned for the duration of the call; Erlang theory can be applied directly. Traffic assumptions, when given, are typically for user traffic, which is equivalent to MSC traffic. All types of traffic are included at this point, regardless of the terminating point.
- **At the Radio Network Controller (RNC).** Resources should be defined for the Iu and Iub interfaces. Erlang theory can be applied directly for the resources on the Iu interface and for the common resources, shown as the combining/splitting function in Figure 3.1. For the Iub resources, the Erlang theory can also be applied, but the amount of traffic should be multiplied by a factor that represents soft handover. At the RNC, additional resources can be defined, but would generally be implementation-dependent. For a generic architecture [1], the resources would still be divided between common (i.e., one per call) and dedicated (i.e., one per Iub) resources.
- **At the Node B.** The main capacity-associated resources to dimension are carriers (deployment of single or multiple frequencies), channel elements, code channels, and power. In addition, the dimensioning should be done in such way that the Uplink noise rise has a limited impact on the cell coverage. Channel elements are the easiest resources to dimension because one is needed per call, per Node B, regardless of the percentage of softer handover. Proper channel elements dimensioning is nevertheless important to ensure that they will not introduce hard blocking in either the Uplink or Downlink. The code channels must be dimensioned according to softer handover; one code channel is required per handover leg. The power required per link depends on the Radio Frequency

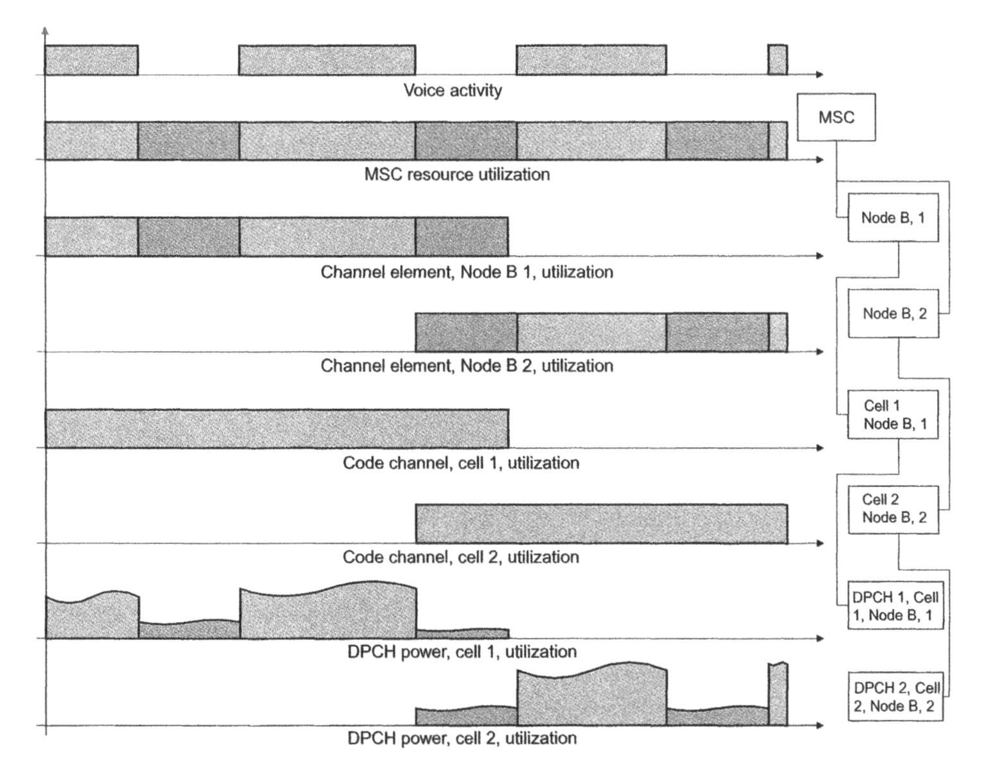

Figure 3.2 Resource utilization and soft/softer handover

(RF) conditions and the handover state. Dimensioning should be done through simulations for a given set of assumptions. Both the code channel and power resources mainly limit the Downlink, while the Uplink noise rise limits the Uplink traffic.

Figure 3.2 illustrates these resource utilization concepts. Cell resource utilization (Dedicated Physical Channel (DPCH) power or Uplink noise rise) cannot be estimated for a given time; instead, it should be based on average resources. For the DPCH power, this average depends on the average channel conditions, the minimum and maximum settings of DPCH power, and the utilization of discontinuous transmission (DTX).

In this section, we shall first estimate user traffic, or the traffic requirements of the system. Then we shall estimate the Uplink and Downlink capacities of a WCDMA system to determine the limiting link for traffic dimensioning.

3.1.1 Capacity Requirements

In the CS domain, the traffic requirement is defined in terms of Erlangs, during the busy hour. However, this may not always correspond to marketing considerations. In addition, such a definition is not practical for network planning because it does not show the distribution of users within the given area.

The approach is based on a generic traffic requirement, adjusted for busy hour requirements and to account for geographical distribution, as required for network planning purposes.

Table 3.1 Example of traffic assumptions used during initial planning

Subscriber	Voice-centric	Data-centric	Corporate
Target population	1000000	1000000	1000000
Target penetration	40%	10%	20%
Per-subscriber monthly Minutes of Use (MOU)	1000	1000	2000
Per-subscriber monthly (MB)	10	100	200

The easiest way to estimate traffic for a commercial application is with monthly Minutes of Use (MoU). Service to end users is commonly sold in terms of MoU quotas. These quotas vary by type of user, as shown in Table 3.1.

Network planning requirements must be expressed for the busy hour because it is the limiting factor in a network. To reconcile coverage-versus-capacity trade-offs when a network is implemented, requirements should be distributed over distinct areas rather than expressed for the entire network.

The geographic distribution of subscribers can be estimated from demographic information. In some countries, such data for populations of entire cities is readily available from the census bureau. In addition, this information is becoming more available from public or commercial databases, sorted by postal code and, possibly, by socio-economic categories.

Because population distribution can be estimated from these data sources, the remaining task is to translate monthly MoU data into busy hour data, based on an estimated service penetration.

Busy hour requirements can be estimated from monthly MoU by calculating the inter-day and intra-day peak-to-averages. The *intra-day* peak-to-average is the difference in traffic loading between the busy hour and the average hour (using system-wide statistics). This typically ranges from 2 to 6. The *inter-day* peak-to-average is the difference between the busiest day and the average day (again using system-wide statistics). This has a narrower range, from 1.2 to 3. With this information, the traffic per subscriber can be calculated using the formula shown in Equation 3.3.

$$\begin{aligned} BusyHourTraffic = {} & \frac{MonthlyTraffic}{30 \times 24} \times \text{Peak-To-Average}_{Inter\text{-}Day} \\ & \times \text{Peak-To-Average}_{Intra\text{-}Day} \end{aligned} \tag{3.3}$$

With the adoption of bundled calling plans, which offer large numbers of minutes for a fixed cost plus steep discounts for nights and weekends, both the intra-day and the inter-day peak-to-averages have reduced considerably in the last few years.

An alternative way to estimate busy hour traffic from monthly MoU is to assume that the traffic is Poisson distributed. This method can be applied to estimate an uneven traffic distribution, even if no actual peak-to-average ratio is known. Using the simplified probability function of the Poisson distribution [2], busy hour loading can be determined by finding n that solves Equation 3.4.

$$\sum_{i=0}^{n} \frac{e^{-\lambda} \times \lambda^{i}}{i!} \leq (1 - GoS) \tag{3.4}$$

In this formula, λ represents average loading, and *GoS* represents the desired GoS of the system, or the probability that a given user experiences blocking. In this case, n is consistent with the unit used for λ, either MoU at busy hour or milliErlangs (mErl) at busy hour.

MoU should be translated into mErl, the value more commonly understood by network planners. This conversion is easy: 1 Erlang corresponds to 1 resource utilized for the entire period of observation. During busy hour, the period of observation is 1 hour, or 60 minutes. Consequently, 1 Erlang = 60 MoU, or 1 MoU = 16.66 mErl.

Figure 3.3 shows a numerical example of the indicated distribution. The example assumes the following:

- All calls are served in random order.
- An infinite number of resources are available.
- Holding time is constant or exponential.
- All blocked calls are cleared.

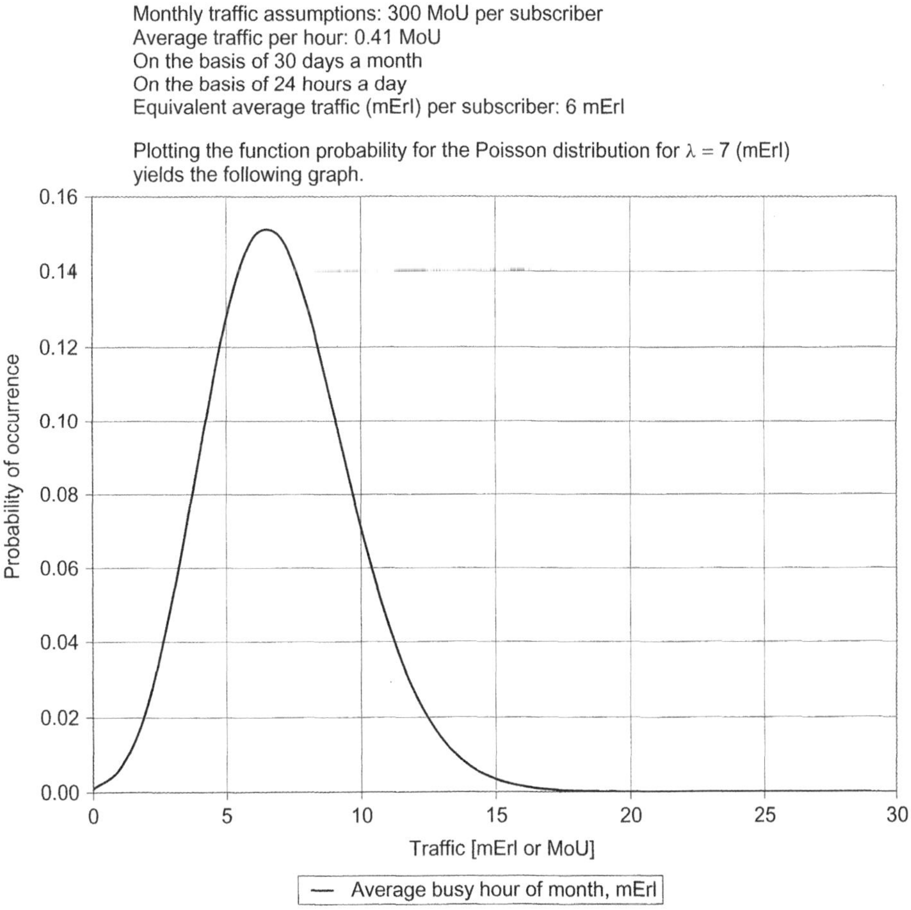

Monthly traffic assumptions: 300 MoU per subscriber
Average traffic per hour: 0.41 MoU
On the basis of 30 days a month
On the basis of 24 hours a day
Equivalent average traffic (mErl) per subscriber: 6 mErl

Plotting the function probability for the Poisson distribution for $\lambda = 7$ (mErl) yields the following graph.

Figure 3.3 Poisson distribution of traffic to estimate busy hour usage from monthly usage

In Figure 3.3, Equation 3.4 is solved for $n = 12$, meaning busy hour traffic of 12 mErl is assumed for network planning. This corresponds to a peak-to-average of 1.7. This example assumes that traffic is evenly distributed across all days. However, it is more accurate to assume that traffic is unevenly distributed over the days of the week [3]. As for other traffic distributions, a Poisson distribution can be assumed. In this case, a two-step process can be followed to determine expected traffic per day and expected traffic at busy hour. On the basis of the above calculation, traffic per subscriber during the busy hour of the busy day is expected to be 19 mErl. This is based on a day-to-day peak-to-average of 1.6, and an intra-day peak-to-average of 1.6.

Another way to determine traffic requirements is to use existing network information. In this case, the entire network traffic requirement per cell, including geographical distribution, is known. For CS (voice) traffic, extrapolating WCDMA traffic from another network is valid as long as charges for service are consistent. For packet switched (PS data) traffic, the extrapolation is less valid because traffic tends to depend on the service offered (maximum data rate).

While GoS depends on specific goals defined by service operators, GoS values for voice services are usually 2% for high-mobility users and around 1% for low-mobility users (Wireless Local Loop (WLL) services). This GoS should not be mistaken for the probability of success for a call; GoS only considers blocked calls (calls for which no resources are available), while the probability of success includes any type of failure.

3.1.2 Uplink Capacity Estimation

For the Uplink, the upper boundary of the capacity (N_{pole}) of a WCDMA carrier can be estimated using the standard Uplink capacity equation [4], as shown in Equation 3.5. This widely accepted formula is derived from early Code Division Multiple Access (CDMA) work [5,6] and applies to WCDMA if variables are set properly. Equation 3.5 estimates the capacity of a single cell from the spreading bandwidth (W), the radio access bearer (RAB) bit rate (R_b), the required Energy per bit-to-Total Noise ratio (E_b/N_t), the activity factor (ν), and the interference factor (α). To estimate the capacity of a Node B, sectorization gain must be included. This gain is typically estimated at 2.55 for a three-sector cell.

$$N_{pole} = \frac{W/R_b}{E_b/N_t \times \nu \times (1+\alpha)} \tag{3.5}$$

The results of this equation vary according to the supplied assumptions and approximations [7]. Before we discuss the limitations of this equation, let us consider the following factors that influence the pole capacity of the system.

- ***W* (Spreading bandwidth).** The spreading bandwidth of the system is fixed by the standard at 3.84 megachips per sec.
- **R_b (Radio access bearer bit rate).** Table 3.2 shows typical bearer bit rates for selected applications. Capacity can be increased only by using applications that require a lower bearer bit rate. An example for voice is using an Adaptive Multirate (AMR) of 7.95 rather than the typical AMR of 12.2. Because this change degrades the perceived voice quality, all choices must be evaluated before applying it.

Table 3.2 Typical bearer bit rate for selected applications

Application	Typical bearer bit rate [kbps]	Comment
Voice	12.2	Other supported codec [8,9] rates (not necessarily implemented) are: 10.2, 7.95, 7.4, 6.7, 5.9, 5.15, 4.75
Video-telephony	64	H.324 video codec [10,11] can work with a bearer rate as low as 32 kbps, but 64 kbps is considered the minimum bearer bit rate for acceptable quality on a QCIF display
PS data service	64 or 128	Higher bearer bit rates typically are not used on the Uplink even if defined up to 384 kbps [12]

Table 3.3 Uplink E_b/N_t requirement for Speech AMR, 12.2

Bearer	Target BLER [%]	Minimum requirement [dB]				
		Static (AWGN)	Case 1	Case 2	Case 3	Case 4
Speech AMR 12.2 kbps	1	5.1	11.9	9.0	7.2	10.2

- **E_b/N_t (Energy per bit-to-total noise ratio).** As mentioned in Chapter 2, E_b/N_t is largely influenced by the data rate, channel conditions, channel coding used, the target Block Error Rate (BLER), and the hardware implementation. The effect of channel conditions and channel coding can be estimated from the Node B minimum performances listed in Table 3.3. The effect of hardware implementation, considered in the minimum performance as an implementation margin, cannot be estimated; therefore, the RF engineer should rely on vendor information for proper planning. The test case conditions shown in this table refer to multipath profiles, shown in Table 3.4. Additional cases are defined in the standard [13], but are not discussed here.
- **ν (Voice activity factor).** The voice activity factor depends on the type of vocoder used, channel coding, and the actual application. For voice, including a one-way monologue, actual voice activity is typically only 85% [14]. Because two-way dialogue is more common, the per-user voice activity is 40 to 45% at the Application Layer. To translate this into the voice activity factor, the vocoder and coding scheme must be included. The vocoder is important because silence and background noise are also coded, but at a lower data rate. The coding scheme is important because the Transport Channel can handle only a finite number of formats, to which the signaling bearer is added. Once this is factored in, the practical voice activity factor is in the range of 58 to 67%, with a possible peak of up to 85% if echo cancelers are not used [15]. Considering this range, mobile-to-mobile voice activity requires a higher voice activity factor, compared to mobile-to-land traffic. For PS data applications, an activity factor, also referred to as efficiency factor, can also be defined and will depend both on the type of application

Table 3.4 Multipath profiles used for performance evaluation, per 25.101

Case, per 25.101	Case 1, speed 3 km/hr		Case 2, speed 3 km/hr		Case 3, speed 120 km/hr		Case 4, speed 3 km/hr	
	Relative delay [ns]	Relative mean power [dB]	Relative delay [ns]	Relative mean power [dB]	Relative delay [ns]	Relative mean power [dB]	Relative delay [ns]	Relative mean power [dB]
First detected path	0	0	0	0	0	0	0	0
Second detected path	976	−10	976	0	260	−3	976	0
Third detected path	NA	NA	20,000	0	521	−6	NA	NA
Fourth detected path	NA	NA	NA	NA	781	−9	NA	NA

(e-mail, web browsing, media streaming) and the channel switching mechanism implemented. Table 3.17 expands on this and shows that the activity factor in the PS domain can range from a few percentage points up to 70%.

- **α (Interference factor).** The interference factor depends mainly on the quality of network planning, because it represents other-cell interference. The interference factor is directly affected by the overlap between cells, and by the ability of a cell to power-control calls. This last point is significant when Active Set sizes are artificially limited, or when handover is delayed (notably through the use of long Time-to-Trigger or high Cell Individual Offset). Values for α range from 0.55 to 0.65 [16]. This range represents an optimized network with minimal cell overlap. Unfortunately, this optimization is not always compatible with the high building penetration loss expected in current networks. Figure 3.4 shows the effect of building penetration loss, antenna tilt, and Active Set size on α where the interference factor is estimated for different configurations, based on regular site orientation, flat terrain, and simulation using a commercially available network planning tool. In commercially available network planning tools, α is not directly available. Usually the quantities available in a network planning tool are either the frequency reuse factor or the frequency reuse efficiency. These quantities can be estimated from the total received power and the same-cell contribution. The quantities can then be correlated with α, as shown in Equations 3.6 and 3.7 [7].

$$\mathrm{F} = \mathrm{Frequency\_reuse\_factor} = 1 + \alpha = \frac{\mathrm{Total\_Rx\_power}}{\mathrm{Rx\_power\_from\_same\_cell}} \tag{3.6}$$

$$\mathrm{F_f} = \mathrm{Frequency\_reuse\_efficiency} = \frac{1}{1+\alpha} = \frac{\mathrm{Rx\_power\_from\_same\_cell}}{\mathrm{Total\_Rx\_power}} \tag{3.7}$$

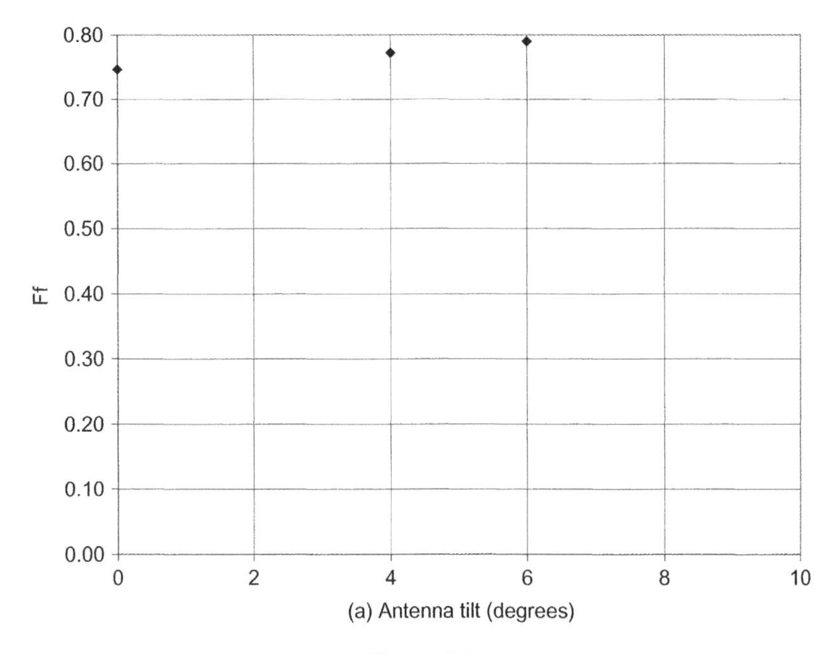

Figure 3.4

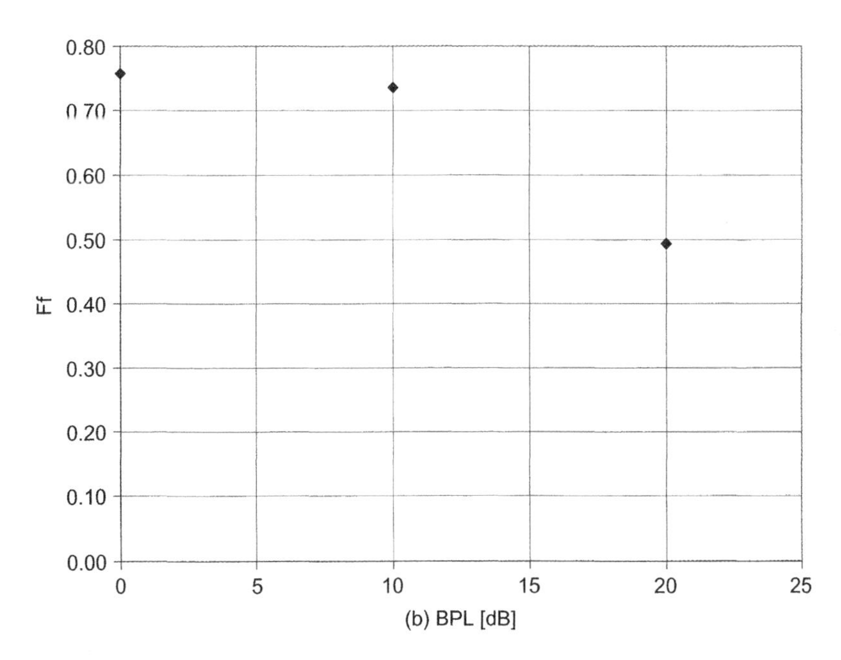

Figure 3.4 How network configuration influences the interference factor: (a) Antenna tilt, (b) BPL, (c) Active Set size

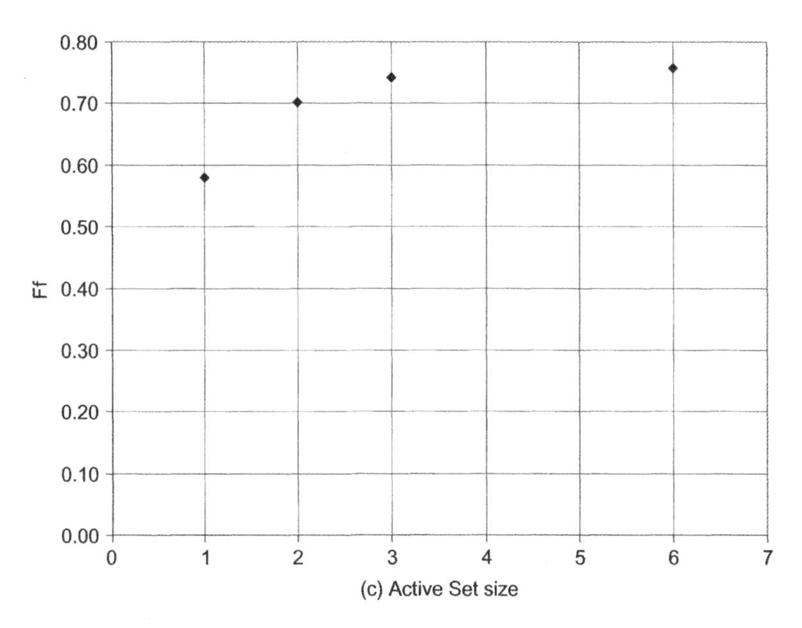

Figure 3.4 *(continued)*

Figure 3.5 illustrates the concept behind Equation 3.5.

The resultant pole capacity has no practical application; each user transmits at maximum power to overcome the noise, while the Uplink coverage is reduced to nothing. From the pole capacity, a practical capacity (N_{user}) for a system can be calculated after the Uplink

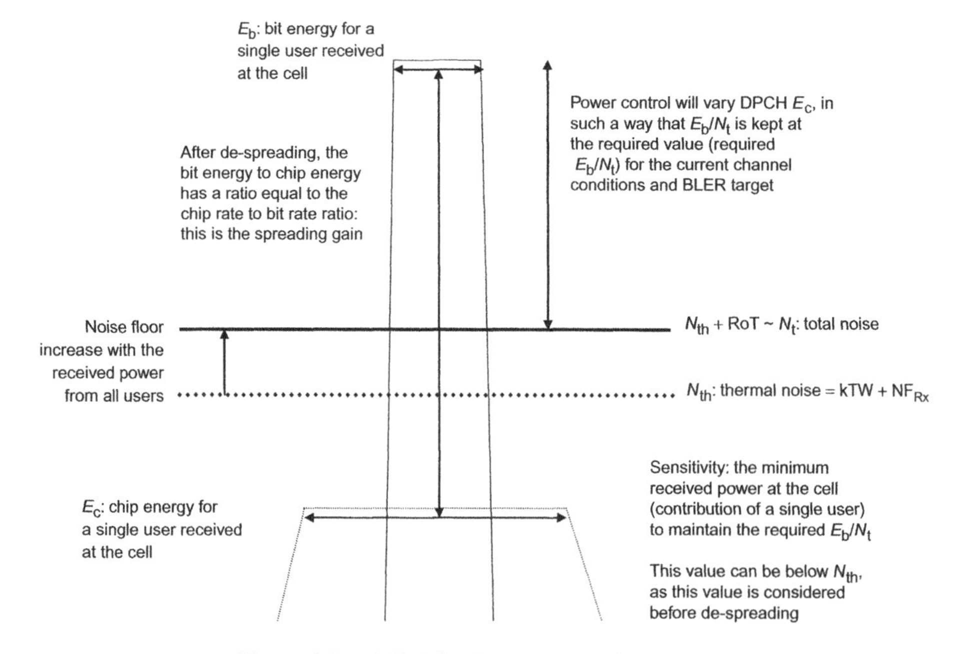

Figure 3.5 Uplink loading and capacity concepts

loading (η) operating point has been determined, as shown in Equation 3.8.

$$N_{\text{user}} = N_{\text{pole}} \times \eta \tag{3.8}$$

The maximum Uplink loading is selected to ensure that the network remains stable and that coverage is not adversely affected. As discussed in Chapter 2, the Uplink Link Budget is affected by the Rise Over Thermal, which is directly dependent on the loading, according to Equation 2.2. The traffic mix also affects the Uplink loading, mainly when PS data is predominant, because PS data services are asymmetric.

The capacity calculated from Equations 3.5 and 3.8 represents the number of resources available on the Uplink radio link, and not the capacity in Erlangs. To derive the capacity in Erlangs, the Erlang B formula is used (Equation 3.1) after setting the GoS.

Other resource dimensioning is considered later in the planning process, after the Uplink and Downlink capacity estimations are known. All calculations are based on these limiting factors.

On the Downlink, code space is a limiting factor, as we will see later. On the Uplink, it is assumed that code space can never be limiting, because users have their own code spaces to distinguish different channels (Dedicated Physical Control Channel (DPCCH) or Dedicated Physical Data Channel (DPDCH)). In addition, every user is differentiated by means of scrambling codes [17]. Therefore, Uplink capacity is limited by scrambling codes rather than by channelization codes. With 2^{24} scrambling codes available, no limitation is expected.

3.1.3 Estimating Downlink Capacity

Unlike Uplink capacity, which can be easily – if not perfectly – modeled analytically, it is better to use simulations to estimate Downlink capacity. Two factors make it difficult to analyze the effect of each user on overall Downlink capacity. First, users experience unique radio conditions, including unique orthogonality, at their various geographical locations. This results in a unique power being consumed by each user, which leads to the second limitation: power is shared among users, making the capacity dependent on the location of, and the RF conditions seen by, the users.

An analytical definition of Downlink capacity estimates the required power per user, because High-Power Amplifier (HPA) is the most limiting Downlink resource. As seen in the Link Budget defined in Chapter 2, DPCH power is mainly affected by system loading (DPCH E_c/I_{or}), interference (expressed in terms of geometry), the required E_b/N_t, the channel conditions that influence the achieved E_b/N_t, and handover state.

Path loss in this Link Budget does not affect capacity directly because DPCH power on the Downlink is set to overcome other-cell interference primarily, and, to a lesser degree, other-user interference. Path loss and building penetration loss in particular affect the signal and the interference equally; thus they have a minimal effect on capacity.

The Link Budget presented in Chapter 2 showed how the variables discussed above affect the DPCH power and, therefore, the capacity. Table 3.5 presents the results.

The following conditions apply to the results shown in Table 3.5.

- **Loading.** Capacity is estimated when loading is set to 100%; loading is expressed as the fraction of HPA power used. As loading decreases, absolute power per channel decreases, but the per-channel E_c/I_{or} remains almost constant.

- **Geometry and handover state.** Handover gain and geometry are considered together. As demonstrated in Chapter 2, geometry depends on the handover state. When handover is disabled, or prevented through the use of parameters, the mean geometry in the system degrades.
- **Channel condition.** Channel conditions affect both the required E_b/N_t and the multipath condition. With AGWN, no multipaths are considered. This improves the achieved E_b/N_t, as discussed in Chapter 2.

Table 3.5 includes an estimated number of users. This number is based on total HPA power, power for all control channels, and the required estimated DPCH power. In this table, a maximum HPA power of 43 dBm and a Common Control Channel (CCCH) power of 36 dBm are assumed, leaving 15 W available for traffic.

The estimated number of users in table 3.5 generally does not include a voice activity factor. Voice activity affects Downlink capacity because transmit power is reduced during DTX. If DTX is used in a voice call, only the control portion of the channels is transmitted during idle periods. However, what is actually transmitted during DTX, as well as the respective power levels of the control and data parts of the transmission are considered. On the Downlink, this concept is important for accurately predicting power resource utilization, as shown in Figure 3.6.

During DTX, average transmit power is reduced according to the power offset and the slot format. For any slot format [18], Equation 3.9 can estimate the power difference between DTX and non-DTX frames, using the number of bits (N_{data}, N_{data2}, N_{TFCI}, N_{Pilot}) and the different power offsets (PO1, PO2, PO3), expressed in linear units.

$$\text{Reduction\_DCH\_Power}_{No\_DTX} = \frac{N_{\text{data1}} + N_{\text{data2}} + N_{\text{TPC}} \times PO2 + N_{\text{TFCI}} \times PO1 + N_{\text{Pilot}} \times PO3}{N_{\text{TPC}} \times PO2 + N_{\text{TFCI}} \times PO1 + N_{\text{Pilot}} \times PO3} \tag{3.9}$$

For the voice example – Case 3 in Table 3.5 [13], with handover and 100% loading – the required DPCH is 27.1 dBm for 31 supported users. With DTX and a voice activity factor of 60%, the average transmit power becomes 25.7 dBm, and the estimated number of users increases to 43. Unlike the Uplink, for which the activity factor can be derived directly from added users, on the Downlink, the relationship between the voice activity factor and the increased number of users is not linear.

Table 3.5 Effect of selected parameters on Downlink capacity

Simulation conditions	Mean DPCH [dBm]	Estimated users per cell
Voice, Case 3, HO, 100% loading	27.1	31
Voice (60% activity), Case 3, HO, 100% loading	25.7	43
Voice, Case 3, HO, 70% loading	25.5	28
Voice, Case 3, HO, 50% loading	24.1	23
Voice, Case 3, no HO, 100% loading	31.9	10
Voice, AWGN, HO, 100% loading	22.6	88

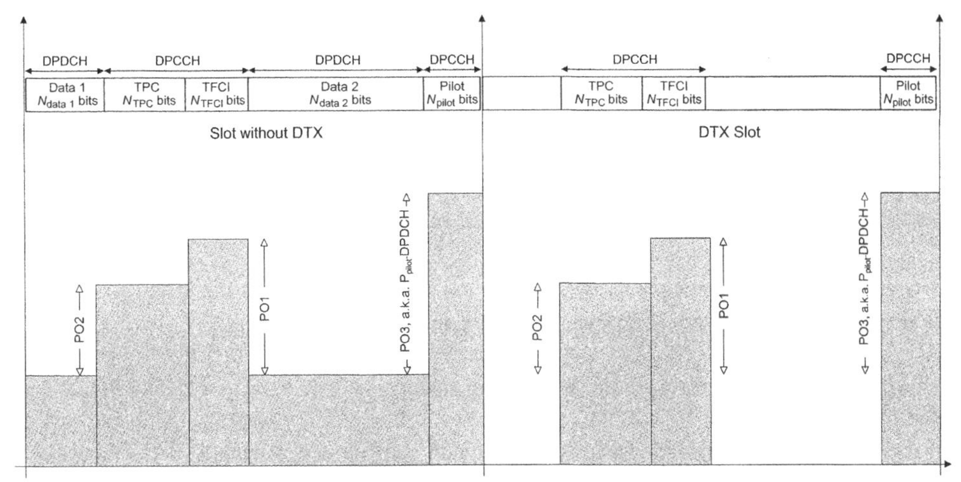

Figure 3.6 Average transmit power during DTX

3.2 Effect of Video-Telephony and PS Data on Traffic Engineering

The previous section discussed how voice services affect capacity engineering. For 3G systems, estimating capacity involves other services as well. Video-telephony and PS data services are important differentiators for 3G carriers. This section compares voice service to video-telephony and PS data services.

3.2.1 WCDMA Traffic Engineering and Video-Telephony

Video-telephony and voice are both CS services; therefore, minimal differences are expected between them. Of the factors that affect voice capacity, the following also affect video-telephony:

- **Target E_b/N_t.** The target E_b/N_t changes for both the Uplink and Downlink, as mentioned in Chapter 2. The BLER target selected for this service affects the required E_b/N_t.
- **Voice activity factor.** During a video-telephony call, the codec adjusts the quality of the transmission (mainly the picture quality) to the available bandwidth. Therefore, the activity factor for video-telephony is expected to be 100% at all times.
- **Bit rate.** Video-telephony services are currently carried over 64 kbps CS RAB, thus influencing the capacity of both the Uplink and the Downlink.

Table 3.6 compares capacity for voice and video-telephony services on the Uplink and Downlink. Uplink capacity is calculated directly from Equation 3.5. Downlink capacity is estimated on the basis of average geometry, as presented in Table 2.9, and following

Table 3.6 Capacity comparison between voice and video-telephony services

Condition	Voice capacity [users per cell]	Video-telephony capacity [users per cell]
Case 3, Uplink, Ff = 0.65	61	15
Case 3, Downlink, HO, 100% loading	43	16

the process discussed in Section 3.1.3. This comparison shows that the reduced E_b/N_t for video-telephony does not compensate for the higher bit rate.

- Overall, compared to voice, video-telephony capacity is reduced by a factor of 2.5 to 4 for the air interface.
- On the Iub and the interface above it, the capacity difference relates to data rate only. In this case, at equal bandwidth, video-telephony capacity is reduced by a factor of 5 to 9, when considering the voice activity factor.

When adding video-telephony service to commercial systems, the capacity for video-telephony as compared to voice services should be weighed carefully. For the air interface alone (usually the costliest of all interfaces), video-telephony capacity could cost two to three times more than voice capacity.

3.2.2 WCDMA Traffic Engineering and PS Data

PS data is bursty. This key characteristic greatly affects WCDMA traffic engineering for PS data services, where resources are not assigned for the entire duration of a data exchange, but are allocated and deallocated dynamically, according to the data. To understand how PS data traffic affects bearer occupancy and how the bearers should be dimensioned, the variables [19] involved in PS data traffic (also shown in Figure 3.7) are first considered.

- **Session occurrence.** If a session is defined as a dataflow between applications [20], several sessions could be open at the same time on a given terminal. A session could last a few seconds or a few days. The definition of a session also depends on what needs to be dimensioned. When the Core Network (CN) and the GPRS Gateway Support Node (GGSN) for General Packet Radio Services (GPRS) must be dimensioned, the session can be defined as the entire time the Packet Data Protocol (PDP) context is active. In terms of the actual occupation of radio resources, the session should be defined as the period during which the end user application is active (used) rather than the time the application is open. Figure 3.7 shows the difference between these session definitions. For either definition, the probability of occurrence of a new session can be considered independently of existing sessions and could be modeled with a Poisson distribution. For a given user, this Poisson distribution would use the average number of sessions per unit of time as the parameter (λ in Equation 3.4).
- **Data volume within the session.** For the entire session space, this is a random variable that could take any value from a few bits to several megabits. The data volume

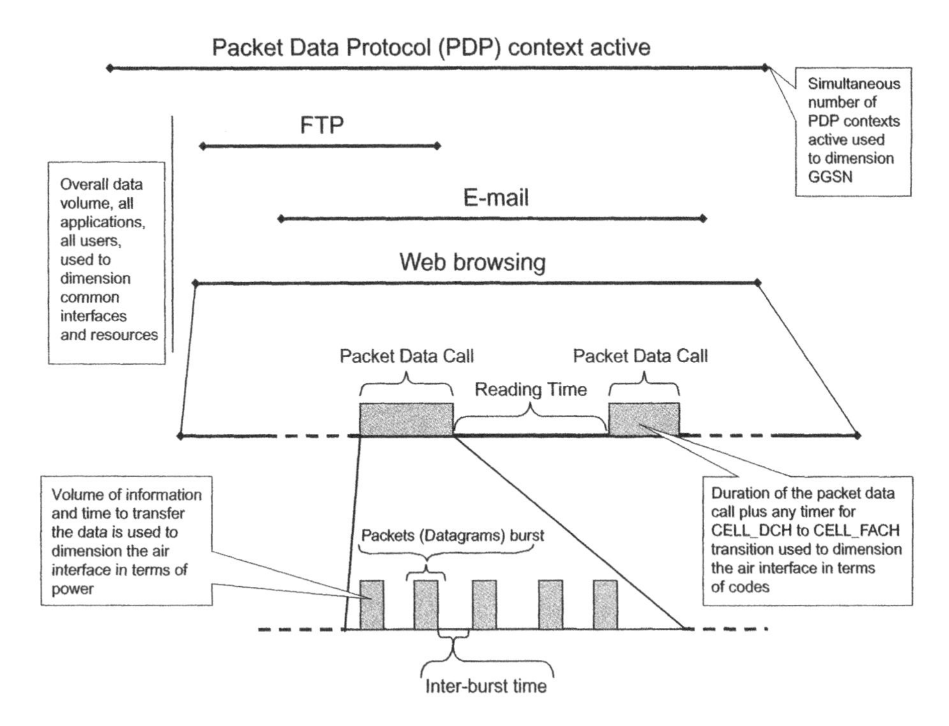

Figure 3.7 Packet data session, call, and associated terminology

per session can then be modeled as a Poisson distribution, with the average volume per session as the parameter; however, this does not fully account for the packet aspects of the service. To account for those, the session is further divided into its basic blocks.

- **Number of packet calls per session.** The number of packet calls within a session can be defined as the distinct period when data is exchanged among the distant terminals. Example: for reading Web-based e-mail, the *session* is the time the e-mail application is open, while each *packet data call* is the data exchanged to open a single message. For Web-based e-mail, reading time is an issue; depending on the application, this time could be longer than the packet data calls. From a dimensioning perspective, the resource associated with a given packet call is the code channel.
- **Idle time between packet call and reading time.** The reading time affects code channel usage. Ideal systems would use no resources during reading time, because no data is exchanged. In current systems, this is controlled mainly by the CELL_DCH to CELL_FACH timer. This timer artificially prolongs the packet data call to minimize the number of transitions and the associated signaling. If the reading time is less than this timer, the entire session is considered a single packet data call, in terms of resource usage.
- **Data volume in a packet call.** This variable can be expressed in terms of the number of packets (or datagrams) in a session and the packet size. Data volume and data rate of the bearer affect the capacity in terms of HPA power. This represent to the WCDMA PS data traffic for most air interface planning purposes.

The above definition of PS data traffic is suited for true packet data traffic but may be too complex for WCDMA, because of limitations on resource assignment. The resources for a true packet data service, such as 802.3 for local area network or High-Speed Downlink Packet Access (HSDPA) for wide area network, are shared among all users and are used only when a packet is transmitted. Such traffic has multiple levels of randomness that can be modeled [21,22].

In WCDMA Release 99, resources are not dynamically assigned. The burstiness of traffic is affected mainly by the CELL_DCH to CELL_FACH transition (or to a lower state), an inherently slow process. To address this limitation, PS data traffic can be dimensioned by using the traffic volume and efficiency factors. The efficiency factors represent wasted capacity (resources), caused by inefficient switching between modes. The following factors apply to busy hour dimensioning.

- **Data volume at busy hour.** This can be estimated from the monthly volume, as was done for CS domain services. This data volume is used to dimension the interfaces, Iu, Iub, and above.
- **Channel elements and code channel.** In the PS domain, the channel elements and code channel utilization are affected by the same variables as in the CS domain. In addition, they are affected by CELL_DCH to CELL_FACH (and CELL_FACH to CELL_DCH) channel type switching. Once a channel efficiency factor is defined, data volume should be increased in proportion to this factor.
- **Power.** When the user equipment (UE) is in CELL_DCH mode but not transmitting data, the DCH power may average to a lower value if DTX is used. Applying the defined channel efficiency factor can decrease the estimated transmit power. Unlike the channel elements and code channel, this effect is not direct because Control Channel information must be transmitted continuously, as mentioned in Section 3.1.3.

The channel type switching mechanism conserves resources during reading time by preserving the PDP context and releasing physical resources. In addition, channel rate switching affects resource usage. This mechanism can be set to either limit the power used by a single user, or adjust the bearer rate according to the needs of the application. The former is more widely implemented. It increases the spreading factor (SF), thus the coding gain, when RF conditions degrade. The latter method uses the Orthogonal Variable Spreading Factor (OVSF) code tree more efficiently by assigning a Radio Bearer rate that corresponds to application requirements.

To circumvent this multivariable process, and because PS data services are commonly implemented as best effort services, a model that accounts for delay has been defined. Latency can be estimated if the data service is an M/M/1/∞ queue (commonly referred to as M/M/1). This model has the following features:

- The channel is considered as a single resource (the 1 in the M/M/1) that serves requests.
- The arrival and departure of the requests can be modeled as Markov processes (the "M"s in M/M/1).

This queuing theory estimates the latency, L, (or delay) to serve a request from the arrival rate, λ, and the serving rate, μ, according to Equation 3.10.

$$L = \frac{\lambda}{\mu - \lambda} \tag{3.10}$$

At the cell level, the arrival rate can be estimated by averaging the number of requests from all the users of a cell. This can be derived from the monthly usage and any peak-to-average or traffic profile information, divided by the number of cells in the network. The serving rate depends on the channel characteristics and the request size. Equation 3.11 estimates the serving rate.

$$\mu = \frac{Channel\_throughput}{Request\_size} \tag{3.11}$$

In Equation 3.11, *channel_throughput* represents the throughput supported by the air interface, regardless of scheduler characteristics or arrival rate of the request. This corresponds to the aggregate throughput of all users performing File Transfer Protocol (FTP), or similar downloads. The *request_size* varies for each application, but an average request size can be determined, on the basis of traffic profile. The bearer efficiency also varies by application, but mainly affects the code channel, channel elements, and upper resources, and not the air interface resources.

Bearer efficiency is defined as the ratio of the arrival rate over the serving rate; however, this definition is not very useful. A better method calculates it using Equation 3.12, to ensure that the delay, L, is acceptable to the end user. This delay measures only the time between a request from the user and the initiation of the download; it does not include the time to complete the download. To estimate the total delay, the download (or data transfer) time is added, which is controlled mainly by the assigned Radio Bearer and the radio conditions.

$$L = \frac{\rho}{\mu \times (1 - \rho)} \tag{3.12}$$

Equation 3.12 can be reformulated as Equation 3.13 to estimate efficiency on the basis of the user requirements (L) and the supported data rate of the channel (μ).

$$\rho = \frac{L \times \mu}{(1 + L \times \mu)} \tag{3.13}$$

The following example on numerical latency calculation clarifies these concepts. From these calculations, we can estimate what serving rate the channel can support (μ) and how much the channel should be loaded (ρ), thus indirectly finding the arrival rate (λ) that a cell can support. Knowing the total number of requests to serve for all users of all cells, and the average arrival rate per cell, we can easily calculate the number of cells.

Table 3.7 Downlink capacity estimation for different PS data rates

Data rate	Downlink [users per cell]	Cell throughput [Mbps]
Case 3, 64 kbps, 100% activity	19	1.2
AWGN, 64 kbps, 100% activity	27	1.7
Case 3, 128 kbps, 100% activity	9	1.1
AWGN, 128 kbps, 100% activity	14	1.8
Case 3, 384 kbps, 100% activity	3	1.1
AWGN, 384 kbps, 100% activity	5	1.9

Numerical Example of Latency Calculation, Single Session

This example assumes an average request size of 480 kbytes over a single WCDMA channel with a channel throughput of 900 kbps.

From the request size and the channel, the serving rate (μ) can be estimated using the following equation:

$$\mu = \frac{Channel\_throughput}{Request\_size}$$

$$\mu = \frac{900[kbps]}{480[KB/request]} = 0.34[request/sec]$$

The utilization efficiency can be calculated to ensure that system delays are less than 5 seconds:

$$\rho = \frac{L\mu}{(1 + L\mu)} = \frac{5 \times 0.34}{1 + 5 \times 0.34} = 0.63$$

From the utilization efficiency and the channel throughput, cell throughput can be calculated as follows:

$$Design\_cell\_throughput = \rho \times Channel\_throughput = 0.63 \times 900 = 567[kbps]$$

In this example, a cell can transfer 567 × 8 × 3600/1024 = 15.9 [MB] of data at busy hour with a maximum delay of 5 seconds to serve a request.

To complete the dimensioning process, channel throughput must be estimated on the basis of the number of simultaneous users, as was done for CS services.

Table 3.7 summarizes the calculated capacities for different Radio Bearer rates.

The cell throughput in column three is estimated from Equation 3.14.

$$Cell\_throughput = Number\_of\_bearers \times bearer\_rate \tag{3.14}$$

In all cases, the code tree is not a limitation. Capacity is limited by DPCH power, which in turns depends on interference. Table 3.7 shows that, for all channel conditions the cell throughput improves when using a higher data rate bearer.

3.3 Multiservice Traffic Engineering

3.3.1 Multiservice Capacity

So far, we have estimated Uplink and Downlink capacities for a single service only. This is a necessary step, but it does not reflect an actual system in which all services are offered simultaneously. Estimating Multi service capacity requires an understanding of how resources are shared when multiple services need different levels of resources.

To understand resource sharing better, the code channel can be considered. Because the code channel does not depend on any other parameters (RF conditions, channel conditions, etc.), it provides a simple illustration of the relationship among services, based the OVSF length. Table 3.8 lists the minimum OVSF length and the number of OVSFs available

Table 3.8 Code channel availability for different services

Service	Minimum OVSF length	Number of OVSF available	Carried Erlangs @ 2% GoS	Trunking efficiency
Voice, AMR	128	125	112	90%
CS 64	32	31	22	71%
PS 64	32	31	22	71%
PS 128	16	15	9	60%
PS 384	8	7	3	43%

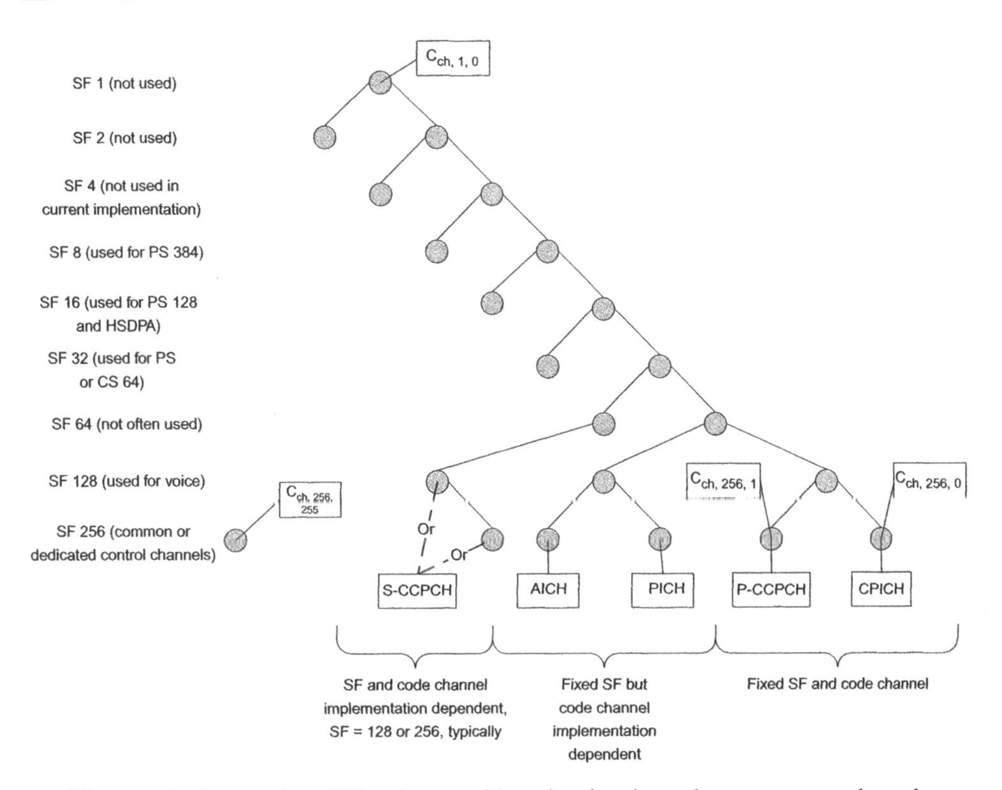

Figure 3.8 Downlink OVSF code tree with optional and mandatory common channels

for each service. For each service type, the carried Erlangs are estimated, assuming a 2% GoS.

This example covers only the Downlink. On the Uplink, each UE has its own code tree, so the code tree is not a limiting factor in that direction. On the Downlink, the number of OVSFs available for each dedicated channel is reduced, because multiple common channels must be supported. Figure 3.8 summarizes the mandatory Downlink channels and the mandatory (or implementation-dependent) values of their OVSFs. It also shows optional Downlink common channels.

In the OVSF code tree structure, one PS 384 connection uses the same resources as four PS 64 connections or 16 voice connections. However, in terms of the SF, the probability of having SF = 8 free channels is not just 4 (or 16) times less than the probability of having one SF = 32 (or SF = 128) free, because the equivalent SF = 32 (or SF = 128) free channels must be contiguous and start at a specific position.

Therefore, the availability of an OVSF of a specific length is determined by the number of OVSFs of same length or shorter that are used, as well as by the number of longer OVSFs used. The OVSF allocation algorithm at the Node B normally manages the availability of consecutive OVSFs. This algorithm also allocates and optimizes the code tree to maximize the availability of shorter OVSFs.

For multiservice dimensioning, it is important that sharing resources among services does not impair the individual service performance, or lead to over-dimensioning. Figure 3.9 helps us understand resource sharing better. For single-service dimensioning (which is simply a visual description of the Erlang theory), it must be kept in mind that traffic may vary during the observation period.

A voice example with a 2% GoS and 125 resources can support 112 Erlangs. If traffic takes an average of 112 resources, the system must be over-dimensioned to 125 to ensure that no more than 2% of the calls are blocked, on the basis of traffic distribution.

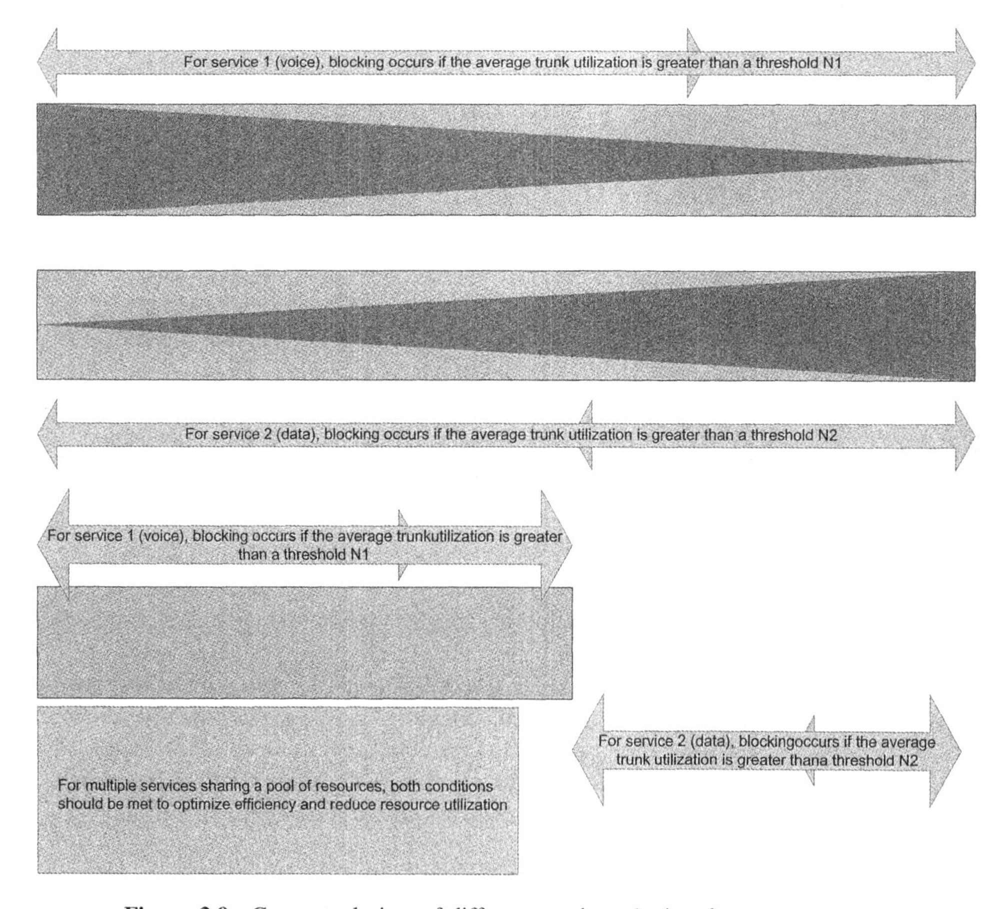

Figure 3.9 Conceptual view of different services sharing the same resources

Using the same example for PS 384, to carry three calls with a 2% blocking probability, the system must be over-dimensioned to seven resources. To provide both services, each must be over-designed to ensure that the blocking probability for both is limited even at peak traffic. It is not practical to assume that one service will be at low traffic to compensate when another service is at peak traffic.

A simple way to address these issues during dimensioning is to divide resources linearly among services. In the case of OVSF, the division is done on the basis of OVSF length: one PS 384 call (OVSF 8) is equivalent to 16 voice calls (OVSF 128), or 4 video-telephony calls (OVSF 32).

Figure 3.10 illustrates this for three services. Using an example of 3 OVSF = 16 to support the PS 384 data traffic, if 10 OVSF = 32 are required for CS 64, then only 41 resources are available to support voice (OVSF = 128). Because of the nature of the code tree, this graph is a step function rather than a line.

The information in Figure 3.10 would be more useful if it were translated into Erlangs, as shown in Figure 3.11. Using the same example, with 3 OVSF = 16 for PS 384, only 0.602 Erlangs can be supported; with 10 OVSF = 32, 5.08 Erlangs are supported along with 31.9 Erlangs for voice. Obviously, the trunking efficiency in this case is greatly reduced, compared to all the resources used for any of the individual services. From this number of available resources, the corresponding carried Erlangs at the desired blocking probability can be estimated, as Figure 3.11 shows. To do the dimensioning, the operating point on Figure 3.11, which presents the traffic ratio among the different services, relative to the traffic ratio at the network level, is determined.

Once the traffic mix is stated in terms of Erlangs, we can locate the operating point on the curve. The *operating point* is the closest point on the curve presenting the same ratio among the services. Because of the curve's discrete nature, for the more demanding resources (in

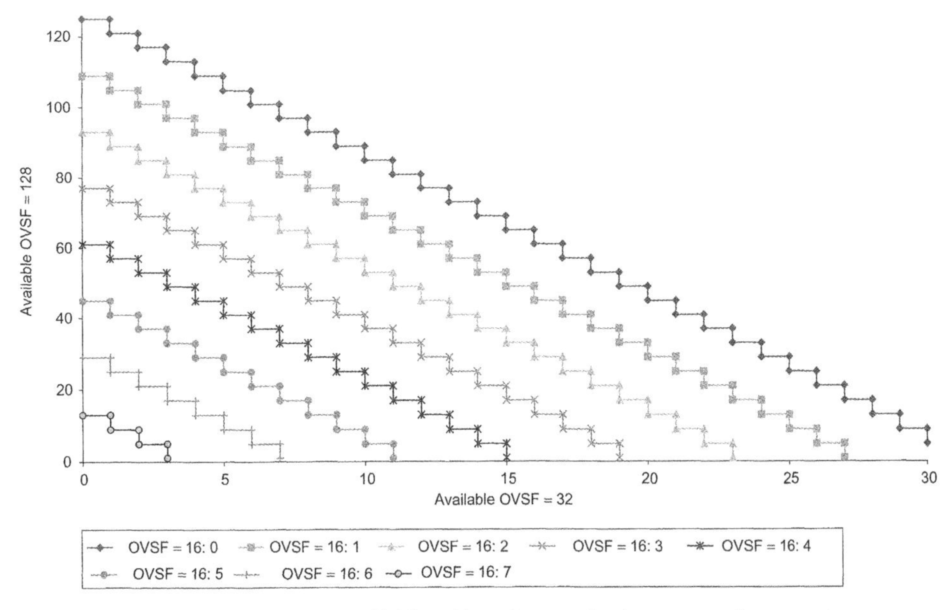

Figure 3.10 Resources available with code tree sharing among three services

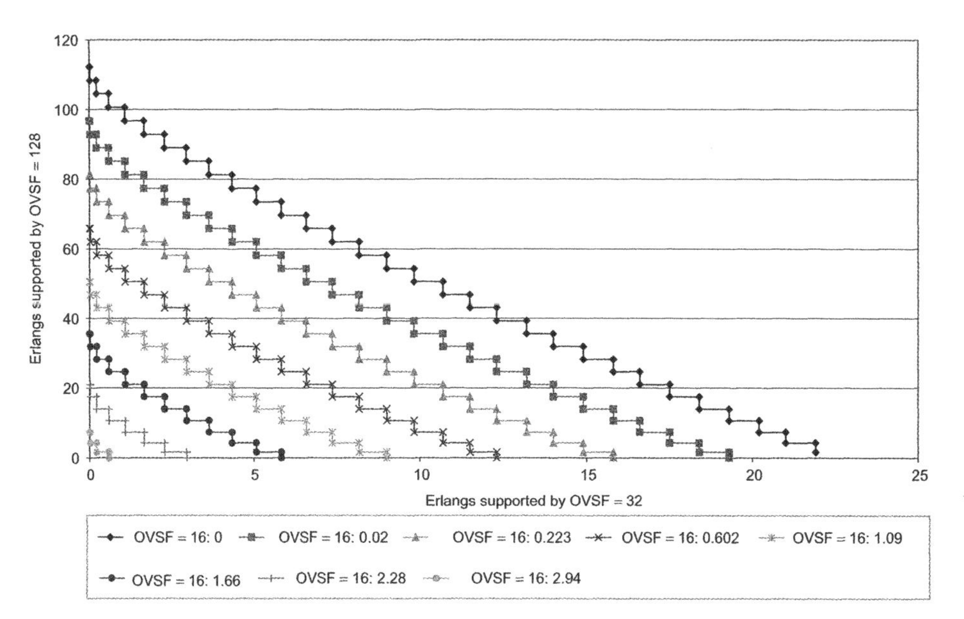

Figure 3.11 Carried Erlangs with code tree sharing among three services, assuming 2% blocking probability

this example, the shorter OVSF) only the closest approximation–not an exact ratio–can be found. Using this graph would assume that the code channel utilization is the limiting factor, rather than power. In most deployment scenario, power, in particular for high data rate service is the limiting factor. To portray this, we can construct the same graph with resources based on the average power used per service and the available power for traffic. Figure 3.12 illustrates this, using the Case 3 channel assumptions defined in Table 3.4.

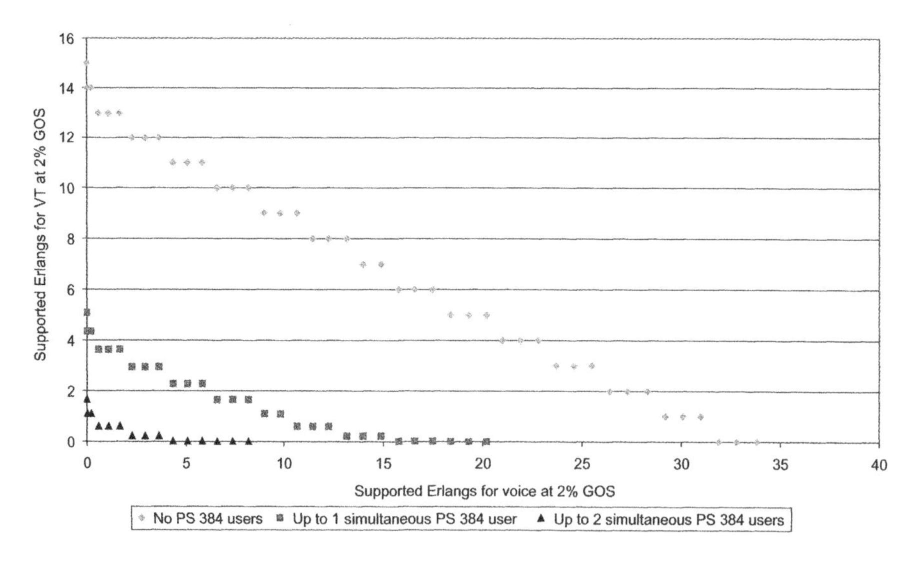

Figure 3.12 Carried Erlangs per cell, with DPCH power sharing among three services

Table 3.9 Multiservice dimensioning example, part 1

Service	Required traffic over entire system
Voice	1000 Erlangs
Video-telephony	300 Erlangs
PS Data	1 GB

Table 3.10 Multiservice dimensioning example, part 2

Service	Required number of cells
Voice	152
Video-telephony	181
Total	181 cells required, for the most limiting service

Let us look at an example of multiservice dimensioning. For a network with the traffic requirements shown in Table 3.9, how many cells are required?

By first assuming that one PS 384 resource will be sufficient, we can estimate the per-cell capacity from Figure 3.12 to be 6.61 Erlangs for voice and 1.66 Erlangs for video-telephony. With this per-cell capacity, Table 3.10 shows the number of required cells.

We can verify the number of cells sufficient for PS traffic. For best effort, one PS 384 resource can carry 168 MB of data at busy hour. For all cells, it can carry 30 GB, which meets the capacity requirement. Dimensioning for data should also consider latency, as described in Section 3.2.2.

3.3.2 Uplink and Downlink Capacity Comparison

So far, we have estimated the Uplink and Downlink capacities for each service, and the effect of multiple services on overall cell utilization. Now we can compare the Uplink and Downlink for a call model rather than for services. In this process, we will also see how the call model (or traffic) affects coverage. Coverage and capacity are interdependent in a WCDMA system.

Table 3.11 summarizes the capacity figures estimated in the previous sections.

For most cases, the Downlink is limiting the capacity of the network. One of the limitations of such estimates, for PS data in particular, is that they are based on 100%

Table 3.11 Uplink and Downlink capacity comparison, per service

Condition	Case 3, Uplink, Ff = 0.65 [users per cell]	Case 3, Downlink [users per cell]
Voice capacity, 60% activity	61	43
Video-telephony capacity, 100% activity	15	16
PS 64, 100% activity	17	19
PS 128, 100% activity	10	9
PS 384, 100% activity	3	3

Table 3.12 Average throughput and activity factor for selected applications

Application	Average throughput Uplink/Downlink [kbps]	Activity factor Uplink/Downlink assuming 64/384 bearer, no channel switching [%]	Uplink/Downlink traffic ratio
Web-based e-mail (example 1)	5/23	8/6	22%
Web browsing	6/29	9/8	21%
Video streaming	6/113	9/30	5%

activity for both Uplink and Downlink. This assumes that channel and rate switching perfectly track user needs, as Chapter 5 explains. Without channel or type switching, activity can be as low as 6%, per the examples given in Table 3.12.

In Table 3.12, the activity factor for selected applications is experimentally determined in the absence of channel rate and type switching. The efficiency factor is estimated using Equation 3.15.

$$Activity\_factor = \frac{Average\_throughput}{Bearer\_throughput} \tag{3.15}$$

A more conservative efficiency factor, when channel switching is implemented, is about 50% in the Downlink. For the Uplink, the stated activity factor is affected by the channel type switching only, because PS data rates lower than 64 kbps are not usually implemented, even though the standard supports them.

To redefine the Uplink and Downlink capacity estimates in terms of simultaneous users, we can assume that an average user has the properties summarized in Table 3.13. On the Downlink, DPCH power is reduced during DTX. The power resource can support 20 users, but in reality it is limited to 15 due to code channel availability (OVSF length 16 required for PS 128). The Uplink is limited by the activity factor and can support over 200 users. Obviously, the Downlink is limiting.

For peer-to-peer service, the Uplink and Downlink potentially have equal traffic, eliminating the imbalance among traffic, capacity, and utilization.

Now let us look at loading. In the previous sections, capacity was estimated for 100% loading, or pole capacity for the Uplink. On the Downlink, pole capacity is not used often because a theoretical limit on capacity cannot be drawn in the Downlink as it can for the Uplink. When estimating coverage for a call model, Uplink loading is a key factor. We

Table 3.13 Typical user profile for PS data services

Property	Value for average user
Average Uplink throughput	6 kbps
Average Downlink throughput	55 kbps
Uplink activity factor	8% (assuming 64 kbps bearer and no channel switching)
Downlink activity factor	40% (assuming 128 kbps bearer and no channel switching)
Uplink to Downlink traffic ratio	15%

can estimate this from Tables 3.11 and 3.13. The results vary by call model, because each service has a different traffic balance. Comparing the Uplink and Downlink capacities for the services, yields the following results:

- **Voice service.** An Uplink loading of 70% is realistic.
- **For PS data.** Loading depends on the assigned Radio Bearer and the application.
- **Video streaming.** If carried over a 64/128 Radio Bearer, the Uplink is limited to 5%.

3.4 Capacity Planning

As described in the previous sections, many factors complicate network capacity planning: the supported services (CS or PS), applications, assigned Radio Bearers, and the radio environment. Furthermore, the network planner must rely on assumptions and estimations.

Network planning tools can help with capacity planning, assuming the tools support 3G. This section explains how to use network planning tools and describes the limitations associated with them. We will first review the inputs required for capacity planning, then discuss how these inputs can be used in a network planning tool.

3.4.1 Input for Capacity Planning

A network planning tool uses traffic modeling to simulate the effects of user and traffic distribution over the area of interest. It then verifies that, for each service, the desired GoS can be achieved. Each service defines its own requirements, but ultimately the collective effect of all the services should be determined.

Because most network planning tools evolved from voice-only tools, they expect the planner to provide the probability of traffic for each basic area, or *Bin*. For voice, this is easily estimated by setting a user distribution and a traffic assumption for each subscriber. However, this basic method of estimating traffic requirements is not well suited for PS data because resources are dynamically assigned and vary for different applications.

The following sections concentrate on the traffic per subscriber for the CS and PS domains, and describe a simple user distribution.

3.4.2 Capacity Planning for the CS Domain

For voice services, or the CS domain in general, traffic can be simulated fairly accurately from user distribution and traffic per user. The distribution of traffic for WCDMA is much more critical than in FDMA/TDMA systems, because WCDMA capacity is limited not by the physical resources but by interference. In other words, it is limited by the geometry.

How traffic per user is defined affects capacity planning when all resources are considered: channel elements, code channels, and power. To estimate all the resources for the CS domain, in addition to the Erlang load per user, the activity factor and handover state should be known. The handover state affects the channel element and code channel resources, but it is not covered here in detail because the network planning tool determines this from the user location and local conditions.

Table 3.14 summarizes how soft Handover Reduction Factor (HORF) and Softer Handover Reduction Factor (Sf_HORF) affect the various physical resources. Traffic puts

Table 3.14 Relationship between average physical resource usage and soft/softer handover reduction factor

Resources	Affected by:	Comments
MSC resource	1 per call	Reference traffic: T
Iu resource	Typically 1 per call	Traffic = T × 1
RNC resource	Vendor implementation, typically 1 per call	Traffic = T × 1
Iub resource	Sf_HORF	Traffic = T× Sf_HORF
Channel element	Sf_HORF	Traffic = T× Sf_HORF
Code channels	HORF	Traffic = T× HORF

additional constraints on dimensioning. For example, RNC dimensioning cannot be limited to carried traffic only; it should also include busy hour call attempts (BHCA), signaling load, and the number of connected Node Bs. These dimensioning considerations vary by vendor implementation and are not addressed here.

Both HORF and Sf_HORFs can be defined as the average number of resources per call, assuming that multiple resources are used during handover. The difference between softer and soft reduction factor supports the estimation of resources that will or will not be shared during a call. Even though the vendor implementation determines the resource sharing at the Node B, we can simplify the approach by stating that channel element and Iub resources are shared only during softer handover, while MSC and Iu resources are always shared. All other resources are never shared; each segment of a call uses a different resource.

Equation 3.16 shows several simple equations to calculate the soft HORF or Sf_HORF as a function of the various handover state probabilities, which can be easily determined from the network planning tool.

$$\begin{aligned}
&Softer_handover_reduction_factor\\
&Sf_HORF = p1 + p2 + p3 + (2 * p4) + (2 * p5) + (2 * p6) + (3 * p7)\\
&Handover_reduction_factor\\
&HORF = p1 + (2 * p2) + (3 * p3) + (2 * p4) + (2 * p5) + (4 * p6) + (3 * p7)\\
&With\\
&p1 : probability(no_handoff)\\
&p2 : probability(2 - way_softer_handoff)\\
&p3 : probability(3 - way_softer_handoff)\\
&p4 : probability(2 - way_soft_handoff)\\
&p5 : probability(3 - way_soft/softer), 3_cells_from_2_node - B\\
&p6 : probability(4 - way_soft/softer), 2_cells_each_from_2_node - B
\end{aligned} \tag{3.16}$$

From these equations, we can deduce that the HORF is always greater than Sf_HORF, or that a call consumes fewer channel element resources than radio links.

The relationship between code channel and power resource utilization cannot be directly established, unless we include the activity factor in the estimation of traffic per user. In this case, if users are distributed using a Monte Carlo simulation, they are defined as follows:

- **Idle.** Not consuming any resources.
- **Connected.** A call is established, thus consuming all resources, but at a reduced power.
- **Connected and active.** A call is established and transmitting at full power, for the given RF conditions.

The ratio of power between a connected user and a connected/active user depends entirely on the slot format and the power offset *(PO1, PO2, PO3)* parameters defined in Section 3.1.3.

CS domain traffic can be defined entirely on the basis of an Erlang value per subscriber and an activity factor, as summarized in Table 3.15. Defining traffic in terms of calls per hour and call duration is identical to defining traffic in mErls. It is valid because network planning tools only perform static simulations and do not account for time.

3.4.3 Capacity Planning for the PS Domain

The PS domain cannot use the same inputs as the CS domain because channel switching affects system capacity and affects each resource differently. The most easily observed effect is on the resource itself; various data rates require different DPCH power, channel element, and code channel resources.

A less obvious effect, and one that is more difficult to estimate or simulate with a network planning tool, is how the achieved data rate at a given location influences user behavior. Would a user employ video streaming as frequently when only a PS 64 RB is available, compared to PS 384? This question cannot be answered here, but it introduces the PS domain concept that a user can either send volumes of information or use an application for a given duration, as summarized in Table 3.16.

It is important to understand the different application types when setting up the user traffic profile with channel rate switching, as shown in Table 3.17.

Table 3.15 CS domain traffic required for network planning

Parameter	Typical value
Calls per hour	Assumption or historical data
Mean holding time	Assumption or historical data
mErlangs	Assumption or historical data mErlangs can be estimated $mErlang = \frac{call\_per\_hour \times mean\_holding\_time[s]}{3600}$
Activity factor	Voice: 60% Video-telephony: 100%

Table 3.16 PS domain application examples

Application	Planning impact
Streaming (audio or video)	Duration-based application; volume of information depends on the achieved quality
E-mail	Volume-based application; duration of session depends on how fast the data is transferred
Web Browsing	Volume- or duration-based application; depends on type of browsing: searching for specific information is volume-based; leisurely browsing is duration-based

Table 3.17 Example of traffic profile with channel rate switching

Application	RB [kbps]	Call per hr	Mean holding time [sec]	Activity factor [%]	Comments
Video streaming	DL 384	1	300	70	Assumes stream encoded at 300 kbps
	UL 64	1	300	16	Assumes TCP ACK generated traffic, MSS 576, each segment ACK'ed
	DL 128	1	300	70	Assumes stream encoded at 100 kbps
	UL 64	1	300	6	Assumes TCP ACK generated traffic, MSS 576, each segment ACK'ed
	DL 64	1	300	70	Assumes stream encoded at 50 kbps
	UL 64	1	300	3	Assumes TCP ACK generated traffic, MSS 576, each segment ACK'ed
E-mail	DL 384	1	26	80	Assumes reception of 1 MB of data
	UL 64	1	26	20	Assumes reception of 1 MB of data
	PS 128	1	80	80	Assumes reception of 1 MB of data
	UL 64	1	80	8	Assumes reception of 1 MB of data
	PS 64	1	160	80	Assumes reception of 1 MB of data
	UL 64	1	160	4	Assumes reception of 1 MB of data
Web browsing	DL 384	1	100	7	Assumes volume-based browsing
	UL 64	1	100	10	Assumes volume-based browsing
	DL 128	1	110	20	Assumes volume-based browsing
	UL 64	1	110	8	Assumes volume-based browsing
	DL 64	1	130	35	Assumes volume-based browsing
	UL 64	1	130	7	Assumes volume-based browsing

The traffic value in Table 3.17 is not absolute; it is a relative value affected by rate switching as follows:

- **Duration-based applications.** For applications that are entirely duration (time) based, the overall volume of data is reduced when the Downlink Radio Bearer rate decreases. This decrease in Downlink traffic also decreases the Uplink traffic, but because the Radio Bearer rate is assumed to be constant, the activity on the Uplink is reduced.

- **Volume-based applications.** For volume-based applications (such as e-mail), activity is expected to be almost constant on the Downlink, mainly due to the slow start and handshake between the client and the server. The high activity factor shown in the example assumes very efficient channel type switching, which uses delayed switching from CELL_FACH to CELL_DCH, and rapid switching from CELL_DCH to CELL_FACH.

In Table 3.17, the activity factors are estimated from the same experiment as in Section 3.3.2, but in this case channel switching is simulated. For Web browsing, also a volume-based application, activity decreases when a higher Radio Bearer rate is used, but improves the user experience. With a low Radio Bearer rate, page loading is delayed, as observed from the increased mean holding time per session.

Using this definition of volume-based or duration-based traffic for PS data, the call model can be developed in two steps. First, the application based on the example in Table 3.17 is defined. Then, the Radio Bearer characteristics are defined. The Radio Bearer characteristics contain the data relative to the target E_b/N_t, bit rate, and power offset.

These inputs for PS data are sufficient and easier to use than packet-type call models. In a packet-type call model, an application is defined in terms of the number of packets, size of the packets, and packet arrival rate. This model is derived from a true PS data medium, such as LAN. It is useful for estimating traffic across interfaces but it cannot accurately model radio resource utilization. Nevertheless, packet-type call models are appropriate for HSDPA planning because resource allocation is dynamic; changes can occur every 2 ms.

Channel rate switching is a key feature of 3G network planning tools. Without it, Monte Carlo simulations become drastically unrealistic, thus invalidating any conclusion drafted by the planner. However, it is not essential that 3G network planning tools include channel type switching. For Release 99 planning, channel type switching is limited to CELL_DCH, to and from CELL_FACH. In the CELL_FACH state, all radio resources are released; therefore it does not affect radio network planning. For HSDPA planning, the activation of high-speed service is best simulated as type switching, but it could be approximated with rate switching.

For completeness, traffic should be distributed geographically. This distribution is more critical for WCDMA than for Global System for Mobiles (GSM), or more generally for FDMA/TDMA systems, because the capacity of a WCDMA system is affected by the locations of the users. As discussed in Section 3.4.4.1, geometry is the main limiting factor for the Downlink. The Uplink is limited mainly by the interference factor, α (see Section 3.1.2), which also could be linked to geometry. Highly accurate traffic distribution can address this limitation, by ensuring that the geometry is optimized for the locations of the users. Several methods can be applied to create the traffic density, with varying degrees of accuracy. These methods are summarized below:

- **From existing network measurements.** Traffic measured on existing networks is accurate in terms of traffic, but must be scaled up or down according to the projected user uptake. Accuracy is limited by the unknown factors that must be applied for different domains: traffic from an existing network is either voice only, or voice and low-rate PS data. The voice traffic per subscriber can be considered constant when migrating to a newer network, but the PS data traffic should be scaled to consider the new services enabled by the higher data rate is considered. Even the traffic distribution may change with the newer services, especially audio streaming and mobile TV. Accuracy is also limited by the number of sites in the existing network. Traffic measured at a specific cell is distributed over the entire cell area; therefore, the resolution of the demand map is limited by the number of cells. In addition, accuracy is limited by the accuracy of the RF model used to determine the coverage. The error is less evident for co-located cells, because both the existing and the new system are subject to the same variations. This method is best suited for microcells or indoor systems, which cover a limited area. Their precision allows the creation of accurate distribution maps, even if microcell or indoor systems were not considered in an early phase of WCDMA deployment.

- **From census data.** This method gets the population density from census data or similar sources. Census data is readily available, and thus can be used even by greenfield operators. On the other hand, it does have some limitations. Census data collects information for large populations over a potentially large area, typically by postal code, or a similar administrative region. In addition, census data provides information only about the population, not subscribers or traffic. Depending on the data source, this type of information might include the residential population only, thus omitting traffic generated by temporary occupants of businesses or commercial areas. The population should be converted into subscribers based on marketing assumptions such as mobile subscriber penetration and market share, or even time of day. From this, one can project traffic per subscriber based on assumptions of wireless habits by socio-economic class. Additional population information, such as income, increases the accuracy of this method.
- **From clutter distribution.** Both of the above methods of estimating traffic density distribution can be supplemented by weighting the data on the basis of clutter, roads, or any geographic information (usually known as *morphologies*). This weighting relies on local knowledge. Different markets or operators have different weights. For example, an operator that targets residential users will assign different weights than an operator targeting business users.

3.4.4 Capacity Planning with a Network Planning Tool

At the network planning stage, capacity planning involves estimating the different resources (channel element, code channel, power) used, and adjusting the network configuration to increase the number of resources. Because network planning tools focus primarily on the radio aspects, not all resources get equal attention. This section examines each resource discussed so far, and explains how to estimate or improve it during the planning phase.

3.4.4.1 DPCH E_c/I_{or}

The DPCH E_c/I_{or} resource captures the portion of HPA power required to sustain a given user in Connected Mode. The resource is defined as a fraction of the total transmit power to provide scalability. In contrast, an absolute value such as W or dBm would not be scalable for different values of total transmit power. Another advantage of using

DPCH E_c/I_{or} is that it illustrates the fact that increasing the HPA power does not increase capacity. Instead, it increases interference, requiring higher DPCH power to meet the E_b/N_t requirement. The following Link Budget parameters affect DPCH E_c/I_{or}:

- **Cell loading.** This value is also known as total transmit power, commonly represented by I_{or}. Total cell power includes both the Common and the Dedicated Channels. This is important because any reduction in Common Channel power increases the power available for DPCH, thus increasing the capacity. One goal of the network planning process is to achieve a balance between Common and Dedicated Channel power. In a fully loaded system, the Common Channel power is adjusted until the call fails because of a failure to satisfy DPCH E_b/N_t or CCCH E_b/N_t, with an equal probability of either reason being the cause. Although this concept is accurate, its use is limited with most network planning tools because only the Common Pilot Channel (CPICH) is modeled, not all the CCCHs.
- **Minimum and maximum DPCH power setting.** These limits on the DPCH power control affect overall cell capacity in opposite ways. For the minimum DPCH power, users in close proximity to a cell will use at least that amount of power, even if the radio condition would allow a lower value. Ultimately, the minimum DPCH power setting sets the maximum limit on cell capacity. The maximum DPCH power, on the other hand, limits how much HPA power a single user, in bad RF conditions, would consume. By limiting the maximum DPCH power, the lower limit of cell capacity is effectively set, because the cell has a limited dynamic range, usually to 25 dB.
- **Cell geometry.** Cell geometry is an important element of network planning. It is affected by the RF configuration, the handover parameter setting, and the traffic distribution, all of which contribute to the loading of each cell. To optimize the Downlink capacity of a cell, the geometry must be optimized, but this data is not directly available to a network planning tool. Instead, the combined CPICH E_c/N_o or the frequency reuse factor variable can be used. The frequency reuse factor was introduced earlier (see Equation 3.6), but the concept of combined CPICH E_c/N_o is new. The combined CPICH E_c/N_o can be estimated by summing the E_c/N_o of all servers in the Active Set, as shown in Equation 3.17. As this equation indicates, the combined CPICH E_c/N_o improves as the Active Set size increases, which is equally true for geometry. Geometry is also affected by the RF configuration and the overlap between cells. Minimizing overlap between cells during planning limits other-cell interference and improves geometry. Improving geometry by optimizing the RF configuration instead of handover parameters provides benefits even when the Active Set size is limited to 1, as in the Idle state or for the HS channels in HSDPA.

$$Combined\_CPICH\_E_c/N_o = 10 \times \log\left(\sum_{i=1}^{ASsize} 10^{\frac{(E_c/N_o)_i}{10}}\right) \tag{3.17}$$

- **Soft handover combining gain.** Soft handover combining gain is the reduction in DPCH power when the Active Set size is unlimited, compared to the DPCH power required for an Active Set size limited to one. Estimating the combining gain from this definition requires running a simulation twice, once with the Active Set size set as expected, and again with it limited to one. Although the combining gain affects the required DPCH, optimizing the combining gain is not a good way to optimize system capacity, because the combining gain monotonically increases with increased Active Set size, while the total DPCH power does not, as shown in Figure 2.23.

- **Target E_b/N_t.** Target E_b/N_t depends on the target BLER and the channel condition, or multipath profile and mobility. The target BLER can be changed for the PS domain because users will detect an increase of the BLER target only as a reduction of throughput, not of quality, assuming that Acknowledged Mode (AM) is used. For CS domain service, Transparent Mode (TM) is normally used; therefore, any increase in the BLER target degrades voice or image quality. For a given BLER target, the target E_b/N_t for any service should be set according to the expected channel conditions in the planned morphology. Therefore, this variable cannot optimize capacity. The only possible optimization is accurately estimating the channel condition (i.e., the required E_b/N_t) as an input to network planning.

Although DPCH power is a good measure of individual cell capacity, it should be balanced with system capacity. Accordingly, total DPCH power serves as a better metric than individual cell DPCH power, because the total considers all the cells in the Active Set. As mentioned above, increasing the Active Set size improves geometry and combining gain, all other parameters being constant. For large Active Set sizes, the combining gain is not constant for each increase of the Active Set size; instead, it decreases. To compensate for this during network planning, given maximum loading, all the influential parameters are adjusted until a balance between performance and capacity is achieved. To help with that balancing act, Table 3.18 summarizes the main parameters and performance metrics that affect capacity.

DPCH power is the most expensive resource to implement. Once the RF configuration and parameters have been optimized, the only way to increase this resource is by adding a new cell, either on the same or a new frequency.

3.4.4.2 Code Channels

Code channel utilization in WCDMA relates directly to the Radio Bearer rate use, and alternately to the activity factor on the assigned Radio Bearer. As it is true for PS domain applications, a very low activity factor exhausts the OVSF while the DPCH power requirement remains low. Because network planning tools usually rely on static simulations, it is impossible to tune the channel switching parameters during planning.

The current release of the standard allows the use of secondary code trees, even though they have not been commercially implemented. In theory, a secondary code tree would mitigate any code tree utilization issues, by doubling the number of OVSF available. However, a secondary code tree would be implemented using Secondary Scrambling Codes (SSCs), which do not have the same orthogonality characteristic as Primary Scrambing Codes (PSCs). This nonorthogonality effectively increases intra-cell interference and ultimately reduces the capacity of the cell.

3.4.4.3 Channel Elements

The concept of channel element varies by vendor implementation. In the first-deployed CDMA system, a channel element was effectively an integrated circuit (IC) capable of

Table 3.18 Main parameters and performance metrics with their impact on capacity

	E_b/N_t target not met	Minimum CPICH E_c/N_o not met	DL power saturation	Code Channel (OVSF) saturation
CCCH power assignment		Insufficient CPICH power to overcome the loading and RF conditions	CCCH power can be reduced to allow more DPCH power	Overall cell coverage should be reduced by reducing both the HPA and CCCH power assignments
Maximum DPCH power setting	Value set too low; insufficient power can be allocated to an individual user		Value possibly set too high; verify if any individual user reached the maximum value	
Minimum DPCH power setting			Value possibly set too high; verify if DPCH is set to the minimum value for multiple users	
Active Set size and HO parameter	Active Set size too small or HO parameter too restrictive; user cannot benefit from HO combining gain		Active Set size too small or HO parameter too restrictive; user cannot benefit from HO combining gain	Active Set size too large and HO parameters too permissive
Incorrect channel switching parameters				Incorrect channel switching leads to low utilization of the RB, with a risk of running out of OVSF before other resources

encoding or demodulating a single code channel. Currently, *channel element* can refer either to a physical IC or to a processing capability. In terms of processing capability, it is easy to understand that a lower SF Radio Bearer uses more channel element resources than a higher SF.

The number of channel elements is not set during planning. Instead, the required number of channel elements is estimated on the basis of the traffic carried by each cell, or each Node B if the channel elements are pooled. In this estimation, it is important to dimension the channel element pool large enough to ensure low blocking probability.

3.4.4.4 Iub Resource

The Iub resource, effectively the bandwidth of the Iub interface, can be determined after RF planning is complete, based on the traffic carried per Node B. Similar to channel element dimensioning, the Iub resource is planned to ensure low blocking probability, usually much lower than the expected GoS.

3.4.4.5 RNC Resources

The ability to plan RNC resources is limited because most network planning tools do not define RNC as a resource. This is not critical, because the RNC resource dimensioning can be done after RF dimensioning is complete. It must be kept in mind that RNC resources are limited by three constraints:

- **Number of ports available.** This limits either the number of Node Bs or the number of equivalent backhauls (usually E1 or T1) that can be physically connected to a RNC.
- **Total carried traffic.** After the number of Node Bs connected to a given RNC is known, the total carried traffic can be estimated by summing the traffic carried by each Node B, discounting the handover traffic.
- **Busy hour call attempts.** BHCA can be estimated from traffic and mean holding time for CS domain calls. For the PS domain, this estimation does not apply because channel type switching also can be considered a call attempt. A more complete call model for the PS domain should consider reading time, an important factor from a user's point of view, which affects the need for type switching.

3.4.5 Microcell Issues

Microcells can enhance overall network capacity, if they are planned properly. Any resulting capacity gain is primarily due to propagation conditions. Capacity improves for both Uplink and Downlink because microcells provide a predominant line of sight, which means a single path is expected. This lowers the E_b/N_t requirements to nearly Additive White Gaussian Noise (AWGN) conditions. From this alone, on the Uplink, the capacity gain is significant even if received diversity is not implemented. On the Downlink, the capacity gain is mainly caused by improved geometry, if proper isolation is maintained between the micro and the macrocells. Improved geometry reduces the required DPCH, thus improving the Downlink capacity. For the microcell, the full capacity gain may not be reached, because the DPCH power settings are likely RNC parameters (in the current implementation) rather than being cell or Node B configurable. With a single power setting and a limited dynamic range for DPCH, typically 25 dB, a large portion of the connection to the microcells would be at the DPCH power floor.

In addition to the DPCH power limitation, how the microcell is deployed affects its capacity and the relative capacity compared to the macro layer. For a microcell, capacity gain on the Downlink relies on isolation between the micro and macro layers. Gain is achieved either by spatial or frequency isolation. *Spatial isolation* refers to increasing the path loss between the macro and the micro layers; *frequency isolation* refers to using different WCDMA carriers for the micro and the macrocells. Table 3.19 summarizes the effects of both.

In Table 3.19, the DPCH E_c/I_{or} represents the capacity of embedded cells, for a regular macro network as shown in Figure 3.13, and assuming even traffic distribution. In this

Table 3.19 Effect of spatial and frequency isolation on cell capacity

	Macrocell	Microcell
	Average DPCH E_c/I_{or} [dB]	
Macrocell only	−18.4	NA
Microcell far from Macro	−18.4	−15.1
Microcell close to Macro	−18.4	−14.1
Microcell middle of Macrocell coverage area	−18.4	−14.5
Microcell on separate carrier	−18.9	−20.1
Microcell indoor (3 cells, close, middle, and far) on same carrier	−18.3	−21.5
Microcell indoor (3 cells, close, middle, and far) on separate carrier	−18.4	−21.7

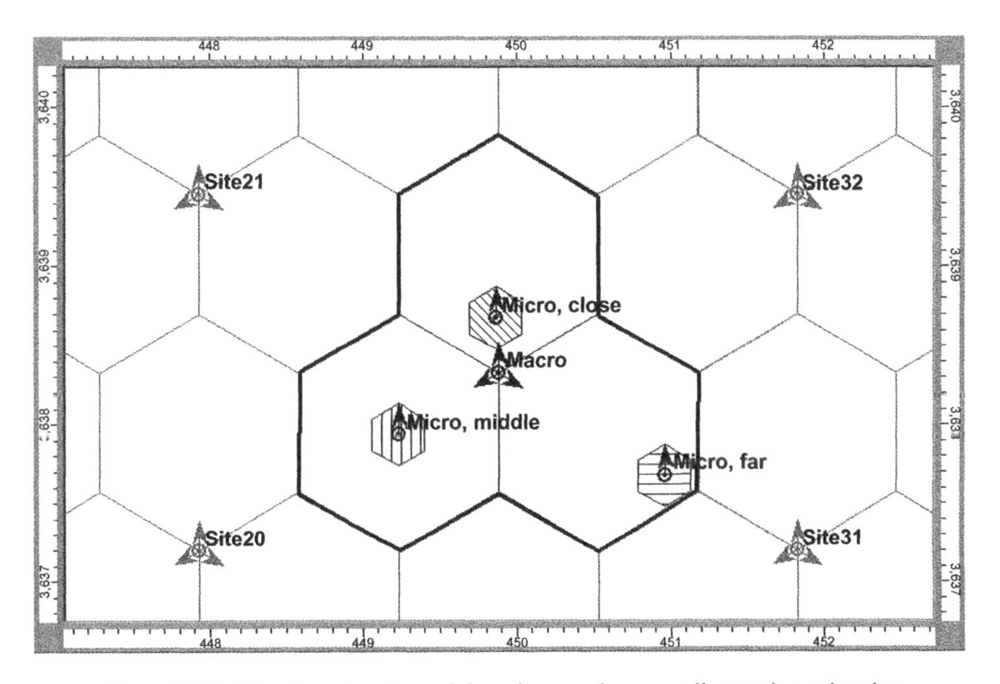

Figure 3.13 Regular network used for micro- and macrocell capacity estimation

simulation, the advantage of representing the capacity by the DPCH E_c/I_{or} is that traffic distribution does not have to be assessed.

In all cases, the macrocell capacity is nearly constant, as long as nearby microcells do not affect the Downlink macrocell's capacity, in terms of power resources. The emphasis should be on power, rather than other affected resources; for example, where a microcell is deployed near a macrocell, microcell users are predominantly in handover, thus affecting the code channel OVSF resources.

The capacity of the microcell cannot be generalized, but depends on the deployment options. When microcells are deployed outdoors at the same frequency, capacity increases only modestly when the microcell is placed at increasing distances from the macrocell.

The 1 dB difference for DPCH E_c/I_{or} between the closest and farthest microcell location translates into a difference of about five users, since 20% of the power is reserved for control channels. In other words, increasing microcell capacity by increasing only the spatial isolation has limited results. This must be considered in light of an actual deployment, especially when microcells are located in coverage holes where the isolation is good.

When microcells are positioned to increase the capacity, frequency isolation provides the best capacity increase. Compared to macrocells, the capacity gain can be almost one and a half times greater. The main advantage here is that the capacity increase does not depend on the achieved spatial isolation. In a multicarrier case, soft handover is not available between carriers and thus has no impact on code channel utilization.

When the microcells are deployed indoors, using a separate carrier does not provide a significant gain over using the same carrier as the macro network. This can be explained easily: the building penetration loss increases the isolation and reduces other-cell interference.

As seen in Table 3.19 and in the following discussion, a microcell's capacity greatly depends on its deployment; thus, accurate placement is very important when planning a microcell. Accurate placement ensures maximal isolation from the macro network and places the microcell in high traffic areas. The relative imprecision of traffic demand maps could produce the following undesirable consequences:

- **Increased handover area between micro and macro cells.** Capacity is affected because the geometry is reduced and line-of-sight conditions are less likely to be present.
- **Geometry degradation.** If a connection to a single server is achieved (e.g., by modifying parameters), geometry is likely to degrade further, affecting both links.

For best results, the microcells and macrocells are planned simultaneously. Without proper coordination between layers, any capacity gain provided by the microcell is hindered by potential handover between the layers, assuming a single carrier is used. To address this, isolation is maintained between the microcells and macrocells.

From the capacity simulation performed, deploying microcells always increases the network capacity, with the greatest gain provided by microcells deployed on a specific carrier, either indoor or outdoor. In a deployment, this capacity advantage should be balanced with overall spectrum utilization strategy: only when a significant portion of the traffic is carried by the microcells can a specific carrier be dedicated to it. In addition, the optimization of a multicarrier network is more complex when carriers are dedicated to specific layers than when all carriers are used for all layers. Inter-frequency reselection and handover should be optimized to ensure that the transitions happen where and when expected.

3.5 Optimizing for Capacity

During initial optimization, it is difficult to estimate network capacity because it involves the radio and resource aspects of a network. However, testing an unloaded network can indicate how much capacity a network can support. This can be achieved by knowing the trade-offs between coverage (Quality of Service) and capacity. Once this is known, an unloaded test can estimate how various changes implemented in a network would affect capacity. After testing, long-term observation of resource utilization will verify these assumptions or reveal any needed adjustments.

3.5.1 Coverage and Capacity Trade-offs

Up to now, we have dealt independently with coverage and capacity. For the dimensioning stage this is reasonable, because the actual site and the subscriber distribution are not known. This is also acceptable for looking at even distributions in the Downlink capacity simulation. During network planning, however, subscriber distribution should be known, or at least approximated. Now the trade-off between coverage and capacity becomes relevant.

We have seen that Downlink coverage is limited initially by the CPICH and other CCCH's power, and secondarily by the maximum DPCH power allocation. Capacity is affected mainly by the average DPCH power per connection and, to a lesser extent, by the available DPCH power. In addition, if there is a high probability of having subscribers at the edge of coverage, the maximum DPCH setting also affects capacity, because a single user could potentially consume all the available power. On the Downlink, the main challenge is setting the maximum DPCH power high enough to ensure coverage (Quality of Service), yet low enough to minimize its effect on capacity.

For voice, which is still the basic service, the maximum DPCH should be set to maintain the minimum E_b/N_t requirement (up to 13 dB for Case 2) at the cell edge, where the geometry could reach −6 dB. If these extreme values were used in the Link Budget defined in Chapter 2, the required maximum DPCH power would be 30.5 dBm, allowing for no handover combining gain. In other words, cell capacity would be exhausted if these extreme conditions were applied to 14 users simultaneously, an unlikely scenario. For voice, planning to provide service over a wide range of channel and geometry conditions does not significantly hinder capacity.

For services that require additional power (such as higher data rate services), Quality of Service and capacity must be more carefully balanced. For example, consider video-telephony service, typically carried by a 64-kbps Radio Bearer in the CS domain. In the same extreme conditions defined previously, the required DPCH power would be over 38 dBm, with a resulting capacity of three users. Limiting the available DPCH power would minimize the risk of running out of capacity, but would greatly reduce coverage. For instance, if the system were optimized such that the worst geometry is 0 dB, then a DPCH power of 32.5 dBm would be sufficient. This would ensure that capacity is never below 10 simultaneous users. This illustrates how vital it is that network planners weigh all of these factors: RF configuration, the maximum DPCH power, and the number of served users.

These concepts apply to the PS domain as well, although the results would be worse because of the reduced spreading gain associated with the low SF used in this domain. However, the PS domain can apply rate switching to limit the required DPCH power. In this case, the balance is not between coverage and capacity, but rather between Quality of Service, expressed in terms of user throughput, and capacity.

3.5.2 Capacity Estimation in a Deployed Network

Tests of an unloaded network can help estimate capacity in several ways, depending on the available datasets. The estimations should be based on geometry, channel conditions from mobile logs, or DPCH E_c/I_{or} from network logs. The accuracy of the estimations produced by each method varies.

The following sections review the accuracy of each of these estimation methods. For all of them, overall accuracy is determined by how and where the supporting data is collected. To achieve the most accurate estimations, the test data over an area that closely matches actual user distribution and usage is collected.

3.5.2.1 Geometry

In a deployed system, geometry can be estimated from RF measurements. Equation 3.18 defines geometry.

$$Geometry = \frac{\hat{I}_{or}}{I_{oc}}. \tag{3.18}$$

To define geometry, the following values must be known:

- Energy received from a cell not in soft handover, other-cell interference (I_{oc})
- Received signal from the active cell or cells ($\hat{I}_{or}$)

These values are not known directly; instead the total received signal power (I_o) and Pilot Received Signal Code Power (RSCP) for the active cell are measured. From the RSCP, $\hat{I}_{or}$ can be calculated, if cell loading is known. In other words, the fraction of HPA power used for the Pilot ($CPICH\_E_c/I_{or}$) must be known, as shown in Equation 3.19.

$$\hat{I}_{or} = \sum_{i=1}^{ASsize} \frac{RSCP_i}{(CPICH\_E_c/I_{or})_i} \tag{3.19}$$

This value can be calculated from the total cell power, which is available from network logging, and the CPICH power setting. Once ($\hat{I}_{or}$) and Received Signal Strength Indicator (RSSI) are known, I_{oc} can be calculated, assuming the thermal noise, N_{th}, using Equation 3.20.

$$RSSI \cong I_o = N_{th} + \hat{I}_{or} + I_{oc} \tag{3.20}$$

Using all the equations in this section, the geometry can be estimated using Equation 3.21.

$$Geometry = \frac{\sum_{i=1}^{ASsize} \frac{RSCP_i}{(CPICH\_E_c/I_{or})_i}}{I_o - N_{th} - \sum_{i=1}^{ASsize} \frac{RSCP_i}{(CPICH\_E_c/I_{or})_i}} \tag{3.21}$$

The accuracy of this calculation relies on the accuracy of the CPICH E_c/I_{or}, which in turn depends on how the total transmit power is measured in the network. Total transmit power may be available in real time, or as a counter collected over a long time-interval. In the latter case, the estimation of geometry is fairly inaccurate, but still usable. In the real-time collection of total cell power, the accuracy of the geometry is affected by the synchronization of both logs, which is not a technical issue and can be addressed procedurally.

If measurements are performed in unloaded conditions, the CPICH E_c/I_{or} can be considered constant over the measurement period and uniquely dependent on the CCCH power settings.

Although difficult to estimate accurately, geometry is still useful when setting the handover parameter because it accounts for the effect of Active Set size on interference and resource utilization.

3.5.2.2 DPCH E_c/I_{or}

If network logging is available, capacity can be estimated from each user's known power usage, instead of from the geometry and a full set of assumptions. When using network logs to determine power usage, power must be expressed as DPCH E_c/I_{or}–not as an absolute value–for scalability reasons.

The only exception is when estimating the total consumed power, as opposed to the per-link power. The total consumed power can be calculated simply as the sum of all per-link power over the entire set. This value is not used directly to estimate the capacity, but to estimate the efficiency of the handover. As the Active Set size increases, because of the combined gain, the per-link power diminishes. On the other hand, the total transmit power initially decreases until the efficiency of the handover reaches the peak [23]; then the total transmit power increases. By estimating this value while optimizing handover parameters (as in the geometry case), the optimal set of parameters can be found by comparing the total transmit power for different sets and in relation to the Active Set size.

3.5.3 Capacity Monitoring for a Deployed Network

When commercial traffic is present in the network, the best way to monitor capacity is to collect performance data from network counters. The implementation of a network counter is vendor-specific. Therefore, this section discusses only the types of information that should be collected and how accurately this information represents actual subscriber traffic and capacity requirements.

3.5.3.1 Total HPA Power

During commercial operation of a network, monitoring the total cell transmit power does not accurately indicate cell capacity; it reflects only the cell's remaining capacity. The total HPA power variable should be used both as a long-term average, over an hour or a day, and as a maximum value. By observing the long-term average, the network engineer can forecast when blocking will occur in the cell, and take action to prevent it.

For total HPA power, as for most resource utilization indicators, we look at the individual cell data and focus on the most used cells. It is also important to look at a cell's value in relation to the other cells. An isolated cell, showing high utilization, is treated differently than a cluster of cells. For an isolated cell, traffic balancing may be a solution; for a cluster of cells, a more radical solution (such as cell splitting or addition of a carrier) may be needed.

3.5.3.2 Code Channel Utilization

In WCDMA systems, code channels are not expected to be a limiting resource unless channel switching for PS domain services is not properly optimized. Therefore, monitoring code channel utilization, even if incomplete, should provide a good indication of the channel switching efficiency. Incomplete indicators reflect resource efficiency, but not the user's experience.

Unless handover states are logged in the network through a different means, code channel utilization can be used to estimate softer handover, by comparing the utilization of the code channel (1 per link) and channel elements (1 per Node B). This method still does not accurately estimate soft handover because data from a common resource (usually at the RNC level) should also be included.

Code channel utilization is a good indicator of the traffic carried by a sector, especially if the utilization report covers different OVSF lengths. The OVSF utilization together with the total transmit power can be used to estimate the average DPCH power and, therefore, can be used to estimate and forecast cell capacity.

3.5.3.3 Channel Element Utilization

In early network implementations, channel elements are often deployed conservatively because of their associated high cost. For this reason, channel element utilization must be monitored along with blocked calls to determine if and when the resource pool should be increased.

3.5.3.4 Number of Calls at the Node B and RNC Level

The number of calls at the Node B or the RNC level can complement the channel element utilization value. The difference between the two sets of numbers indicates the probability of handover. These numbers vary by vendor because they can account for only the number of call originations, or they count each handover request also as a call. Only the second method is useful in estimating the handover overhead factor.

3.5.3.5 Average DPCH Power

Average DPCH power can be estimated from code channel utilization and total transmit power, if it is not available directly. To forecast capacity, this value is considered relative to the total transmit power. If taken as an absolute, the value would rise with increased traffic demand because both the same-cell and other-cell interference would rise.

During the commercial operation of the network, comparing the average DPCH power for all the cells is a good way to identify areas that need RF optimization. In those areas where RF configuration is not optimal, interference increases, and each call requires additional power to fulfill the E_b/N_t requirement.

3.5.3.6 Handover State

When monitoring the handover state, soft handover (more than softer) does not need to be monitored to forecast the capacity offered by the radio resource. Other counters such as DPCH power, channel elements, and code channel more directly address the radio resources.

Monitoring the handover state is mainly used to ensure that the Iub interfaces are appropriately dimensioned, using an accurate handover overhead factor. This proper dimensioning is particularly important for PS domain services because the Iub interface requirement is at a maximum.

References

[1] Kaaranen H, Ahtiainen A, Laitinen L, Naghian S, Niemi V. *UMTS Networks*. New York: Wiley; 2001.
[2] Downing D, Clark J. *Statistics the Easy Way*. Barron's; 1997.
[3] Andriantiatsahoniaina LA, Trajokovic L. Analysis of user behavior from billing records of a CDPD wireless network. *IEEE Conference on Local Computer Networks*; Tampa, Florida 2002.
[4] Kim KI. *Handbook of CDMA System Design, Engineering and Optimization*. Prentice Hall; 1999.
[5] Gilhousen KS, Jacobs IM, Padovani R, Viterbi AJ, Weaver LA, Wheatley CE. On the capacity of a cellular CDMA system. *IEEE Transaction on Vehicular Technology* 1991; 40: 303–312.
[6] Viterbi AM, Viterby AJ. Erlang capacity of a power controlled CDMA system. *IEEE Journal on Select Areas in Communications* 1993; 11(6): 892–999.
[7] Vanghi V, Damnajanovic A, Vojcic B. *The CDMA2000 System for Mobile Communication*. Prentice Hall; 2004.
[8] 26.071. Mandatory Speech Codec speech processing functions, AMR Speech Codec; General Description. 3GPP; 1999.
[9] 26.103. Speech Codec list for GSM and UMTS. 3GPP; 2002.
[10] 26.110. Codec for Circuit Switched Multimedia Telephony Service; General Description. 3GPP; 2001.
[11] 26.111. Codec for Circuit Switched Multimedia Telephony Service; modifications to H.324. 3GPP; 2000.
[12] 34.108. Common test environments for User Equipment (UE), conformance testing. 3GPP; 2003.
[13] 25.101. User Equipment (UE) radio transmission and reception (FDD). 3GPP; 2004.
[14] Gruber JG. A comparison of measured and calculated speech temporal parameters relevant to speech activity detection. *IEEE Transactions on Communications* 1982; COM-30(4): 728–738.
[15] Falsafi A, Bruemmer K, Deschennes J-H. *Characterizing Backhaul Traffic in 3G Networks Using Real-World Speech*: IEEE Communication Society; 2004.
[16] Holma H, Toskala A. *WCDMA for UMTS: Radio Access for Third Generation Mobile Communication*: Wiley; 2002.
[17] 25.213. Spreading and modulation (FDD). 3GPP; 2002.
[18] 25.211. Physical channels and mapping of transport channels onto physical channels (FDD). 3GPP; 2002.
[19] *Selection procedures for the choice of radio transmission technologies of the UMTS*. ETSI; 1998.
[20] Newton H. *Newton's Telecom Dictionary:* CMP Books; 2001.
[21] Ben Slimane S, Le-Ngoc T. *A Double Stochastic Poisson Model for Self-similar Traffic*. IEEE; 1995.
[22] Anagnostou ME, Sanchez-P J.A, Venieris IS. A multiservice user descriptive traffic source model. *IEEE Transactions on Communications* 1996; 44(10): 1243–1246.
[23] Vanghi V., Chevallier C. WCDMA Handover Parameters Optimization *IEEE International Conference on communication (ICC)*, 2004, Paris, France, Vol. 7, 20–24 June; 4133–4137, Digital Object Identifier 10.1109/ICC.2004.1313326..

4

Initial Parameter Settings

Christopher Brunner, Andrea Garavaglia and Christophe Chevallier

4.1 Introduction

Network parameters play an important role in determining the correct behavior of the system and in achieving target performance. This chapter focuses on the initial settings for system parameters, to guarantee reasonably good operation of the network at the beginning of the optimization process or upon the friendly launch of services. For further insight into optimizing and fine-tuning the parameters, see Chapter 5.

As in many telecommunication systems, the parameters are intended to give the operator enough flexibility to (re)configure the network according to a specific strategy and development stage, without having to modify the corresponding network element software. In UMTS systems, the system parameters are typically stored in the RNC and Core Network (CN) databases and can be managed via the Operation and Maintenance Center (OMC). The parameters are then distributed with appropriate signaling to the network elements, as well as to the UE (user equipment).

Most of the signaling that involves system parameters is done through Layer 3 signaling. From a UTRAN point of view, the Radio Resource Control (RRC) protocol [1] plays a major role. The RRC has a control interface for each of the layers and sub-layers belonging to the Access Stratum (AS) with the capability to start, stop, and configure all of them. Other protocols used to set AS parameters are the Medium Access Control (MAC) protocol [2], the Radio Link Control (RLC) protocol [3], the Packet Data Convergence Protocol (PDCP) [4], and the Broadcast/Multi-cast Control (BMC) protocol [5]. Non-Access Stratum (NAS) parameters are usually defined in the CN [6] and distributed to the different network elements of the UTRAN by the RNC over the Iu [7], Iur [8], and Iub [9] interfaces using the corresponding protocols.

4.1.1 Broadcast of System Information

The RRC is the overall controller of the AS of UMTS, responsible for configuring all the involved layers and providing the signaling interface to the NAS layer. The RRC is also

WCDMA (UMTS) Deployment Handbook: Planning and Optimization Aspects QUALCOMM Incorporated

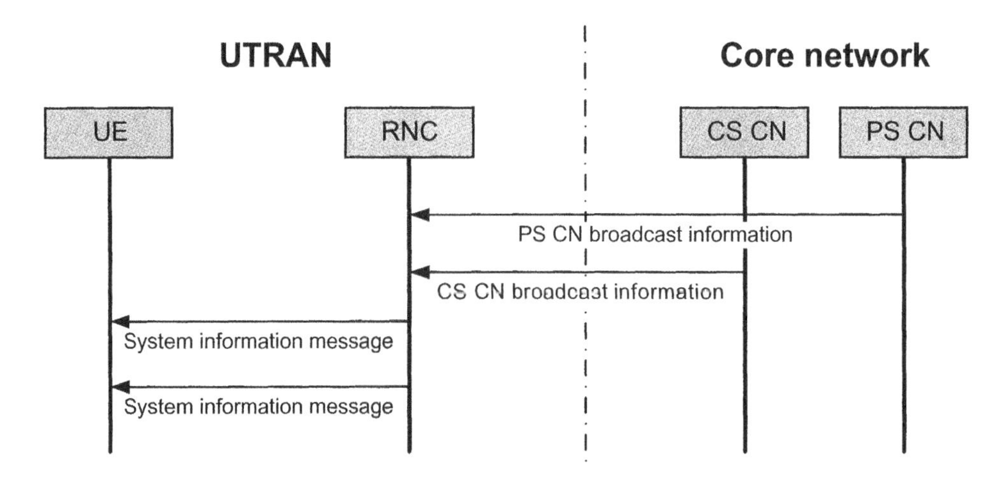

Figure 4.1 System information processing in UMTS

responsible for a set of procedures that establish, route, retain, and release the connection between the UTRAN and the UE.

One of the key RRC procedures for setting parameters is the *Broadcast of System Information*. This broadcast transfers the corresponding system parameters to all the UE present in a cell. As shown in Figure 4.1, system information includes Information Elements (IEs) from both the AS (originated from the UTRAN) and the NAS (originated from the CN). Both kinds of information are collected by the RNC and distributed to the UE as System Information messages, using the RRC protocol. The messages are sent on a logical Broadcast Control Channel (BCCH), which is mapped to either the BCH or the FACH Transport Channel.

The system information elements are broadcast in System Information Blocks (SIBs). A SIB groups together system information elements of the same nature. Different SIBs may have different characteristics in terms of scope (PLMN or cell), repetition rate, and requirements on the UE to read the SIB. Fast-changing (dynamic) parameters and more static parameters are grouped into different SIBs to prevent frequent reading of the same information.

To accommodate SIBs of very different sizes, a System Information message can carry either several complete SIBs or only a part of a SIB, while always fitting into the BCH or FACH transport blocks. A total of 11 combinations are allowed, according to the standard [10], with SIBs that can either be aggregated together in the same message or segmented into several messages, depending on their size.

Figure 4.2 shows the organization of the SIBs. A single Master Information Block (MIB) contains the scheduling information for the SIBs. One or two additional Scheduling Blocks may optionally be included in the MIB to provide schedule information for the rest of the SIBs. Some IEs are duplicated in different SIBs, for example Cell Reselection parameters in SIB3 and SIB4 (also grouped in Figure 4.2). SIB3 is used when the UE operates in Idle Mode, while SIB4, if transmitted, is used in Connected Mode. In this example, the transmission of SIB4 is signaled by a flag in SIB3, indicating that the UE needs to read additional information from (in this case) SIB4. To avoid any conflict or

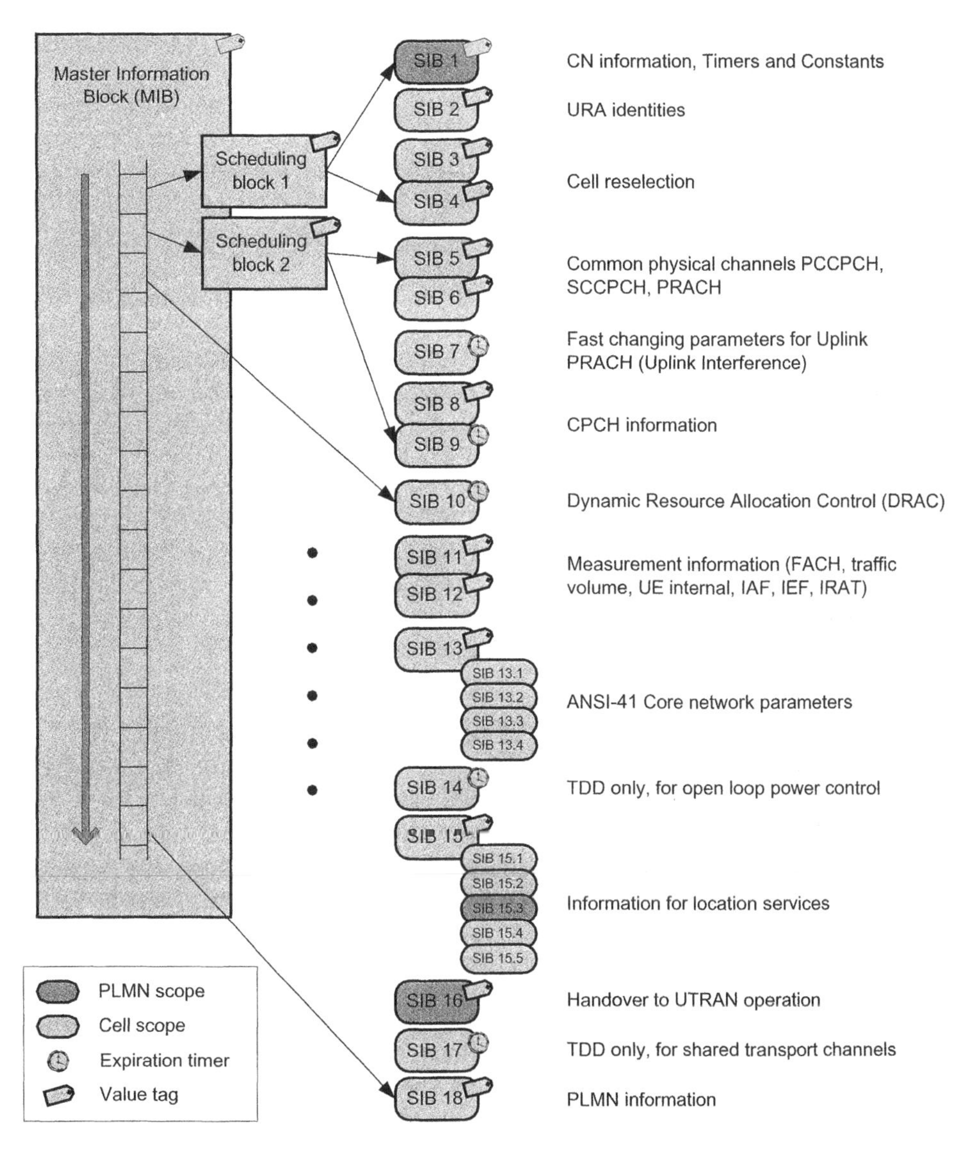

Figure 4.2 Structure of system information blocks according to 3GPP standard [10]

undefined behavior, the standard also defines the default values when specific SIBs are missing. In the previous example, if SIB4 is not transmitted, the UE defaults to SIB3.

It can be noted in Figure 4.2 that the legend (bottom left) provides the scope and the change control method for the different SIBs.

When a UE first camps on a cell, it must read all of the broadcast SIBs for that cell according to the schedule in the MIB, thus retrieving the parameter values. Because of this process, the SIB scheduling affects the reselection performance, since the reselection can only be effective after all the SIBs have been read. To mitigate this effect, the UE may optionally store System Information messages for a given cell so that they can be used later when returning to that cell, without having to read them all from the BCH–assuming none of the parameters has changed since the UE last read the SIBs.

Each SIB has a defined method of change control to allow the UE to quickly determine if the content has changed. SIBs with more static parameters use a *value tag* for change control, distributed by the MIB. When the UE reads the MIB, it compares the value tags for the scheduled SIBs with the corresponding value tags previously stored; all the SIBs with new value tags are reread. The MIB also contains a value tag that is sent regularly by the UTRAN on a static schedule.

More dynamic SIBs use *SIB-specific timers* for change control, so rereading the SIB can be triggered by timer expiration. The UTRAN can also inform the UE of system information changes with a Paging Type 1 message (for the Idle, CELL_PCH, and URA_PCH states) or with a System Information Change Indication message (when the UE is in the CELL_FACH state). Finally, the UE must reread the SIBs whenever their scope has changed, because of moving across cell or PLMN boundaries.

SIBs are not the only way to modify system parameters. In Connected Mode or during the RRC connection process, the UTRAN may change the parameter settings by means of RRC messages. An example is the Neighbor List and, more generally, handover parameters. In Idle Mode this information is carried by SIB11 and, potentially, SIB12. Once the connection is established, the Measurement Control Messages can provide new information to the UE, for a more dynamic control of the parameters.

4.1.2 Translation between Information Element Values and Engineering Values

The RRC protocol uses a set of messages that are described in Ref [10]. The messages are encoded by means of Abstract Syntax Notation One (ASN.1) and include a list of IEs with the corresponding values. For each IE, the standard specifies the characteristics in terms of validity range and the presence of the element in a given message, and provides a one-to-one mapping between the encoded values and the engineering values that will be applied to the system.

To analyze performance and verify the actual settings, it is important to extract the correct engineering values for the parameters of interest from the encoded messages, as illustrated in the following example. Consider a Measurement Control Message sent by the RNC to request the UE for intra-frequency measurement reports for Event 1a (a primary CPICH enters the reporting range). In Figure 4.3, part of the original encoded message is reported on the left, and the parameters of interest for the example are highlighted in bold.

By applying the mapping as specified by the standard (see Table 4.1), it is possible to extract the real engineering values of the parameters from the message, as indicated in the right side in Figure 4.3. Some tools that are used to analyze and set parameters may be able to convert directly from IE to engineering values, on the user interface.

4.1.3 Over-the-Air Parameter Verification

After completing the initial parameter settings, it is a good practice to verify if what is transmitted over-the-air corresponds to the desired values set in the databases. At first glance, this seems superfluous; however, practical experience shows that there are always cases where the real parameters transmitted by the network elements differ from

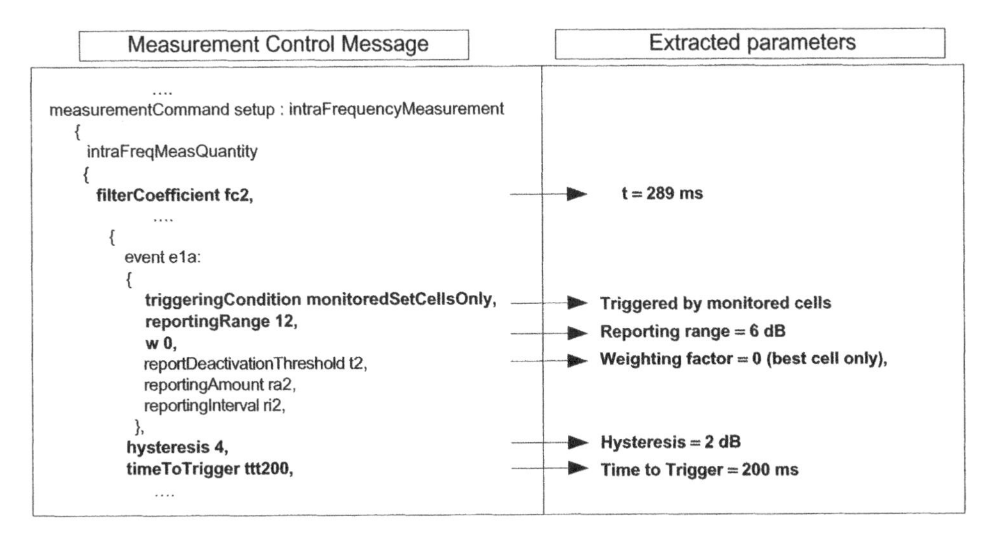

Figure 4.3 Sample mapping between IEs and engineering values for Event 1a parameters

Table 4.1 Specified mapping between IEs and engineering values for Event 1a parameters [10]

	IE value	Engineering units
Triggering conditions	Enumerated {Active Set cells, Monitored Set cells, Active Set cells and Monitored Set cells, Detected Set cells, Detected Set cells and Monitored Set cells}	Active Set cells, Monitored Set cells, Active Set cells and Monitored Set cells, Detected Set cells, Detected Set cells and Monitored Set cells
Reporting range	Integer (0–29)	(0–14.5) dB, in 0.5 dB increments
Filter coefficient	Integer (0, 1, 2, 3, 4, 5, 6, 7, 8, 9, 11, 13, 15, 17, 19)	0 = no filtering Otherwise the time constant of the low-pass filter is calculated as $\tau = -T/\ln(1-2^{-k/2})$ T = 200 ms indicates the intra-frequency measurement period
Weighting factor	Integer (0–20)	(0–2), in 0.1 increments
Hysteresis	Integer (0–15)	(0–7.5) dB in 0.5 dB increments
Time-to-trigger	Enumerated {0, 10, 20, 40, 60, 80, 100, 120, 160, 200, 240, 320, 640, 1280, 2560, 5000}	0, 10, 20, 40, 60, 80, 100, 120, 160, 200, 240, 320, 640, 1280, 2560, 5000 [ms]

the values set in the OMC. Several factors could contribute to this situation: frequent parameter updating for various network elements, an upgrade of software versions, or human error when setting parameters.

A mismatch between desired and actual parameter values could affect a limited number of cells, entire Radio Network Subsystems (RNS), or even the entire network. The RNS mismatch case would produce an inconsistency between the values present in a small group of cells and the rest of the network, which could cause performance degradations or, in more severe cases, procedure malfunctions.

Here is a suggested way to verify over-the-air parameters:

- Collect measurements and signaling traces with the UE, in Idle and Connected Mode, in different areas of the network.
- Use suitable tools to extract the parameter values for all visited cells.
- Check the extracted parameters on a cell basis for inconsistencies with the selected settings.

During this process, it is important to match the proper parameter set to the proper cell. The difficulty is that, in Idle Mode, the UE must read SIBs from both the camping cell and from any candidate for reselection, but only the identity of the camping cell is consistently and uniquely reported by optimization tools. To clarify this uncertainty, it is necessary to rely on low level information from the test UE: either information that the reselection attempt has started, or information about Rake finger assignment. In the first case, once the reselection process starts, any SIB reading should be attributed to the reselection target cell. In the second case, because the Active Set size is limited to 1 in Idle Mode, whenever fingers are assigned to more than one Primary Scrambling Codes (PSC), a reselection attempt should be suspected and a SIB reading should be attributed to a non-active PSC. In either case, once possible errors are identified, the affected parameters can be adjusted via the OMC terminal.

4.2 Physical Layer Parameters

Before discussing the main parameters that control the operation of the UE in its different states, we must address parameters related to Layer 1. Here, we include only the parameters generally considered to be the most important for network planning and optimization.

4.2.1 Frequency Selection and Management

In WCDMA, the channels are defined in terms of UTRA Absolute Radio Frequency Channel Number (UARFCN). The relationship between the UARFCN and the frequency is defined in the standard [11]. UARFCN can be based on a center, Downlink, or Uplink frequency.

$$Frequency = UARFCN * 200 \text{ kHz} \tag{4.1}$$

According to Equation 4.1 and the range allowed for the UARFCN (from 0 to 16383), WCDMA could be deployed over any frequency band lower than 3.2 GHz. The flexibility to deploy WCDMA in the mobile frequency band of any country is enhanced by the

Table 4.2 Frequency allocation for land mobile operation

Frequency band UL/DL [MHz]	Typical frequency duplex	Used in:	Defined in the standard?[a]
1920–1980 2110–2170	190	IMT2000 band, Europe, China, Korea, Japan	Yes
1850–1910 1930–1990	80	North America	Yes
1710–1785 1805–1880	95	Europe	Yes
824.02–848.97 869.04–893.97	45	North America	No
872.01–914.99 917.01–954.99	45	Europe	No
887.1–924.99 832.1–869.99	55	Japan	No
1750–1780 1840–1870	90		No
452.5–455.73 462.5–465.73	10	Europe	No
776–794 746–764	30	North America	No

[a] The standard does not prevent deploying WCDMA in frequency bands not defined in [11].

option of setting the UARFCN Uplink and Downlink independently, which would adjust the duplex frequency for bands that are not yet entirely defined; see Table 4.2.

As mentioned before, the channel raster for WCDMA is 200 KHz, but the bandwidth is typically 5 MHz. With the channel raster, the separation between adjacent channels can be adjusted in a practical range of 4.4 to 5.6 MHz. This adjustment can reduce the adjacent channel interference when deployment configuration does not allow spatially separating two different network carriers that border each other.

4.2.2 PSC Planning

PSCs differentiate cells on the Downlink. Codes are generated [12] to ensure that PSCs are independent of one another. This greatly simplifies PSC planning because the only requirements are to prevent cells with the same PSC from overlapping and avoid repetition of a PSC in the Neighbor List. Overlapping should be stringently defined: the E_c/N_o (or Received Signal Code Power [RSCP]) from another cell of the same PSC must be significantly below the E_c/N_o (or RSCP) threshold above which the Rake searcher assigns a finger to the path. In practice, this value depends on the UE implementation and is not defined in the standard. Suggested values defined in Table 4.3 ensure that no overlap between cells is observed during the PSC planning stage.

To prevent PSC repetition when Neighbor Lists merge (Active Set >1), the condition should be verified both on neighbors and second-order neighbors; that is, the Neighbor List of each member in the Neighbor List.

Table 4.3 Cell overlap threshold for PSC planning

Quantity	Minimum performance	Suggested threshold	Comment
E_c/N_o	−20 dB	−25 dB	Typical UE implementations do not assign fingers for path below −22to −25 dB
RSCP	−117 dBm	−127 dBm	Margin in line with LNF margin, for (1 − reliability) target

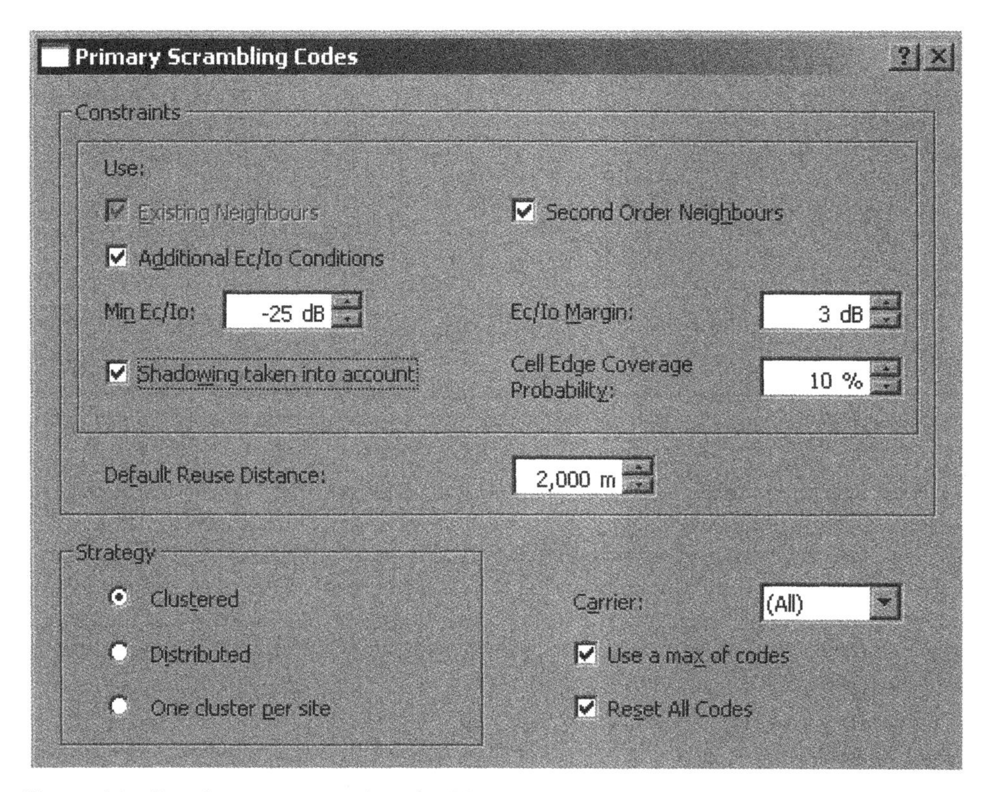

Figure 4.4 Sample parameter settings for PSC planning in a commercial network planning tool

Such a threshold might not be directly usable in a network planning tool, depending on the tool implementation. For reference, Figure 4.4 shows an example of PSC planning input.

4.2.3 Power Allocation

Power allocation encompasses power assignments for both common channels and dedicated channels in the Downlink. In contrast to common channels, dedicated channels are power controlled, so power assignments to dedicated channels set upper (and lower) transmit power limits.

Common and dedicated power assignments must be set jointly to ensure that both sets of assignments meet coverage goals at maximum projected load conditions, as discussed in Chapter 2. This also implies that if the coverage for common and dedicated channels is not comparable, power and capacity is wasted.

Of all the power assignments, only the CPICH, AICH, and Page Indicator Channel (PICH) transmit power values are signaled to the UE.

The transmission of the CPICH transmit power value is for open loop power control. If the CPICH transmit power value is known at the UE, the Downlink path loss can be calculated from the measured RSCP value and used to determine the Access probe power in the Uplink, or simply to report path loss when this measurement quantity is set.

The transmitted value of AICH and PICH Tx power is not used directly by the UE but is available for verification of the assigned power. Both power assignments ensure that the Downlink indicators are properly decoded [13]. The main difference between them is that the AICH has a set information rate, but the number of page indicators can vary from 18 to 144. The recommendations listed in Table 4.4 assume 18 page indicators per frame; if more page indicators per frames are set, the PICH power must be increased by 3 dB for each doubling of the page indicator number, to compensate for the reduced spreading gain.

Chapters 2 and 3 discussed dedicated power settings to fulfill coverage or capacity requirements. Table 4.5 presents DPCH power assignments that yield an acceptable trade-off between these two requirements.

Table 4.4 AICH and PICH power offset settings

Parameter	Setting	Comment
AICH power offset	−5 to −7 dB	Setting is in reference to CPICH
PICH power offset	−6 to −7 dB	Setting is in reference to CPICH for 18 PI per frame offset Increase by 3 dB when PI per frame doubles

Table 4.5 Typical power assignments for the main Radio Access Bearer (RAB)

Bearer	Setting	Comment
Speech AMR, 12.2 kbps	31.5 dBm	Reference coverage, in line with CPICH coverage
Circuit switched data, 64 kbps	36.5 dBm	Bearer typically used for video telephony Match the coverage of voice service
Packet switched data, 64 kbps	36.5 dBm	Coverage continuity over the entire cell area is provided by channel rate switching
Packet switched data, 128 kbps	37 dBm	
Packet switched data, 384 kbps	37 dBm	

4.3 Intra-frequency Cell Reselection Parameters

4.3.1 Introduction

In UMTS systems, one of the key performance metrics is standby time. In Idle Mode, standby time depends on the implemented cell reselection mechanism and its corresponding parameters. To differentiate their service from others, operators can adjust standby time and camping cell quality performance by optimizing operator-configurable cell reselection parameters.

This section investigates the impact of these parameters, on the basis of diverse field data from different characteristic RF environments collected in commercial networks (for example: outdoor, indoor, low/high speed, Pilot polluted/not Pilot polluted). In this section, we will compute performance metrics (camping cell quality, relative standby time) for different parameter sets, using a simulation platform that efficiently employs over-sampled channel measurements to improve reliability and includes a standby time model described below. Simulation results illustrate the trade-offs for different parameter settings in various RF environments. Interestingly, camping cell quality and standby time do not always move in opposite directions.

The intra-frequency cell reselection mechanism is also used in Connected Mode in all states except for CELL_DCH. Section 4.3.7 briefly addresses how the trade-offs in Connected Mode, especially in the CELL_FACH state, compare to the trade-offs in Idle Mode.

4.3.2 Overview of the Intra-frequency Cell Reselection Procedure

In UMTS, according to the cell reselection criterion [14], the UE shall regularly search for a better cell to camp on. This mechanism ensures acceptable quality and strength of the camping cell and is necessary to achieve the desired call setup performance. A very reactive cell reselection mechanism can guarantee a better quality of the camping cell at the expense of frequent reselections, which decreases standby time. The reselection criterion is based on parameters provided by the network.

In Idle Mode, the UE operates in Discontinuous Reception (DRX) to improve its standby time. At the beginning of each DRX cycle, the UE wakes up, reacquires the camping cell, and reads its PICH. Depending on the measured quality of the camping cell, the UE may trigger intra-frequency measurements and evaluate the cell reselection criterion. In particular, measurements are triggered if the common Pilot signal-to-noise ratio (CPICH E_c/N_o) of the camping cell falls below $Qqualmin + S_{\text{intrasearch}}$. The cell reselection criterion is then evaluated by comparing the quality of the camping cell with that of the monitored cells. The measurement quantity can be either the CPICH E_c/N_o or the CPICH RSCP. The reselection criterion [14] ensures that the UE will reselect to a new cell if the quality of the new cell is at least $Qhyst2_s + Qoffset2_{s,n}$ dB better for Treselection seconds than the camping cell quality, assuming that E_c/N_o was chosen as a measurement quantity.

In GSM, Hierarchical Cell Structures (HCS) have been introduced to minimize signaling. Fast-moving UEs camp on large cells or boomer cells. Slower-moving UEs transition to smaller cells (macrocells or even microcells). The number of reselections over a fixed duration allows the UE to decide if it is moving at high or low speed. Small cells offer the advantage of providing more capacity. And, since only a small share of users move at higher speeds, only a small fraction of the bandwidth must be dedicated to boomer cells.

What works well in GSM, though, is less straightforward in WCDMA. Several operators have only two (wideband) carriers (as opposed to many narrowband GSM carriers). Therefore, dedicating one wideband carrier to boomer cells significantly reduces capacity. The standard also supports HCS on a single carrier (boomer cells and microcells on one carrier). However, that comes at the expense of a significant increase in interference. Hence, using HCS in WCDMA is not recommended, especially over a single carrier, as detailed in Chapter 3.

4.3.3 List of Intra-frequency Cell Reselection Parameters

Table 4.6 lists all the parameters provided by the standard to tune the intra-frequency cell reselection mechanism.

If the quality measure for cell selection and reselection is set to CPICH E_c/N_o, $\text{Qoffset1}_{s,n}$ and Qhyst1_s affect only inter-system reselection. Otherwise, $\text{Qoffset1}_{s,n}$ and

Table 4.6 Intra-frequency cell reselection parameters

Parameters	Range, units	Definitions
Quality measure	CPICH E_c/N_o – CPICH RSCP	Choice of measurement to use as quality measure Q for ranking candidate FDD cells for cell reselection
DRX cycle	80 ms to 5.12 sec in discrete increments	Periodicity of the paging occasion corresponding to a given UE in Idle Mode
Treselection	0 to 31 sec	Cell reselection timer value
Qrxlevmin	−115 to −25 dBm, in 2 dB increments	Minimum quality level in the cell expressed in terms of CPICH RSCP
Qqualmin	−24 to 0 dB	Minimum quality level in the cell expressed in terms of CPICH E_c/N_o
$S_{intrasearch}$	−32 to +20 dB, in 2 dB increments	Measure other intra-frequency cells if CPICH E_c/N_o < Qqualmin + $S_{intrasearch}$
Qhyst1_s	0 to 40 dB, in 2 dB increments	Hysteresis: used for TDD and GSM cells and for FDD cells if the quality measure for cell selection and reselection is set to CPICH RSCP
$\text{Qoffset1}_{s,n}$	−50 to 50 dB	Offset between the two cells: used for TDD and GSM cells and for FDD cells if the quality measure for cell selection and reselection is set to CPICH RSCP
Qhyst2_s	0 to 40 dB, in 2 dB increments	Hysteresis: used for FDD cells if the quality measure for cell selection and reselection is set to CPICH E_c/N_o
$\text{Qoffset2}_{s,n}$	−50 to 50 dB	Offset between the two cells: used for FDD cells if the measurement quantity is CPICH E_c/N_o

Qhyst1_s would be shared by intra-frequency and inter-system reselection, not allowing separate settings for both mechanisms.

If camping cell measurements fall below Qrxlevmin or Qqualmin, procedures are initiated that may lead to an inter-system cell reselection [14]. If another system such as GSM is available (typically the case), Qrxlevmin or Qqualmin qualify as both intra-frequency and inter-system parameters and, thus, should be optimized for both schemes, as discussed in Chapter 6. If no other system is available, Qrxlevmin and Qqualmin should be optimized to avoid premature out-of-service declarations [14] as opposed to maintaining the minimum cell quality and strength that guarantee the desired call setup performance.

For the intra-frequency cell reselection simulations, we suggest the following values: Qqualmin = −18 dB and Qrxlevmin = −113 dBm. With these settings, both Qrxlevmin and Qqualmin are set slightly lower than the expected E_c/I_o and RSCP at the cell edge for a loaded system, as calculated in Chapter 2. This discrepancy is necessary to offer some added protection against out-of-service conditions. For instance, because of weak coverage, the noise component may be larger than in the calculation shown in Chapter 2.

In the initial release of the standard, Treselection was not only used by intra-frequency cell reselection but was also shared with the inter-frequency and the inter-system cell reselection schemes as well. So long as UE implementations are based on the latest standard release, which allows scaling of Treselection for inter-frequency and inter-system cell reselection, this limitation must be taken into account.

4.3.4 Intra-frequency Cell Reselection Metrics

Camping on a strong cell is only the first step in a successful call setup.

- UE-terminated calls start with a page on the Downlink, followed by an Access attempt on the Uplink before proceeding with call setup.
- UE-originated calls start with an Access attempt on the Uplink before proceeding with call setup.

Accordingly, the UE must be able to do the following: receive pages on the Downlink for UE-terminated calls, access the Random Access Channel (RACH) in the Uplink, and successfully demodulate the ensuing call setup messaging. To facilitate the entire process, several repetition mechanisms are in place. The pages are repeated on the Downlink if the UE does not respond to the initial page, either because the page was lost due to a weak Downlink Channel or because the UE was not in the paging area (i.e., Location, Registration, or Routing) or was turned off. The Access probe is repeated with increased power if the Node B is not able to detect the first probe. Finally, the probing mechanism can be repeated if the call setup messages are not successfully exchanged.

Because cell reselection serves call setup, the high-level performance metrics of the entire call setup scheme–consisting of cell reselection, paging, Access, and call setup messaging–are the following: UE standby time, call setup performance (latency and success rate) for UE-originated and UE-terminated calls, and the capacity required for the common control channels involved in the call setup.

If we concentrate on cell reselection, the metrics of interest are UE standby time, channel strength, and quality during a paging occasion. Section 4.2.3 briefly discussed common channel power settings related to paging. Section 4.4 discusses Access-parameter settings.

To use simple performance metrics in a simulation context, we evaluate camping cell CPICH RSCP and E_c/N_o:

- Camping cell CPICH RSCP approximately correlates to the ability to close the Uplink (Access, Uplink call setup messaging).
- Camping cell E_c/N_o correlates to the ability to close the Downlink (successful paging occasions, Downlink cell setup messaging).

The UE standby time model should be flexible enough to apply to any UE implementation. The model described below fulfills this requirement as it considers relative standby time [15].

The model compares each set of parameters to a reference case, examining the ratio between the two standby times. The reference case is a stationary UE (i.e., reselection rate equal to zero) with a DRX cycle length of 2.56 sec.

Let α be the ratio of the current being drawn when the UE is awake and asleep. Let f_a be the fraction of time when the UE is awake. The ratio between the standby times of the two sets of parameters can then be expressed as:

$$\gamma = \frac{ST_2}{ST_1} = \frac{1 + f_{a,1}(\alpha - 1)}{1 + f_{a,2}(\alpha - 1)} \tag{4.2}$$

In our simulations, we assumed $\alpha = 100$. The fraction of time the UE is awake, f_a, is the average awake time per cycle, T_{awake}, divided by the DRX cycle length:

$$f_a = \frac{T_{awake}}{T_{DRX}} \tag{4.3}$$

For T_{awake}, we used the following model:

$$T_{awake} = (30 + 250 R_{rslct}) ms/cycle \tag{4.4}$$

R_{rslct} denotes the reselection rate, defined as number of reselections per DRX cycle. In Equation 4.4, the first term models the average awake time per cycle needed by the UE to reacquire the camping cell, to monitor the PICH, and to evaluate the cell reselection criterion. The second term models the additional awake time due to reselection. In fact, each time a reselection takes place, the UE must decode the system information broadcasted by the new cell, thus extending its awake time.

Using an average awake time ignores the fact that the UE can store SIBs of cells already camped on and, therefore, can reduce the cell reselection delay if already camped on the cell. This approximation is valid if a new parameter set does not dramatically alter the following ratio: UE-reselected cells that it has already camped on over UE-reselected cells that it has not yet camped on.

4.3.5 Intra-frequency Cell Reselection Trade-offs in Idle Mode

Section 4.3.4 discussed the different phases of call setup, including overall call setup optimization goals: standby time, call setup performance, and the capacity required by paging and Access channels. Trade-offs among these goals are possible. For instance, if we want to improve standby time at the expense of capacity and retain call setup performance, we can increase the DRX cycle and have the UE wake up less frequently, but assign more power to the paging and Access indicator channels to compensate for weaker channels due to reselections taking place less frequently. This approach cannot be taken to extremes, though, because the maximum UE transmit power is limited; this limits how much we can improve Access performance and subsequent call setup messaging on the Uplink. Also, to save resources, Idle Mode coverage should not exceed Connected Mode coverage.

To maximize the success rate of paging and Access, the channel must be strong during paging occasions. The rest of this section emphasizes the trade-offs between standby time and camping cell strength and quality. To illustrate and quantify these trade-offs in realistic scenarios, we passed field measurements through a simulation platform that accurately models UE behavior to determine the above-mentioned performance metrics for different parameter settings (see reference [16]).

4.3.5.1 Field Measurements

We collected field measurements from three different commercial UMTS networks in Europe, using a QUALCOMM test phone. In each market, we performed several runs of data collection along a metric route. We selected routes that best characterized the specific deployment scenarios. In Table 4.7, the measurements are classified in terms of radio frequency (RF) environment and mobility. Markets X and Z are examples of macro-deployment scenarios in suburban and urban areas. Market Y is an example of an in-building micro-deployment scenario. The metric route leads through Pilot-polluted areas with several Pilot signals of similar strength, areas with fragmented coverage and frequent changes of best server, and areas dominated by one Pilot.

4.3.5.2 DRX Cycle Length versus Treselection

Let's first study the interaction between DRX cycle length and Treselection. To do this, we tested the DRX cycle length values of 0.64 sec, 1.28 sec, and 2.56 sec in combination with the Treselection values of 0 sec, 1 sec, and 2 sec. This produces nine sets

Table 4.7 Classification of field measurement data

Market (number of runs)	RF environment	Mobility
X (4)	Outdoor; fragmented coverage; Pilot pollution	Vehicular; 40 min mean time/run; 35 km/hr mean speed/run; highly variable speeds from run to run
Y (3)	Indoor; some Pilot pollution	Pedestrian; 20 min mean time/run; 4 km/hr mean speed/run
Z (1)	Outdoor; some fragmented coverage	Vehicular; 30 min mean time/run; 25 km/hr mean speed

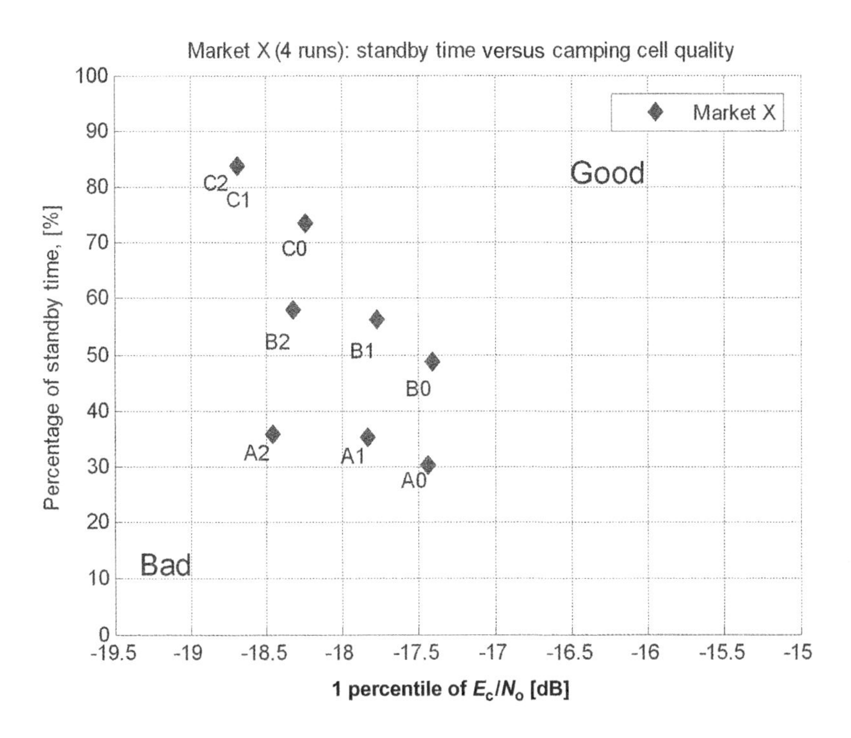

Figure 4.5 DRX cycles versus treselection for market X; Qqualmin + $S_{intrasearch}$ = 10 dB; Qhyst2_s + Qoffset2_n = 3 dB

of parameters to test in the simulator: A0, A1, A2, B0, B1, B2, C0, C1, and C2. The letter indicates the DRX cycle length value: A, B, and C stand for 0.64 sec, 1.28 sec, and 2.56 sec, respectively. The number indicates the Treselection value: 0, 1, and 2 stand for 0 sec, 1 sec, and 2 sec, respectively. The measurement-triggering threshold is set to 10 dB (i.e., Qqualmin + $S_{intrasearch}$ = 10 dB) and the hysteresis is set to 3 dB (i.e., Qhyst2_s + Qoffset2_n = 3 dB).

Figure 4.5 shows the performance of each set of parameters for Market X. A DRX cycle of 1.28 sec (set B0, B1, and B2) improves the UE standby time by about 20 to 25% with respect to a DRX cycle of 0.64 sec (set A0, A1, and A2), surprisingly without compromising camping cell quality. Of course, the UE wakes up more frequently with a DRX cycle of 0.64 sec and can adapt more quickly to deteriorating channel conditions. However, more deteriorating CPICH E_c/N_o measurements are captured if the reselection process is delayed. As expected, a DRX cycle of 2.56 sec (set C0, C1, and C2) instead of 1.28 sec (set B0, B1, and B2) improves UE standby time by about 25% but decreases cell quality by 0.5 to 1 dB[1]. The performance of sets C1 and C2 coincide because the Treselection parameters (1 sec and 2 sec) are smaller than the DRX cycle period of 2.56 sec. Simulations based on field data from Market Y and Z yield similar results.

[1] The DRX cycle is much larger than the reselection delay, producing results in line with expectations.

4.3.5.3 Effect of Velocity and RF Environment

Next, let us focus on field data from Markets X, Y, and Z for a DRX cycle equal to 1.28 sec (sets B0, B1, and B2) and different Treselection values (see Figure 4.6). A Treselection of 1 sec (set B1) instead of 2 sec (set B2) improves the camping cell quality by about 0.5 to 1 dB at the expense of standby time, which is reduced by less than 2%. A Treselection of 0 sec (set B0) instead of 1 sec (set B1) improves the camping cell quality by about 0.5 to 0.7 dB at the expense of roughly 5% reduction in standby time. From Figure 4.9, we can also observe that the UE speed impacts the reselection process. As the speed increases, for a given parameter, the camping cell quality (as measured by the 1 percentile of E_c/N_o) distribution degrades and the stand-by time degrades.

To analyze the effect of velocity on cell reselection performance, we compare the fastest and the slowest run from Market X (average speeds of 45 km/hr and 25 km/hr, respectively), as shown in Figure 4.7. Velocity degrades the reselection performance both in terms of standby time (by 2%, higher speeds increase the number of reselections over time) and cell quality (by 0.5 dB). Different velocities do not produce different trade-off curves. The trade-off curve depends mainly on the RF environment.

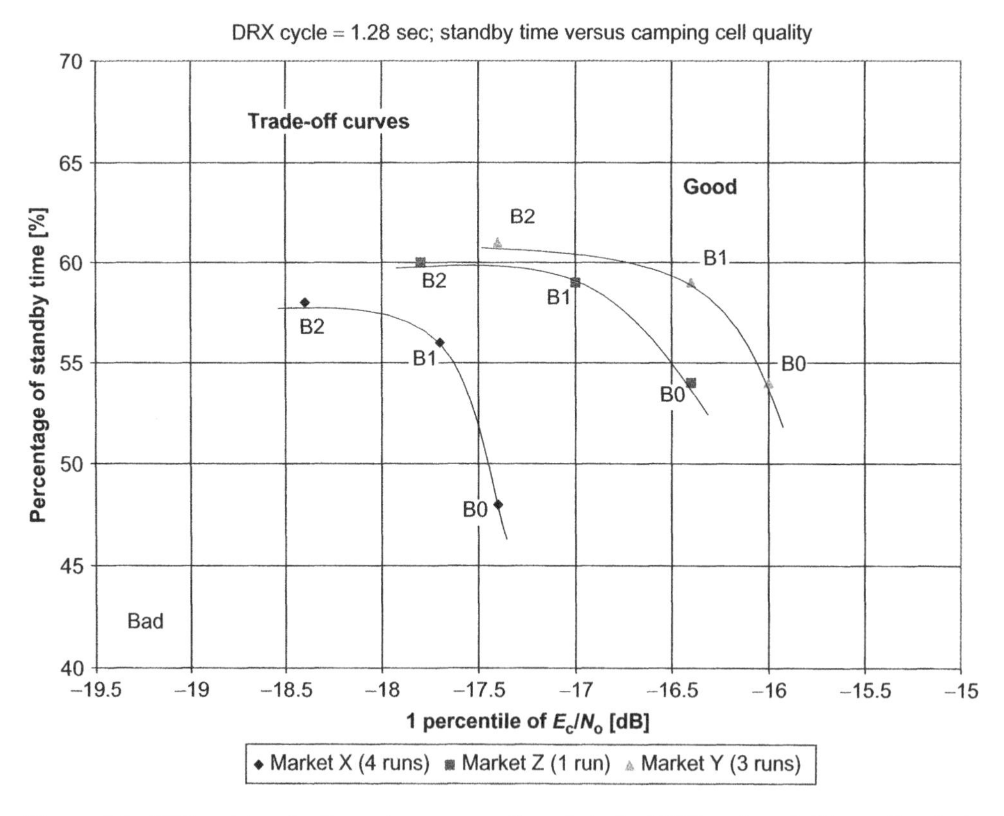

Figure 4.6 Different treselection values across all markets for DRX cycle = 1.28 sec; Qqualmin + $S_{intrasearch}$ = 10 dB; $Qhyst2_s + Qoffset2_n$ = 3 dB

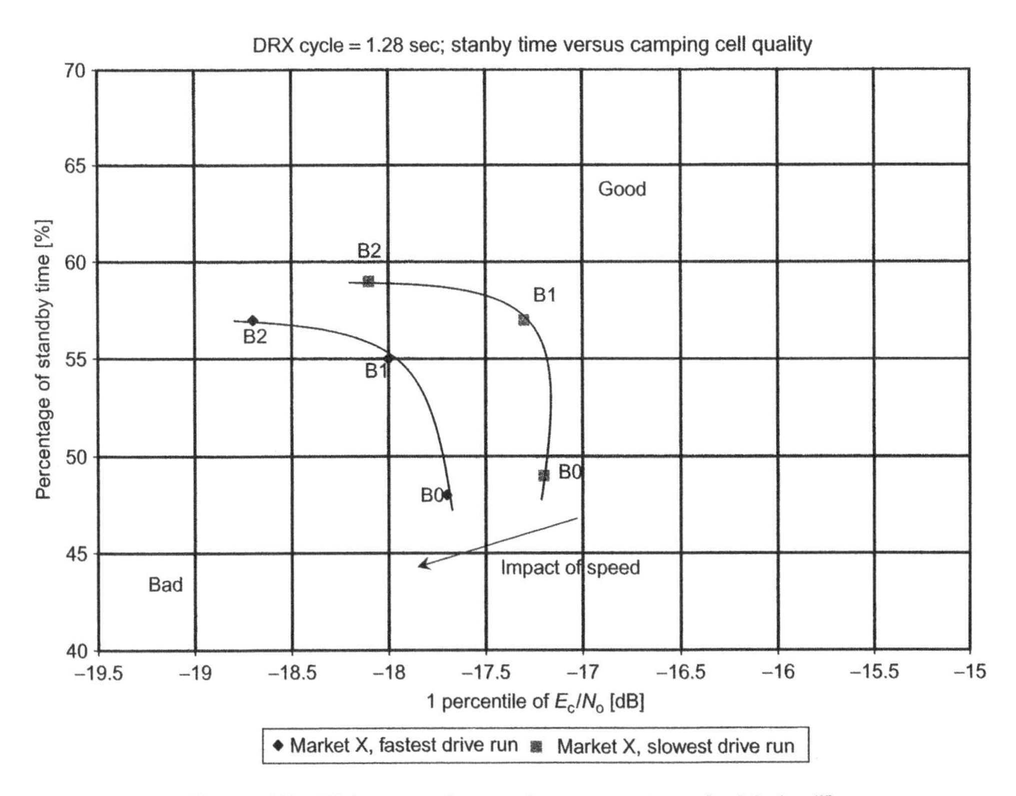

Figure 4.7 Highest speed versus lowest speed run for Market X

4.3.5.4 Treselection versus Hysteresis

Treselection and hysteresis (Qhyst2_s+ Qoffset2_n) are inter-dependent in controlling the trade-off between cell quality and standby time. This section studies the interaction between these two parameters, with a focus on DRX cycle length of 1.28 sec. The Treselection values of 0 sec, 1 sec, and 2 sec are tested in combination with different hysteresis values. In particular:

- The Treselection value of 0 sec is tested with hysteresis values from 1 to 13 dB, in 2 dB increments.
- The Treselection values of 1 sec and 2 sec are tested in combination with hysteresis values from 1 to 7 dB, in 2 dB increments.

Figures 4.8 and 4.9 show the performance of each set of parameters for Markets X and Y, respectively. In each plot, the three curves correspond to the three tested values of Treselection. The data points on each curve represent hysteresis settings. Each data point is labeled with a letter "h" followed by a number that indicates the hysteresis value (for example, "h7" represents "7 dB hysteresis").

Increasing the hysteresis always reduces the reselection rate. While this always improves standby time, it does not necessarily degrade cell quality. On the contrary, for a given

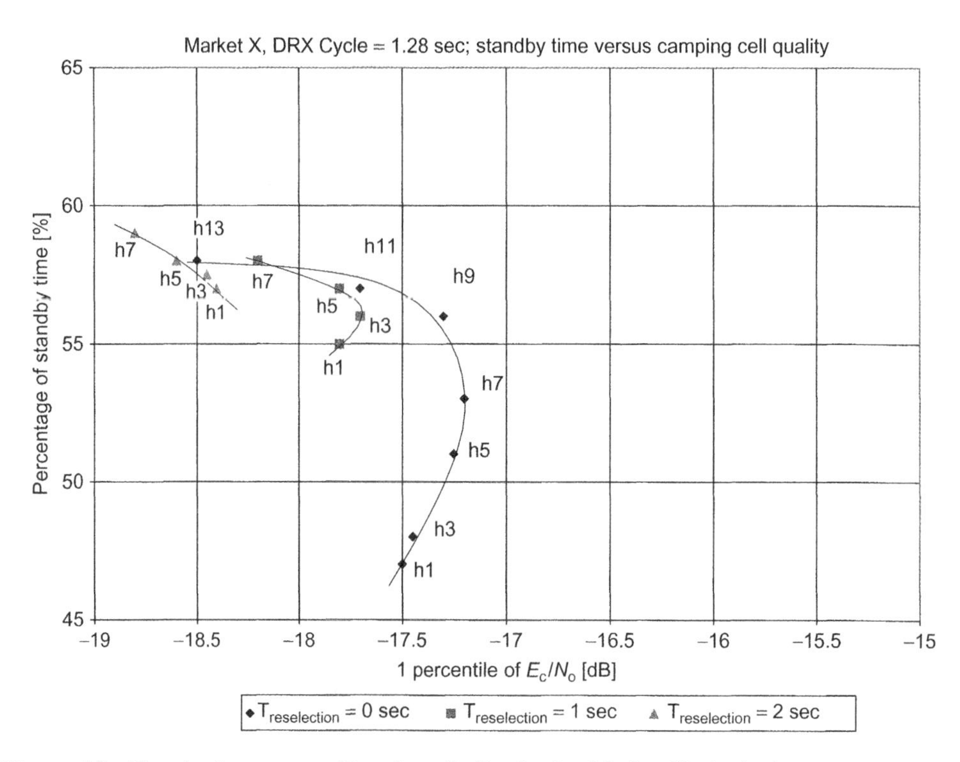

Figure 4.8 Treselection versus Qhyst2$_s$+ Qoffset2$_n$ for Market X; in both cases: Qqualmin + $S_{intrasearch}$ = 10 dB

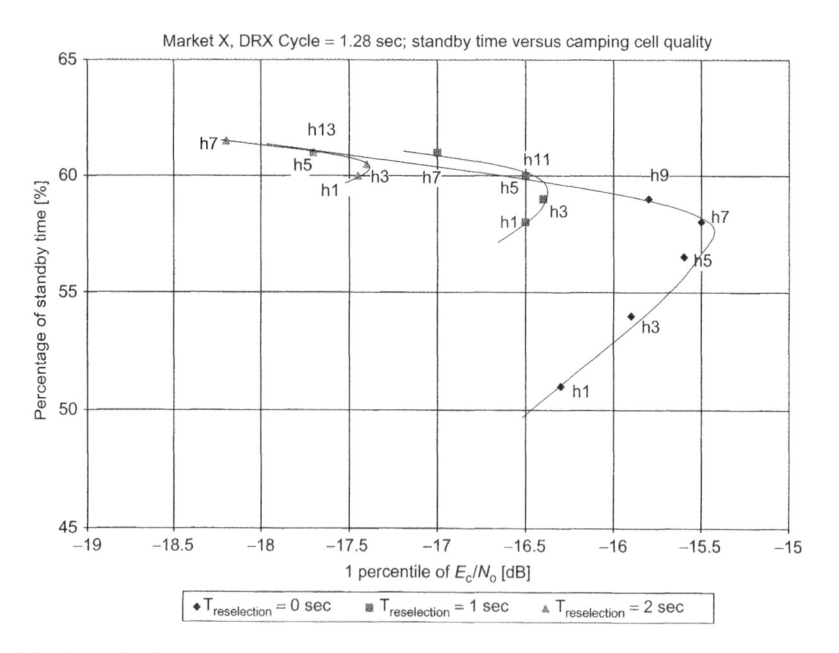

Figure 4.9 Treselection versus Qhyst2$_s$+ Qoffset2$_n$ for Market Y; in both cases: Qqualmin + $S_{intrasearch}$ = 10 dB

Treselection, there is an optimal amount of hysteresis that maximizes cell quality by preventing the UE from making poor reselection decisions in response to large signal fluctuations; as expected, higher hysteresis is required for smaller Treselection values.

On the basis of simulation results, the minimum recommended hysteresis settings are 7 dB and 3 dB for a Treselection of 0 sec and 1 sec, respectively. Higher hysteresis values can be used to improve camping cell quality, at the expense of decreased standby time. Our preferred setting is a Treselection of 0 sec with a hysteresis of 7 to 9 dB, because this maximizes cell quality without significantly degrading the standby time performance. However, as mentioned in Section 4.3.3, Treselection is shared with the inter-system reselection functionality. Because inter-system reselection is more expensive, a Treselection larger than 0 sec is suggested. Accordingly, our recommendation combines a Treselection setting of 1 sec with a hysteresis of 3 dB.

4.3.5.5 $S_{intrasearch}$ and Qqualmin

The measurement-triggering threshold (i.e., $Qhyst2_s + Qoffset2_n$) should be set to initiate measurements when the camping cell quality starts to deteriorate and a cell reselection is desirable. On the other hand, unnecessary measurements should be avoided because of their negative impact on standby time[2]. In the previous simulation, the measurement-triggering threshold was set to −10 dB. Let's compare four different threshold levels: −8 dB, −10 dB, −12 dB, and −14 dB. Two different combinations of Treselection and hysteresis are used: Treselection of 0 sec with 7 dB hysteresis, and Treselection of 1 sec with 3 dB hysteresis. The DRX cycle length is set to 1.28 sec.

Figure 4.10 shows the simulation results. A threshold of −10 dB maximizes the cell quality. Threshold levels lower than −10 dB can be used to reduce the reselection rate, and thus trade camping cell quality for standby time.

4.3.6 Intra-frequency Cell Reselection Parameter Recommendations for Idle Mode

Table 4.8 shows our recommended set of parameters for intra-frequency reselection. This parameter set works well in all scenarios and achieves a good compromise between camping cell quality and UE standby time.

Simulation results also show that the cell reselection metrics depend on the deployment scenario. To achieve the best performance, the parameters can be fine-tuned for a specific scenario or, even more aggressively, on a per-cell basis.

4.3.7 Intra-frequency Cell Reselection in CELL_FACH State

In Connected Mode, the UE can be in four states: CELL_DCH, CELL_FACH, CELL_PCH, and URA_PCH. These states, mainly used in the PS domain, allow trade-offs among resource utilization, UE power consumption, and data latency. For instance, CELL_FACH is activated if the data rate is too low to justify the CELL_DCH state but too high for any other state. In all states except CELL_DCH, the cell reselection mechanism manages mobility.

[2] The standby time model previously presented does not consider the intra-frequency measurement rate, because the power required for intra-frequency measurements is negligible compared to the power required for intra-frequency cell reselection.

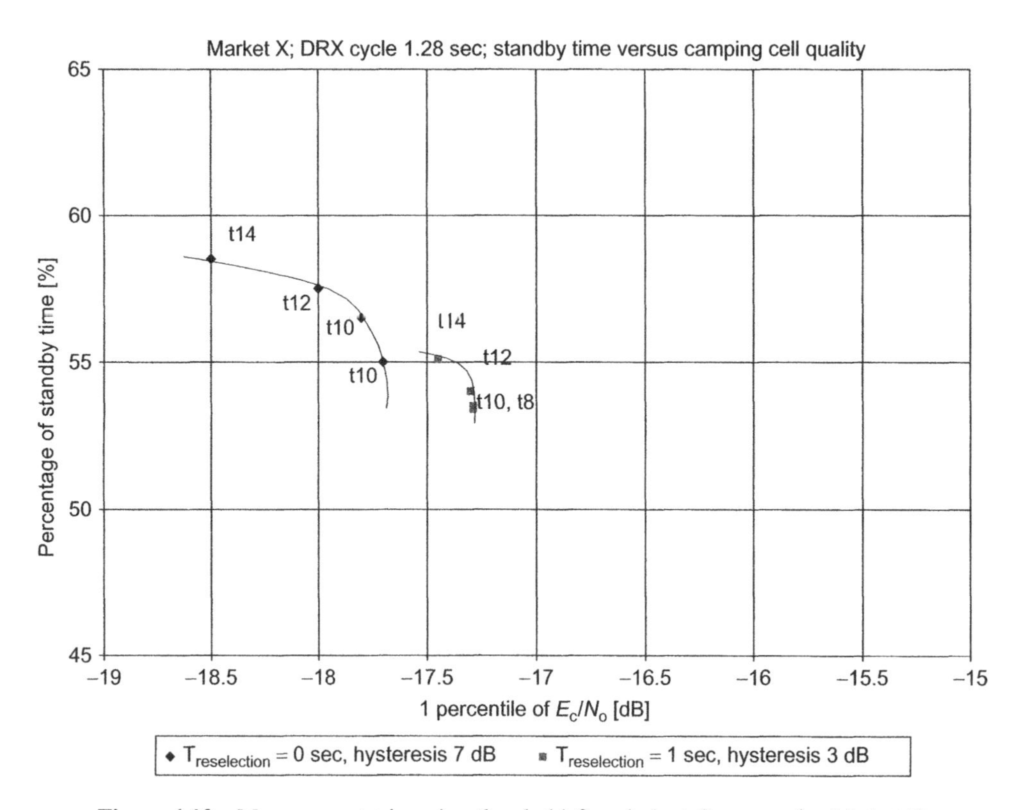

Figure 4.10 Measurement-triggering threshold Qqualmin + $S_{intrasearch}$ for Market X

Because the UE is in Connected Mode for much less time than it is in Idle Mode, and because it does not switch to Sleep Mode (DRX) in CELL_FACH, standby time is of less importance in the selection of DRX cycle length. Cell quality is important in Idle Mode, but even more important in CELL_FACH, CELL_PCH, and URA_PCH states. The standard addresses these constraints by supporting different parameter settings for Idle Mode and Connected Mode.

However, we have shown in previous simulations that reducing the DRX cycle from 1.28 to 0.64 sec does not significantly improve performance, in terms of cell quality. Therefore, the Idle Mode parameters should be sufficient in all Connected Mode states.

4.3.8 Inter-frequency Cell Reselection Considerations

The inter-frequency cell reselection scheme shares most of the parameters with the intra-frequency cell reselection scheme. The only exception is the threshold parameter, below which to measure other cells. The intra-frequency cell reselection parameter is $S_{intrasearch}$ and the corresponding parameter for inter-frequency cell reselection is $S_{intersearch}$.

Inter-frequency cell reselections are less optimized in terms of power consumption. Also, the reselection delays are longer than intra-frequency cell reselection delays. Therefore, inter-frequency cell reselection should take place less frequently. This can

Table 4.8 Intra-frequency cell reselection parameter recommendations

Parameters	Recommendation	Comments
Quality measure	CPICH E_c/N_o	According to [17], the noise and interference term N_o of the CPICH E_c/N_o measurement quantity is the same for different cells at the same location. Therefore, intra-frequency cell reselection performs the same for both measurement quantities. $Qhyst1_s$ and $Qoffset1_{s,n}$ are not shared and thus are solely available for inter-system cell reselection
DRX cycle	1.28 sec	A DRX cycle length of 1.28 sec is preferred over a DRX cycle length of 0.64 sec, because it improves the standby time by ~25% without reducing camping cell quality
Treselection	0 sec or 1 sec	If there are no inter-system considerations to take into account, 0 sec outperforms 1 sec
Qrxlevmin	−113 dBm	Typical setting for intra-frequency only. Different settings may be used, depending on inter-system aspects; see Chapter 6
Qqualmin	−16 to −18 dB	Typical setting for intra-frequency only. Different settings may be used, depending on inter-system aspects; see Chapter 6
$S_{intrasearch}$	Qqualmin + $S_{intrasearch}$ = 10 dB	
$Qhyst1_s$	−2 dB	Typical setting for intra-frequency only. Different settings may be used, depending on inter-system aspects; see Chapter 6
$Qoffset1_{s,n}$	−1 dB	Typical setting for intra-frequency only. Different settings may be used, depending on inter-system aspects; see Chapter 6
$Qhyst2_s$	If Treselection = 1 sec, $Qhyst2_s$ + $Qoffset2_n$ = 3 dB	Recommendation is given for Qhyst2 + Qoffset2. Individual settings of Qhyst2 and Qoffset2 are not considered
$Qoffset2_{s,n}$	If Treselection = 0 sec, $Qhyst2_s$ + $Qoffset2_n$ = 7 dB	

be achieved by setting the threshold defined by Qqualmin + $S_{intersearch}$ lower than the intra-frequency threshold defined by Qqualmin + $S_{intrasearch}$.

Inter-frequency cell reselection serves two purposes. If a second carrier addresses only hotspots, reselection to the first carrier must take place when the UE moves out of second carrier coverage. If the load on one carrier is high, reselection to another carrier should take place to achieve load balancing. Inter-frequency cell reselection implicitly takes care of load balancing between carriers, if the measurement quantity is E_c/N_o. If one carrier is not loaded, its E_c/N_o will be significantly larger than that of a loaded carrier and reselection to the unloaded carrier will take place. Load balancing can be further managed during access using implementation-dependent parameters.

Inter-frequency cell reselection considerations should go hand in hand with inter-frequency handover considerations. Section 4.5.7 discusses inter-frequency handover considerations.

4.4 Access Parameter Recommendations

Access parameters are used for call access, either call origination or termination, and also any time the UE communicates on the RACH. This includes any signaling while in Idle Mode, typically Location Area Update, Routing Area Update, and Cell Update, or any data transfer while in Connected (but not Dedicated) Mode, typically CELL_FACH.

Access takes place on the RACH. The UE sets the RACH preamble initial transmit power according to Equation 4.5.

$$\begin{aligned} Preamble\_Initial\_Power = {} & Primary\ CPICH\ TX\ power - CPICH\_RSCP \\ & + UL\ interference + Constant\ Value \end{aligned} \tag{4.5}$$

The first two terms on the right-hand side represent an estimate of Downlink path loss. Solving in terms of the *Constant Value*, and assuming that Downlink and Uplink path losses are equal, one simply finds that *Constant Value* represents the RACH preamble received E_c/N_o. Therefore, the *Constant Value* IE should be set to the preamble target E_c/N_o.

To allow for occasional decorrelation between Downlink and Uplink, a conservative[3] approach is to set *Constant Value* to the lowest estimated RACH detector, E_c/N_o.

The preamble duration is 4096 chips, corresponding to a processing gain of ~36 dB. When using an energy detector with coherent integration time equal to 4096 chips to detect the preamble at the Node B, the detection threshold that achieves sufficiently low probability of false alarm (~ 1 false alarm per second, or 1/750 = 0.0015%) is equivalent to $E_c/N_o = -25$ dB[4]. Assuming that the Node B employs such a threshold and has dual receive antennas, achieving a probability of detection of 50% or better in an Additive White Gaussian Noise (AWGN) Channel requires the preamble received E_c/N_o to be −26 dB or −19.5 dB for 0 km/hr and 150 km/hr, respectively[5].

Figure 4.11 illustrates the Access procedure. Probes are sent at increasing power if an ACK is not received. After the first set of probes is sent, further sets can be sent to exploit time diversity. The parameters *PwrRampStep, PreambleRetransMax*, and *Mmax* are typically operator-configurable.

There is a power offset between the preamble and the RACH message control part, as shown in Figure 4.12. This offset is necessary because the preamble can be received with more or less power, due to the detection and demodulation mechanism at the Node B.

[3] "Conservative" in this context refers to avoiding excess Uplink interference. The Access procedure (power ramping) compensates for a potential underestimate.

[4] This is an estimate of the E_c/N_o required to detect that a probe was sent (not necessarily to decode the probe successfully).

[5] The UE speed is relevant because it affects the required detection integration interval.

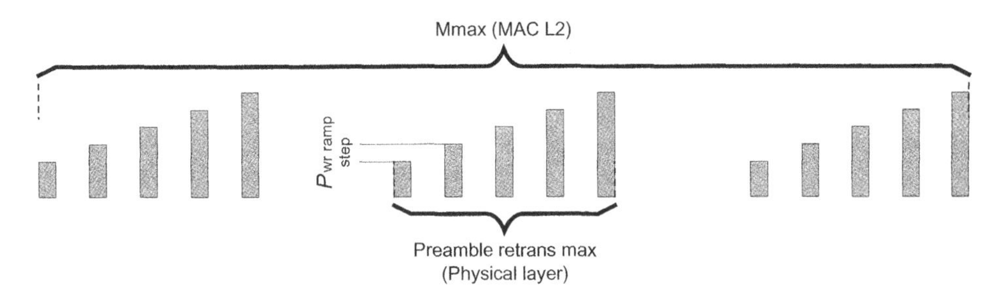

Figure 4.11 Access procedure on RACH

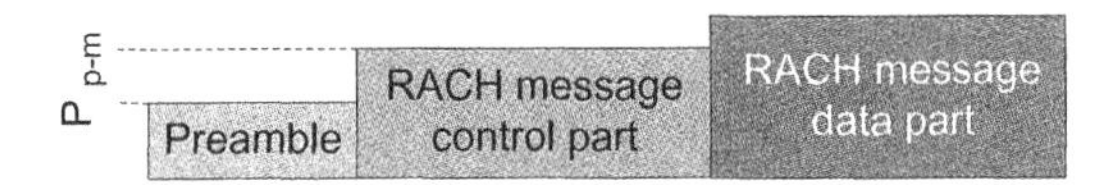

Figure 4.12 Power offset P_{p-m}

The trade-off in Access is latency versus Uplink interference. Because the Uplink channel is not known, probes are sent with increasing power until received by the network.

In unloaded conditions, no more than two probes should be required. In loaded conditions, this number may increase, especially if the Uplink interference value, transmitted in SIB 7, is not accurate. If the network is loaded, the average number of probes close to 1 indicates that initial power settings are too high. If the network is unloaded and the first probe is never successful, clearly the initial power settings are too low. Given typical decorrelation between Uplink and Downlink and a noise rise in the Uplink due to load limited to 6 dB (75% loading), a step size of 2 to 3 dB ensures that Access can be quickly completed and interference kept to a minimum, even if the Uplink interference value is not updated in real time.

In addition to the Layer 1 Access mechanism, a Layer 3 mechanism has been defined to take over if Layer 1 fails. If an RRC Connection Setup message has not been received within T300 after sending a RRC Connection Request message, the Access procedure is repeated up to N300 times.

Table 4.9 lists the main access-related parameters. The Comments column explains the reasoning for the recommended settings.

4.5 Intra-frequency Handover Parameters

4.5.1 Introduction

In UMTS, intra-frequency handovers are assisted by Measurement Reports from the terminals triggered by Active and Monitored Set Channel measurements. Triggering points depend on measurement-reporting parameters. Performance metrics (call quality, capacity, Active Set size, measurement-reporting rate) are determined for different parameter sets, using field data to illustrate the effects of different parameter sets.

In contrast to intra-frequency cell reselection, intra-frequency handover is used only in the CELL_DCH state of Connected Mode.

Table 4.9 Access parameter recommendations

Parameters	Range	Recommended setting	Comments
Constant value	−35 to −10 dB	−26 to −24 dB	This parameter mainly affects the average number of preambles, because it controls the power of the first one [10].
Power ramp step	1–8 dB	2–3 dB	UE transmit power increases between subsequent transmissions on the RACH when no acquisition indicator is received. Larger values cause the final preamble to be at very high power while smaller values increase the number of unsuccessful probes.
Preamble retrans max	1–64	6–10	Maximum number of preamble transmissions in one preamble ramping cycle. Preamble retrans max * power ramp step should allow the UE to reach the maximum allowed transmit power, based on all the other Access parameters.
Mmax	1–32	3–6	Maximum number of preamble ramping cycles. Increasing the number of preamble cycles within an attempt does not significantly affect the overall call setup time. In rapidly changing conditions, a higher setting may be beneficial compared to higher layer retransmission (T300 and N300).
Power offset P_{p-m}	−5 to +10 dB	2–4 dB	Optimization of this parameter mainly depends on cell detection threshold. Increase the value if ACKs are detected, but RRC_connection requests fail.
T300	100 ms to 8 sec in discrete increments	1 sec	Maximum time from transmission of RRC Connection Request message to reception of RRC Connection Setup message. Compensate for low settings with a high number of repetitions (N300) to benefit from time diversity.
N300	0–7	5	Maximum number of retransmissions of the RRC Connection Request message. From a user's point of view, a higher number of repetitions is preferable to longer delays.

4.5.2 Intra-frequency Handover Procedure

Once the UE is in the CELL_DCH state of Connected Mode, the UTRAN sends the UE a Measurement Control Message [10], which contains the parameters controlling the intra-frequency, inter-frequency (see Section 4.5.7), and inter-system (see Chapter 6) handovers. These values update those previously received in SIB11 or SIB12.

For each type of handover, the UE is informed of the measurement objects (cells), measurement quantity (CPICH E_c/N_o, CPICH RSCP, path loss, UTRA Carrier RSSI), reporting quantities (separately for Active Set cells, Monitored Set cells, Detected Set cells), reporting criteria (periodical or event-triggered), and Reporting Mode (acknowledged or unacknowledged). When the reporting criteria are fulfilled, the UE responds by sending a Measurement Report Message to the UTRAN. Upon receipt of these measurements, the RNC may update the Active Set (and send an Active Set Update message) accordingly, depending upon resource availability and other internal considerations.

Measurement reporting is controlled by a set of rules defined in the Layer 3 RRC protocol [14]. For instance, the triggering condition for Event 1a, which indicates that a cell should be added to the Active Set, is as follows.

Event 1a triggering condition is satisfied whenever the quality M of a new cell, that is for example its CPICH E_c/N_o, satisfies the following condition:

$$10\log M_{\text{new}} + CIO_{\text{new}} \geq W10\ \text{Log}\left(\sum_{i=1}^{N_A} M_i\right) + (1-W)10\ \text{Log}\ M_{\text{best}} - \left(R_{1a} - \frac{H_{1a}}{2}\right) \tag{4.6}$$

Where:

- $\boldsymbol{M}_{\text{new}}$ is the measurement result of the cell entering the reporting range.
- $\boldsymbol{CIO}_{\text{new}}$ is the individual cell offset for the cell entering the reporting range if an individual cell offset is stored for that cell. Otherwise it is equal to 0.
- $\boldsymbol{M}_{\text{i}}$ is the measurement result of a cell not forbidden to affect reporting range in the Active Set.
- N_A is the number of cells not forbidden to affect reporting range in the Active Set.
- $\boldsymbol{M}_{\text{best}}$ is the measurement result of the cell not forbidden to affect reporting range in the Active Set with the highest measurement result.
- $\boldsymbol{W}$ is the weighting factor, as further defined in the next section.
- $\boldsymbol{R}_{1a}$ is the reporting range constant, in dB.
- $\boldsymbol{H}_{1a}$ is the hysteresis parameter for Event 1a.

Figure 4.13 shows an example of Event 1a reporting, to illustrate the conditions and involved parameters for a radio link addition. The figure represents a non-zero *Time-to-Trigger* (*TTT*) and a weighting factor $W = 0$. (These are not necessarily recommended parameters but are used to illustrate the process.)

In Figure 4.13, the quality M of Cell 3 (PCPICH3) increases enough to meet the triggering condition for Event 1a. After the condition is observed for a time equal to TTT, the UE starts reporting Event 1a (repeatedly with a time period equal to *Reporting Interval* up to a number of *Reporting Amount* event reports). Presumably, Cell 3 is not added to the Active Set. Eventually, the quality M of Cell 3 no longer meets the triggering condition and the UE stops reporting Event 1a.

Event 1b is the counterpart to Event 1a. If triggered, the RNC is asked to remove that cell from the UE's Active Set.

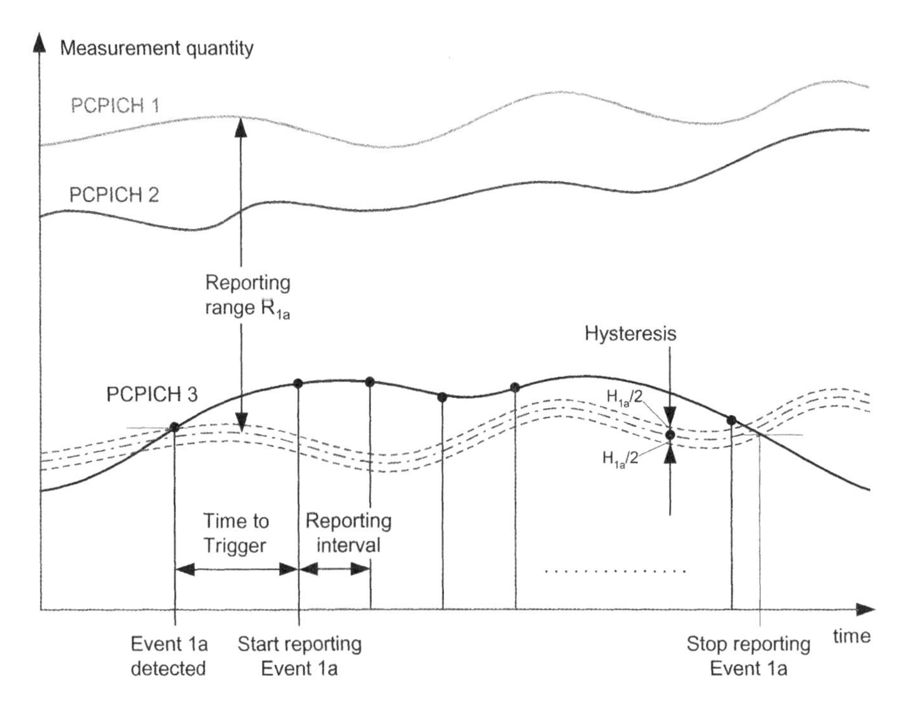

Figure 4.13 Radio link addition (Event 1a)

Event 1c combines the functionality of Event 1a and 1b by replacing an Active Set cell. The triggering condition occurs if:

$$M_{\text{New}} \geq M_{\text{InAS}} + H_{1c}/2 \tag{4.7}$$

Where:

- $\boldsymbol{M}_{\text{New}}$ is the measurement result of the cell not included in the Active Set.
- $\boldsymbol{M}_{\text{InAS}}$ is the measurement result of a cell in the Active Set.
- $\boldsymbol{H}_{1c}$ is the hysteresis parameter for Event 1c.

Figure 4.14 shows an example of Event 1c reporting to illustrate the conditions and involved parameters for a radio link replacement. The figure represents a non-zero *Time-to-Trigger*. (This is not necessarily recommended, but helps to illustrate the process.) PCPICH 1 is replaced by PCPICH 4 in this example.

4.5.3 Intra-frequency Handover Parameters

Table 4.10 lists all parameters provided by the standard for tuning the intra-frequency handover mechanism. This table covers only the most commonly used events for intra-frequency handover:

- **Event 1a.** a server is within a dynamic threshold of the best server(s).
- **Event 1b.** a server is below a dynamic threshold of the best server(s).
- **Event 1c.** a server is within a dynamic threshold of the cells in the Active Set.

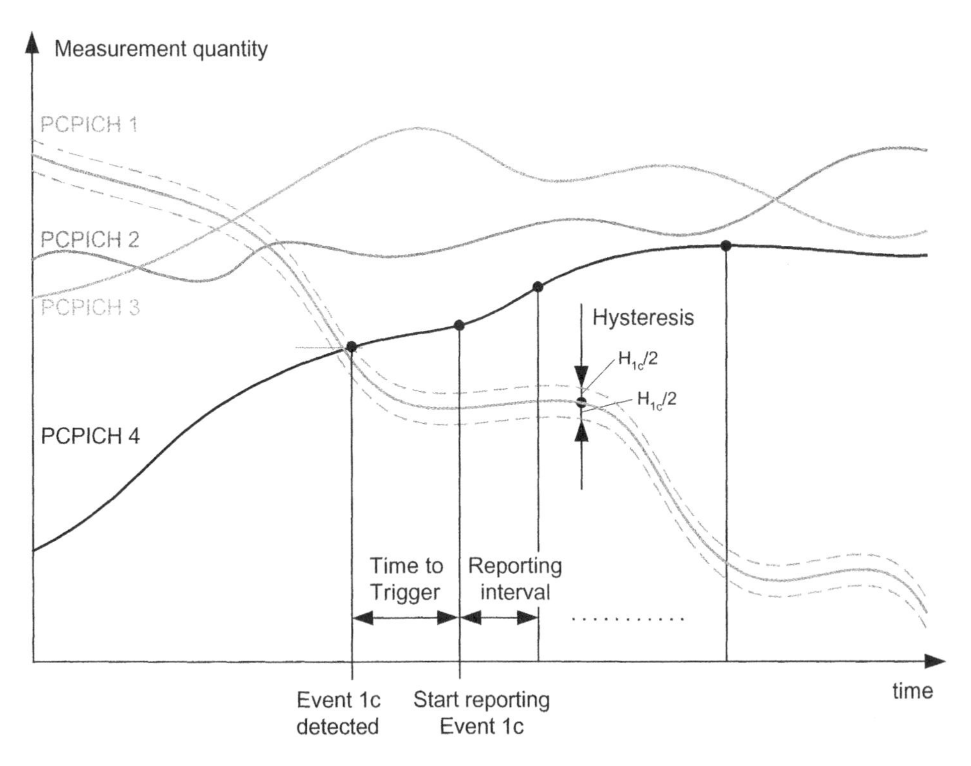

Figure 4.14 Radio link replacement (Event 1c)

4.5.4 Intra-frequency Handover Metrics

In intra-frequency handover optimization, the following four metrics are commonly evaluated:

- **Call quality.** Call quality is covered by evaluating the tail of the CPICH E_c/N_o distribution. An increase indicates a reduced probability of BLER due to sharp drops in signal quality and a reduced probability of reaching the maximum transmit power.
- **Resource utilization (codes, channel elements).** Resource utilization is captured by calculating the mean Active Set size.
- **Signaling (of measurement reports, requires RNC processing power).** RNC processing requirements are addressed by evaluating the mean measurement-reporting rate.
- **Capacity.** To obtain a simple capacity metric, we calculate the total power across different cells required to serve the terminal; then we normalize that total power and take the inverse.

Let us discuss the capacity metric in more detail. It applies only to the Downlink. First, we assume that in soft handover Node Bs are transmitting to a UE with the same power. Equation 4.8 corresponds to UE receive power. If the strongest link is serving the UE, the left term applies. If the UE is being served by an Active Set larger than 1, the right term applies. The receive power must be the same, no matter how many links are serving the UE.

Table 4.10 Intra-frequency handover parameters

Parameters	Range, units	Definitions
Measurement Report Transfer Mode	Acknowledged Mode RLC, Unacknowledged Mode RLC	Indicates the type of Measurement Report Transfer Mode.
Periodical Reporting/Event Trigger	Periodical reporting, Event Trigger	Indicates periodical or Event Trigger.
Measurement quantity	CPICH E_c/N_o, CPICH RSCP, path loss	The quantity the UE shall measure for intra-frequency measurement.
Filter coefficient	0, 1, 2, 3, 4, 5, 6, 7, 8, 9, 11, 13, 15, 17, 19	Equivalent time constant of the low-pass filter applied to the CPICH quality measurements, that is, CPICH E_c/N_o. Filtered quality measurements are used by the UE to evaluate the event triggering condition and are also reported in the *Measured Results* IE of the Measurement Report message.
Reporting Range (Event 1a)	0–14.5 dB in 0.5 dB increments	Constant in the inequality criterion that must be satisfied for an Event 1a to occur.
Hysteresis (Event 1a)	0–7.5 dB, in 0.5 dB increments	Hysteresis between the condition to activate and deactivate Event 1a reporting.
Time-to-Trigger (Event 1a)	0, 10, 20, 40, 60, 80, 100, 120, 160, 200, 240, 320, 640, 1280, 2560, 5000 ms	Period of time for which the triggering condition must be satisfied before transmission of the Measurement Report message can occur.
W (Event 1a)	0–2, in 0.1 increments	Weighting factor applied to the cell quality in the inequality criterion that must be satisfied for an Event 1a to occur.
Report Deactivation Threshold (Event 1a)	0 (= not applicable), 1, 2 – 7 cells in Active Set	Maximum number of cells allowed in Active Set in order for Event 1a to occur.
Reporting Amount (Event 1a)	1, 2, 4, 8, 16, 32, 64, Infinity	Maximum number of Measurement Report messages sent by the UE for periodic reporting triggered by an Event 1a.
Reporting Interval (Event 1a)	0, 250, 500, 1000, 2000, 4000, 8000, 16000 ms	Transmission period of Measurement Report messages sent by the UE for periodic reporting triggered by an Event 1a.
Reporting Range (Event 1b)	0–14.5 dB, in 0.5 dB increments	Constant in the inequality criterion that needs to be satisfied for an Event 1b to occur.
Hysteresis (Event 1b)	0–7.5 dB, in 0.5 dB increments	Hysteresis between the condition to activate and deactivate Event 1b reporting.

Table 4.10 (*continued*)

Parameters	Range, units	Definitions
Time-to-Trigger (Event 1b)	0, 10, 20, 40, 60, 80, 100, 120, 160, 200, 240, 320, 640, 1280, 2560, 5000 ms	Period of time the event triggering condition must be satisfied before transmission of the Measurement Report Message can occur.
W (Event 1b)	0–2, in 0.1 increments	Weighting factor applied to the cell quality in the inequality criterion that needs to be satisfied for an Event 1b to occur.
Reporting Range (Event 1c)	0–14.5 dB, in 0.5 dB increments	Constant in the inequality criterion that needs to be satisfied for an Event 1c to occur.
Hysteresis (Event 1c)	0–7.5 dB, in 0.5 dB increments	Hysteresis between the condition to activate and deactivate Event 1c reporting.
Time-to-Trigger (Event 1c)	0, 10, 20, 40, 60, 80, 100, 120, 160, 200, 240, 320, 640, 1280, 2560, 5000 ms	Period of time the event triggering condition must be satisfied before transmission of the Measurement Report Message can occur.
W (Event 1c)	0–2, in 0.1 increments	Weighting factor applied to the cell quality in the inequality criterion that needs to be satisfied for an Event 1c to occur.
Replacement Deactivation Threshold (Event 1c)	1, 2–7 cells in Active Set	Minimum number of cells allowed in the Active Set in order for Event 1c to occur.
Reporting Amount (Event 1c)	1, 2, 4, 8, 16, 32, 64, Infinity	Maximum number of Measurement Report Messages sent by the UE for periodic reporting triggered by an Event 1c.
Reporting Interval (Event 1c)	0, 250, 500, 1000, 2000, 4000, 8000, 16000 ms	Transmission period of Measurement Report Messages sent by the UE for periodic reporting triggered by an Event 1c.

$$P_{TX}(t) \max_i \frac{1}{L_i(t)} \stackrel{!}{=} \sum_{i \in ASET(t)} \frac{P_{SHO\_TX}(t)}{N(t)} \frac{1}{L_i(t)} \tag{4.8}$$

Where:

- L_i denotes the path loss.
- $P_{TX}(t)$ denotes the transmit power from a single Node B.
- $P_{SHO\_TX}(t)$ denotes the transmit power from each Node B of the Active Set.
- $N(t)$ denotes the Active Set size.

According to Equation 4.8, the least power is required if the Active Set size is small, but it also implies that cells can be instantly swapped as soon as another cell becomes stronger. The equation is an approximation, because swapping, adding, and deleting cells from the Active Set requires approximately 300 ms in existing UMTS networks. If the delayed Active Set change is taken into account, an Active Set size larger than 1 has the advantage of providing macro-diversity gain, because signals are received from different sites, but Tx power is wasted if the links have different path loss.

Equation 4.8 captures the relative capacity loss due to links of different path loss. Because it is very tedious to determine the capacity gain resulting from macro-diversity (this depends on time and on frequency diversity, which in turn depends on the channel and on the coding and interleaving schemes), the equation takes into account only the capacity loss caused by links of different strength. This is accurate for line-of-sight channels and/or stationary conditions but, of course, is only an approximation for non–line-of-sight channels in mobile conditions.

The relative loss in capacity at a given point in time, $Q(t)$, is defined as the ratio of the required transmit power for the best link to the required transmit power across all soft handover links. It can be easily computed if the Pilot quality of each link is available, as shown in Equation 4.9:

$$Q(t) = \frac{1}{N(t)} \frac{\sum_{i \in ASET(t)} CPICH_i \frac{E_c}{I_o}(t)}{\max_i CPICH_i \frac{E_c}{I_o}(t)}$$

$$\text{With } Q(t) = \frac{P_{TX}(t)}{P_{SHO\_TX}(t)} \text{ and } CPICH_i \frac{E_c}{I_o}(t) \sim \frac{1}{L_i(t)} \tag{4.9}$$

To obtain the relative loss in capacity over a drive route, integration over $Q(t)$ is necessary.

4.5.5 Intra-frequency Handover Trade-offs

In early deployments, the goal is to maximize call quality and minimize signaling; that is, to reduce measurement reporting at the expense of capacity and resource usage. Signaling requires RNC processing power, which may be limited. If several Node Bs are connected to one RNC, the problem is aggravated. Call quality may suffer from insufficient coverage, especially indoors, and from large delays between sending Measurement Reports and receiving Active Set Update messages. As deployments become more mature, load increases and capacity becomes more important.

This section quantifies those trade-offs, allowing the selection of the best solution in view of the optimization goals.

4.5.5.1 Field Measurements

To illustrate the impacts and trade-offs for different intra-frequency measurement-reporting parameter sets, we imported channel measurements from a typical route into a simulation tool, and then calculated a set of performance metrics for different parameter sets.

The selected route presents different conditions, all characteristic of vehicular traffic. Some sections of the route are relatively well optimized, while others are subject to Pilot pollution and overshooting cells. We drove the route four times in different traffic

Table 4.11 Selected handover parameter sets

Set	W	R (1a)	R(1b)	H(1c)	TTT (1a)	TTT (1b)	TTT (1c)
1	1	5.0	7.0	3	0	0.1	0.1
2	1	4.0	6.0	3	0	0.1	0.1
3	0	4.0	6.0	3	0	0.1	0.1
4	0.5	4.0	6.0	3	0	0.1	0.1
5	0	3.0	5.0	3	0	0.1	0.1
6	0	2.5	4.0	2	0.1	0.64	0.1
7	0.5	2.5	4.0	2	0.1	0.64	0.1
8	0	3.0	4.5	2	0.1	0.64	0.1
9	0	2.5	4.0	2	0	0.64	0.1
10	0	2.5	4.0	1	0.1	0.64	0.1
11	0	2.5	4.0	2	0.1	0.32	0.1
12	0	3.0	4.5	3	0.0	0.64	0.1

conditions and at differing speeds. After evaluating the complete route, parameter settings that work well when averaged over all conditions can be determined. However, the RF environment affects the trade-offs (e.g., quality, signaling, capacity).

4.5.5.2 Parameter Sets

Handover parameters that are not listed in Table 4.11 remain constant.

The following additional parameters were used for the simulation:

- Filter Coefficient K: 3
- Deactivation Threshold: 2
- Replacement Activation Threshold: 3
- Reporting Amount (Events 1a and 1c): ∞ (infinity)
- Reporting Interval (Events 1a and 1c): 1 sec

4.5.5.3 Call Quality versus Capacity

Figure 4.15 illustrates the trade-off between call quality and air-interface capacity for the parameter sets listed in Table 4.11. The call quality is approximated by the 5th percentile of the CPICH E_c/N_o distribution, calculated by combining the cells in the Active Set (ASET).

For the same TTT 1b value, there is a straightforward trade-off between both metrics: if TTT 1b is increased, the capacity is negatively affected, given the same call quality. Weak cells remain in the Active Set longer, contribute less to the overall received channel quality, and interfere with other users.

Sets 1 and 3, and sets 2 and 5 (shown in the circles) perform similarly. In both cases, the parameter W has been changed from 1 to 0 and the reporting ranges have been adapted: a larger reporting range is used for $W = 1$ than for $W = 0$.

4.5.5.4 Mean Measurement-Reporting Message Rate versus Mean Active Set size

Figure 4.16 shows the relationship between the average rate of Measurement Report Messages (MRM) and the average size of the Active Set in the UE for the parameter sets in Table 4.11. A low value for both variables is preferred.

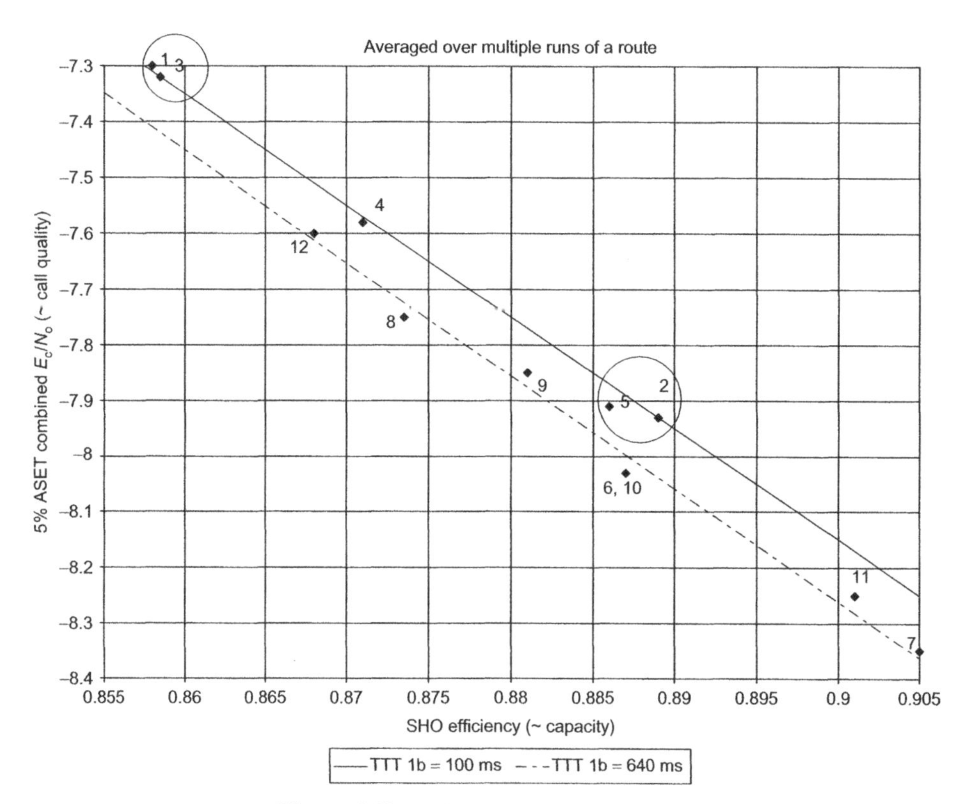

Figure 4.15 Call quality versus capacity

The metrics for parameter sets 1 and 3, and sets 2 and 5 (circled) differ significantly in Figure 4.16, while in Figure 4.15 they show very close performance. The MRM rate is considerably higher for sets 1 and 2, which are characterized by $W = 1$, with a benefit of a slightly reduced mean Active Set size.

An increased TTT 1b provides a significantly reduced measurement-reporting rate. This is relevant if the processing power of the RNC is limited.

It may seem surprising that $W = 1$ leads to much more measurement reporting than $W = 0$. The reason is that when $W = 1$, the sum over all Active Set cells is compared to the strongest (Event 1a) neighbor cell (refer to Equation 4.6), or the weakest (Event 1b) Active Set cell. Therefore, a cell may move in and out of the Active Set even if its quality or strength fluctuates less than the difference between Events 1a and 1b reporting ranges. This is not the case for $W = 0$, which compares the best Active Set cell to the strongest neighbor cell or the weakest Active Set cell.

4.5.6 Intra-frequency Handover Parameter Recommendations

In light of the simulation results shown in the previous two figures, the recommendations in Table 4.12 follow the values defined as parameter set 12 in Table 4.11.

4.5.7 Inter-frequency Handover Considerations

Three mechanisms control a transition from one carrier to another:

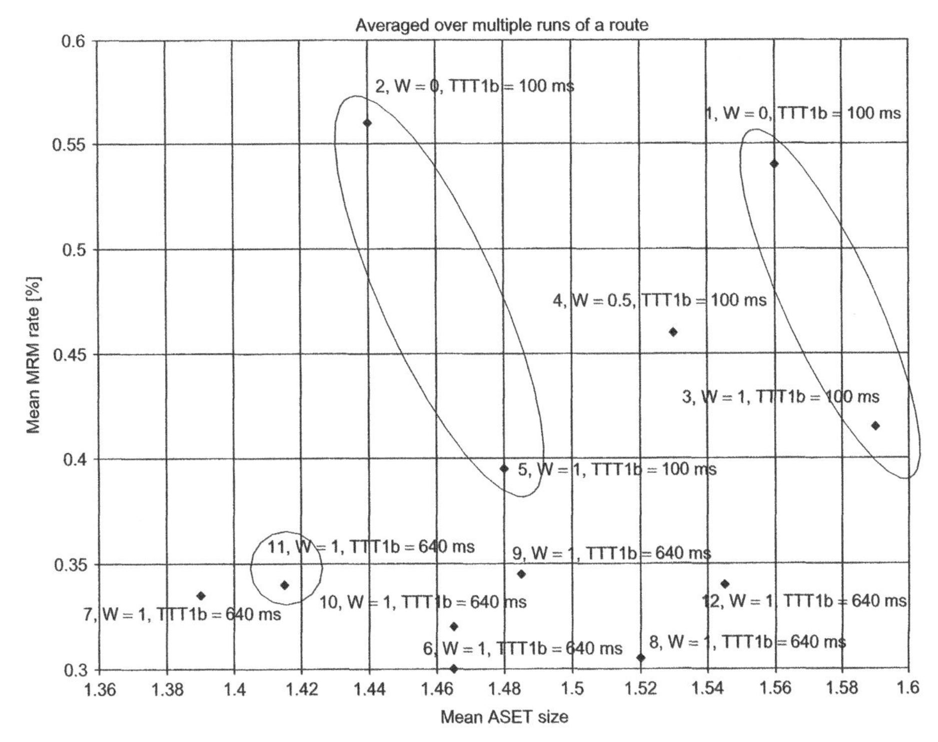

Figure 4.16 Mean Measurement Report Message (MRM) rate versus mean Active Set (ASET) size

- Inter-frequency cell reselection
- Redirected RRC setup during call setup: call setup takes place on another carrier
- Inter-frequency handover

The first mechanism, inter-frequency cell reselection, was covered in Section 4.3.8. It takes care of load balancing and loss of coverage on the current carrier.

The second mechanism, redirected RRC setup during call setup, is not equivalent to inter-frequency handover because the call is set up on another carrier as opposed to being handed over. To achieve load balancing, it may not be sufficient to rely solely on inter-frequency cell reselection. For instance, the UE is unaware of the resource utilization. If interference conditions are still acceptable but most of the codes are reserved, an inter-frequency cell reselection would not take place. Therefore, during RRC Connection Setup, the infrastructure can reply with RRC Connection Reject and propose another carrier.

The third mechanism is inter-frequency handover. This occurs if a UE leaves coverage of that carrier and if it is a real-time service.

If it is not a real-time service, it may not be necessary to support inter-frequency handover in Connected Mode. Instead, it may be more straightforward to allow the call to drop, in which case cell reselection would move the UE to the stronger carrier and the call would reestablish. A specific example would be a situation that offers complete UMTS Rel-99 coverage on the original carrier and limited HSDPA coverage on a second

Table 4.12 Intra-frequency handover parameters

Parameters	Recommendations	Comments
Measurement Report Transfer Mode	Acknowledged mode RLC	Acknowledged mode measurement reporting ensures quick retransmission of erroneously transmitted or lost measurement reports, enhancing reliability at the expense of more signaling and processing at UE and RNC.
Periodical reporting/Event Trigger	Event Trigger	In a loaded network, periodical reporting generates a high load of measurement reports on the Uplink, thereby consuming processing power in the UTRAN. Unless these measurement reports are used to optimize the system, the Event Trigger Reporting Mode is preferred, because reports are sent only after the UE observed the triggering condition.
Measurement quantity	CPICH E_c/N_o	According to [17], the noise and interference term N_o of the CPICH E_c/N_o measurement quantity is the same for different cells at the same location. Therefore, intra-frequency handover performs the same for either CPICH E_c/N_o or CPICH RSCP. Because Pilot power settings are sometimes used to change cell sizes, for example, for load balancing, path loss is not recommended as a measurement quantity.
Filter coefficient	2 or 3	The time constants of coefficients 2 and 3 correspond to 289 ms and 458 ms, respectively. Performance at higher velocities degrades less if the filter coefficients are lower. If the RNC can process frequent measurement reports and the delay between sending a Measurement Report Message and receiving an Active Set update is about 300 ms, a filtering coefficient of 2 is recommended. Otherwise a coefficient of 3 is recommend.
Reporting range (Event 1a)	3 dB	If the parameter value is set too large, cells of relatively poor quality may be included unnecessarily in the Active Set, thereby degrading Downlink capacity. If the parameter value is set too small, cells of relatively good quality may not be included in the Active Set, leading to poor call quality if (Uplink) power maxes out (Pilot pollution, cell edge), or if Active Set updates are delayed because of RNC issues. Also, capacity may be wasted by not tapping into macro-diversity gains.

Table 4.12 (*continued*)

Parameters	Recommendations	Comments
Hysteresis (Event 1a)	0 dB	Can be set in combination with reporting range. However, the only advantage is a very minor reduction in measurement reporting.
Time-to-Trigger (Event 1a)	0 ms	If necessary, measurement reports shall be delayed by using a larger filtering coefficient as opposed to increasing the TTT. This ensures that cells with quickly increasing power are taken into account quickly, not after a fixed time.
		A small reduction in mean Active Set size can be achieved at the expense of a significant increase in measurement reporting if *W* is increased from 0 to 1 and Events 1a and 1b reporting ranges are adapted to maintain capacity and call quality. Given the limited processing power of current RNCs, this trade-off is frequently viewed as unfavorable.
W (Event 1a)	0	For a larger *W* and increased Events 1a and 1b ranges, the second cell is added more quickly but the third, less quickly. In Pilot-polluted areas, the strength of the third cell may rise quickly but it does not become part of the Active Set until the RNC has processed the Measurement Report Message and issued an Active Set update, thus reducing call quality.
		On the other hand, a larger *W* may lead to capacity increases in scenarios with large macro-diversity gains (e.g., no LOS, no time diversity = low velocity, low BLER targets) because the second cell is added more quickly.
Report deactivation threshold (Event 1a)	>=3	If the parameter value is set too small, a cell of relatively good quality may not be included in the Active Set. Including that cell would have been otherwise possible (and desirable) given the selected Event 1a triggering conditions. The recommended value may be further limited by the maximum Active Set size of the UTRAN.
Reporting amount (Event 1a)	∞(Infinity)	If the parameter value is set too small, the cell addition procedure may unnecessarily fail, for example because of lack of network resources that is only temporary but extends beyond the reporting period.

(*continued overleaf*)

Table 4.12 (*continued*)

Parameters	Recommendations	Comments
Reporting interval (Event 1a)	500 ms or 1000 ms	The interval should exceed the average time between sending a measurement report and receiving an Active Set update. If the interval is too large, an addition may be unnecessarily delayed.
Reporting range (Event 1b)	4.5 dB	Must be set larger than the reporting range (Event 1a). Cells can be retained in the Active Set by choosing a larger reporting range (Event 1b) and/or a large TTT (Event 1b). A large TTT has the advantage of keeping the cells in the Active Set if the fluctuation rate of the signal increases because of higher velocities.
Hysteresis (Event 1b)	0 dB	Can be set in combination with reporting range. However, the only advantage is a very minor reduction in measurement reporting.
Time-to-Trigger (Event 1b)	640 ms	Measurement reporting can be reduced significantly by increasing the TTT of Event 1b. In quickly fluctuating environments (Rician or Rayleigh fading at medium or large velocities), it is detrimental to quickly drop a cell (due to a low TTT for Event 1b). The link might suffer until the RNC includes the cell after an Event 1a is reported. The reduction in measurement reporting comes at the expense of a small loss in capacity as weak cells remain in the Active Set. If calls drop because cells are not being added quickly enough in fast fading scenarios, this value can be increased.
W (Event 1b)	0	Here, the same argument as for W (Event 1a) applies.
Hysteresis (Event 1c)	3 dB	If the parameter value is set too large, cells of relatively good quality may not trigger an Event 1c and will not replace cells of relatively poor quality, thus degrading Downlink capacity. If the parameter value is set too small, cells of only marginally better quality than that of an Active Set cell may trigger an Event 1c, increasing signaling load without appreciably improving the combined Active Set cells' quality.

Table 4.12 *(continued)*

Parameters	Recommendations	Comments
Time-to-Trigger (Event 1c)	0 ms	If necessary, measurement reports shall be delayed by using a larger filtering coefficient as opposed to increasing the TTT. This ensures that cells with quickly increasing power are taken into account more quickly, not after a fixed time.
Replacement deactivation threshold (Event 1c)	1+Report deactivation threshold	If the parameter value is set larger than 1+Report Deactivation Threshold IE, Event 1c is prevented from ever occurring. If the parameter value is set smaller than 1+Report Deactivation Threshold IE, a cell of relatively low quality, which is not deemed strong enough for Event 1a triggering and inclusion in the Active Set, may be included in the Active Set following Event 1c triggering.
Reporting amount (Event 1c)	∞(Infinity)	If the parameter value is set too small, the cell replacement procedure may unnecessarily fail, for example because of temporary lack of processing capability or in the event of signaling loss.
Reporting interval (Event 1c)	1000 ms	The interval should exceed the average time between sending a measurement report and receiving an Active Set update. If the interval is too large, an addition is unnecessarily delayed.

carrier. Cells on the second carrier would not have to provide continuous coverage. It would be sufficient to deal with hotspots by deploying isolated cells.

If inter-frequency handover is necessary, Event 2a can be configured to indicate change of best frequency and Event 2b to indicate that the current frequency is below a threshold and an alternative frequency is above a threshold. If the UE has only one RF front-end, Compressed Mode must be activated to measure other carriers. Event 2d can be configured to activate Compressed Mode and Event 2f to deactivate Compressed Mode. Because inter-frequency handover is a hard handover, as opposed to a soft intra-frequency handover, it may affect a real-time application. As a result, it might be preferable to deploy clusters of Node Bs with a second carrier rather than with isolated cells.

References

[1] 3GPP R2-050703; TSG-RAN WG2. Meeting #4. Discussion on cell selection and reselection parameters. Scottsdale, Arizona, USA; Feb. 2004.
[2] TS 25.321. Medium Access Control (MAC) protocol specification. 3GPP; 2005.
[3] TS 25.322. Radio Link Control (RLC) protocol specification. 3GPP; 2004.
[4] TS 25.323. Packet Data Convergence Protocol (PDCP) specification. 3GPP; 2002.

[5] TS 25.324. Broadcast/Multicast Control (BMC). 3GPP; 2004.
[6] TS 24.008. Mobile radio interface Layer 3 specification; core network protocols; stage 3. 3GPP; 2005.
[7] TS 25.413. UTRAN Iu interface Radio Access Network Application Part (RANAP) signaling. 3GPP; 2003.
[8] TS 25.423. UTRAN Iur interface Radio Network Subsystem Application Part (RNSAP) signaling. 3GPP; 2004.
[9] TS 25.433. UTRAN Iub interface Node B Application Part (NBAP) signaling. 3GPP; 2004.
[10] TS 25.331. Radio Resource Control (RRC) protocol specification. 3GPP; 2004.
[11] TS 25.101. UE Radio transmission and reception (FDD). 3GPP; 2004.
[12] TS 25.213. Spreading and modulation (FDD). 3GPP; 2002.
[13] Vanghi V, Sarkar S. Performance of WCDMA downlink access and paging indicators in multipath rayleigh fading channels. *Proceedings of PIMRC 2003. 14th IEEE Proceedings on Volume 1*, Beijing, China, 7–10 Sept. 2003. pp 331–335.
[14] TS 25.304. User Equipment (UE) procedures in Idle Mode and procedures for cell reselection in Connected Mode. 3GPP; 2004.
[15] Sarkar S, Tiedemann E, cdma2000: battery life improvement techniques. *Proceedings of PIMRC 2000. The 11th IEEE International Symposium on Volume 2*, London, UK, 18–21 Sept. 2000. pp 954–958.
[16] Flore D, Brunner C, Grilli F, Vanghi V. Cell reselection parameter optimization in UMTS. *Proceeding of ISWCS 2005. 2nd International Symposium on Wireless Communications Systems 2005*, Sienna, Italy, Sept. 2005.
[17] TS 25.215. Physical layer – measurements (FDD). 3GPP; 2005.

5

Service Optimization

Andrea Forte, Patrick Chan and Christophe Chevallier

5.1 KPI and Layered Optimization Approach

5.1.1 Main KPI Definitions

Key Performance Indicators (KPIs) are a minimum set of metrics for tracking system progress toward a performance target. During network deployment and optimization, several sets of KPIs can be defined. A different set could be used for each optimization phase described in Chapter 2, as summarized in Figure 5.1.

From this optimization process, three categories of KPIs can be distinguished. Each is discussed below.

- **RF KPIs.** Used during initial RF optimization or each time RF conditions change significantly; for example, when a new site is added or a cell is reoriented. RF KPIs are discussed in detail in Chapter 2 and are summarized in Table 5.1.
- **Service KPIs.** Obtained through drive tests and used during initial optimization to estimate system performance. Service KPIs should track all aspects of user experience, from call origination to call release. Table 5.2 summarizes the relevant service KPIs. The rest of this section discusses service KPIs for access failures, poor quality, and call drops. The KPIs in Table 5.2 represent a minimum performance level to support the design target defined in Chapter 2: a network with cell edge coverage probability of 90% (97% coverage area, assuming $\sigma = 8$ dB). For retention and access performance, the probability of success should correspond to the coverage area. For example, the call success rate target should be 97%. Expecting higher success would be equivalent to expecting service in an area where coverage is not guaranteed. Results depend on how the tests are performed. The case discussed here assumes that the tests are run over the entire predicted coverage area.
- **Operational KPIs.** Used once sufficient commercial traffic is carried by the network, when the actual user experience can be evaluated and network growth can be

WCDMA (UMTS) Deployment Handbook: Planning and Optimization Aspects QUALCOMM Incorporated

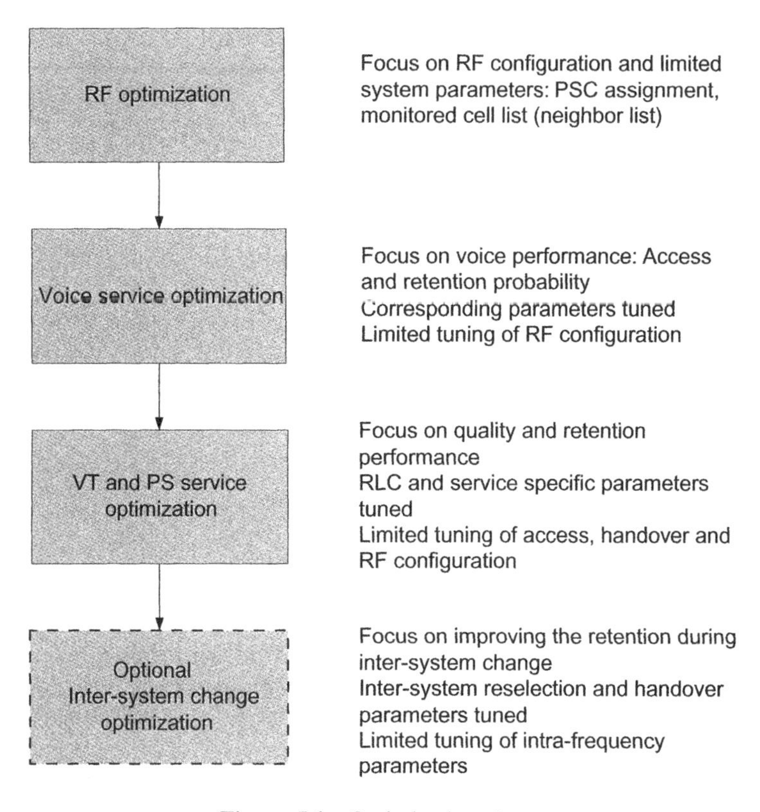

Figure 5.1 Optimization phases

Table 5.1 RF KPI examples

Task	KPI	KPI target example
RF Optimization	Measured RSCP	> −88 dBm over 97% of area, measured outdoors
	Measured E_c/N_o	> −9 dB over 95% of area
	Cell overlay	<3 cells within the reporting range, Event 1a (r1a) over 95% of area
	Qualitative distribution	No cell overshoot Minimal cell fragmentation Minimal change in best server Minimal differences between prediction and measurements

planned. Operational KPIs should be measured and trended over an extended period of time. They should be measured with network counters or with a protocol analyzer attached to a network interface (usually Iub or Iu). Network counters are generally easier to use than protocol analyzers, but they are limited by each vendor's implementation.

Table 5.2 Service KPI examples

KPI	Definition	Target value
Set up success rate MO	Access failure: call fails to reach alerting state Call setup success ratio = (number of calls reaching alerting state)/(number of calls initiated by user)	>97%
Set up success rate MT	Access failure: call fails to reach alerting state (B party) after initiation (A party) Call setup success ratio = (number of calls reaching alerting state)/(number of calls initiated)	>95%
Drop rate	Call drop: unintended release of the radio link after reaching alerting state Call drop ratio = dropped calls/(number of initiated calls – access failure)	<3%
Set up latency MO	Delay between user-originated call and alerting state	<6 sec for 90% of the calls
Set up latency MT	Delay between first page and reaching alerting state (terminated side)	<8 sec for 90% of the calls

When evaluating the KPIs mentioned above, it should be understood that confidence levels in the measured performance greatly depends on the number of samples collected. The number of samples is less important during the initial round of optimization, which focuses on discovering and resolving the main issues in the network rather than on the statistical validity of the measurements. In later rounds of optimization, as the network approaches commercial operation, the number of samples is important to verify that the target performance is met and to determine whether the measured performance, typically over a limited drive route, is an accurate statistical representation of the performance of the entire network. Determining the number of samples needed to statistically validate the measurements involves statistical theory [1]. Without reviewing the complete mathematical justification, the number of samples to satisfy a required confidence level can be estimated from Equation 5.1.

$$n = \max\left[z_{\alpha/2}^{2} \times \frac{p(1-p)}{e^{2}}, \frac{5}{(1-p)}\right] \tag{5.1}$$

where

n is the number of samples, estimated from the following:

- $z_{\alpha/2}$ is the standard normal distribution, typically 1.96 for 95% confidence level or 2.576 for 99% confidence level
- p is the expected probability of success
- e is the acceptable error in the measurement

The second term considered for the maxima ensures that a sufficient number of samples are reached, allowing the necessary simplification for the first term to be valid.

5.2 Voice Service Optimization

5.2.1 Adaptive Multirate Codec

An Adaptive Multirate (AMR) codec [2] is the main speech codec for a number of services, including voice telephony, Video-Telephony (VT), and Multimedia Messaging Service (MMS). Its popularity is due to its performance-to-data-rate trade-off and its compatibility with other popular speech codecs such as Global System for Mobile (GSM) Enhanced Full Rate (EFR) codec (AMR 12.2 kbps mode), North American Time Division Multiple Access (TDMA) IS-641 EFR codec (AMR 7.4 kbps mode), and ARIB Personal Digital Cellular (PDC) EFR (AMR 6.7 kbps mode).

AMR can operate in eight different modes, supporting the variable bit rates shown in Table 5.3. Choosing among AMR modes involves a compromise between voice quality and network capacity. Higher system capacity can be achieved by lowering the data rate of each voice user. Similarly, if there is sufficient network capacity, each user may use a higher data rate AMR mode for better quality, providing flexible resource management.

In an ideal implementation, network operators should be able to control the balance between capacity and quality under different situations, but typically only the 12.2 kbps option is implemented. Even when multirate is implemented, a call would be set up in one mode and remain unchanged for the duration of the call. An exception to this is Tandem Free Operation mode, in which a Wideband Code Division Multiple Access (WCDMA) mobile and a GSM mobile are engaged in a call. In this scenario, supported by the standard in Release 4 and higher, the AMR mode adapts to the GSM link quality and can change upon command as frequently as every 20 ms speech frame [2–4].

The details of AMR coding are discussed in Refs [5–7]. At a high-level, three streams of bits are generated with different error-protection requirements: Class A, Class B, and Class C. The Class A bits, necessary to attempt frame decoding, are the most important and so are convolutionally coded at rate 1/3 and are also Cyclic Redundancy Check (CRC) protected. The Class B bits do not have a CRC, but are convolutionally coded at rate 1/3. Class C bits, the least important, do not carry a CRC, and are coded at rate 1/2.

If the CRC for the Class A bits does not check, the entire bad frame is discarded. An error concealment mechanism such as voice muting should then be implemented to

Table 5.3 List of AMR codec modes

AMR codec mode	Source codec bit-rate
AMR_12.20	12.20 kbps (GSM EFR)
AMR_10.20	10.20 kbps
AMR_7.95	7.95 kbps
AMR_7.40	7.40 kbps (IS-641)
AMR_6.70	6.70 kbps (PDC-EFR)
AMR_5.90	5.90 kbps
AMR_5.15	5.15 kbps
AMR_4.75	4.75 kbps
AMR_SID	1.80 kbps

minimize perceived poor quality. The perceived voice quality depends on the Block Error Rate (BLER) and the implemented error concealment mechanism.

5.2.2 AMR Service

Call failure events for voice services can be classified into two categories: call access failures and call drops. Call access failure occurs when a call cannot be successfully established. A call drop is the abnormal disconnection of an established call. During network optimization, both categories should be measured independently because call access failures are counted in call delivery KPIs while dropped calls are counted in call retention KPIs.

Some qualities of access failures and dropped calls are specific to voice service, while others are generic and apply to any service. For simplification, any failure associated with the Signaling Radio Bearer (SRB) can be considered generic, while failures associated with the Radio Bearer (RB) or Radio Access Bearer (RAB) should be considered specific. A Radio Resource Control (RRC) connection is required before the RB carries the voice payload, as described in Chapter 1.

Compared to other services such as VT, AMR payload encompasses three distinct subflows, each of which is mapped to a logical Dedicated Traffic Channel (DTCH), as shown in Figure 5.2.

Each DTCH is mapped to a Dedicated Channel (DCH) with Transmission Time Interval (TTI) of 20 ms, sent in Radio Link Control Transparent Mode (RLC TM). RLC TM is used for the three AMR DTCHs because retransmission of lost or corrupted AMR frames would not be compatible with the low latency required for voice services, where latency over 200 ms is perceptible.

At the Physical Layer, the transport blocks from each transport channel, user payload, and SRB are multiplexed and mapped onto the Coded Composite Transport Channel (CCTrCh) to allow the transmission of all channels over a single coded channel.

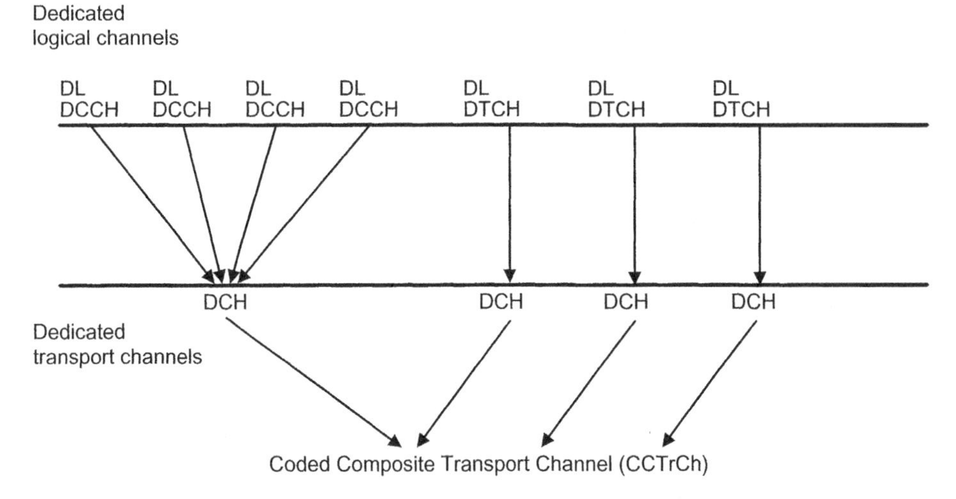

Figure 5.2 AMR mapping of dedicated logical channels onto dedicated transport channels

Because call access failures and dropped calls are counted differently, it is important to establish a unique delineation in the signaling flow to distinguish the call setup stage from the call established stage (conversation stage). Two messages can be used to make this distinction:

- **Call Control (CC): Connect ACK.** End-to-end connection is fully established. This message would accurately depict the user's perception of access performance, but it relies on a user action of accepting the call. Because user action is required, which may affect the call setup latency, this message is not preferred.
- **Call Control (CC): Alerting.** End-to-end connection is almost established and a ringing tone signal is sent to the terminating phone. Because no user action is required, this is the recommended Layer 3 message for distinguishing between the two types of failures. However, it may not be valid when an auto-answer number is called.

Now that we understand the differences between these two types of messages, we can explore how access failure and dropped calls relate to signaling. The troubleshooting section (Section 5.2.6) discusses how these types of failures are affected by RF conditions.

5.2.3 Call Setup, Events, and Signaling

There are two main types of calls:

- **Mobile-Originated (MO) call.** User Equipment (UE) initiates a call to another party.
- **Mobile-Terminated (MT) call.** UE is called by another party.

On the basis of AMR call setup, the Layer 3 signaling flow varies for each call type. Figure 5.3 shows the Mobile Originated (MO) call flow and Figure 5.4 shows the Mobile Terminated (MT) call flow.

Table 5.4 compares both types of calls. The main difference is how the call is detected in a call flow, or in a log collected with a test UE. Test UE used for MO calls should indicate when a user initiates a call; otherwise, the optimization engineer would have to rely on random access. MT calls are initially detected by receipt of a paging message; however, detecting a paging message can be challenging, as we will see later.

While call detection differs for each call type, they use similar messaging to establish and set up calls, even though the source and target are different.

The following sections discuss each step of the MO and MT call flows: system access, RRC Connection Setup, Core Network (CN) negotiation, RB setup, and end-to-end connection.

5.2.3.1 System Access

System access is achieved by means of the Random Access Channel (RACH) procedures [8], during which RACH preambles are sent by the UE and acknowledged on the Downlink Acquisition Indicator Channel (DL AICH) before the RACH message can be sent on the Uplink Physical Random Access Channel (UL PRACH), as illustrated in Figure 5.5.

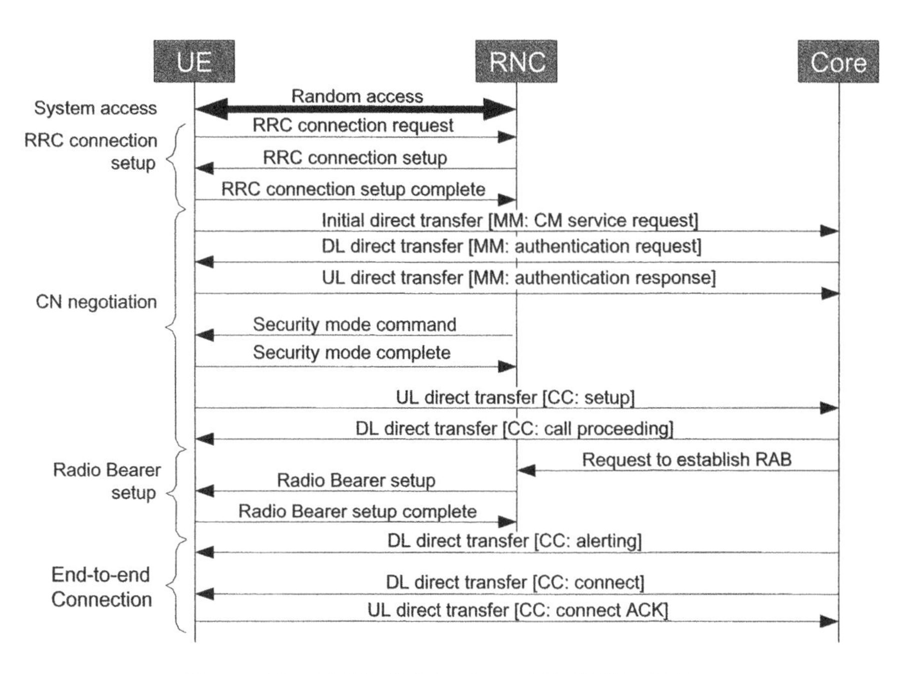

Figure 5.3 Mobile Originating (MO) AMR call flow

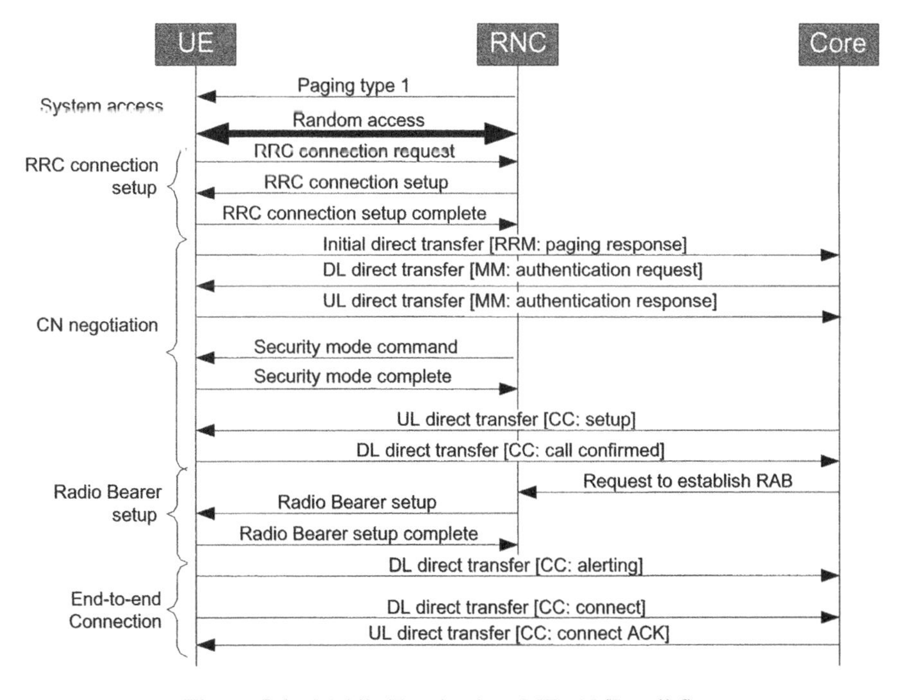

Figure 5.4 Mobile Terminating (MT) AMR call flow

Table 5.4 Layer 3 signaling comparison of MO and MT AMR call types

Messages	MO	MT
Paging type1	NA	UE ← RNC
RRM: Paging response	NA	UE → CN
CM Service request	UE → CN	NA
CC: Setup	UE → CN	UE ← CN
CC: Call proceeding	UE ← CN	NA
CC: Call confirmed	NA	UE → CN
CC: Alerting	UE ← CN	UE → CN
CC: Connect	UE ← CN	UE → CN
CC: Connect ACK	UE → CN	UE ← CN

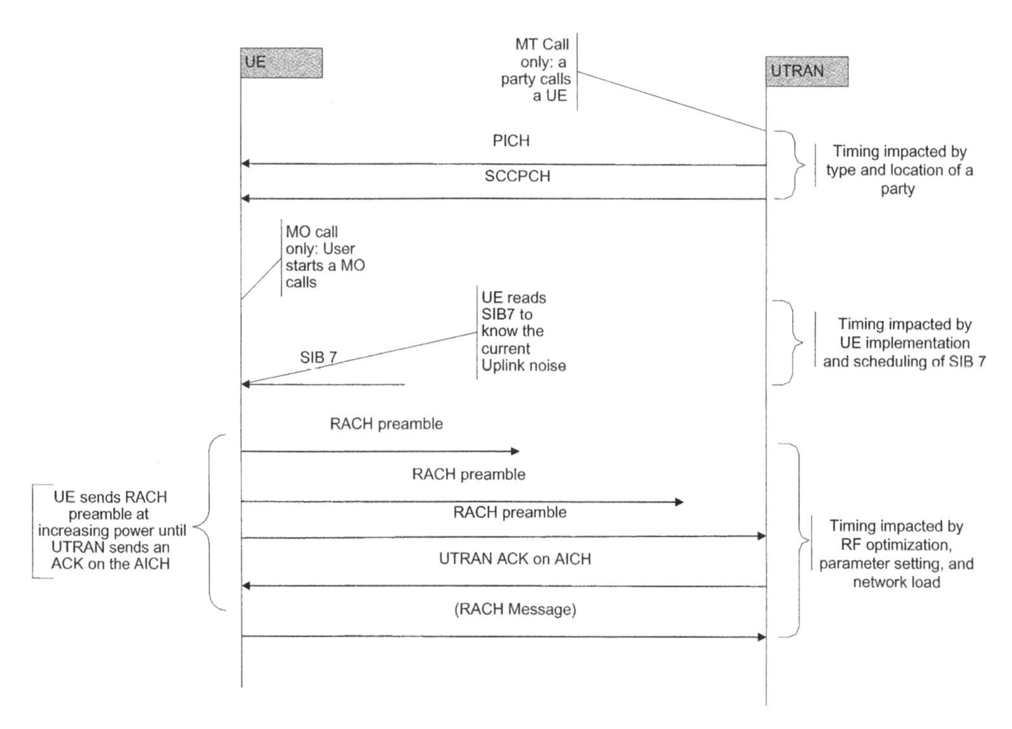

Figure 5.5 MO and MT call access process

For an MT call, the RACH attempt starts after the incoming paging indication on the Paging Indicator Channel (PICH) and the paging information is received on the Paging Channel (PCH, transmitted over the Secondary Common Control Physical Channel (SCCPCH)). Paging Type 1, or general paging, is the Layer 3 message that appears in most MT call logs. Paging Type 2, or dedicated paging, is used if the UE is already in Connected Mode. It sets up a concurrent call when the UE is in the CELL_DCH or CELL_FACH state. Because Paging Type 2 is sent on Dedicated Control Channel (DCCH), only Paging Type 1 is discussed here.

Paging Type 1 identifies the start of a call setup procedure for an MT call, assuming that only a UE log is available and that the paging message was captured by the logging device. For a thorough analysis, the standard paging cause for AMR MT should be "Terminating Conversational Call."

Successful reception of the paging message is complicated by the fact that the UE operates in discontinuous reception (DRX) mode. In DRX mode, the network must send the page indication and page only in the slot in which the UE is awake. This process conserves the battery life of the UE, but it requires that the UE quickly and accurately set up its receiver to demodulate the PICH, and then the SCCPCH. During this stage, the UE is camped on a single cell, as defined in the cell reselection process [9]. Proper RF optimization is necessary to ensure sufficient E_b/N_t for the detection of PICH and SCCPCH at all times. This is achieved by using proper cell reselection parameters, as discussed in Chapter 4, and by setting up clean boundaries between cells to support fast and accurate cell reselection for the UE, which would allow the UE to camp on the strongest cell.

A more robust way to determine the start of an MT call is a network log. For a mobile-to-mobile call, the log of the calling UE shows a Non-Access Stratum (NAS) setup message as soon as the MSC starts to connect the MT UE.

For MO calls, detection of the paging message is not an issue because it does not indicate a call. The only way to use messaging to detect a call attempt is to look for a RACH attempt. This method is not ideal because a RACH sequence could be present in the messaging for other reasons, such as Location Area Update (LAU) or Routing Area Update (RAU). To identify an MO call, the RACH message must contain the RRC connection request message with an establishment cause of "Originating Conversational Call."

For MO or MT calls, the RACH process can start only after the UE has decoded the proper parameters to ensure that the access preamble will be at the proper power. One of these parameters, UL interference, is set dynamically by the network on the basis of the Uplink load. This message is sent over System Information Block (SIB) 7. Therefore, decoding SIB 7 is a good starting point to detect call access if a UE internal message is not available. The drawback of decoding SIB 7 before making any attempts is that scheduling this SIB can delay call setup. According to the standard [10], SIB 7 can be scheduled within a period of 320 ms (minimum) up to several seconds. Since the UE must read SIB 7 before starting the RACH process, any increase in SIB 7 scheduling prolongs call setup. Other parameters of interest in the RACH process affect the open loop power control. Section 5.2.6 discusses these parameters.

5.2.3.2 RRC Connection Setup

The RRC Connection Setup establishes SRBs to carry dedicated signaling. The RRC connection request contains the UE identity, optional cell measurement results, and the establishment cause. For call access, or speech AMR service, this cause is recorded as either "Originating Conversational Call" or "Terminating Conversational Call" [10]. The "Terminating Conversational Call" cause could lead to false event reporting if it is not properly verified in call analysis, because the UE also performs Periodic Registration/LAU/RAU as part of the Mobility Management, which must be ignored when calculating call setup statistics.

Successfully establishing the RRC connection is the most challenging part of call setup. This can be attributed to two factors: the admission control implementation and the size of the RRC Connection Setup message. The latter is the main challenge. During admission control implementation, an RRC connection reject is sent if no resources are available for allocation, or if the call should be redirected to a different system or carrier. After successful resource allocation, the RRC Connection Setup message–which contains SRB information including the mapping details of dedicated logical, transport, and physical channels–is sent on the Forward Access Channel (FACH) (over [Downlink Secondary Common Control Physical Channel] [DL SCCPCH]). This RRC Connection Setup message contains a significant amount of information, and it spans multiple frames while not yet operating in closed loop power-controlled condition. This makes it difficult for the UE to receive the message, especially if the SCCPCH power allocation is not set to accommodate low geometry.

After the RRC Connection Setup message is received, the UE can set up the low data rate DCH according to the RRC Connection Setup message. First, only the PDCCH-containing Transmit Power Control (TPC) and Pilot bits are sent to allow the inner loop power control to converge. Afterwards, the RRC Connection Setup Complete message is used to acknowledge the setup message and send UE-capability information to the network. At this point, the UE should have transitioned from Idle state to CELL_DCH state.

At this time, the connection is power-controlled and may support handover, depending on the Universal Terrestrial Radio Access Network (UTRAN) implementation. Both features improve the reliability of the connection.

5.2.3.3 Core Network Negotiation

The low data rate DCH facilitates upper layer signaling with the NAS layer, which performs all authentication and security procedures along with additional processes in the CN to establish the end-to-end connection.

To initiate the connection request to the upper layers, MO calls use the Connection Management (CM) Service Request and MT calls use the paging response, as shown in Table 5.4. In the CM Service Request message, the CM Service Type field indicates "Mobile originating call establishment" as the cause for a MO AMR call [11]. The Paging Response message does not need to carry service information because the NAS layer already knows what service is being set up for the MT call.

To perform two-way authentication, the UE checks the Authentication Token (AUTN) and the network checks the Signed Authentication Response (SRES), which is calculated from the RAND number from the network. Depending on the supported security capabilities, ciphering and integrity protection are switched on to enable encryption of user data and signaling messages.

For MO calls, the UE sends main parameters such as bearer capabilities and dialed digits to the network using the Call Control (CC) Setup message. For MT calls, the network uses the Setup message to send the same, or similar, information to the UE. The next step is similar; MO calls use Call Proceeding messages (UTRAN to UE), while MT calls use Call Confirmed messages (UE to UTRAN) as a Layer 3 acknowledgment.

5.2.3.4 Radio Bearer Setup and Reconfiguration

After the CN negotiation, the UE can establish the DCH(s) for the requested service. Existing DCHs also may be reconfigured to meet the requirements, depending on the current configuration. Before sending the RB Setup message, call admission control must be checked and resource allocation performed for every resource involved in a call, as discussed in Chapter 3.

For AMR voice, the UE typically sets up three dedicated logical/transport channels mapped onto one CCTrCh. Information in the RB Setup message is similar to the RRC Connection Setup:

- **NAS Layer 2.** Radio Bearer/logical/transport channel mapping
- **Layer 2.** Channel coding, Radio Link Control (RLC) parameters, TTI, BLER targets, Transport Format Combination Set (TFCS)
- **Layer 1.** Spreading Factor (SF), OVSF code, Scrambling Code, frame offset, power control parameters

At this point, the radio link is completely established; however, the end-to-end connection is not yet fully established.

The RB Setup message can also be used as a reconfiguration message because the existing SRBs are, from this point forward, multiplexed with RAB onto a single physical dedicated channel.

5.2.3.5 End-to-End Connection

The Alerting message is sent from the network to the UE for MO calls, or from the UE to the network for MT calls. The directions for Connect and Connect ACKnowledge (ACK) are reversed, depending on the call type. As shown in Figures 5.3 and 5.4, the Connect message is sent by the UTRAN for MO calls, but by the UE for MT calls.

To avoid interruptions or restarts of the process, the timing of events is important for the end-to-end connection. If the end-to-end connection delay is too long, the network clears the call to avoid unnecessary allocation of network resources. Figure 5.6 shows the timers. They are typically broadcast and found in SIB 1 (T300) or are hard coded (T303, T310, and T313).

- **T300 and N300.** T300 is the UE timer for repeating the RRC connection request; N300 affects the number of RRC connection requests at Layer 3. T300 and N300 impact the duration of call setup. These parameters should be optimized with other call setup parameters at Layer 2 and Layer 1, such as mMax, preambleRetransMax, powerRampStep, and other open loop power control parameters.
- **T303.** Used by the UE to track the CN negotiation (security procedures) between CM Service Request and Call Proceeding.
- **T310.** Used by the UE to track the RB setup time between Call Proceeding and Alerting.
- **T313.** Used by the CN to track the final completion of the end-to-end connection between Connect and Connect ACK.

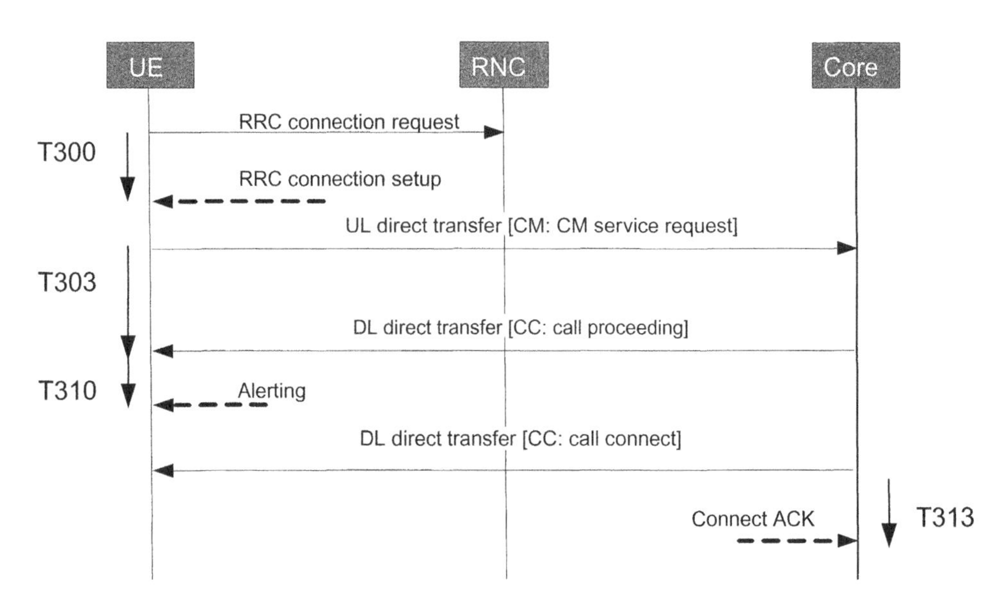

Figure 5.6 Main timers in AMR call flow

5.2.4 Call Retention Event and Signaling

In Connected Mode, CELL_DCH, call mobility is accomplished by handover, which replaces the cell reselection performed in Idle Mode. Before a handover can take place, the UE must assess all the cells with PSCs listed in the Monitored Set. The latter is signaled either on SIB 11, SIB 12, or in a Measurement Control Message(MCM). The network constructs and controls the list, and communicates it to the UE in two ways. Initially, the UE uses the same Neighbor List as broadcast in SIB 11 or 12 [10]. Once a MCM is received, the values received from the MCM supersede the stored information elements. This addresses the fact that each UE may have different cells in the Active Set, and various combinations of these Active Sets require different Neighbor Lists.

As defined in the standard [12], the UE must be capable of measuring up to 32 intra-frequency cells, 32 inter-frequency cells (on up to 2 additional carriers), and 32 GSM cells. During soft handover, the network must merge and combine the Neighbor Lists of all the cells in the Active Set. If the number exceeds the imposed limit [12], the network uses an infrastructure vendor-proprietary algorithm to truncate the list by discarding some PSCs from the Measurement List to be put in the MCM. The UE reports the measured PSC, including timing information, according to the threshold or periods defined in the Measurement Report Messages (MRMs). The network analyzes the MRMs, and then changes the Active Set in the Active Set Update (ASU) message. All network resource allocation changes must be implemented before an ASU is sent. Call admission and congestion control are involved during the decision process, and the internal algorithms are implementation dependent. It is not necessary to have a one-to-one matching of ASU to Measurement Report. It does not constitute a failure to have one ASU in response to multiple measurement reports. The delay between the first report of a strong server and its inclusion in the Active Set is still a useful indicator of quality when evaluating a system; long delays could indicate suboptimal queuing of MRM at the RNC. This MRM queuing, as well as discarding and processing, is RNC implementation dependent. For example, some implementations may send a single ASU when Events 1a and 1b MRMs are received, while others process each message and eventually send an ASU for each, slowing the handover process.

An ASU message can be sent in RLC Acknowledged Mode (AM) or Unacknowledged Mode (UM). The UE responds with a Layer 3 acknowledgment message–either ASU Complete (ASUC–Success) or ASU Failure (ASUF–Failure). These messages are always sent in Acknowledged Mode. ASU Failure can have a number of causes: full Active Set, deletion of a nonactive link, addition of a link with a different SF, or deletion of all the active links. Therefore, it is not a good indicator of handover performance in a system.

For successful ASUs, the UE modifies the radio links according to the ASU message and performs the Layer 1 synchronization. After the network receives the UE acknowledgment of the ASU message, it should update the final Neighbor List to be used by the UE for future measurements. These Neighbor List changes are sent to the UE in the subsequent MCM. If the combined Neighbor List needs no changes, no MCM is needed. Therefore, the absence of an MCM after ASCU does not necessarily indicate a problem. Figure 5.7 shows the entire handover signaling cycle.

Successful and timely handover is a key factor for reducing call drop occurrences. Handover optimization is a three-step process, of which the optimization engineer can control only two:

- **RF configuration optimization.** Minimize cell overlap to reduce the need for handover. This step involves Neighbor List optimization, because it notifies the UE which cells overlap.
- **Handover parameters.** These are discussed in Chapter 4. For a given RF configuration, handover parameters are set to balance resource utilization and call quality by reporting only the strongest interferer to be included in the Active Set.
- **RNC processing of MRM.** This process is implementation dependent and cannot be controlled by the optimization engineer.

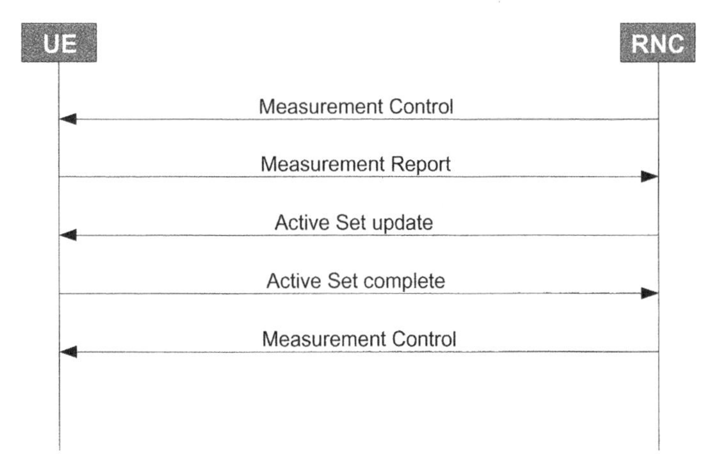

Figure 5.7 Handover signaling

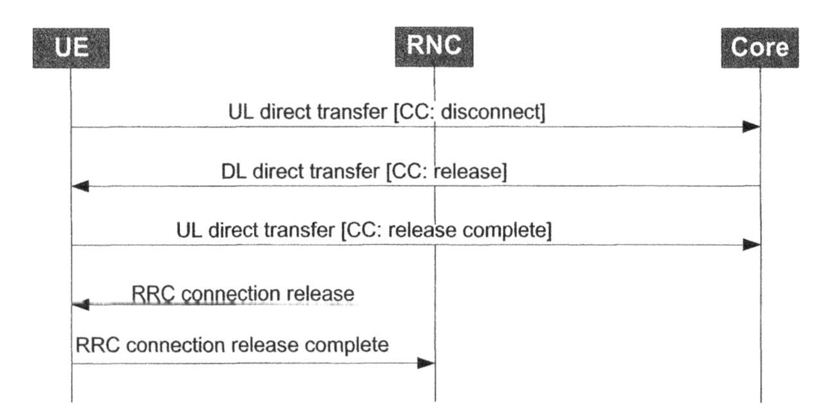

Figure 5.8 Normal AMR call disconnection call flow

Unlike other call events such as call origination, handovers, or call release, a dropped call is not defined by a signaling signature but rather by an absence of signaling. Figure 5.8 shows a normal AMR call disconnection.

Ultimately, the Disconnect message must be checked for the following cause value: Normal Call Clearing. For AMR voice service, the absence of such a signature can indicate a dropped call, as shown in Figure 5.9.

In Figure 5.9, the top example shows a normal release with RRC Connection Release and RRC Connection Release Complete messages before entering Idle Mode. The bottom example shows that the UE makes several successive measurement reports without receiving any ASU and finally transitions to Idle Mode. This is a typical signature of a

Friendly Viewer
File Edit Filters Help

Log Code	Message	Call Flow	Category
0x412F	WCDMA UL_DCCH Message: rrcConnectionReleaseComplete	BS <<<< MS	WCDMA_RRC_UL_DCCH_MESSAGES
0x412F	WCDMA UL_DCCH Message: measurementReport	BS <<<< MS	WCDMA_RRC_UL_DCCH_MESSAGES
0x412F	WCDMA DL_DCCH Message: rrcConnectionRelease	BS >>>> MS	WCDMA_RRC_DL_DCCH_MESSAGES
0x412F	WCDMA UL_DCCH Message: rrcConnectionReleaseComplete	BS <<<< MS	WCDMA_RRC_UL_DCCH_MESSAGES
0x412F	WCDMA UL_DCCH Message: rrcConnectionReleaseComplete	BS <<<< MS	WCDMA_RRC_UL_DCCH_MESSAGES
0x412F	WCDMA BCCH FACH Message: systemInformation	BS >>>> MS	WCDMA_RRC_BCCH_BCH_MESSAGES
0x412F	WCDMA BCCH FACH Message: systemInformation	BS >>>> MS	WCDMA_RRC_BCCH_BCH_MESSAGES
0x412F	WCDMA BCCH FACH Message: systemInformation	BS >>>> MS	WCDMA_RRC_BCCH_BCH_MESSAGES
0x412F	WCDMA BCCH FACH Message: systemInformation	BS >>>> MS	WCDMA_RRC_BCCH_BCH_MESSAGES
0x412F	WCDMA BCCH FACH Message: systemInformation	BS >>>> MS	WCDMA_RRC_BCCH_BCH_MESSAGES

Friendly Viewer
File Edit Filters Help

Log Code	Message	Call Flow	Category
0x412F	WCDMA UL_DCCH Message: activeSetUpdateComplete	BS <<<< MS	WCDMA_RRC_UL_DCCH_MESSAGES
0x412F	WCDMA DL_DCCH Message: measurementControl	BS >>>> MS	WCDMA_RRC_DL_DCCH_MESSAGES
0x412F	WCDMA UL_DCCH Message: measurementReport	BS <<<< MS	WCDMA_RRC_UL_DCCH_MESSAGES
0x412F	WCDMA UL_DCCH Message: measurementReport	BS <<<< MS	WCDMA_RRC_UL_DCCH_MESSAGES
0x412F	WCDMA UL_DCCH Message: measurementReport	BS <<<< MS	WCDMA_RRC_UL_DCCH_MESSAGES
0x412F	WCDMA UL_DCCH Message: measurementReport	BS <<<< MS	WCDMA_RRC_UL_DCCH_MESSAGES
0x412F	WCDMA BCCH FACH Message: systemInformation	BS >>>> MS	WCDMA_RRC_BCCH_BCH_MESSAGES
0x412F	WCDMA BCCH FACH Message: systemInformation	BS >>>> MS	WCDMA_RRC_BCCH_BCH_MESSAGES
0x412F	WCDMA BCCH FACH Message: systemInformation	BS >>>> MS	WCDMA_RRC_BCCH_BCH_MESSAGES
0x412F	WCDMA BCCH FACH Message: systemInformation	BS >>>> MS	WCDMA_RRC_BCCH_BCH_MESSAGES

Figure 5.9 Comparing normal release to a call drop

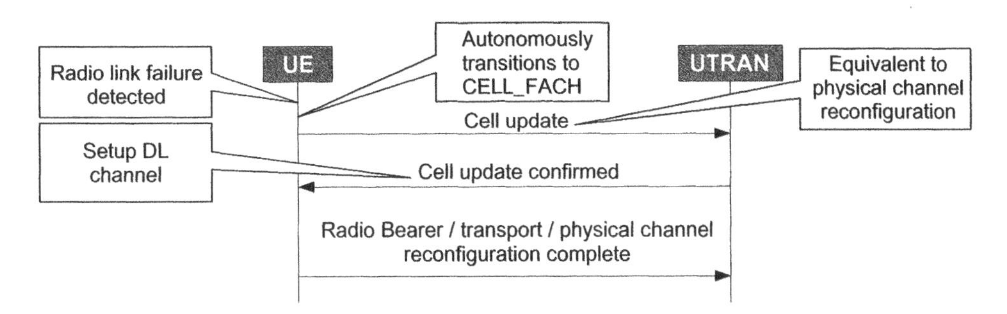

Figure 5.10 Call reestablishment signaling flow

dropped call. If a call drops because of a radio link failure, and if network settings allow call reestablishment, the UE can reestablish the call connection through the cell update procedure [10].

After the drop, a suitable cell is reselected and the UE sends a cell update, as shown in Figure 5.10. This procedure requires the radio condition to recover quickly from the radio link failure; otherwise higher layers on the UTRAN will clear the call.

After a suitable cell is found, the UE transitions to CELL_FACH. The UE sends a Cell Update message using a random access procedure, the normal procedure for radio link establishment. In this procedure, the network can send a Cell Update Confirmed message to instruct the UE to return to the CELL_DCH state with new RB, transport channel, and physical channel information (with new assigned dedicated channel information). This is similar to the procedure used in channel-type switching (from CELL_FACH to CELL_DCH) during a packet switched call, as detailed in Section 5.4.3. The UE then responds with one of the following acknowledgment Layer 3 messages: RB Reconfiguration Complete, Transport Channel Reconfiguration Complete, or Physical Channel Reconfiguration Complete.

If the connection is successfully reestablished, the dropped call could be a *system-perceived call drop* rather than a *user-perceived call drop*, because the user does not have to manually intervene to reestablish the connection. System-perceived call drops and user-perceived call drops should be counted separately during network analysis. It can take up to T315 seconds to complete the link reestablishment procedure, during which the UE transitions to CELL_FACH, recovers from the radio link failure, reads all the SIBs, sends the cell update message, and receives the cell update Complete message with new channel information. During this time, the conversation sounds are muted to the user.

5.2.5 Connection Supervision and Link Quality Indicators

In CELL_DCH Mode, Uplink and Downlink quality can be estimated by means of several metrics:

- **Block Error Rate.** BLER should be close to the signaled BLER target during a call. Increased Downlink BLER indicates that the required E_b/N_t cannot be fulfilled. This could indicate incorrect outer loop power control or that the required Dedicated Physical Channel (DPCH) power is not available, either due to the load on the network or because

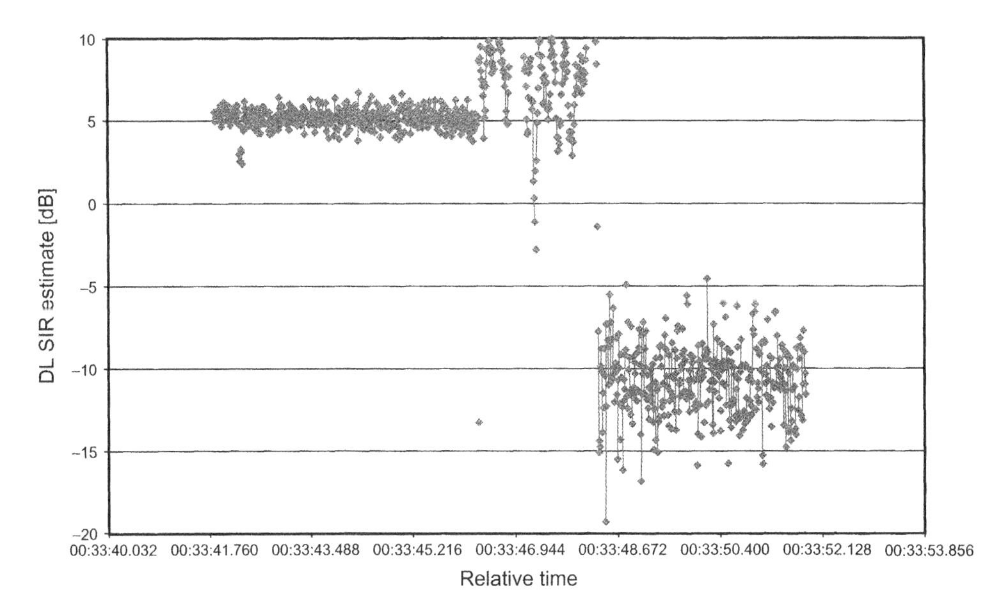

Figure 5.11 Downlink SIR during a dropped call

the maximum allowed power has been reached. Measuring the Downlink BLER is a more common practice than measuring the Uplink BLER. More tools are available for logging and processing Downlink BLER, while only vendor-specific infrastructure tools can be used to log and process Uplink BLER data.

- **Downlink Signal-to-Interference (DL SIR).** DL SIR indicates Downlink quality. During a call, the SIR varies between the implemented lower and upper boundaries. Lower SIR denotes favorable RF conditions, as shown in Figure 5.11. Before 00:33:46, the signal can be demodulated with limited power assigned to the DPCH. However, a very low (negative) SIR could indicate a disabled DPCH, as depicted after time 00:33:48. This behavior may be observed after a sudden increase of the Downlink SIR, which indicates that the power control algorithm tried to maintain the BLER target until the DPCH is disconnected.
- **Uplink Signal-to-Interference (UL SIR).** UL SIR can be used to estimate the RF conditions of the Uplink channel in the same way as Downlink SIR. As with the Uplink BLER, extracting this information requires access to UTRAN vendor-specific tools for logging and for parsing logs.
- **Transmit Power Control (TPC).** Commands history ("0" or "1") for the Downlink (transmitted by the UE on the Uplink) and Uplink (transmitted by the UTRAN on the Downlink). During a call, the TPC should present a random arrangement of "0" and "1" based on the power control algorithm. In either direction, a higher occurrence of "1" (power up) indicates degradation of the radio link. Power control tries to compensate by requesting additional transmit power. For the UL TPC, the sequence of bits has specific meaning in out-of-sync conditions. For example, when the UE transmitter is disabled, the Uplink SIR drops and the Qin condition is not fulfilled. Node B starts the out-of-sync TPC pattern [8]. The Node B uses various TPC patterns depending on the Active Set size.

- **UE Transmit Power.** Can also indicate Uplink RF conditions, if power control works perfectly. An increase of UE transmit power correlates with high Uplink BLER or high Uplink interference. It is easier to examine UE transmit power than the Uplink quality metrics BLER or SIR because the UE transmit power information can be extracted directly from the UE log and does not require additional post-processing or trace synchronization.

In addition to these commonly available metrics, the UTRAN and UE (Layer 1) constantly monitor the Uplink and Downlink for synchronization through Qin and Qout, which are in-sync and out-of-sync primitives, respectively. It is important to understand this process because it is the source of dropped calls that can have misleading signatures.

Downlink out-of-sync is reported with each frame using the CPHY-out-of-sync-IND primitive, which checks to see whether *either* of the two following quality criteria is true:

- Downlink Dedicated Physical Control Channel (DL DPCCH) quality over the previous 160 ms is worse than Qout [13]. For this situation, Qout is not defined formally in the standard, but is commonly implemented using DL SIR.
- All of the last 20 transport blocks received have CRC errors and, of the CRC-protected blocks, all transport blocks received in the last 160 ms have CRC errors.

If N313 successive out-of-sync indicators are detected at Layer 1, the UE waits for the T313 timer to expire before declaring a radio link failure. Both N313 and T313 are broadcast in SIB 1. T313 is implemented to allow the link to recover. If N313 successive in-sync indicators are detected, the entire out-of-sync process is reset.

Downlink in-sync is reported in every radio frame (10 ms), using the CPHY-sync-IND primitive. This process checks to see whether *both* of the following quality criteria are met:

- DL DPCCH quality over the previous 160 ms is better than Qin [13].
- At least one correct CRC is received in a TTI ending in the current frame. Only CRC-protected blocks are considered. The criterion is assumed to be fulfilled if no CRC-protected blocks are transmitted.

When the T313 timer expires, the UE declares radio link failure (see Figure 5.12). After Qout is detected, the UE Power Amplifier (PA) is turned off. Because the Downlink cannot be demodulated reliably, power control information is not received; thus the PA is turned off to avoid generating interference on the Uplink. If the Qout condition is maintained for N313 frames, the UE declares CPHY-out-of-sync. The UE then starts a process similar to initial acquisition of the radio link, because the system timing is considered lost at this point. If the acquisition process does not succeed within T313, the link is considered lost and radio link failure is declared.

The N313 and T313 parameters directly influence how long a call can be maintained in bad RF conditions. If these parameters are too short, many calls may be prematurely dropped under rapidly changing RF conditions. On the other hand, setting these times too long allows more time for calls to recover but may affect call quality and resource utilization.

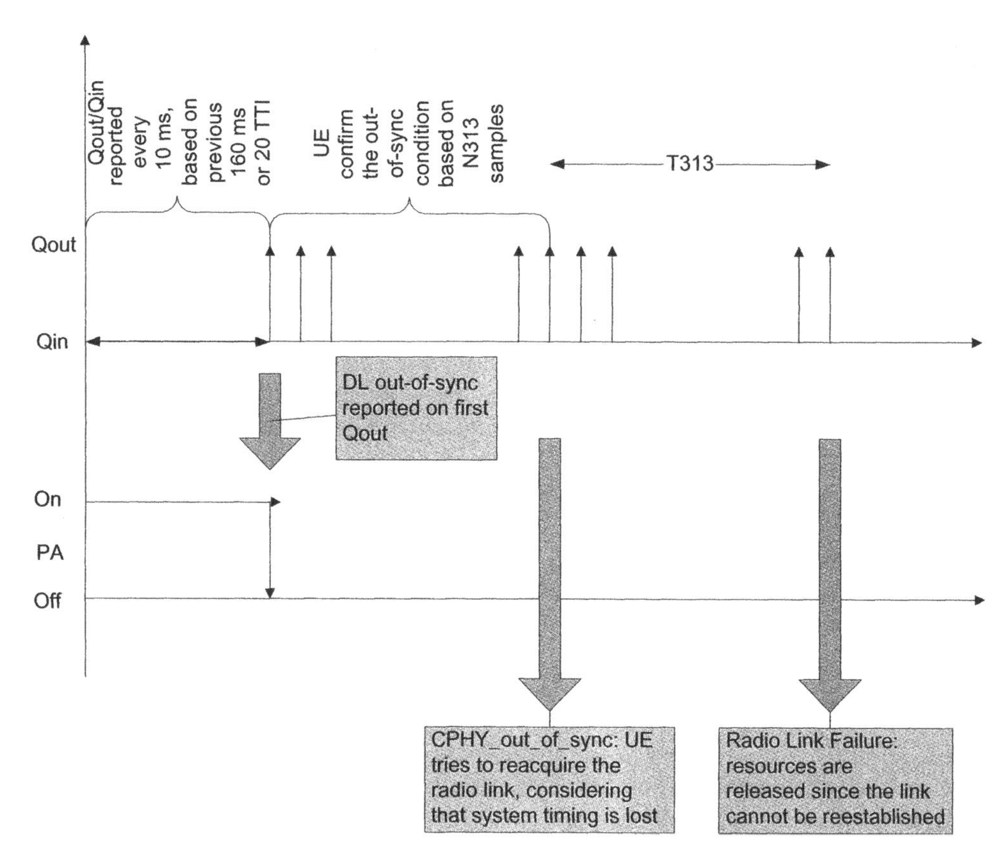

Figure 5.12 Call drop due to loss of Downlink synchronization

For simplicity, this example assumes that the UE PA turns off after the first Qout. The actual process is more complex: 160 ms after physical channel establishment, the UE turns its transmitter on or off according to the Downlink DPCCH quality criteria, as follows:

- The UE turns off its transmitter when it estimates that the DPCCH quality over the last 160 ms is worse than a threshold Qout.
- The UE can turn its transmitter on again when it estimates that the DPCCH quality over the last 160 ms is better than a threshold Qin. When transmission is resumed, the power of the DPCCH is the same as when the UE transmitter was turned off.

This allows the UE to turn off its PA after 160 ms, but turn it back on only after 10 ms if the Qin threshold is lower than Qout. It is not uncommon for the UE to turn power on and off under bad RF conditions in response to synchronization loss and recovery.

For the Node B, a similar process is available through the CPHY-sync-IND or CPHY-out-of-sync-IND primitives. The Node B monitors the Uplink synchronization [8] and reports a CPHY-sync-IND or CPHY-out-of-sync-IND primitive to the RL Failure/Restored triggering function. The Uplink synchronization criteria are not specified in the standard, but could be based on similar measurements–CRC and/or DPCCH quality. With soft handover, a call could be supported by several radio links; therefore, some vendors have implemented separate timers for individual links and the overall connection. If a radio link fails, only the link-specific timer expires and only that specific link clears. Alternatively, if the connection timer expires, the connection would drop.

5.2.6 Troubleshooting AMR Failures

During voice service optimization, it is important to analyze the quality metrics and resolve the issues observed during testing to determine what actions will improve the metrics. We have developed a simple troubleshooting guide to facilitate the problem-solving process. Its methodology is based on the analysis of UE logs only because network-based logs are not as readily available as commonly as UE logs.

We begin with a description of the optimization processes for call access and retention. Then we review examples of some common failures for each of them.

5.2.6.1 AMR Analysis Process

The flowchart in Figure 5.13 can be used to analyze AMR call delivery performance. The specific quality metrics for AMR call delivery, call setup success ratio and call

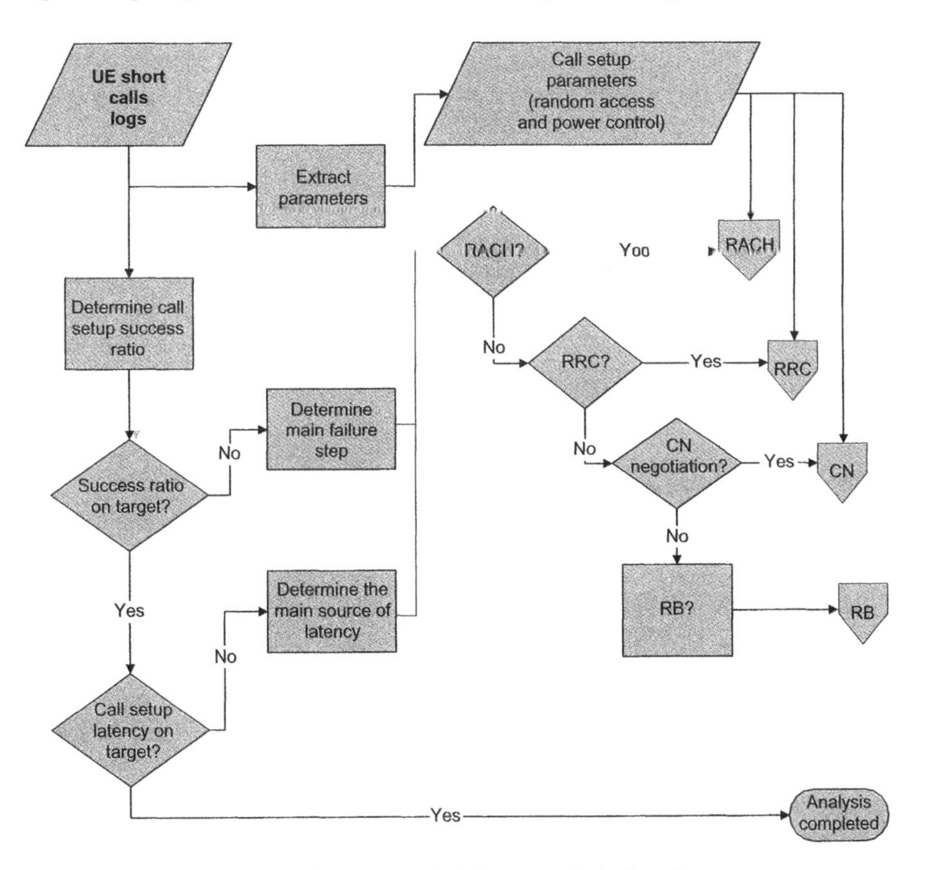

Figure 5.13 AMR call delivery analysis flowchart

setup latency, are discussed in Section 5.2.8. System parameters should be extracted from parameter audits.

The analysis in Figure 5.13 follows the basic steps defined in the signaling flow discussed in Section 5.2.3:

- RACH (Figure 5.14) corresponds to the system access step.
- RRC (Figure 5.15) corresponds to the RRC Connection Setup step.
- CN negotiation (Figure 5.16).
- RB (Figure 5.17) correspond to the RB setup step.

Compared with Section 5.2.3, end-to-end connection is not included in this analysis. Issues during this stage can be analyzed as dropped calls, as shown in Figure 5.18.

To perform both analyses efficiently, the network planner must verify that all the nodes and features in the system are available, as described in Section 2.2.2. The analysis should account for any nodes or features that were unavailable during testing, as this affects recommended changes. The network planner must also know the current parameter settings, for the expected and the implemented parameters, as described in Section 4.1.3.

The process shown in Figure 5.13 is more appropriate for initial optimization. During this phase it is important to remember the goal: solving all problems is not economically feasible. We must focus on the critical KPIs to know when they have been met. This way, we can optimize the network to an acceptable quality level with the least effort. The analysis in Figure 5.13 concentrates on the main milestones at which a call can fail. An alternative would be to focus on the area in which most problems are observed.

5.2.6.2 AMR Call Delivery Analysis: System Access

Figure 5.14 shows the access process. The first step in examining the access process is to determine if an ACK of the preamble is received.

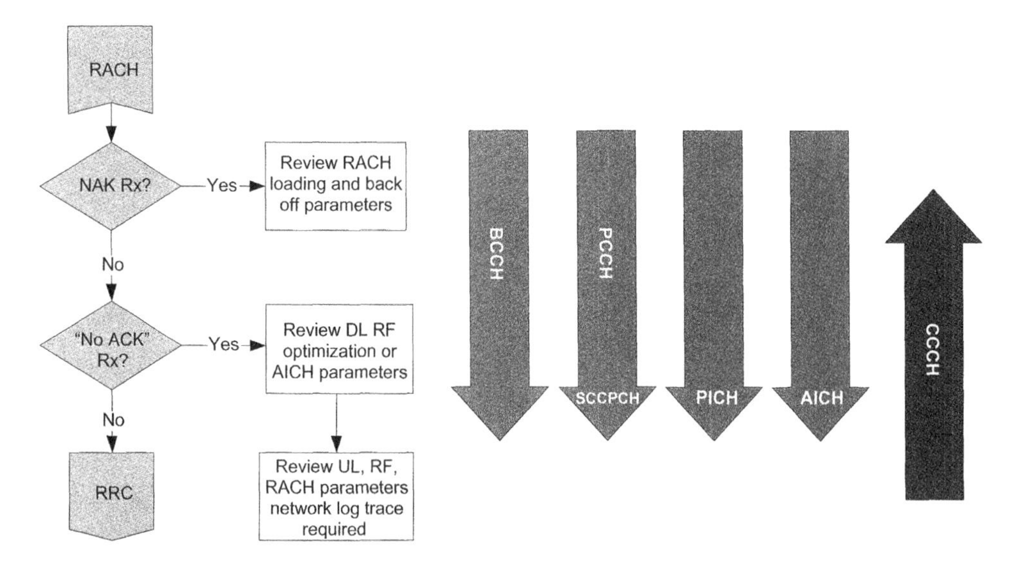

Figure 5.14 AMR call delivery analysis flowchart: system access step

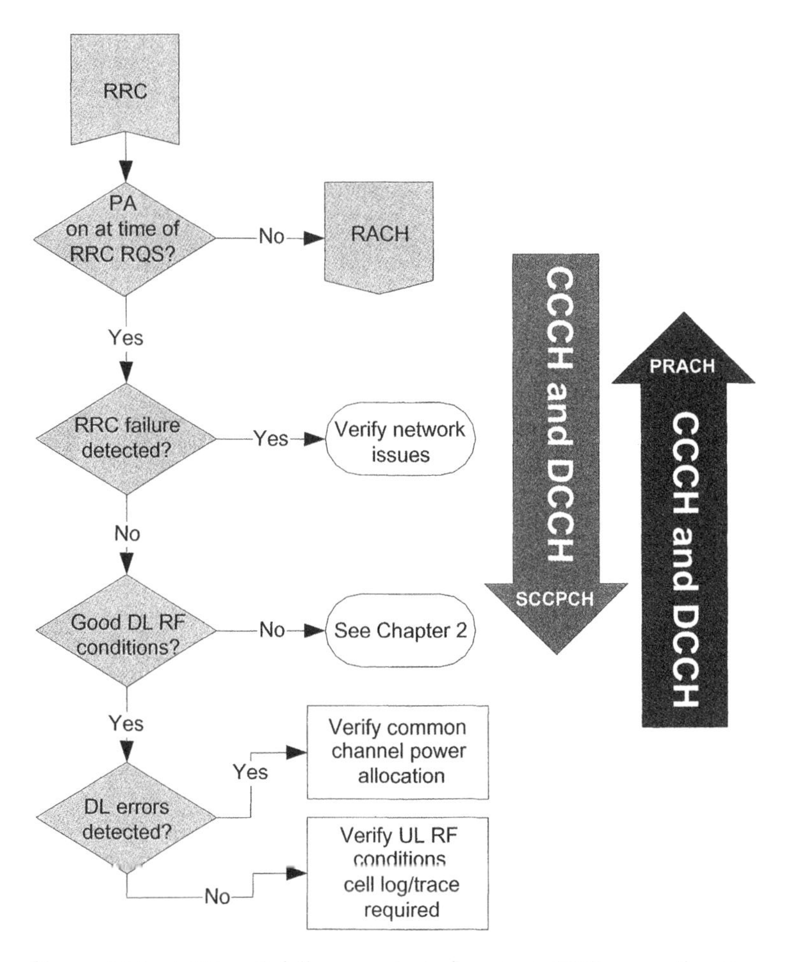

Figure 5.15 AMR call delivery analysis flowchart: RRC connection step

Assuming the UTRAN is error free, the absence of an ACK could indicate that the Node B never received the preamble, or the UE did not correctly decode the Acquisition Indicator Channel (AICH). To accurately determine if the preamble was received, network logging should be available. In the UE log, the transmit power provides the only indication that the preamble might not have reached the cell. If the access sequence ends before the maximum allowable transmit power is reached, this indicates that the preamble was not received. In this case, we carefully review the access parameters described in Section 4.4, including the power allocation of the AICH. Insufficient power allocation would prevent the Acquisition indicator from being demodulated by the UE in the whole range of geometry.

5.2.6.3 AMR Call Delivery Analysis: RRC Connection

Figure 5.15 shows the RRC connection step. During RRC connection, the first step is to verify that the RRC Connection Request message (RRC RQS) was sent.

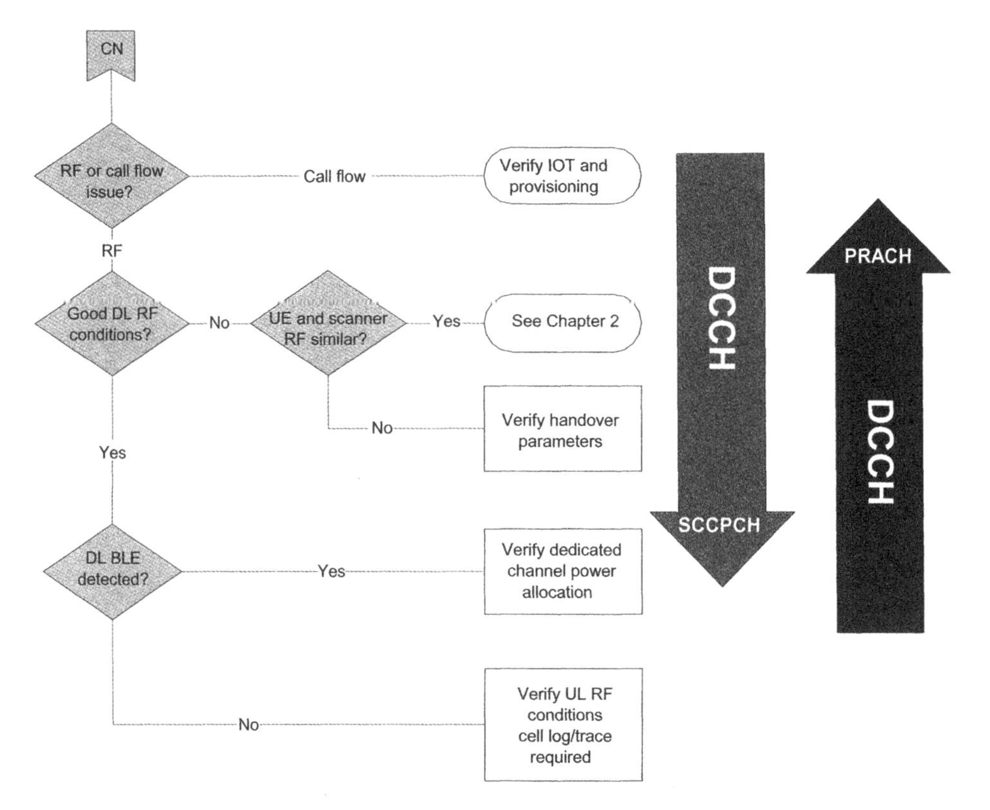

Figure 5.16 AMR call delivery analysis flowchart: Core Network negotiation step

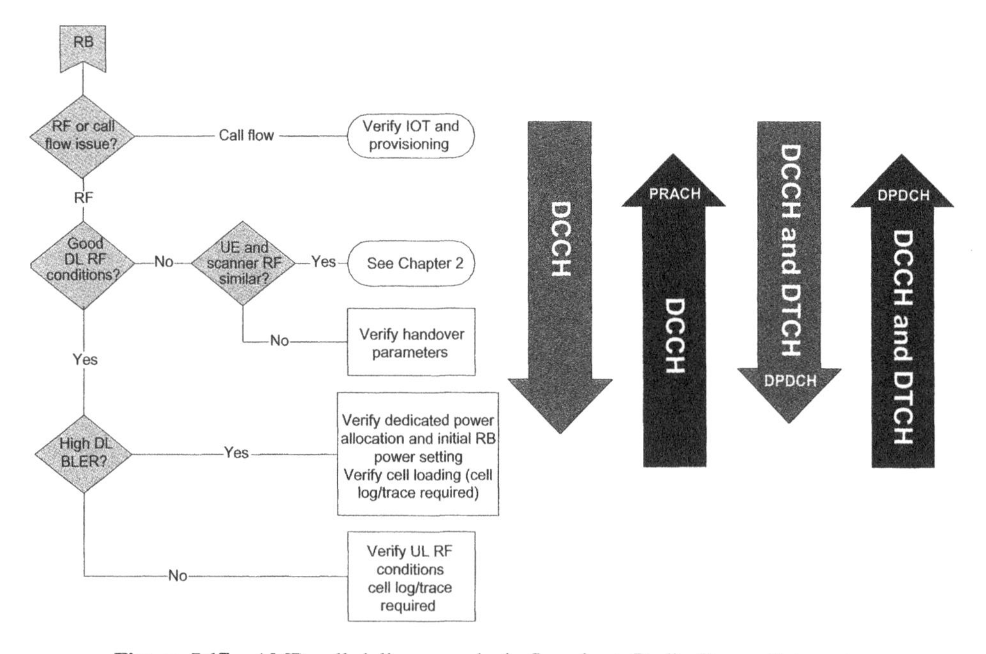

Figure 5.17 AMR call delivery analysis flowchart: Radio Bearer Setup step

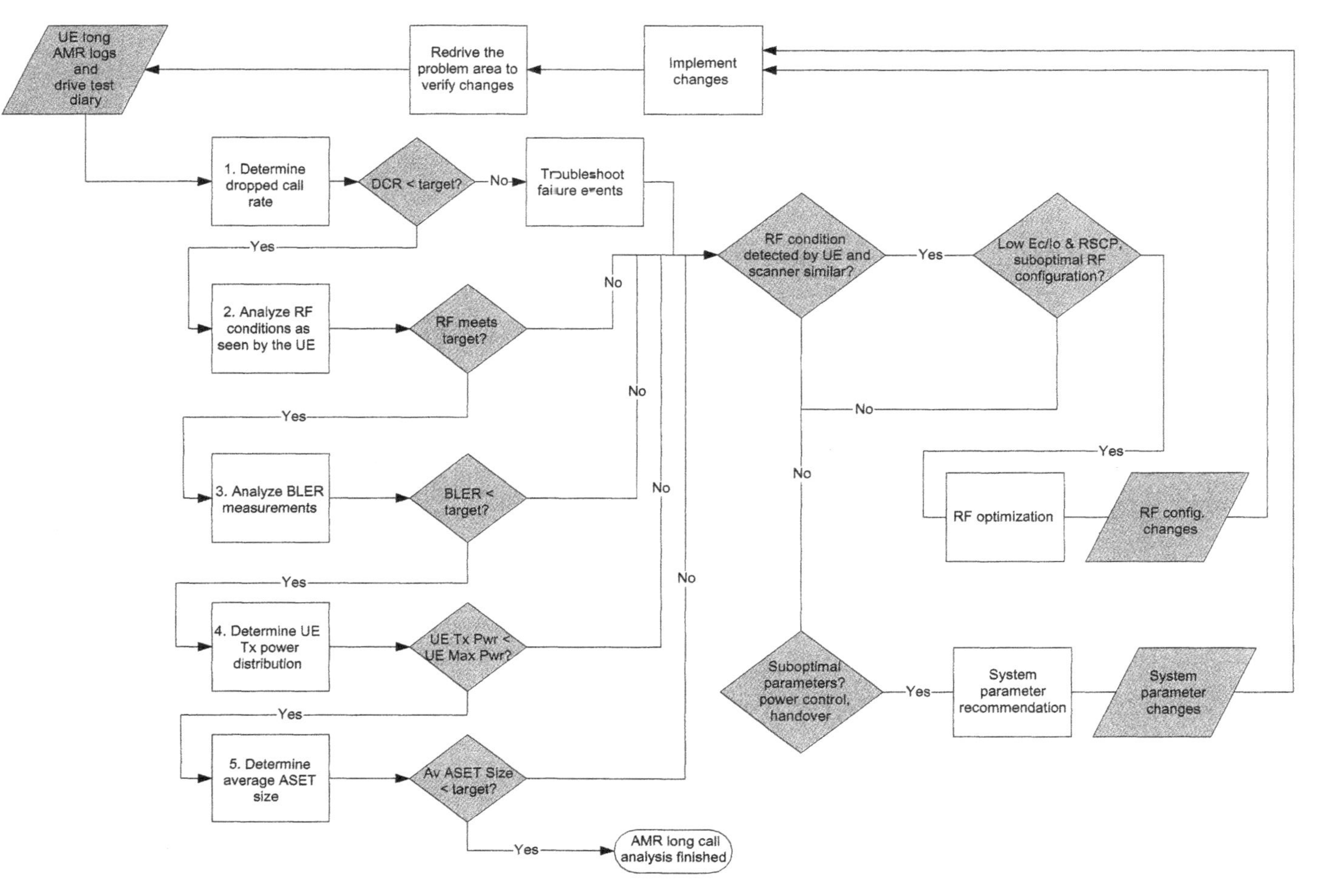

Figure 5.18 AMR call retention and quality analysis flowchart

In the UE logs, the RRC RQS message is detected when it is sent from Layer 3, not when it is actually sent from Layer 1. Therefore, detection of the RRC RQS message is not a definitive indication that the message was actually sent. It must be verified if the access step was successful. After the message is sent, the only way to reliably know if it was received at the Node B is to observe the network logs. In the absence of such logs, we can look at the Downlink RF condition to see if a RRC Connection Setup message was received. Acceptable RF conditions must be defined in terms of RCSP and CPICH E_c/N_o according to the RF KPI defined in Section 2. If the Common Pilot Channel (CPICH) quality is good, power allocation of the SCCPCH is analyzed carefully.

If a failure indication (e.g., RRC connection reject) is received, the call should be considered blocked rather than failed. The main reasons why calls are blocked are that the admission control threshold was reached, or equipment malfunctioned. We must review the network logs to determine why the call was blocked.

5.2.6.4 AMR Call Delivery Analysis: Core Network Negotiation

Figure 5.16 shows the CN negotiation step. After a successful RRC connection, dedicated SRBs are established. Failure after this milestone can easily be linked to RF conditions or protocols. Check the power control information (TPC bits), presence of Block Error (BLE), and CPICH quality to identify RF issues. Review the signaling flow and timing, as described in Section 5.2.3, to detect protocol issues.

5.2.6.5 AMR Call Delivery Analysis: Radio Bearer Setup

From an RF perspective, the RB setup step shown in Figure 5.17 is similar to the RRC Connection Setup: new RBs are added, for which resources need to be available. Once the resources are allocated, the main cause of failure in the Downlink is either insufficient power allocation or inefficient power control.

5.2.6.6 AMR Call Retention and Quality Analysis

Once the RBs are set up, the call can be analyzed for retention, as shown in Figure 5.18. The specific metrics are Dropped Call Rate (DCR), UE perceived RF performance (E_c/N_o, [Received Signal Code Power] [RSCP], etc.,), BLER measurements, UE transmit power, and Average Active Set Size (Section 5.2.8). These metrics must be compared to the target KPIs for performance benchmarking. Failure events must be studied and matched to typical failure symptoms, as discussed above.

The following sections explain different failure symptoms and suggest resolutions, starting with RF performance issues and evolving to more advanced issues.

5.2.6.7 Bad RF Performance (Downlink, Uplink, or Both)

Even when RF is optimized, RF issues could still be detected during service optimization. This could happen, for example, if RF optimization was based on a statistical analysis that overlooked some specific areas. Evaluating RF performance requires identification of RF quality metrics including CPICH E_c/N_o, CPICH RSCP, UE transmit power (UE Tx Pwr), and signal quality of the Uplink and Downlink for dedicated channel RF (TPC history, SIR, and BLER). A complete diagnosis requires a multidimensional lookup matrix.

Table 5.5 RF failure symptom matrix

		E_c/N_o			
		HI		LO	
		UE Tx Pwr		UE Tx Pwr	
		LO	*HI*	LO	*HI*
RSCP	*HI*	No RF issue	Limited UL	DL interference	UL interference
RSCP	LO	DL network edge	UL network edge	DL coverage hole	Lack of coverage (DL and UL)

Table 5.5 shows a simplified lookup matrix for symptoms covering only the main metrics: CPICH E_c/N_o, CPICH RSCP, and UE Tx Pwr. In most cases, these are sufficient to identify RF issues; Uplink and Downlink signal quality could be used to complement the data.

RF failure could be caused by any of the conditions listed below:

- **No RF issue.** When no RF issues are detected but dropped calls are still found, analyze the UE and network logs for protocol issues or equipment failures.
- **Limited UL.** Networks should be designed for balanced Uplinks and Downlinks; therefore, finding limited Uplink coverage without Downlink issues should be rare. If detected, verify either Uplink interference or equipment issues. In particular, carefully consider the Uplink received chain, including the equipment. The first kind of equipment to be investigated are Tower Mount Amplifiers (TMA), if installed. Equipment malfunction is not the only reason for limited Uplink. Uplink power control equalizes all the signal levels. In the case of Uplink interference, especially external interference, the Uplink power control will force an increase in UE transmit power. Confirm this condition by checking the transmitted Uplink interference value in SIB 7, or the noise rise at the cell level, available on the Operation and Maintenance (O&M) terminal.
- **DL interference.** Pilot pollution is a generic term that actually refers to Downlink interference. This could be caused by multiple cell overlap, which is the origin of the term Pilot pollution–when more servers or paths are detected in a given area than can be accommodated in the Active Set or by the Rake receiver. Missing neighbor or missed handover is detected in the same way: the UE cannot utilize the co-channel signal and is, thus, interfering. Downlink interference can also be caused by external interference, which can be easily identified by performing a channel sweep using a spectrum analyzer. In all the Downlink interference cases, the reported E_c/N_o for the active server drops, while the total received power stays about constant or even increases. In this condition, in Connected Mode, BLER is likely to increase because DPCH power is normally set to overcome only a limited range of interference.
- **UL interference.** This case is similar to the limited UL case previously discussed, except that CPICH E_c/N_o is also low, and both Uplink and Downlink interferences are detected. In this situation, investigate cell overlap.

- **Unbalanced links (Uplink/Downlink).** An AMR call is a symmetric service that requires both Downlink and Uplink coverage. An indication of unbalanced links is that the UE transmit power reaches the maximum values even in good Downlink coverage (good CPICH E_c/N_o or low BLER). Low UE transmit power when DL BLER or CPICH E_c/N_o is high indicates an unbalance in favor of the Uplink. Check TPC history and SIR figures to evaluate the link qualities of the dedicated channels when link imbalance is suspected. When this is observed, assess the following conditions:
  - **DL network edge.** Downlink network edge is detected when the Uplink has TMA, high UE transmit power (class 3 or above), or an exceptionally sensitive Node B. Downlink network edge condition is also likely to occur if the Node B transmit power is artificially reduced; for example, as an attempt to solve DL interference.
  - **UL network edge.** Lack of UL coverage is indirectly detected by observing the UE transmit power. Insufficient UL coverage and link imbalance have similar signatures. UL network edge can be remedied by decreasing the UL path loss, typically by adding a TMA in the received branch.
  - **Coverage hole, UL, and DL.** Lack of coverage in DL and UL for both common (i.e., CPICH) and dedicated (i.e., DCH) channels. It is mainly characterized by low RSSI (lower than −95 dBm) or low RSCP (lower than −105 dBm) as well as high transmit power. In this case, adding a Node B or a repeater is the only long-term solution, even if an expensive one.

The checklist in Table 5.6 summarizes recommendations from the previous discussion for addressing Uplink or Downlink limitations.

5.2.6.8 Missing Neighbors

The previous section mentioned that missing neighbor relationships present a signature similar to Downlink interference. This section explains that and provides a solution.

Table 5.6 Solving Downlink or Uplink coverage limitations

Steps to resolve limited Downlink coverage and high Downlink interference
Check the transmit antenna system (including the antennas and High-Power Amplifier: HPA) and power settings, and check for external interference, using a spectrum analyzer
Reduce Pilot pollution by reducing the number of dominant servers in the problematic area; for example, downtilt antennas, reduce transmit power levels, change antenna directions, or reduce feeder cable loss
Increase the dominance of the best server(s), by reducing antenna downtilts, increasing transmit power levels (installing a High-Power Amplifier), or changing antenna directions
Install additional cell sites to address inadequate DL coverage area
Steps to resolve limited Uplink coverage and high Uplink interference
Check the receive antenna system, power settings, and the UL noise rise (SIB 7)
Try to use diversity (if not used) or install Low-Noise Amplifiers (LNA) to reduce the Uplink path loss
Install additional cell sites to address inadequate UL coverage area
Check infrastructure, vendor-specific implementations for UL interference handling

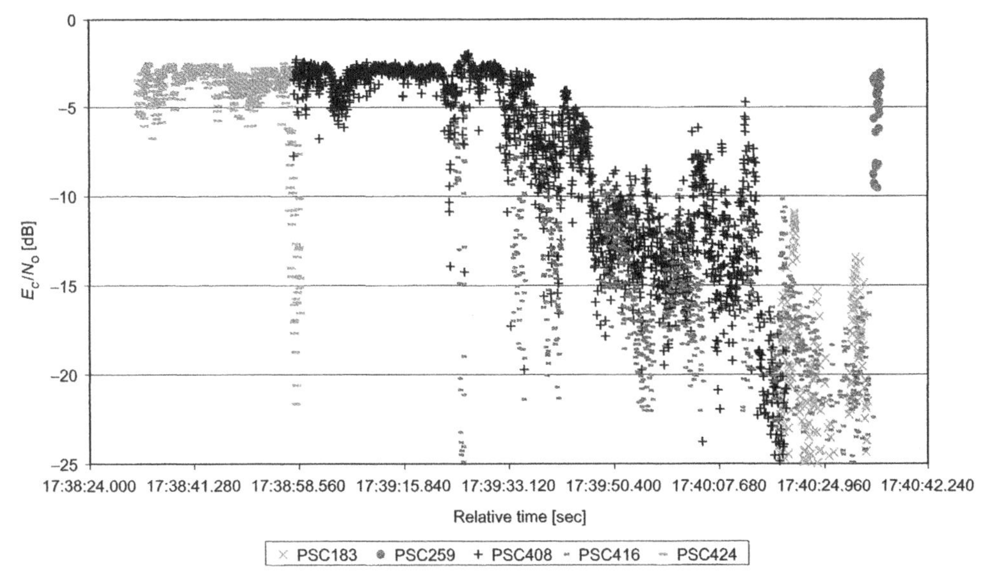

Figure 5.19 Missing neighbor signature

- **Failure symptoms:**
  - High interference identified by low CPICH E_c/N_0 due to missing neighbor relation definitions.
  - Soft handover cannot take place because a strong suitable cell is not in the Neighbor List.
  - Frequent cell reselections or handovers to weak cells. Because the strong cells are not in the Neighbor List, they cannot be reselected or handed over to. This leads to large CPICH E_c/N_0 fluctuations, as shown in Figure 5.19.
  - After the failure, or after a delayed handover, the CPICH E_c/N_0 of the best server improves significantly, as shown by PSC 259 in Figure 5.19.
- **Steps for Resolving Missing Neighbors:**
  - Verify that all sites are operating and have the desired parameter settings.
  - Verify that the best server, including the one detected after the recovery, can cover the test area.
  - Correct overshooting cell(s) first, as they might behave like missing neighbors. As mentioned in Chapter 2, limiting cell overlap minimizes the need for handover and long Neighbor Lists.
  - Add the missing neighbor to the Neighbor List for the best serving cell, if justified by the network plan. Instead of adding the missing neighbor relation as a quick fix to the problem, consult the network plan to estimate whether the addition will be beneficial in the long term, or whether a change of RF configuration should be considered.
  - Verify that neighbor relation additions are symmetrical.
  - If the resulting Neighbor List is long (more than 14 to 16 entries), consider further RF optimization and reduce cell overlap.
  - Rank all neighbors, if required or deemed necessary by the UTRAN (vendor-implementation dependent). Missing neighbors may arise from Neighbor List trun-

cation due to overly long Neighbor Lists from Active Set cells in soft handover. Check the Neighbor List merging algorithm to reduce the probability that important neighbors are deleted or truncated from the list.

5.2.6.9 Incorrect Network Settings and Max RLC Reset Reached

For Signal Radio Bearers, if the ACKs from the AM RLC Protocol Data Unit (PDU) are lost and successive retransmission attempts fail, RLC Reset may take place. When the maximum allowed number of RLC resets is reached, radio link failure is reported. This failure is rare for AMR calls because they can occur only on the basis of the traffic sent on the Signal Radio Bearers. Since the user payload is in transparent mode (RLC TM) no acknowledgment is used for this type of traffic. If radio conditions are bad, synchronization is lost before Max RLC Reset occurs.

- **Failure symptoms:**
  - Low call retention performance and/or high BLER in good RF conditions.
- **Steps for Resolving Incorrect Network Settings and Reaching Max RLC Reset:**
  - Check whether the Max RLC Reset is correlated with bad RF performance. Interoperability testing may help indicate if the Signal Radio Bearer and the user Radio Bearer are not performing equally.
  - Check all equipment for proper setup and calibration.
  - If failures occur during paging, preamble ramping, and initial network access, check and correct the Paging and Access parameters, and open loop power control parameters.
  - If failures occur in the middle of an established call, check and correct the Measurement Reporting Control message and parameters for handover, power control, and power allocation.
  - If failures occur in a WCDMA boundary area near the WCDMA-GSM borders, check and correct the Compressed Mode Operation and Inter-RAT parameters.
  - If failure is related to Maximum RLC Reset Reached, check the AM RLC parameters.

5.2.6.10 Faulty Hardware Equipment

If equipment appears to be defective or incorrectly set up, check the following possible causes:

- **Network infrastructure.** The RNC, Node B, backhaul, antenna system, or equipment may be faulty even when they are newly installed. Check the installation and put system/equipment-monitoring systems in place to report potential faulty equipment alarms.
- **Cell barred or reserved.** Some parts of the network may be reserved or barred for operational maintenance and upgrades. Monitor operational conditions of the network nodes throughout the drive tests.
- **Incorrect network parameters.** If parameter settings are suboptimal, the UE could initiate calls on the wrong cell, could be unable to quickly reselect a better cell, and preamble/message power could be insufficient in good RF coverage. In Connected Mode, the UE may not handover to the most appropriate cells and closed loop power control may be inefficient.

- **User equipment (USIM, mobile equipment).** Mobile terminals may be another problem source. Different UE brands/types and software versions could affect UE performance, which may result in poor network performance as measured by these UE devices. It is important to choose a good and stable UE device for testing, benchmarking, and performance comparison (to evaluate parameter changes). In addition, available USIMs may not support certain service features.

- **Failure symptoms:**
  - Unexplained failures, often in good RF conditions.
  - Call release or rejection with abnormal causes (unspecified, pre-emptive release, congestion, reestablishment reject, user inactivity, directed signaling, and connection reestablishment) while the UE is camping on or connected to given cells.
- **Steps for resolving faulty hardware:**
  - Check the RF conditions.
  - Check all equipment for proper setup and calibration.
  - Examine UE and network logs to understand the issue from both sides.
  - Verify network loading at the time of the event.
  - Test with known equipment that operates properly, and use several different UE brands.
  - Try to isolate the problem from other parts of the network.

5.2.6.11 Loss of Synchronization (Downlink, Uplink, or Both)

Synchronization can be lost during handover or call establishment, as described in Section 5.2.5. In fact, loss of synchronization is the main reason for dropped calls and it leads to radio link failure. In a strict sense, most call drops should be tagged with "Loss of Synchronization." However, this type of classification is not sufficiently meaningful. Synchronization loss can have many root causes, including bad RF performance and missing neighbor relations. The category defined here addresses unexplained synchronization loss between the network and the UE. For example, frequency and/or timing drift (network and/or UE) may cause loss of link synchronization.

- **Failure symptoms:**
  - Failure occurs when a radio link is added or modified.
  - TPC logs may present a typical pattern of all ones ("111 ... 111") or alternating ones and zeros ("0101 ... 011").
- **Steps for resolving loss of synchronization:**
  - Check the RF conditions, especially link balance.
  - Check for power control problems.
  - Check all equipment for proper setup and calibration.
  - Apply both UE and network logging to understand the issue from both sides.
  - Test with known equipment that performs reliably, and use several different UE brands.
  - Check for sudden drops in SIR estimates (>15 dB), which indicates that the network may have switched off the DL DPCH.
  - Try to isolate the problem from other parts of the network.

5.2.7 Parameter Optimization

Before discussing different system parameters for AMR voice service optimization, it is important to stress that system parameter optimization should be performed only *after* completing all RF configuration optimization. This ensures that system parameters are not used to address RF issues, which could affect system capacity. In other words, even if system parameters improve system performance, they should not be used to replace RF configuration optimization.

From a more practical point of view, optimizing the RF configuration before optimizing system parameters allows us to concentrate on one set of parameters at a time. It is important to know which system parameters to use during RF optimization. A good starting point is a set of generic system parameters, as presented in Chapter 4. This approach is similar to solving a linear optimization problem: it maximizes performance with one variable before setting another variable, thus discovering the optimal point of maximum system performance.

The same approach is valid for AMR and Circuit Switched (CS) or Packet Switched (PS) data optimization. Performing AMR optimization prior to CS or PS data optimization improves system performance for procedures that are common to all services, such as call setup and handovers.

This does not prevent further optimization of the RF configuration during service optimization. It only shifts the focus to concentrate on system parameters during service optimization.

Of the system parameters presented in Chapter 4, the following are most important for satisfactory AMR voice service:

- **Cell reselection parameters** are the first set of parameters to optimize after RF optimization. With proper reselection parameters, the UE will always camp on the best suitable cell, limiting the required power of the access preambles or the Downlink common channels involved in access (SCCPCH, Primary Common Control Physical Channel (PCCPCH), PICH, and AICH). These parameters are especially important to optimize in cases of high access (origination or termination) failure rates or when multiple RRC connection requests are observed because of poor RF conditions.
- **Call setup parameters** (Section 4.4) should be optimized for the best performance of random access and power control. After the reselection is optimized, set the access parameters to ensure that the UE can access the system with the minimum number of preambles and minimum transmit power. These two requirements conflict; a high transmit power for the initial preamble ensures access is detected by the Node B, but also increases the Uplink interference. To address this, focus on the number of preamble optimizations. In an unloaded system, the average number of preambles should be optimized as close to 1.5 as possible. Alternatively, the optimal setting can be found by calculating the total transmit power of all preambles. This minimum will correspond to the lower amount of Uplink interference.
- **Power allocation parameters** (dedicated channels) can be set to favor coverage (call retention) or capacity (number of supported calls). For the long term, power should be allocated at the minimum level to maintain the desired probability of service. With this in mind, optimize the RF first to ensure that the geometry is limited, optimize the handover parameters to ensure that the necessary links are in the Active Set, then

adjust the power allocation. In addition to drive tests, power allocation can be based on statistical data collected at the cell level. In this case, reduce the maximum DPCH power allocation only after considering the effect of loading. Increasing the maximum DPCH power allocation affects the loading and capacity of the system. Consequently, increase it even more carefully than reducing it.

- **Call retention parameters**, that is handover parameters (see Section 4.5), should be optimized so that handover can be performed smoothly and in a timely manner. Parameter tuning after RF optimization can yield surprising improvements in system performance for both call retention and capacity. As with proper RF optimization, when capacity is improved, the Active Set size decreases. Improvements in call retention performance might not be obvious at first glance. However, rapid RF environment changes can potentially drop a call if handover is not quick enough. This problem is called *Slow Handover* and can occur in areas with high WCDMA cell density. On the other hand, network infrastructure (typically RNC) may have problems with too many or too frequent measurement reports from many users in the network. Analyze this on a case-by-case basis.
- **Neighbor Lists** can be seen as both system and RF parameters. They should be optimized so the UE can perform cell reselection and handovers to the appropriate cells. This affects both Idle Mode and Connected Mode performances. RF cleanup is always recommended before adding too many neighbor relations. Reducing long Neighbor Lists can shorten the searching time and increase the search frequency of important neighbor cells.

5.2.8 Call Quality Metrics and Test Process

The main quality metrics for AMR calls were reviewed in Section 5.1.1. During service optimization, their corresponding KPIs can be augmented with more specific quality metrics:

- **High call setup success rate.** UE can successfully perform random access and set up a call with high probability (Section 5.2.3). MT calls have an additional step of paging; thus, the target call setup success rate is typically lower for MT calls because of the possibility of lost paging occasions. Attempts can be counted on the basis of UE internal events (i.e., an event triggered by a key press), the first access preamble, the first RRC connection request, or SIB 7 reading for MO/MT calls. For MT calls, we can also use the Paging Type 1 message.
- **Low call setup latency.** Call setup latency should be as short as possible. It may take longer if a cell reselection occurs just before or shortly after the RRC connection request. If a connection is dropped after RB establishment, call reestablishment may occur, which further lengthens the user-perceived call setup latency.
- **Low call drop rate.** The likelihood of an unintended radio link release after the Alerting message should be low. With the call reestablishment feature, call drops can further be categorized into system-perceived and user-perceived call drops. If call reestablishment is successful, it can be considered a system-perceived call drop; if unsuccessful, it can be considered a user-perceived call drop. However, unlike PS data calls, voice call reestablishment is perceived by the user to a greater extent because conversation is lost (call muted) for several seconds.

- **Excellent voice quality.** Excellent subjective voice quality is desirable and is highly correlated with the radio link quality. The radio link quality can be assessed from the BLER; low BLER is desired. Analysis of mean BLER by itself has limited value; it must be complemented by the measured probability density function of the BLER. Because of closed loop power control, mean BLER should approach the target BLER set up by the system when the network is loaded. High short-term BLER, on the other hand, can directly affect voice quality, causing noticeable degradation over short periods. BLER measurement provides an easy, repeatable test to quickly assess the radio link condition, and can reliably estimate voice quality.

For AMR service, BLER is sufficient to estimate voice quality because the vocoder was selected for adequate quality (Mean Opinion Score (MOS) of 4 or above), even in the presence of errors. MOS scoring is especially important as Voice over Internet Protocol (VoIP) services become available. With VoIP, the quality of the voice cannot be defined entirely from BLER, because PS data service can be subject to delays in addition to lost frames. The length and distribution of these delays affect voice quality differently, depending on when they occur during a conversation.

To test the quality metrics listed above, a test methodology and process must be defined. Rather than define an entire drive test methodology, we will concentrate only on specific tests that are required for AMR evaluation.

Two call types can be used to measure AMR performance: short calls and long (continuous) calls. Short calls (20 to 40 sec) measure call delivery performance (call setup success rate and latency). Long calls measure call retention and quality performance (call drop rate and voice quality).

Figure 5.20 shows a schematic example of UE_1 making a long call that is not to be released by the user manually. It is restarted only when the call drops.

To evaluate this in terms other than the actual test duration, call retention performance is expressed through the equivalent call drop ratio given in Equation 5.2.

$$Equivalent\_call\_drop\_ratio = \frac{Number\_drop\_calls}{(Total\_call\_time/Mean\_holding\_time)} \tag{5.2}$$

In addition, BLER measurements can be collected along the entire drive route, except for brief instances when the call is dropped.

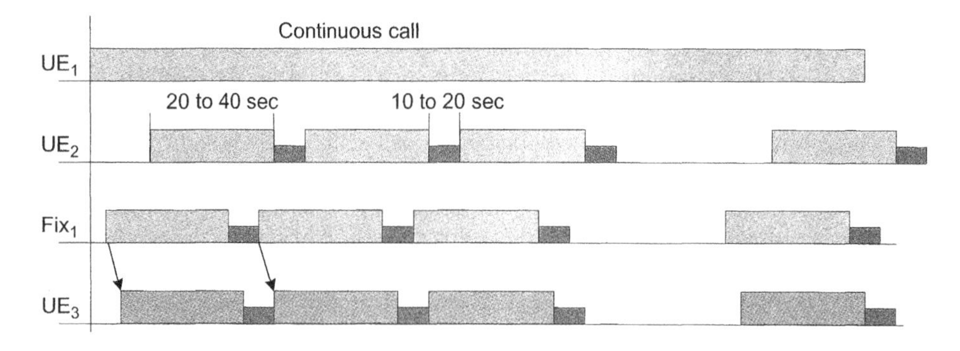

Figure 5.20 Call pattern testing – AMR Speech

In Figure 5.20, UE_2 is making short calls. If the tests are done in parallel, the short calls of UE_2 should start later than the long call of UE_1 to separate the two AMR attempts and insulate the call delivery test. Preferably, UE_2 should dial a wired-line number with auto-answer turned on. The calls should be held for 20 to 40 sec. The wired-line number avoids introducing a second radio link in the end-to-end connection, which could affect the performance evaluation. In this example, UE_2 is used to evaluate the MO call delivery performance on the Uplink.

To measure the MT call delivery performance on the Downlink, UE_3 receives the MT AMR calls from the wired-line phone. Alternatively, a mobile phone could be substituted for the wired-line phone. If so, this mobile phone should remain stationary in an area with good coverage and must not interfere with the drive testing (preferably by using a different PLMN, such as GSM). The mobile phone can also be logged to capture all call access attempts and any other important events for verification. A 10- to 20-sec idle period between short calls must be inserted, to provide sufficient time to release all network resources so the UE can return to Idle Mode.

5.3 Video-Telephony Service Optimization

VT is one of the major new services offered by Universal Mobile Telecommunications Systems (UMTS) and is the main service using the symmetrical bidirectional 64 kbps CS data bearer. It is perceived by the end user as an improvement over voice calls and is therefore regarded as the second service to be optimized. This section focuses on optimization of the VT service and the underlying CS data bearer. Optimizing a CS data bearer for other applications (data modem service notably) is not considered here because, in terms of efficiency, these applications would be best supported in the PS domain. The Third Generation Partnership Project (3GPP) standard recommends CS VT for Release 99 and PS VT for Release 5 and later 3G architectures. They are based on standards derived from the ITU-T H.324, commonly called *3G-324M*. In addition, although VT service is available between a mobile phone and a wired-line phone, we only consider the UE-to-UE case. In a typical optimization scenario, this allows use of the same tools for both the MT and the MO parts of the call.

A video call contains a voice and a video stream in each direction, where more data is used to transmit the video than to transmit the audio. The 3GPP standard offers different techniques to compress both video and audio that does not affect the transmission, metrics, or the test process. A point to keep in mind is that both the VT applications and the coder/decoder affect video quality and call establishment performance.

Figure 5.21 shows a conceptual presentation of the standards involved [14] in VT applications. For video, it is mandatory to support the H.263 video codec. Although, many terminals already support the MPEG-4 simple profile, support for MPEG-4 simple profile and H.261 is optional. For audio codecs, the AMR is considered mandatory and G723.1, optional. For control purposes, the H.245 standard is used. The choice of which video codec to pair with the audio codec is not constrained; it is agreed upon by the applications during session establishment.

In principle, all codecs used for VT compress video information the same way. Motion is captured by sending a sequence of snapshot pictures (called frames). The rapid exchange of these frames gives the impression of smooth movement. Compression [6] is achieved through the two processes described below.

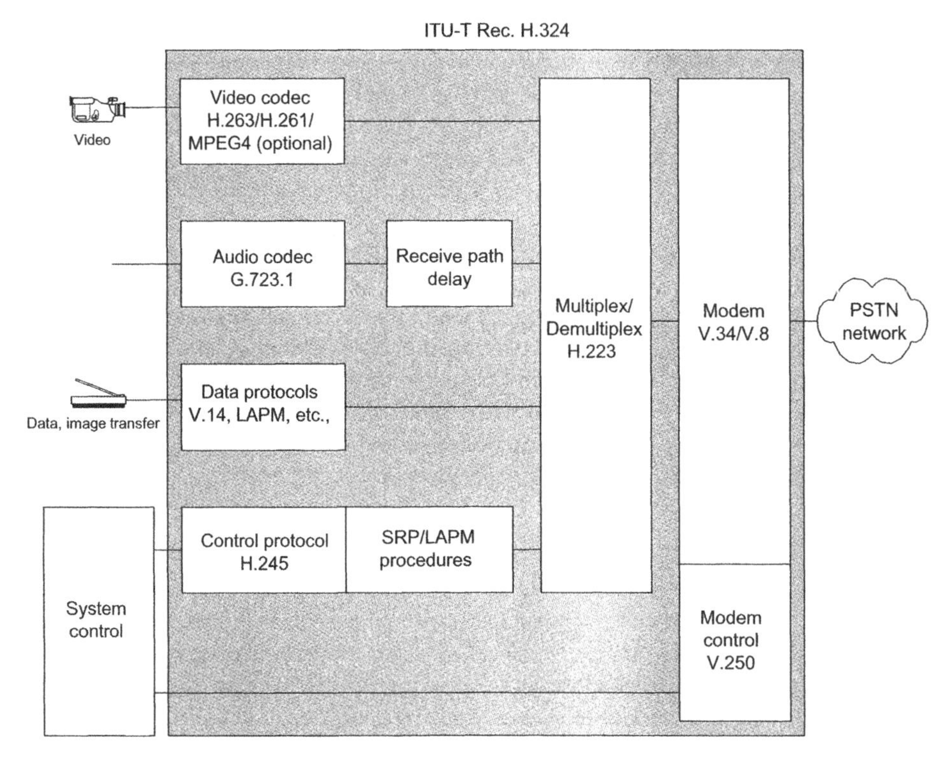

Figure 5.21 Standard codecs involved in video-telephony

In the first process, compression is achieved by sending only the differences between successive images, rather than every complete image. The codec periodically sends complete frames, called "I-frames." They are coded without any reference to previous frames. I-frames are interlaced with "P-frames," which contain only the pixels that differ from those of the previous frame. The I-frames start the picture rendering and recover from situations in which information was lost. With P-frames, the amount of information is dramatically reduced, especially for the assumed typical scenario for VT: a person talking in front of a steady background (like a TV newscast). Hence, I-frames are more important than P-frames. Unfortunately, the network layers used for transport cannot distinguish between I-frames and P-frames. The defined BLER target for VT applications should consequently be based on the worst case: losing the I-frame. This ultimately drives the BLER target requirement very low (less than 1% or even less than 0.5%).

The second process that helps achieve a low data rate is degrading the image quality. If the codec recognizes that it cannot send the complete video information over the available bandwidth, it will reduce the detail level of the video frames either by not sending or delaying smaller objects in the frame, or by reducing the frame transmission rate. This results in alterations (errors) of the encoded sequence, caused by codec limitations and not by radio or signaling effects. On the positive side, the limited resolution of the UE display reduces the needed bandwidth. In 3G-324M VT, standardized frame pixel sizes of 176 × 144 QCIF (Quarter Common Intermediate Format) and 128 × 96 SQCIF (Sub-QCIF format) are the most common.

For VT service, the bandwidth must be guaranteed and the end-to-end delay must be reasonably low and constant. These properties are inherited directly from the CS domain and are very important for good user perception of a video call. As with other CS services, the CS data bearer is associated with the *conversational* Quality of Service (QoS) class and RLC TM (transparent mode). Both ensure real-time delivery of the transmitted data. For UMTS, the only parameter that can be tuned is the target BLER of the associated CS data bearer. Possible settings in the terminal that influence the coding algorithms are beyond the scope of this book and not accessible to the optimization engineers.

5.3.1 Video-Telephony and Voice Comparison

VT multiplexes video, audio, and in-band signaling into one stream at the Application Layer. Synchronization of the parallel audio and video streams (lip synchronization) is a challenge. A short, stable Round Trip Time (RTT) is necessary to ensure a positive user perception; any variation in RTT introduces jitter. Because of the real-time nature of the service, there is no acknowledgment or retransmission mechanism for the audio and video information throughout the network. Corrupted data is discarded and results in quality degradation. The control messages between the VT applications are handled with an acknowledgment mechanism at the application level.

Figure 5.22 compares the call establishment procedures of voice and VT services.

As Figure 5.22 shows, a video call is established similar to a voice call. In this simplified flow, the messaging is identical to that used for voice, up to the Alerting message (not presented for clarity). The RRC Connection is established from the originating UE. The UE then requests the desired service (voice or video call) from the network. This triggers the RAB establishment on the originating side and the RRC Connection and RAB establishment on the terminating side, as well as the security procedures (authentication and ciphering) on both sides. When everything is set up correctly, the destination terminal sends an Alerting message to the originating terminal, indicating that the call was delivered and ready to be accepted. Up to this point, the differences between voice call and video call establishments are limited to the type of service requested and the type of bearer setup. After that point, the two message flows differ, as shown below the horizontal line in Figure 5.22. For a voice call, the voice path is switched and the two users can exchange voice information right away. For a VT call, the CS data path is connected and, after a synchronization phase, the two applications can exchange information to establish a multimedia session. This includes exchanging terminal capabilities, defining which audio and video codecs to use, opening the various logical channels, and exchanging multiplexer tables before the multimedia data can be exchanged.

When performing end-to-end network optimization, special attention is paid to the initial synchronization procedure between the two VT applications. As soon as the CS data bearer is established, the two VT applications start listening to incoming bit streams and attempt to isolate the initial and final flags from the multiplexed frames. These bit patterns must be unique in the stream for the receiver to detect them. During that phase, transmission errors must be very low; otherwise, there is a risk of long synchronization phases, resulting in a long call delay for the user or, even worse, call establishment failure if the synchronization cannot be achieved. This requirement also reduces the BLER target of the CS data bearer.

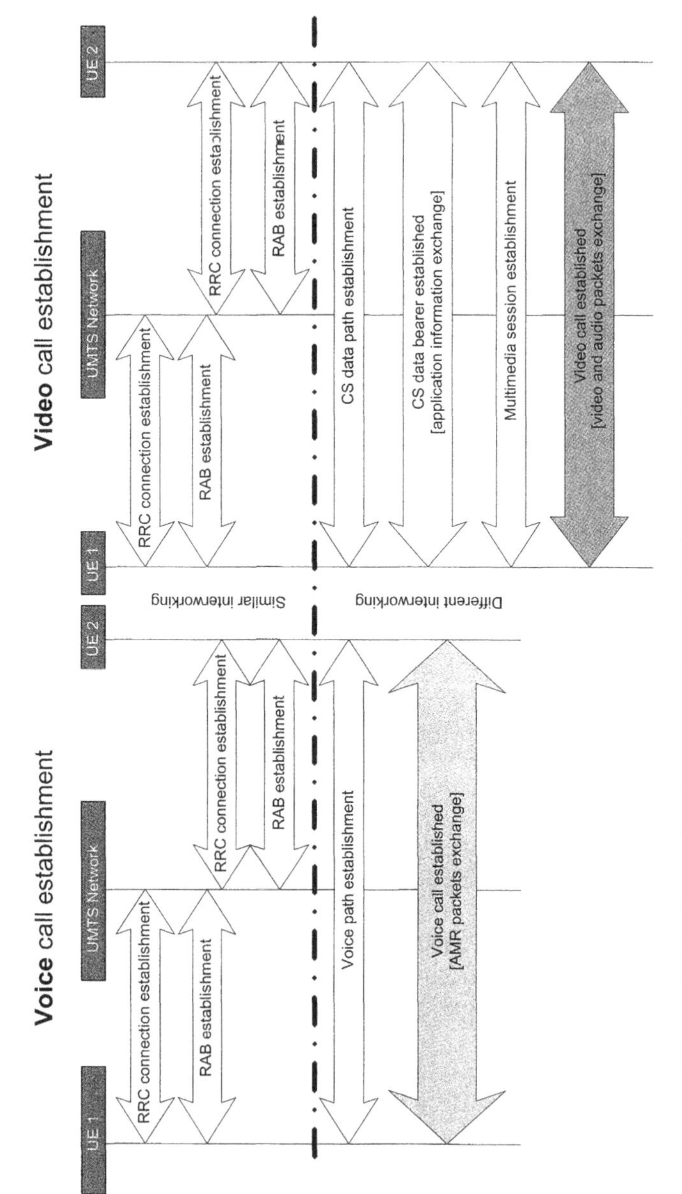

Figure 5.22 Comparison between voice and VT simplified call establishment procedures

During VT call setup, the Application Layer is initialized after the call has been accepted; this lasts several seconds, which the user can perceive negatively. The setup time varies depending on the type (brand) of terminals involved and the (mostly proprietary) synchronization algorithm used. For this reason, it is best to use two UEs of the same band for network optimization testing.

Audio and video streams have redundant error correction information. The amount of redundancy introduced is constant for the voice stream, but not for the video stream. The video codec calculates the optimum amount of redundancy (coding rate) to achieve the specified net transmission rate, on the basis of the amount of information to be transmitted in relation to the available bandwidth. This involves a compromise between video resolution (detail level) and error correction capability, which affects P-frames more than I-frames.

Another important difference between voice call and VT is that a voice call user is silent more than half the time and the AMR coder transmits *silent frames*. These special frames reduce the amount of data transmitted and help reduce the required power on the radio interface, reducing system interference and increasing system capacity and battery life. With VT, even if the user is not talking and everything in the video is still, the video codec uses the extra bandwidth to improve error correction, increase the resolution of the picture, or simply increase the rate of I-frames. This translates into the radio interface being equally loaded throughout the entire video call. This should be kept in mind when setting the admission control and congestion control parameters.

Admission and congestion control algorithms verify the OVSF or channel element availability and the power and interference levels. The power (DL) or interference (UL) levels are most likely to be limiting in a macro network. The two algorithms, interference in the Uplink and power in the Downlink, operate in similar ways. When radio interface resources are requested at the cell level (e.g., for a new service or new soft handover leg), admission control checks to see if the needed resources are available. This verification takes into account both the current resource usage, that is, current transmit power or Uplink interference, and the expected consumption of the new channel. These values are compared to an operator-settable threshold. The congestion control verifies that the current measured situation is under the given threshold. Both power and interference vary greatly, as a function of the data is being transferred. For voice or PS data services, the allocated channel is not always used to its fullest extent. This influences the measured situation. We should remember this fact when defining the thresholds for the two algorithms on the basis of the target quality and target capacity. We should also remember that the allocated channel for VT is always at maximum capacity.

It is also important to consider user expectations when planning VT service. Users expect VT to be as robust and reliable as voice calls, and to have the same voice quality. Many studies have shown, however, that user expectations for video quality are variable. This is partially because the service is new to the broad public (VT over wired telephone lines has, as yet, very limited acceptance). In addition, carriers often lack experience and the tools to understand which quality they should target. Thus, for VT, the acceptable minimum quality that people are willing to pay for remains, to some extent, unclear.

5.3.2 Video-Telephony: Test Process and Metrics

5.3.2.1 Test Process

The test process for VT differs from other AMR or PS data service tests because of the amount of post-processing needed to determine the metrics, sometimes at the frame level.

One of these metrics, subjective quality, varies during playback depending on rendering conditions (terminal, monitor settings, ambient lighting, etc.,).

Data collection for VT requires an equipment setup that can log the RF environment seen by the terminal as well as the signal and video received from the issuing party, while keeping the time references aligned. The received multimedia files must be saved exactly as they are presented to the user, capturing freeze frames, errors, and audio disturbances. In addition, each video frame should be time stamped to allow correlation with the signaling and RF measurements. This can be accomplished by recording the video screens of the transmitting and receiving UEs with high definition video cameras that support time stamping. The process can be simplified by water-marking each video frame with a unique sequence number before transmission. This enables direct identification and analysis of impaired or dropped frames. According to the time stamp, the corresponding RF and signaling situation can be investigated.

A mobile-to-mobile call introduces additional challenges. When detecting an impairment in the received stream, it is not possible to distinguish whether the error was introduced on the first radio interface (transmitting UE to network) or the second radio interface (network to receiving UE), while reasonably assuming that the internal UMTS network is error free. To address this, we keep one terminal in an environment with good RF conditions, and one in an environment where the RF varies. This does not completely remove errors from the radio link on the terminal in the good RF environment, but it minimizes and decorrelates them, allowing us to focus on one interface only.

Figure 5.23 shows a possible test setup, where the VT test process can be divided into three phases: preparation, data collection, and post-processing.

For this test process, we assume that a recorded clip can be sent and recorded over the air interface, rather than during a live session. This is necessary to simplify the post-processing, as both source and received streams must be compared. This process also focuses exclusively on the video aspect, excluding the audio metric.

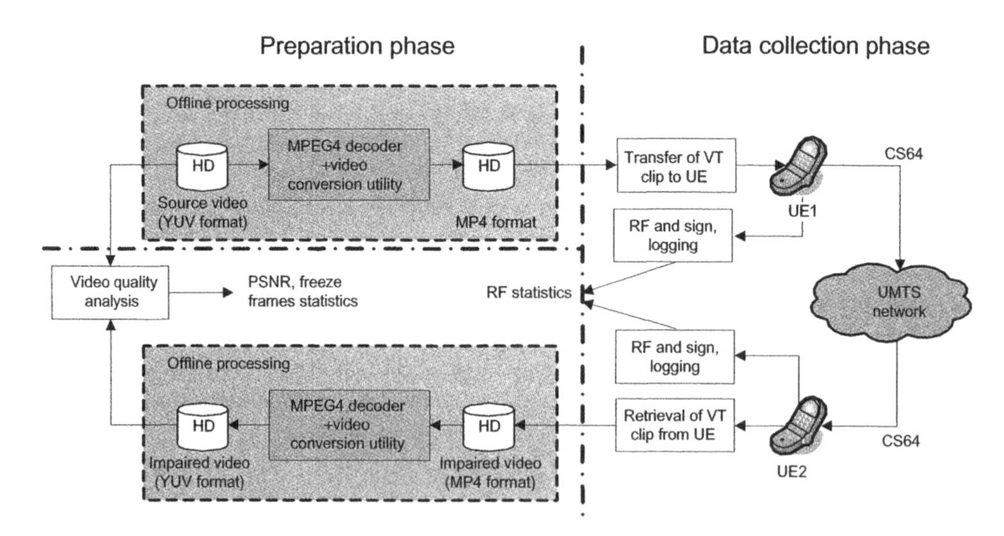

Figure 5.23 Video-telephony test process

- **Preparation phase.** In the preparation phase, select the video clips to use during the data collection phase and set up the test equipment. Select the video clips on the basis of the expected user behavior in terms of the subject of interest (e.g., close-up or wide area, static or high motion, light levels), and take into account the limitations of the codec and the available bandwidth. As explained previously, low bandwidth leads to degradation of video quality, even in the absence of transmission errors. Verify beforehand that the selected clips have sufficient quality, to avoid biasing the measurements with intrinsic errors that cannot be improved by network optimization. After selecting the clips, adapt them to the resolution of the test terminal and watermark them with the time stamp mentioned above. Video clips are usually stored in an uncompressed YUV format and need to be compressed (e.g., into MPEG format) before they can be uploaded to the terminal. Storing clips in uncompressed mode and performing the coding at the time of transmission would be more realistic, but would probably exceed the memory available in the test device. The codec used for video compression should be the same as the codec implemented in the UE. If this is not possible, choose a codec that is commonly used in commercial phones.
- **Data collection phase.** During data collection, video calls are generated in an environment that matches the desired target scenario. The scenario conditions, signaling information, and multimedia received should be recorded with synchronized time stamps on both the sending and receiving terminals for later analysis. It is important to keep all sources and received (impaired) clips in uncompressed YUV format for offline comparison. In contrast to other video coding algorithms, YUV format enables a simple comparison of lines and pixels, which is fundamental for some of the video metrics described in Section 5.3.2.2.
- **Post-processing phase.** During the analysis and evaluation phase, the received (impaired) compressed video format must be decoded back to the uncompressed YUV format. The two uncompressed versions can then be compared and the different metrics computed (see Section 5.3.2.2). Lastly, skilled testers can complete a subjective evaluation to determine the video MOS of the sequence.

5.3.2.2 Metrics

Table 5.7 lists the main metrics for evaluating the quality of VT. The sections that follow provide more details on these metrics.

Table 5.7 Video-telephony quality metrics

Type	Metric	Represents
Video	Peak Signal-to-Noise Ratio (PSNR)	End-to-end degradation of quality between source and received image
	Delta PSNR	Degradation due to transmission process
	Freeze or dropped frames	Major transmission errors
	Video MOS	Overall video quality, as perceived by the user
Audio	Audio MOS	Overall audio quality, as perceived by the user
Synchronization	Lip Sync	Time difference between audio and video stream

5.3.2.2.1 Average Peak Signal-to-Noise Ratio ($PSNR_{avg}$)

Using video coding, we can estimate the quality for a video *MxN* pixels in size by computing the average Peak Signal-to-Noise Ratio ($PSNR_{avg}$) of a luminance, based on 8-bit coding. Equation 5.3 shows this method.

$$\mathrm{MSE} = \frac{\sum\limits_{M,N} [f(m,n) - F(m,n)]^2}{M \times N} \tag{5.3}$$

where

$$\mathrm{PSNR_{avg}} = 20 \times \log_{10}\left(\frac{255}{\sqrt{\mathrm{MSE}}}\right)$$

$f(m,n)$ is Source Video Image Luminance
$F(m,n)$ is Impaired Video Image Luminance

The metric calculation is based on the luminance component of a pixel because the human eye is more sensitive to luminance differences than to color component differences. This calculation is done for every received frame in the impaired video, compared with the same frame in the original video (i.e., the same frame number). For a dropped frame, the PSNR is calculated between the current expected frame and the previous nonfrozen received frame, which corresponds to the error concealment techniques used by decoders. The PSNR calculation reflects the effect of dropped frames. After the PSNR values are calculated for the entire video sequence, the average and the distribution are checked. For a video broadcast, PSNR values above 35 dB are considered as good values. Because VT has limited data rates and transmission errors, PSNR values of compressed clips range from 20 dB, for high motion (sports clips), to 40 dB, for low motion (head-and-shoulder newscasts).

Although PSNR is easy to calculate, it does not correlate perfectly with user perception. In particular, the PSNR calculation does not reflect localized frame corruptions or spots with color shifts. Localized errors are poorly reflected in the PSNR because the average of the entire picture is taken. Even the distribution of the pixel errors is not reflected in the PSNR, meaning that changing a defined amount of pixels, irrespective of their location, produces the same PSNR. However, human eyes perceive error patterns within a frame differently, depending on where the patterns occur. Are the pixels grouped into a single, bigger spot? Or are they spread over the entire picture with only isolated error pixels? For example, poor quality would be detected if several erroneous pixels were concentrated in an area of visual focus (e.g., the eyes of someone speaking). However, the same number of erroneous pixels distributed equally within the frame might go unnoticed.

The PSNR metric calculation is based on the luminance component of a pixel. Consequently, errors in the color components of a pixel do not affect the PSNR. For example, if the sky of a picture is rendered light yellow instead of light blue (assuming the same luminance), the human eye perceives an error, but the PSNR value is not affected. This type of impairment is not expected to be an issue during testing, because of the coding used.

Figure 5.24 Image degradation examples: (a) original image, (b) blockiness, and (c) blurriness

5.3.2.2.2 Delta PSNR

The PSNR metric includes coding and transmission errors, which are perceived differently by the human eye. With coding errors, degradation affects the entire image, adding either blockiness (pixelization) or blurriness, as shown in Figure 5.24. Coding errors depend on the complexity of the image, both in terms of spatial complexity (amount of details) and temporal complexity (amount of movement).

In contrast to coding errors, transmission errors are perceived as random deformations of the image, as shown in Figure 5.25.

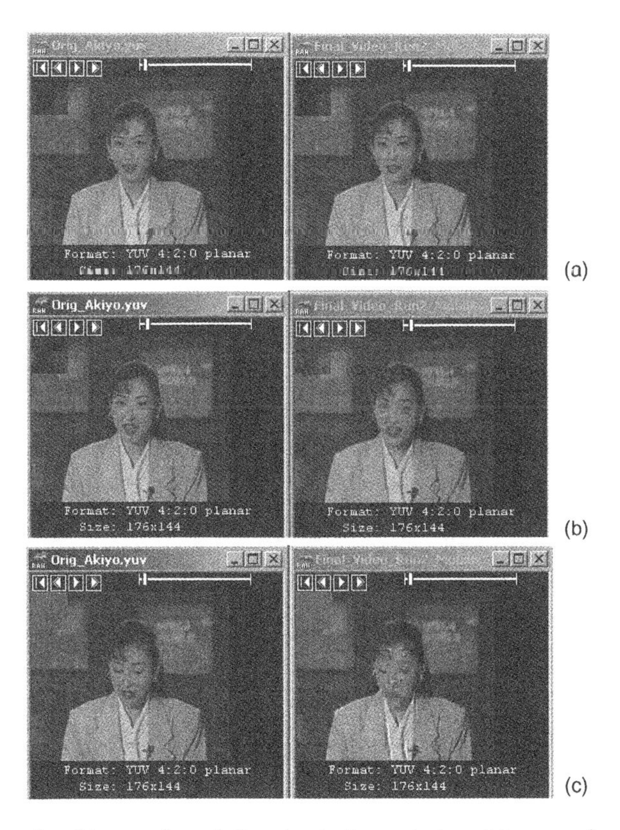

Figure 5.25 Example of image degradation due to transmission: (a) source image and error-free transmission: PSNR = 40.19 dB, (b) source image and minor degradation, PSNR = 30.64 dB, and (c) source image and major degradation PSNR = 23.39 dB

Delta PSNR keeps both effects separate. To estimate delta PSNR, we first estimate the PSNR resulting from encoding alone, and then estimate the PSNR from encoding plus transmission. The delta PSNR is the difference between both PSNRs and reflects the degradation due to transmission errors.

Different software packages can calculate average PSNR values of video clips by comparing the original and the impaired versions in YUV format (or a similar un-encoded format).

5.3.2.2.3 Frozen or Dropped Frames

Another metric that is easy to calculate is the number of frozen or dropped frames. It measures the amount of video information lost. If a video frame is erroneous and cannot be recovered, it is discarded and the previous frame is not replaced (*frozen frame*). A still video is perceived negatively by the user, disproportionate to the duration of the freeze. In case of missing P-frames, the video freezes partially because the difference from the previous frame is affected. The obvious target for this metric is to have no dropped frames.

5.3.2.2.4 Subjective Quality or Video MOS

Variations of average PSNR between samples are of limited value because human perception weights degradation differently, based on its spatial and temporal occurrence. This is sometimes referred to as *recency effect*, in which the perceived opinion is most strongly influenced by the last few seconds of the sample sequence. Hence, an automated video quality evaluation tool must consider the distribution of the PSNR over the entire clip, as well as the intensity and duration of the degradation. Because of its inherent complexity, the choice of quality thresholds is outside the scope of this book. The idea behind this process is to determine the PSNR time distribution, but only when the PSNR is below defined thresholds. If we consider the example given in Figure 5.26, assuming that PSNR

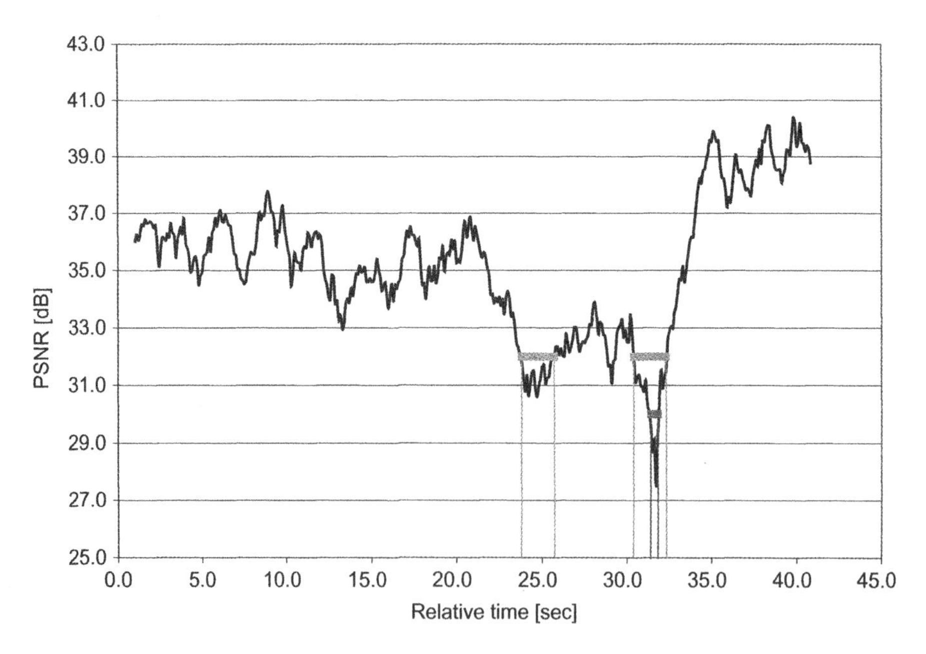

Figure 5.26 Subjective estimation through PSNR mask

thresholds are set to 32 dB for minor degradation and to 30 dB for major degradation, the PSNR falls below the minor degradation threshold 10% of the time, and below the major degradation threshold 1% of the time.

In addition to determining the quality threshold, it is necessary to determine an acceptable time distribution threshold beyond which the quality is not acceptable. For a given clip, these thresholds can be determined by a group of experienced users. The main purpose of this group is to determine thresholds that correlate to MOS scoring for a given clip. The PSNR threshold could be defined in an absolute value, if both coding and transmission are rated. If only transmission is rated, delta PSNR can be used; this value is more often used during optimization.

5.3.2.2.5 Audio MOS

Because the audio data is coded with the standard AMR coder, the default AMR quality metrics can be used. In this case, the objective metrics align well with the subjective perception. Of primary interest is the MOS, as defined in the ITU-T P.800. There are several ways to obtain this: Perceptual Evaluation Speech Quality (PESQ) as defined in ITU-T P.862, Perceptual Analysis Measurement System (PAMS), and Perceptual Speech Quality Measurement (PSQM), as defined in ITU-T P.861. Several software packages are available that can calculate these metrics.

When testing voice only, audio MOS is not normally performed because of the correlation between MOS and BLER that was established during the selection process of the vocoder. For VT, if AMR is used, this correlation can be extended. An understanding of audio MOS is most needed when other Vocoders are used, or when VoIP is considered.

5.3.2.2.6 Lip Synchronization

The easiest way to rank the synchronization between audio and video is called *lip sync*. As the name suggests, it uses the observation that certain events on the video should occur synchronously with their associated sounds (e.g., lip movement synchronized to the sounds of the words). This is usually a manual process, or done with clips that have specific encoding. The test output could be the number of occurrences, the duration for which the media is not in sync, or the average time difference between streams. Special software and content are required for a precise frame-by-frame calculation that determines the amount of video delay compared to the associated audio. Eventually, reference points (audio and video) must be embedded in the original clip. Because of this, a specific test is used to estimate lip sync, instead of the test used to estimate video quality.

5.3.3 VT versus AMR Optimization

This section compares only the differences between VT and AMR call optimizations for access and retention scenarios. The discussion of voice in Section 5.2 is still valid, but it is not needed if the voice optimization was successfully completed before performing VT optimization.

- **Call access optimization:**
  - VT calls require more resources than AMR calls because the lower SF requires higher DPCH power and higher bandwidth, and thus a higher number of channel elements and transport network resources. The admission control algorithm triggers earlier, potentially leading to higher call failure rates, especially in a loaded network.

— The handshake required before starting the video transmission increases the end-to-end call setup latency. It may also increase the overall end-to-end failure rate.

- **Call retention optimization:**
  — The higher amount of required resources affects call retention, just as it does for call access. This can be attributed to the congestion control algorithm, as handover may be blocked in loaded conditions.
  — The higher DPCH power required for VT may increase the call failure rate, or at least cause the target BLER not to converge in low geometry. This is most perceptible when the maximum DPCH power is set to favor capacity rather than coverage. This is further observed when target BLER for VT is much lower than for AMR.

For these reasons, and to simplify optimization by reducing the number of variables, VT testing should be done after the target for voice service has been met. This allows us to concentrate only on RB-specific parameters (target BLER, maximum DPCH setting, admission control, and congestion control) rather than on the full range of 3GPP parameters for both applications at the same time.

5.4 PS Data Service Optimization

Optimizing PS data services is the last step in the service optimization process. It can be divided into three substeps:

1. ***PS call throughput optimization*** – Focuses on a single call in an unloaded cell environment.
2. ***PS cell throughput optimization*** – Focuses on the overall throughput of a cell serving several parallel users.
3. ***PS application optimization*** – Tries to evaluate the expected user experience of specific applications.

Substeps 1 and 2 optimize the same parameters in different ways, so they must be iteratively repeated until the desired balance is reached. During the PS data service optimization process, the operator must decide whether to favor call optimization or cell optimization. Substep 3 explores the interactions between the protocol layers, using relevant traffic scenarios. As we will see, different applications have different QoS profiles. Some parameters (e.g., RAB Inactivity Timer) apply generally to PS data services, while others (e.g., RLC parameters) are QoS-dependent. Optimizing a single application's throughput can be regarded as optimizing the QoS profile associated with the application (e.g., streaming, File Transfer Protocol (FTP), e-mail). A point to keep in mind is that many applications, for example e-mail and Web browsing, could eventually share the same QoS profile due to implementation constraints.

5.4.1 PS Data versus AMR Optimization

In a typical voice or VT call, the user triggers both the call establishment and the call release. Negative events are perceived when the call establishment is not successful, or if the RB is interrupted, since the RB is allocated for the entire call duration. In a packet data session, the user interacts with applications running on a Personal Computer (PC), PDA, or UE. The protocol stack in the terminal ensures that call establishment and release occur with the correct attributes as soon as they are needed. The user has no direct control over this process and may not even notice it, except when the service is initially established and some type of authentication must be provided (e.g., user name and password). As with a PC connected to a Local Area Network (LAN), the user does not know, and is not expected to know, when data is sent over the network.

Another significant difference between voice and data is the asymmetry of the bearer for PS data. In CS services, the path is symmetrical because both users are expected to have the same amount of information to send. This does not generally apply in the packet domain because data transfers from applications are highly asymmetric. A good example is accessing a Web page. A few bytes of data are transferred on the Uplink when a Uniform Resource Locator (URL) is entered, but the page downloaded on the Downlink could contain a large amount of data, from a few kilobits to several megabits.

The Web page example illustrates another difference between voice and packet data: how traffic evolves over time. CS services (e.g., AMR voice or VT) are supposed to generate a constant amount of data over time. For this reason, a constant bandwidth is allocated and granted for the entire duration of the call. In a PS data session, the transmitted traffic varies almost continuously over time. For better resource utilization, it makes sense to adapt the RB to match the actual data rate requirement of the transmitting side. This network behavior is called channel switching; Section 5.4.3 describes it in more detail.

Most PS domain applications require error-free data transmission. Over an error-prone radio link, this implies that data will have to be retransmitted. These retransmissions are handled by the RLC layer and require specific parameter settings, as discussed in Section 5.4.5. This leads to a higher system load and eventually decreases the perceived throughput as the retransmissions are handled by a mechanism transparent to the user. In case of retransmission, the user only perceives stalls or delays in the progress of the task (e.g., FTP download, Web page display, e-mail download), rather than transmission errors.

5.4.2 Typical PS Data Applications and QoS Profiles

During a voice or VT call, the network allocates a fixed bandwidth and must ensure a low, nearly constant end-to-end transmission delay. These are basic requirements for all CS services. In the packet domain, this is not always true because some services require real-time transport while others are delay-tolerant yet more sensitive to transmission errors. QoS classes specify the requirements of the services. Table 5.8 shows the main characteristics of the QoS classes.

The application in the data terminal makes a request to the protocol stack in the PC. Depending on the application, the protocol stack maps the request to a specific QoS profile and routes it accordingly. During the Packet Data Protocol (PDP) context setup, the terminal and the network agree to a QoS profile that matches the QoS characteristics required by the terminal, the subscribed QoS characteristics stored in the HLR, and the resources currently available in the network (i.e., capacity constraints due to current load). Once the PDP context is set, it retains its characteristics until released. Neither the PC nor the user has any influence on the bearer management. Because the bearer attributes

Table 5.8 Quality of Service classes: main attributes

QoS classes	Real-time traffic?	Highly delay sensitive?	Preserves payload content?	Priority
Conversational	Yes	Yes	No	High
Streaming	No	Yes	No	Medium–high
Interactive	No	No	Yes	Medium
Background	No	No	Yes	Low

may affect the monetary cost paid by the user, it is important that the user or the PC request a PDP context that matches the application and its corresponding QoS profile. A practical example of a typical user scenario will help clarify this point.

Assume a user turns on a laptop and plugs in a UMTS data card. The card powers on and registers with the *ATTACH* procedure to the network. The user starts an e-mail client to download mail from a POP3 server. As the application sends the login to the mail server, the protocol stack in the PC recognizes that a connection needs to be established. On the basis of the QoS profile associated with e-mail reading (background class), the PC automatically sends an appropriate PDP Context Request to the network. After the PDP Context and the associated PS data bearer are set up, the e-mail can be downloaded. After the POP3 session is complete, the PDP context remains available.

The user starts reading e-mail and finds a Web link in one message. As the Internet browser starts and requests the HyperText Transfer Protocol (HTTP) page, the PC notices that this request does not have a corresponding PDP for browsing (interactive class). The PC triggers a second, parallel PDP context with the interactive class profile for Web browsing with its associated PS data bearer. After it is established, the PC routes the HTTP request and subsequent dialogs through this PDP.

Now assume that the Web page contains some text, HTTP links, and a small video clip that is played by a streaming player embedded in the Web page. As soon as the streaming player sends the request for the associated video, the PC recognizes that there is no PDP context with the streaming QoS class needed by the player. The PC sets up a third PDP context to handle the request from the video player with its associated PS data bearer. After the video clip has finished playing, the PC still has three active PDP contexts, even though no data is being sent or received.

While reading the Web page, the user decides to download a document from the page. This task is handled by the download manager, an application similar to an FTP client (background class). The request is routed through the first activated PDP context, which is still active and has a background class QoS profile. As the previous example shows, a PDP context and its associated bearer are managed autonomously by the PC. Their establishment or release is fully transparent to the user. Which PDP context (i.e., which QoS profile) is needed for a given request is determined by the requesting application.

As described earlier, different applications map to different QoS profiles. This mapping structure is applied to the entire UMTS network and is divided into individual subsystems [15], enabling the bearer to adapt to the needs of each application and save resources at the same time. The use of QoS profiles makes it easier for the end user to buy the needed network performance from the operator, and operators can charge users differently, on the basis of the attributes of the requested PDP context. In addition to the QoS profile class (maximum and minimum/guaranteed), the PDP context is also defined by the bandwidth (maximum and minimum/guaranteed), allowing charges for data services to be customized to the needs of each user.

5.4.3 Channel Reconfiguration and Resource Planning

Efficient resource utilization is always a major topic for telecommunication networks. In wireless networks, the most critical point for resource constraints is the air interface (Uu interface in UMTS). Because PS data traffic is bursty, it is good to adapt the associated RB to the instantaneous throughput requirements. This is achieved with rate and channel-type switching, which are completely controlled by the network. Both the Uplink and Downlink of the RB can be configured to accommodate different throughputs (e.g., 8, 16, 32, 64, 128, and 384 kbps) independently in the Uplink and Downlink over dedicated (DPCH) or common (FACH) channels.

Optimizing rate and channel-type switching algorithms are critical for optimal network capacity and a positive user perception of the PS data service. However, reconfiguring channel types is time consuming. For the best user perception, reconfiguration decisions should be determined by the amount of data to be transmitted, that is, by the RLC buffer allocation that is periodically reported to the upper layers. From the operator's point of view, reconfiguration should be implemented to minimize resource usage, which encompasses the usage of the OVSF code tree, the DPCH power, and the transport network resources. The optimization goal is to free the unused resources as fast as possible without creating a virtual, user-perceived bottleneck. Retrieval of RLC buffer allocation is easy on the Downlink because those buffers reside in the RNC. For the Uplink, the RLC buffer allocation is retrieved from the RRC measurement procedure, which is optional for traffic volume.

Channel-type switching optimization involves signaling overhead, delay introduced by the switching, user experience, and resource availability. Clearly, the user would prefer the network to never limit bandwidth, while the operator would like to limit the resources allocated to a single user as much as possible to improve the service offered to all users. These two conflicting requirements drive the need for very fast rate switching, which allows operators to dynamically allocate and release the bandwidth according to changing user demands.

Unfortunately, the switching time and the delay introduced by switching cannot be reduced on demand. If rate switching is controlled by buffer allocation, the buffer must be filled up to the defined threshold to trigger the up-switching procedure. Lowering the threshold too much produces a lot of false triggering and dramatically decreases its efficiency. Similarly, if only the buffer occupancy is considered, without considering the time, a single request could trigger the switching, even if it is not needed later on. Before the higher rate is available, a small reconfiguration delay is introduced, because the network and the terminal need some time to reconfigure the channel. The down-switching case is similar, but here the delay is not perceived by the user because it has more bandwidth than needed. Lesser channel-type switching also reduces the imposed signaling overhead as well as the computational performance requirements of the involved RNC and Node B hardware.

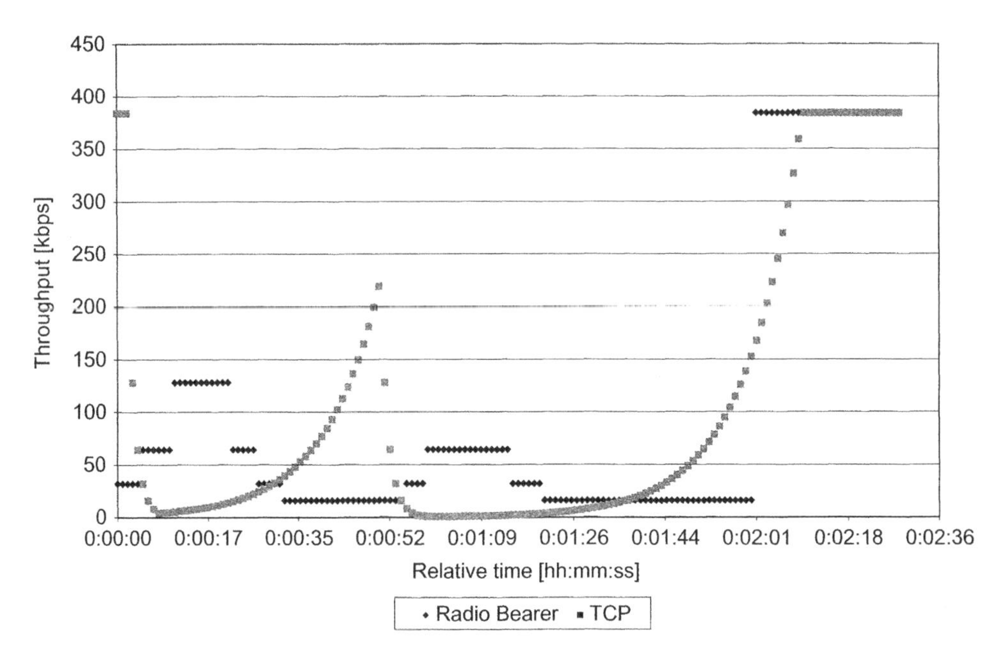

Figure 5.27 Rate switching and TCP interaction

Probably the most important argument for limiting the amount of rate or channel-type switching is the interaction needed with the upper layer protocols, notably Transmission Control Protocol (TCP). Let us demonstrate this with an example of a file being downloaded during an FTP session, as shown in Figure 5.27.

In this scenario, the allocated RB at the beginning of a file transfer is very low (e.g., 32 kbps UL/32 kbps DL). This is reasonable because only a small amount of data has been transferred (user login, directory change, etc.,) before the FTP download started. A point to keep in mind is that rate switching is applied only to the RB. On the Iu interface and in the CN, the maximum bandwidth is assigned during the PDP context establishment (e.g., 64 kbps UL/384 kbps DL). As the FTP protocol begins to transfer data, the buffer fills up because the arrival rate (384 kbps in this example) is larger than the serving rate (32 kbps). The rate switching algorithm should try to increase the RB rate. Let us assume that this process takes 5 seconds. During this time, the transmit buffers are filled, which is seen as congestion on the TCP layer. The TCP congestion avoidance algorithm then exponentially decreases the RAB data rate until the buffer stops overflowing, that is, the allocation pointer drops below a predetermined threshold. Once the congestion condition clears, the RAB rate increases again, but much more slowly than it originally decreased. This process is called *TCP slow start*. Frequent rate switching could lead to situations in which an RB higher than 128 kbps is never observed. If a 384-kbps RB is selected right away, instead of a gradual RB increase, the oscillating bandwidth behavior is not observed, as shown in Figure 5.27 after time 00:02:00.

In practice, not one but several switching thresholds are used to prevent congestion. On the basis of the vendor implementation, the target rate can be defined according to different utilization thresholds.

Channel-type switching and resource utilization in particular can be triggered by other means. An example is a down-switch triggered by bad RF quality. In this case, rather than triggering the down-switching directly from the reported RF condition, a simpler implementation would trigger it from DPCH power. This could produce a positive user experience because higher user throughput is achieved with a lower RB speed but with less retransmission. The positive effect on resource management is that a lower RB requires less DPCH power. This type of rate switching indirectly considers the interference situation (geometry) as well as the radio channel condition (fading and multipath effects). Figures 5.28 and 5.29 illustrate this concept for a DL scenario, both leading to different DL DPCH usage.

Consider the impact of geometry as shown in Figure 5.28. Assume the DPCH power is limited to -7 dB E_c/I_{or}. That is, for a 43 dBm HPA, only a maximum of 36 dBm can be allocated to a single DPCH. In this condition, for a geometry lower than 3 dB, a PS 384 bearer cannot be maintained error free, as shown in Figure 5.28 by the curve leveling off. In this case, the bearer rate should be decreased to PS 128 to limit the number of retransmissions. The same process should be considered when the geometry reaches -3 dB, at which point the bearer rate should be further decreased to 64 kbps.

Similarly, we can consider how channel conditions influence the PS data bearer. The example shown in Figure 5.29 uses the standard channel condition for UE testing [13]. By simulating several cases, we can estimate the required DPCH power and the achieved throughput. If we limit the DPCH E_c/I_{or} to -7 dB for all the cases except the Additive White Gaussian Noise (AWGN) channel, the power setting is not sufficient to maintain an error free PS 384 RB. This can be addressed with rate switching to reconfigure the RB to a lower rate. In an actual implementation, the rate switching would not know the channel

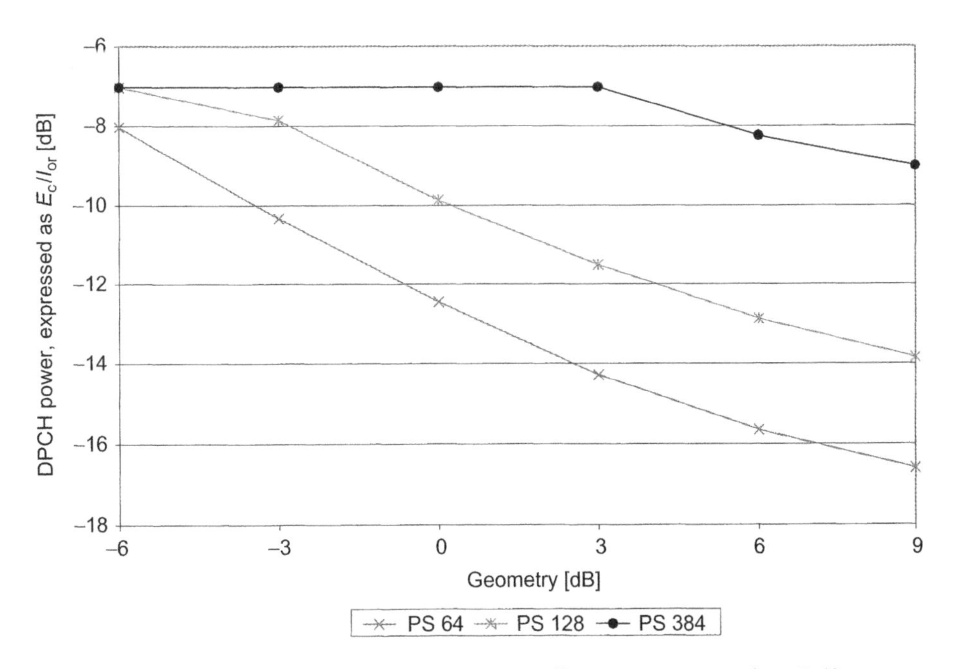

Figure 5.28 Impact of geometry on DL DPCH power, expressed as E_c/I_{or}

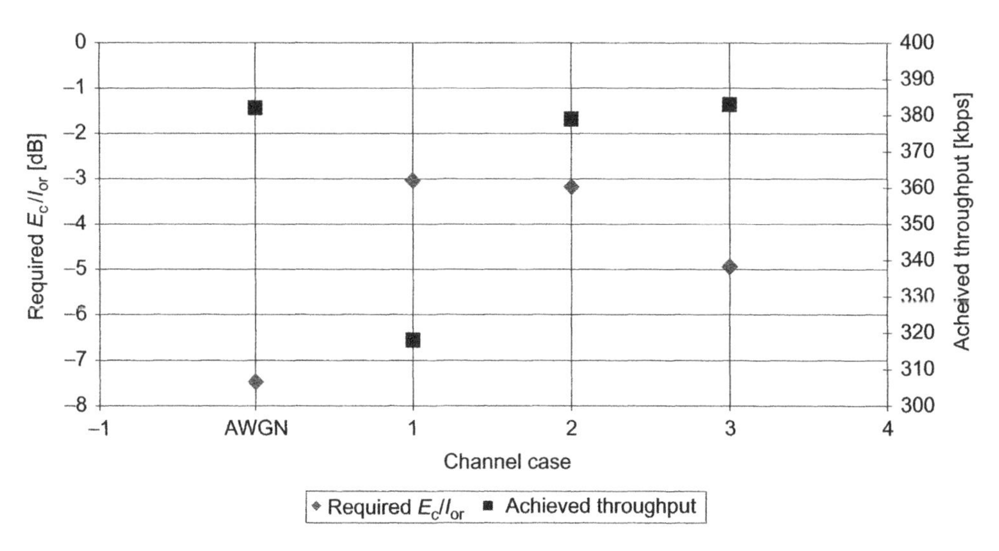

Figure 5.29 Impact of a multipath profile on DL DPCH power, expressed as E_c/I_{or}

condition. Switching would be based only on DPCH power, or Downlink measured SIR, if it was reported by the UE through MRM.

A final situation to consider for this example is long periods of inactivity. In this case, dedicated resources can be released to the terminal of the radio interface while retaining the bearer on Iu and the CN. This is done by switching either the RRC state or the channel type. Figure 5.30 shows the various states associated with the Radio Resource Control protocol (RRC) [10], in a simplified form.

This figure omits the transitions to GSM, which are explained in Chapter 6. The RRC state mainly influences the radio interface. In fact, the CN is informed about RRC state

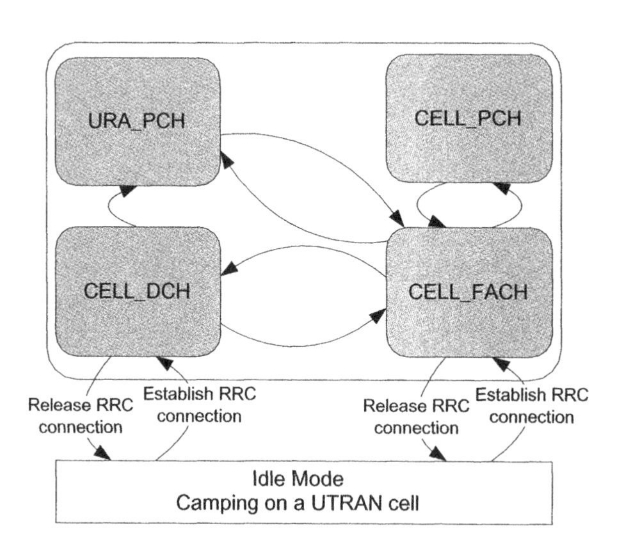

Figure 5.30 Radio Resource Control protocol states

switching only when the UE transitions from Connected Mode to Idle or vice versa. The RNC controls the different Connected Mode states; the CN is not involved. As soon as the UE has dedicated resources, it is in CELL_DCH state. In this state, the UE operates on the Dedicated Channel and the network knows its position at the cell level.

If the UE has less data to transmit or receive than what the dedicated resource can support, it can be moved to another Connected Mode state without informing the CN. If there is still a small amount of data to be transmitted, the UE can transition to the CELL_FACH state, in which only the FACH and RACH/AICH channels are active on the Downlink and Uplink, respectively. If the transmission pause is long enough, the UE can be moved to either the CELL_PCH or URA_PCH state. While in this state, the UE remains in Connected Mode and has at least one PDP context active, but does not have any physical channel resources assigned to exchange data with the network. This conserves both Node B resources and UE battery life. The selection of URA_PCH or CELL_PCH states is dictated by the operator strategy and the network layout and implementation.

From a user's perspective, the RRC state transitions should be optimized so they are fully transparent. The user is affected not only by throughput, but also by the latency (or delay) associated with each state. For example, RBs are set up when in CELL_DCH; thus latency is caused only by the TTI and transmission delays. In CELL_FACH, latency is also affected by the RACH process.

An example of how rate switching affects the user's experience is connecting to a Virtual Private Network (VPN) to access a corporate network from the Internet. This is usually set up using IP-Secure (IP-Sec), which has a built-in keep-alive mechanism by which both ends of the connection periodically check that the link is uninterrupted and uncompromised. This background activity always takes place, even when nothing is being transmitted. The small amount of data exchanged should not prevent the terminal from moving to CELL_FACH; otherwise, the dedicated resources will be wasted. Once the UE is operating in CELL_FACH, the latency introduced in the data exchange could cause the handshake to fail. The situation would worsen if the UE operated in URA_PCH or CELL_PCH, which introduce even longer delays.

Optimizing channel switching is difficult because throughput fluctuates during a PS data session and during a packet exchange within the session. In addition, user expectations vary for different applications supported in the PS domain. To handle higher data volumes and increasingly bursty traffic, future UMTS releases will introduce new features: High Speed Downlink Packet Access ([HSDPA], discussed in Chapter 7) and High Speed Uplink Packet Access (HSUPA). These permit higher data rates using common shared channels.

Introducing these new channels in the UMTS network will affect optimization of PS data services on dedicated channels. Handling traffic with the HSDPA channels will reduce the need for frequent switching. On the other hand, these new channels will require further optimization for the new features and overall system management.

Available bandwidth on the Iub interface is another important optimization consideration. Allocation strategies and constraints are vendor-dependent and usually cannot be tuned during network optimization. To address these constraints, we must ensure that sufficient capacity is available on this interface.

To summarize, optimizing channel-type switching is important to minimize resource utilization. It is also important to consider user perception of the service quality. Perceived quality varies by application, as illustrated by the main three application categories described below.

- **E-mail.** E-mail applications can be divided into two main categories: server-based and client-based. Each generates very different traffic. In a server-based application, the user is always connected to a server, which updates the local interface with either the header only or a full copy of the messages. When just headers are downloaded to the client, the traffic requires very low bandwidth until a message is opened and the entire message is downloaded, including any attachments. To support a positive user experience for such applications, the peak data rate and rapid switching to the maximum data rate are of utmost importance. In a client-based e-mail application, the full messages are downloaded as they become available on the server. A reliable, low data rate service could support this. A high-speed bearer is needed only for login and logout, because a large amount of information may be handled at these times for folder synchronization.
- **Streaming.** For UE-based streaming, the user downloads information at a constant rate for a relatively short period of time (a few minutes). User perception is influenced mainly by the quality of the audio and/or video information received. Video streaming metrics are the same as for a video-telephony application. Audio streaming metrics (bandwidth, dynamic distortion, and so on) should correspond to those defined for home audio appliances. Video and audio streaming both need a continuous information stream, without interruption due to rebuffering. For this type of application, rate switching is of primary importance, because the signal quality degrades at a lower bearer rate.
- **Web browsing.** During a typical Web browsing session, the user starts from a specific Web page. Afterwards, action alternates between downloading and reading. For this application, the type of browser affects the traffic. A mini-browser, as built into a UE, typically downloads less complex information than a browser intended for a PDA or a personal computer. In addition, the content of the downloaded pages (graphics or text) determines the amount of data transferred. This affects reading time and the amount of channel-type switching needed. A realistic model for Web browsing tests should allow the download of multiple pages in a repeatable way and should accurately model user behavior.

5.4.4 Quality Metrics and Test Process

For PS data services, the recommended test setup for collecting optimization data is similar to that used for other services. The main difference is that a data collection tool must be available to collect information, primarily about throughput, at different layers. Tracing capabilities should measure RF and collect signaling and application logs (if available). A network sniffer should run on the UE and the network side. Network internal interface logs are also helpful. A Global Positioning System (GPS) receiver should provide location information and precise time synchronization of the logs, which significantly improves the analysis process.

To optimize call or cell throughput in PS data services, FTP file transfer tests are sufficient. FTP tests ensure that the air interface (Uu) is constantly loaded; however, they do not cover all possible user scenarios. More complex tests are necessary to process all the channel-type switching and scheduling mechanisms involved in user applications such as e-mail, streaming, or Web browsing. In fact, the major difference between these applications and FTP is that FTP tries to continuously load the bearer as much as possible, while other applications are much burstier.

For the best results, we first optimize the PS data bearer throughput by employing FTP tests, then use other applications (such as the ping utility) to optimize channel-type switching and scheduling. One of the advantages of using ping commands to optimize channel-type switching is control over the payload size.

One last point to keep in mind when performing FTP tests is that the transferred file should be uncompressible (such as a binary file) and large enough to provide periods of constant load. To keep the required download and upload times manageable, 5 to 10 MB is used for a 384 kbps Downlink bearer, and 1 to 5 MB is used for a 64/128 kbps Uplink bearer.

Table 5.9 summarizes the key performance indicators for PS data service optimization.

Table 5.9 PS data service KPIs

PS data	Definition	Target value
	PS data call quality metrics	
PS Call throughput	Preferred: avg. throughput measured at Application Layer for FTP transfer in an unloaded system Alternative: throughput measured with network sniffer or Microsoft Windows XP™ performance monitor (Perfmon)	Download only (RAB 384): >240 kbps in mobility; Upload only (RAB 64): >48 kbps in mobility
PS Call setup success rate	Successful PS data call: PS data call reaching PDP context activation accepted PS setup success ratio: number of calls reaching PDP context activation accept/total number of originations	>97%
PS Call setup time	Delay between origination and PDP context activation accept	90% <10 sec
PS Call drop rate	PS dropped call: release of PDP context.	<3%
Round trip time or latency	Return transfer time for a packet through the network	Indicative: 90% <250 ms Target depends on core network architecture and server location
	PS data cell capacity quality metrics	
Single-cell throughput	Preferred: single-cell throughput measured with multiple users in stationary condition	For DL (RAB 384): >800 kbps with user evenly distributed through the cell
Single-cell user capacity	Preferred: total number of simultaneous users (specific RAB type) in stationary condition	For DL (RAB 384): >3 users under near cell condition

The most significant metric is the maximum FTP session throughput, because it is the "real" benchmark that users experience. This can be tested in two different scenarios:

- A single user accesses a cell alone (single-user throughput test).
- Multiple users access a cell at the same time (cell throughput test).

Use both scenarios to test how users affect each other. This may be most visible when optimizing the target BLER for PS data service. The available throughput, sometimes called *goodput*, depends on the number of retransmissions, which in turn depends on the target BLER. For example, if measurements show that a PS 384 bearer with 10% BLER cannot support user throughput higher than 345 kbps, decreasing the target BLER would improve the single-user experience at the expense of other users. As mentioned in Chapter 2, a lower BLER target leads to a higher E_b/N_t target, which results in higher required DPCH. In other words, improving performance for a single user affects system capacity and all other users as the system becomes loaded.

Another important metric is the RTT, which defines the minimum time in which a sent packet can be echoed by the network, for example the time it takes a receiver to acknowledge a packet to the transmitter across the UMTS network. The RTT can be obtained easily with the ping utility. This utility can send a variable number of packets of different sizes, and the address of the target server can change. Pinging can isolate the latency of each node. However, we must remember that the ping utility gives the RTT for the packets, and not the actual one-way delay, or latency. Thus, the asymmetry of the PS data bearer affects the overall result. For large packet sizes, the Uplink delay will be longer than the Downlink delay because the transmission may require multiple TTIs. The measured values depend on the implementation of the CN, which makes the ping test a troubleshooting tool as much as a performance evaluation tool. Areas with abnormally long delays can be isolated and subdivided using the addresses of different nodes along the route.

5.4.5 PS Data Parameters

5.4.5.1 Channel-Type Switching Parameters

Channel-type switching can be based on measurements of buffer allocations for a given duration and can be triggered by reporting events that the UE sends to the network. On the Downlink, this procedure is controlled by vendor-specific parameters rather than standard, over-the-air parameters. On the Uplink, in addition to vendor-specific solutions, the same process can be implemented by setting up traffic volume measurement reporting. The latter can be achieved through MCM controlling Events 4a and 4b. Event 4a is sent when volume exceeds a threshold; Event 4b is sent when it falls below another threshold.

Once traffic measurement reporting is activated, a MRM is generated. The triggering of MRM is based on the following parameters:

- **Traffic volume.** The number of bytes in the transmit buffer.
- **Time-to-Trigger.** A buffer that prevents reporting if the traffic volume is exceeded for only a short duration.
- **Pending Time-after-trigger.** Prevents sending multiple reports if different conditions are met, to limit the amount of signaling.

Table 5.10 Traffic volume measurement parameter trade-offs

Parameter	Setting trend	Impact
Event 4a, *reporting threshold* (or DL equivalent)	Low	Enables quick up-switching, at the expense of resource utilization
	Medium to high	Limits up-switching for high volume data transfers, which improves resource utilization at the expense of user perception
Event 4a, *Time-to-Trigger*	Long	In conjunction with a low reporting threshold, limits the number of up-switches
	Short	Can be used in conjunction with a medium or high reporting threshold to prevent further delays in up-switching
Pending time-after-trigger	Short	Allows rapid reporting to dynamically adapt to traffic volume, at the expense of increased signaling
	Long	Limits the amount of signaling at the expense of less dynamic channel switching
Event 4b, *reporting threshold* (or DL equivalent)	Low	Limits down-switching during periods of low traffic activity, at the expense of higher resource utilization
	Medium to high	Allows for quick down-switching and saves resources at the risk of frequent switching and more signaling
Event 4b, *Time-to-Trigger*	Long	In conjunction with low reporting threshold, further reduces down-switching at the expense of higher resource utilization
	Short	Can be used in conjunction with any reporting threshold to prevent further delays in down-switching

Channel-type switching optimization should be based on both expected user experience and capacity considerations. Table 5.10, which summarizes the main traffic volume measurement parameters, shows the trends and trade-offs.

In addition to traffic-based factors, switching can be triggered by resource utilization either on the Downlink (using a vendor-proprietary implementation) or on the Uplink (using a vendor-proprietary solution or standard event reporting parameters). Optimization decisions vary according to the goals of the operator: for instance, whether user perception or system capacity is most important.

5.4.5.2 RLC Parameters

Most PS data service optimizations–either call or cell throughput–involve RLC protocol parameters. Because PS data services operate in AM, where errors are handled by retransmissions, RLC settings are essential. RLC is a protocol in UMTS networks that allows a lost or erroneous packet to be retransmitted by the lower layer. Table 5.11 summarizes the most important parameters of the RLC protocol, as well as generic recommendations.

The RLC parameters determine whether RLC PDUs are retransmitted or discarded, to speed up data transfer. These parameters are valid only when the RLC is operating in AM. If a PDU is reported in error (Negative Acknowledgment (NAK)) or not acknowledged, it must be retransmitted by the source. Retransmissions can be repeated up to the number of times specified by *MaxDAT*. If the transmission fails, the RLC reset procedure is triggered. This procedure is initialized by sending the RLC Reset PDU. RLC RESET can be triggered after a defined period of time, or after a specified number of retransmissions have been attempted. To reset the RLC state machine, the data currently in the buffer must be discarded, depending on the SDU discard mode [16]. If the reset procedure succeeds, the PDUs that were not sent are handled according to the parameter settings (i.e., discarded or retransmitted), then the transmission continues. If the reset procedure fails, the transmission is terminated and the upper layers are informed (unrecoverable error).

The challenge in optimizing the RLC protocol is to find the best balance between retransmission delays and resource utilization. The following example illustrates this concept.

Under normal RF conditions, transmission errors are rare, their probability of occurrence being set by the BLER target setting. Therefore, most retransmissions are single events. For retransmissions, RLC settings are not important as long as sufficient retransmissions are supported. However, under prolonged challenging RF conditions, retransmissions become more complicated. In the case of a fast rising cell signal or a deep fade, power control might not have enough time to react, or the maximum DPCH could be reached, causing the error rate to increase dramatically (possibly up to 100%). This could lead to a situation in which no PDU, or no error-free PDU, is received that can be acknowledged. If this happens, the RLC will retransmit the erroneous PDUs, that is, the PDUs that were not acknowledged or were reported in error (NAK'ed). To prevent infinite retransmissions, a retransmission counter is maintained. When this counter reaches *MaxDAT*, retransmissions are stopped. The duration of this procedure is set by the *Tx_Window_Size* argument of the *MaxDAT* parameter, and is inversely proportional to the bearer rate.

If a PDU is retransmitted *MaxDAT* times, the RLC reset procedure is triggered. The transmitting side sends a reset and waits for an acknowledgment. If the fading event continues, preventing the RLC Reset PDU from reaching its destination, a second Reset PDU is transmitted after the *Timer_RST* timer expires. This procedure is repeated up to *MaxRST* times. After this, an unrecoverable error is reported to the upper layers, causing the channel to be released and the user's call to be dropped.

At first glance, we might want to prolong the time from the very first failure to release of the channel, in hopes that the deep fading event will terminate and the connection will be reestablished. On the other hand, we must realize that during this entire period the channel is useless for the connection but still generating interference for other terminals that are active in the area. In addition, from the moment the link is released, the UE can eventually acquire another (good) cell and continue the data transfer on that new cell. This reestablishment feature, called *cell update*, can minimize the user's perception of dropped calls.

Table 5.11 RLC parameters

Parameter name	Description	Recommendation	Comment
PDU discard mode	Defines the behavior for unacknowledged PDUs	No discard	Transmission of PDU will be attempted MaxDAT −1 times before the link will be reset
MaxDAT	Maximum number of PDU retransmissions before a RLC Reset is triggered	30	High number allows transmission of the PDU even during temporarily poor radio conditions
Last (re-) transmission PDU Poll	Toggle trigger indication if sender is to transmit a poll after the last PDU available for transmission has been transmitted	Set	Avoids transmission stall after the last PDU is sent
Missing PDU Indicator	Toggle trigger indicating if the UE is to send a STATUS report upon detection of a missing PDU	Set	Ensures retransmission of missing PDU as soon as possible
Timer_RST	Reset timer for RLC resets	> RTT	Ensures that sufficient time elapses between consecutive RST for the RST_ACK to be received
MaxRST	MaxRST defines the maximum number of RLC RESET PDUs retransmissions	8 to 12	Allows the link to be maintained during temporary radio channel degradations
Tx/Rx_Window_Size	Maximum allowed transmitter/ receiver window size	Number of PDU equivalent to 3 × RTT	Ensures that transmission does not stall if PDU is received on the first retransmission
Timer_Status_Prohibit (TSP)	Minimum time between consecutive Status (ACK) reports	~RTT + 2 TTI	Setting TSP > RTT prevents the multiple retransmissions of PDU
Timer_Poll_Prohibit	Minimum time between polls	2 to 3 TTI Should be lower than TST	Trade-off CPU load (large number of polls) and throughput (several PDU to be resent)

An aggressive RLC parameter setting allows only a few retransmissions, providing a short period of time to recover the RLC layer while relying on the upper layers to maintain transmission quality. This results in more unrecoverable errors but allows the UE to change cells more dynamically and recover faster.

An unrecoverable error does not necessarily lead to errors at the Application Layer. If the UE reacquires a cell quickly enough, it could reestablish the connection before the application timer expires. When a new channel is established, the content of the buffers is lost and a retransmission at the Application Layer is required. This introduces a considerable delay from the user's perspective. Therefore, we must consider a more tolerant RLC retransmission setting to allow more retransmission and reset attempts, to keep the channel alive. If the channel recovers, the buffer content may be preserved and the connection can continue. The disadvantage of this approach is the additional interference generated by the retransmissions and resets.

RLC parameters are associated with a channel type or a PDP context. The RLC parameters could be changed at each reconfiguration or activation of a PDP context. As a result, the number of optimization scenarios increases exponentially.

References

[1] Downing D, Clark J. *Statistics the Easy Way*. New York: Barron's Educational Series; 1997.
[2] 26.071. AMR speech Codec; General description. 3GPP; 1999.
[3] 26.093. AMR speech Codec; Source Controlled Rate operation. 3GPP; 2000.
[4] 26.103. Speech codec list for GSM and UMTS. 3GPP; 2002.
[5] Tanner R, Woodard J (eds). *WCDMA Requirements and Practical Design*. New York: Wiley; 2004.
[6] Myers DJ. *Mobile Video Telephony for 3G Wireless Networks*. New York: McGraw-Hill Professional; 2005.
[7] ETSI EN 301 704. Adaptive Multi-Rate (AMR) Speech Transcoding. ETSI; 2000.
[8] 25.214. Physical layer procedures (FDD). 3GPP; 2002.
[9] 25.304. UE Procedures in Idle Mode and Procedures for Cell Reselection in Connected Mode. 3GPP; 2004.
[10] 25.331. Radio Resource Control (RRC) protocol specification. 3GPP; 2004.
[11] 24.008. Mobile radio interface Layer 3 specification; Core network protocols; Stage 3. 3GPP; 2004.
[12] 25.133. Requirements for support of radio resource management (FDD). 3GPP; 2004.
[13] 25.101. UE Radio transmission and reception (FDD). 3GPP; 2004.
[14] 26.111. Codec for Circuit-Switched Multimedia Telephony Service; Modifications to H.324. 3GPP; 2000.
[15] 23.107. Quality of Service (QoS) concept and architecture. 3GPP; 2002.
[16] 25.322. Radio Link Control (RLC) protocol specification. 3GPP; 2004.

6

Inter-System Planning and Optimization

Andrea Garavaglia, Christopher Brunner and Christophe Chevallier

6.1 Introduction

UMTS was designed to meet the anticipated demand for advanced wireless services. The introduction of the WCDMA air interface and the standardization of improved architecture provided high-speed mobile Internet and data connections alongside existing wireless technologies such as GSM/GPRS and Enhanced Data rates for GSM Evolution (EDGE).

Initially, WCDMA was deployed primarily in city centers and business districts, because the higher subscriber density and concentrated demand for new services gave an early return on investment. As a result, a typical deployment consisted of islands of WCDMA coverage in an ocean of GSM cells. Imperfections within WCDMA-covered regions resulted in coverage holes or limited indoor coverage that might trigger inter-system transitions. In that situation, ensuring service continuity at the WCDMA coverage boundaries requires effective Inter-System Handover (ISHO) and cell reselection, along with dual-mode terminals.

Expanding a WCDMA network requires careful planning and optimization of the inter-system boundary regions to provide smooth transitions and seamless mobility for the users. Initially, inter-system transition design is influenced by coverage. In later phases, load sharing and service segregation can also be addressed by inter-system procedures.

This chapter discusses inter-system issues. In particular, it recommends planning and optimization methods to effectively design inter-system boundaries and to set related parameters. Section 6.2 introduces common scenarios and related boundary planning. Sections 6.3 and 6.4 discuss procedures and system parameters for inter-system transitions in Connected and Idle Mode, respectively. Section 6.5 covers system parameter tuning. The chapter concludes with a discussion on multiple carriers and load distribution involving inter-system transitions.

WCDMA (UMTS) Deployment Handbook: Planning and Optimization Aspects QUALCOMM Incorporated

6.2 Inter-System Boundary Planning

Traditionally, mobile communication systems based on different radio access technologies are planned independently of one another, and transitions between them are optimized later, to the best possible extent. This greatly simplifies planning, but may result in poor initial performance of inter-system changes such as reselection, handover, or cell changes.

This section first introduces a formal method of inter-system boundary planning, and later discusses the related optimization process.

6.2.1 Inter-System Borders

Planning the inter-system boundary between WCDMA and GSM is primarily constrained by the existing GSM network. To support current users without disrupting service, the new network must be deployed without changing the existing operational network. In addition, WCDMA network planning emphasizes the primary targeted coverage area, which usually is well within the inter-system boundary, as shown in Figure 6.1.

The difference between the targeted coverage area and the inter-system boundary is the added margin in the design to provide in-building coverage and to increase connection reliability. These margins are usually not at the boundaries; instead, the boundary depends on the actual signal detected, usually at street level. Moreover, because the precise locations of the inter-system boundaries usually are not considered when the system is planned and designed, they must be tuned and optimized at a later phase of network deployment.

To minimize the capital expenditure for a WCDMA network, border cells may be configured to extend their coverage as far as possible. As Figure 6.2 shows, this increases

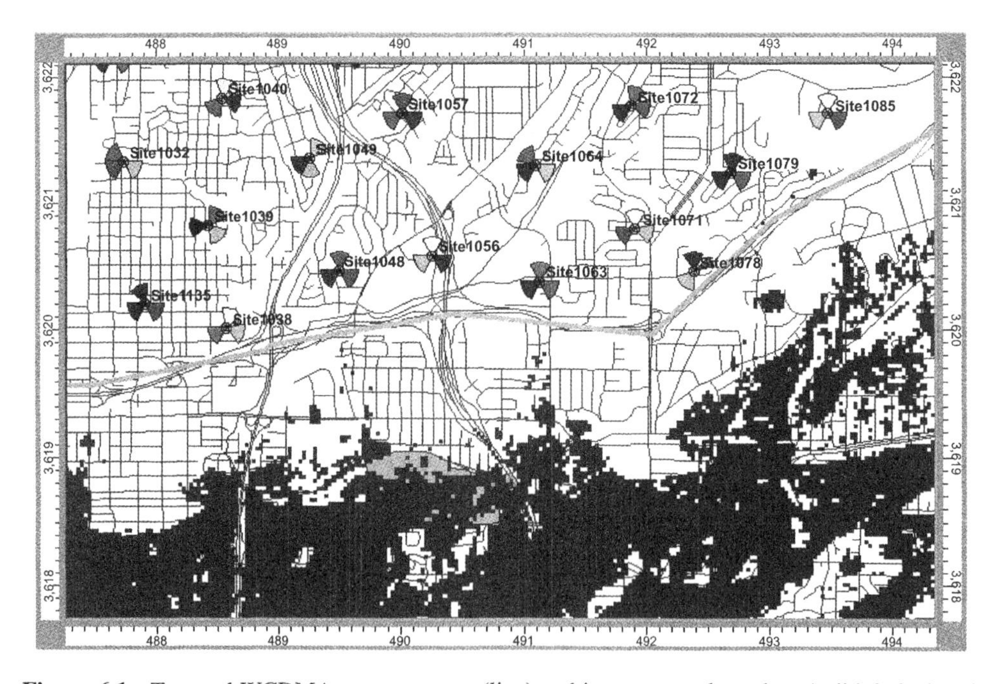

Figure 6.1 Targeted WCDMA coverage area (line) and inter-system boundary (solid dark shape)

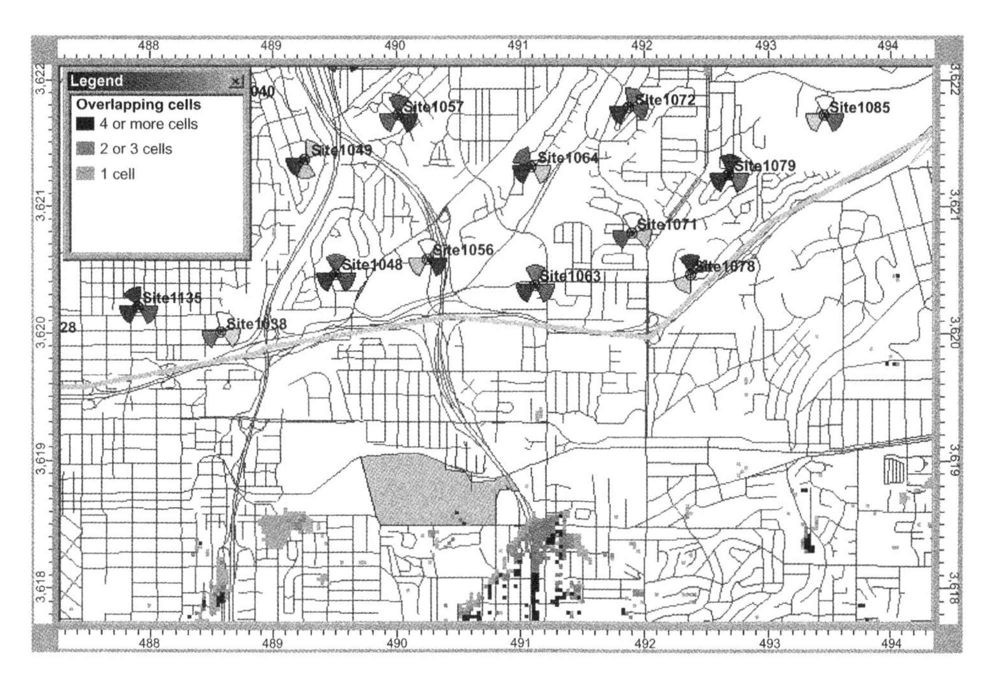

Figure 6.2 Effect of network edges on cell overlap

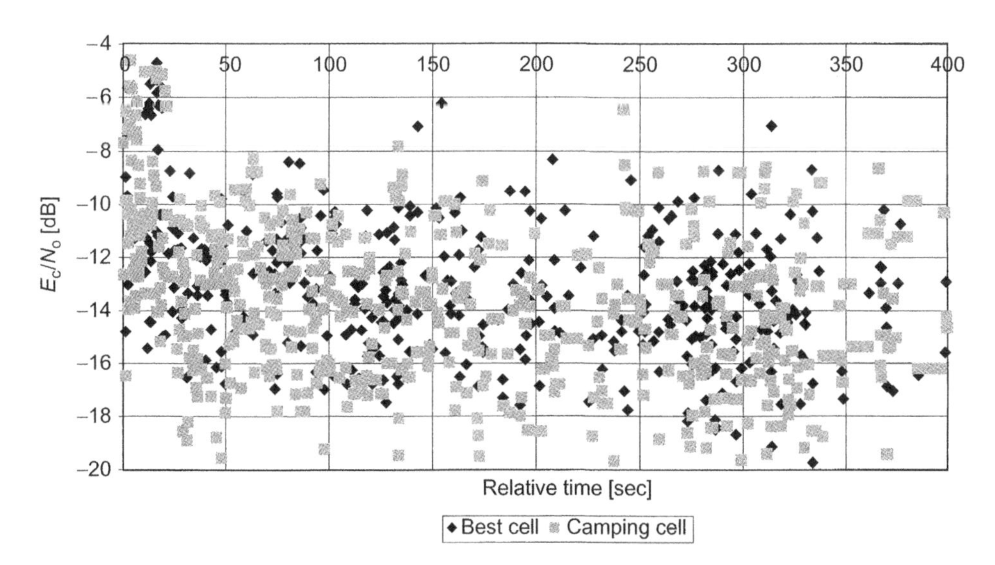

Figure 6.3 E_c/N_o measurements and cell selection suitability criteria

the number of overlapping cells at inter-system borders, which affects both Inter-System Cell Reselection (ISCR) and ISHO.

Cell reselection is affected by the preference given to intra-frequency reselection over inter-frequency or inter-system reselection. Figure 6.3 shows an example of measured

CPICH E_c/N_0 over a drive route at the Network Boundary for the network shown in Figures 6.1 and 6.2.

With multiple cells overlapping in the boundary area, the UE reselects from one weak cell to other equally weak cells until the suitability criteria eventually fail and the UE starts searching for the best neighbor from intra-frequency, inter-frequency, or inter-system lists.

Preventing this problem while maintaining the RF configuration would result in a very high ISCR threshold. For example, the CPICH E_c/N_0 threshold in Figure 6.3 would need to be set at approximately -15.5 dB, assuming suitability criteria of about -18 dB, and 4 sec for the inter-system reselection to take place. If these values were set and the load increased on the WCDMA network, the inter-system boundary would change significantly, as shown in Figure 6.4, and could be well within the original targeted coverage area, depending on the quality of the network plan.

The same results would apply if RSCP were the reselection criteria for intra-frequency. The effect would not be noticeable during load changes, but it would limit the indoor areas where the UE was previously camping on WCDMA. This might be desirable when an indoor solution is available only on GSM/GPRS, not on WCDMA.

To optimize this kind of deployment, the network plan should ensure that a single dominant server is present when the inter-system changes are expected to occur. Adding a single cell or a repeater to the previous example would increase the dominancy over the highway area where the majority of the inter-system changes are expected.

Without adding any more nodes, the inter-system boundary can still be optimized by controlling the coverage of all border cells in the region where most of the inter-system changes are expected. Any of the commonly used RF coverage optimization options could accomplish this.

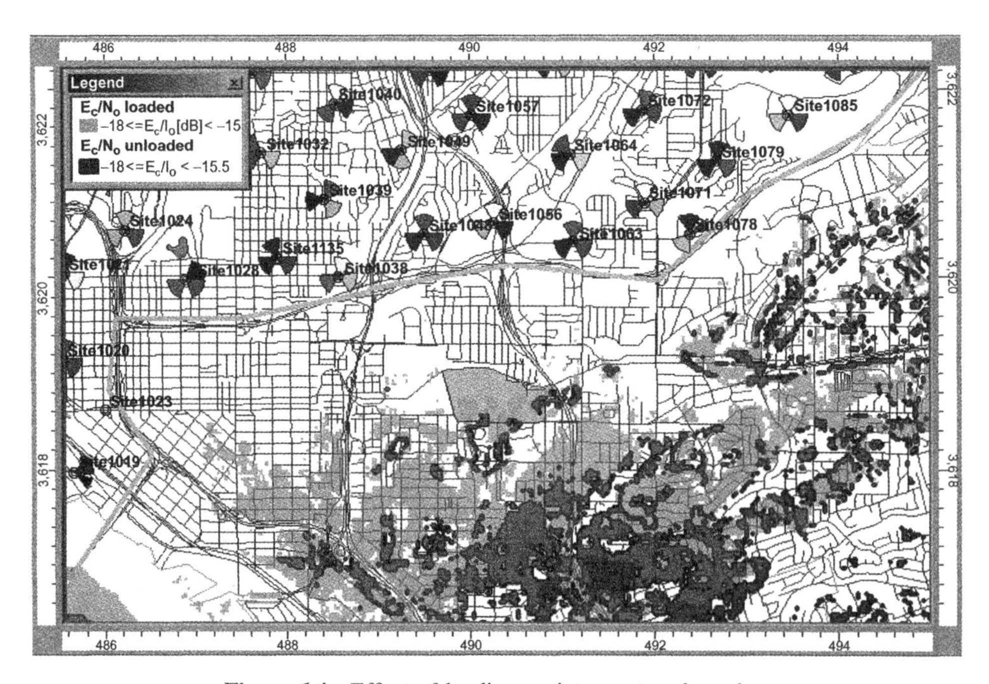

Figure 6.4 Effect of loading on inter-system boundary

6.2.2 Typical Inter-System Scenarios

The examples in the previous section concerned only the inter-system boundary at the WCDMA network edge. This case is the most common and is likely to remain so until the WCDMA footprint reaches a level comparable to that of GSM. Because of the propagation differences between GSM at 900 MHz and WCDMA in the IMT-2000 band (2.1 GHz), GSM coverage is likely to remain dominant until WCDMA can be deployed in lower frequency bands such as 850 or 900 MHz (the actual MHz varies by country). Once this deployment occurs, continuity among the different frequency bands can be achieved through inter-frequency changes rather than inter-system changes.

Given the coverage limitations in the current deployment of WCDMA in the IMT-2000 band, inter-system changes may occur in coverage-hole areas or inside buildings. Indoors, relying on inter-system changes for voice is reasonable until dedicated indoor solutions become widely available. Outdoors at street level, on the other hand, relying on inter-system changes to alleviate coverage deficiencies is not desirable. To prevent inter-system change in these areas, the network is planned properly and the WCDMA layer is optimized.

In-building inter-system scenarios may persist for some time, because of the scarcity of indoor nodes that support WCDMA. Until these are commonly deployed, as suggested by the guidelines in Chapter 8, inter-system changes can extend services indoors. This particular deployment scenario uses the parameter settings detailed in Section 6.6 and is similar to the network edge scenario: inter-system change performance improves if a unique WCDMA best server is present. At the ground level, or at low elevations, it is fairly easy to find such a best server, though a macro network planning tool cannot plan it reliably. Two main factors affect the reliability of network planning tool predictions:

- The exact shape of the building that provides directive attenuation, based on walls and openings.
- The surrounding buildings that reflect or shadow the signal.

On higher floors within the building, especially at heights above the average clutter, it is difficult to have a single best server. The relationship between E_c/N_0 and RSCP for a given server is different at higher elevations than at ground level because more servers (i.e., more interference) are detected. This inconsistency between E_c/N_0 and RSCP makes it difficult to decide which measurement quantity to use for inter-system changes, since either quantity could fall below the set threshold. This is further complicated when suitability criteria are applied. For example, the RSCP suitability threshold might be reached before the E_c/N_0 reselection threshold.

6.2.3 Boundary Determination

When a network planning tool is used to determine the inter-system boundary, the result largely depends on which measurement quantity is chosen (E_c/N_0 or RSCP), and whether Idle Mode (i.e., cell reselection) or Connected Mode (i.e., handover) boundaries are planned. The static nature of such tools affects the predictions they generate. An actual UE implementation usually employs both thresholds and timers to define inter-system changes; this results in a transition region rather than a sharp border.

A reasonable approach would be to plan Connected Mode boundaries to meet call retention requirements, then adjust the Idle Mode boundaries to be slightly larger than the Connected Mode boundaries. This approach meets WCDMA performance requirements for both access and retention performance.

In Idle Mode, only E_c/N_o measurement quantities are effectively available; therefore the boundary region is delimited by the parameters *Qqualmin and* $S_{searchRAT}$, plus any additional offset and hysteresis (as detailed in Section 6.4.4). These values are important for boundary planning because they define suitability limits. They also determine when to start measuring, to ensure that the UE has enough time to reselect to the other system. Estimated distances and a typical speed are used to evaluate how long a UE will stay in the boundary area. That time is compared to $T_{reselection}$ and to the normal cell reselection execution time to verify border region planning. The execution time is critical for verifying whether WCDMA would remain available if the reselection fails and the UE needs to return to its original system.

In Connected Mode and Idle Mode, it is better to estimate a boundary area instead of a single border. Determining the outside boundary is harder for handover than for cell reselection. The reason? ISHO is generally based on events triggered from CPICH measurements, while dropped calls are caused by failure to sustain a predetermined Signal-to-Interference Ratio (SIR) for the selected service. Therefore, the handover boundary can be estimated from the intersection of two plots: the CPICH E_c/N_o (or RSCP, depending on the selected measurement quantity) and the achieved DPCH E_b/N_t. To be comprehensive, the DPCH E_b/N_t for all the services are estimated, not only for voice.

As with cell reselection, the size of the handover area is compared to the typical handover execution time (that is, to the inter-system change delays described in Section 6.3.2).

For both ISCR and handover, we must remember that the defined assumptions limit the accuracy of predictions. For RSCP, assumptions include the propagation model and the margins; E_c/N_o, uses the same assumptions but for multiple links and their simulated load.

6.3 Inter-System Transitions in Connected Mode

ISHO is used in Connected Mode to maintain the desired quality and guarantee service continuity when WCDMA coverage fades out. In WCDMA, the ISHO is a hard handover, usually based on UE measurements. The UTRAN instructs the UE to monitor the quality of the ongoing connection and, if quality falls below a specified threshold, to monitor the signal strength of the other system (such as GSM/GPRS).

The ISHO procedure depends on the UE's domain–circuit switched (CS) and/or packet switched (PS)–as described in Section 6.3.1. In the CS domain, UMTS implements a seamless inter-system hard handover. In the PS domain, the procedure is based on a cell reselection process that cannot maintain the Quality of Service (QoS). In the PS domain, it would be more precise to refer to the process as a Cell Change Order (CCO) instead of ISHO, as used in the CS domain. A better term, when a common term is needed for both domains, would be inter-system change. In the rest of this chapter, *inter-system change* and ISHO are used interchangeably.

The CS and PS domains have two distinct phases for inter-system changes: measurement and execution. In the measurement phase, the UE measures the target system and reports the results to the network. In the execution phase, the UE executes the ISHO procedure. When both phases are present, the handover is called *non-blind*, because the target cell is measured and identified before execution. In *blind handover*, no measurements are made on the target system and the target cell is retrieved from stored database information (see Section 6.3.1.3). However, measurements are made on the current system, according to intra-frequency measurement criteria.

6.3.1 Inter-System Change Procedures

Figure 6.5 shows a simplified inter-system change state diagram, including both ISHO and cell reselection (see Section 6.4) between WCDMA and GSM/GPRS, for CS and PS services.

For CS calls, a handover is performed in Connected Mode. For PS sessions, inter-system transitions are handled either by a CCO or a cell reselection. The following two sections discuss the differences between the CS and PS cases (see Refs [1–3]).

6.3.1.1 Inter-System Handover for Circuit Switched Services

CS connections are used for applications that are sensitive to delays, such as voice or video calls. Because GSM/GPRS currently does not support video-telephony, this section focuses only on voice services.

As indicated at the top of Figure 6.5, ISHO is defined for Connected Mode of CS services in both directions: WCDMA to GSM and GSM to WCDMA. For voice services, however, ISHO in the WCDMA-to-GSM direction is prioritized because good GSM coverage is assumed, which obviates the need to return to WCDMA. The UE reselects back to WCDMA after the call is released. This avoids unnecessary ISHOs, which can

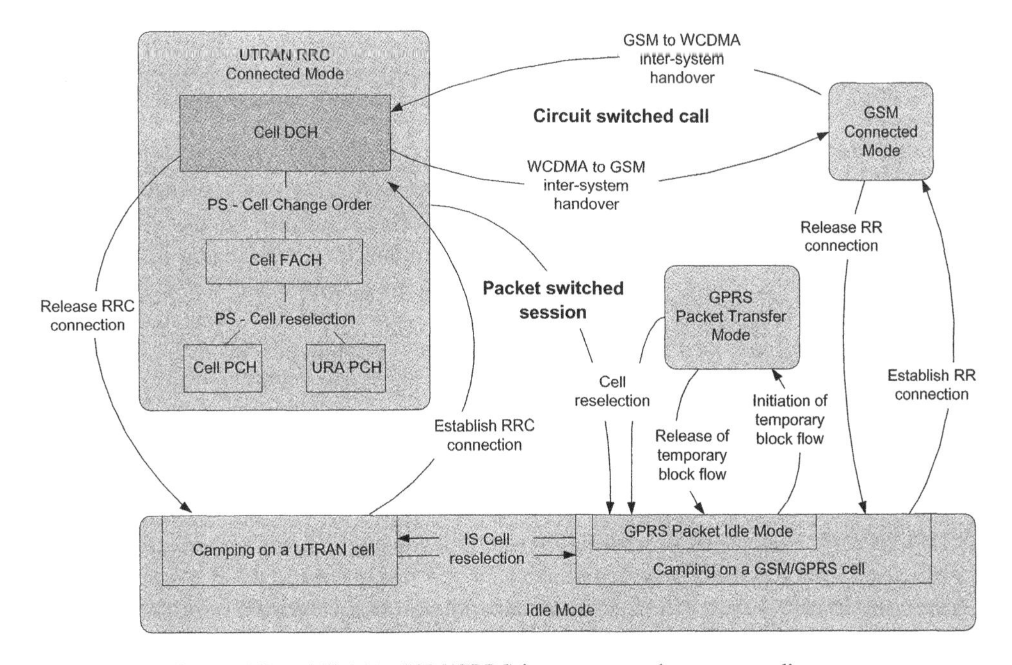

Figure 6.5 WCDMA-GSM/GPRS inter-system change state diagram

increase the risk of dropped calls. As networks become more stable and traffic increases, this process may change to utilize resources better, distribute users between the two networks, and support more advanced services.

In WCDMA, the UTRAN specifies several measurements to monitor the quality of the connection while the UE is connected. If the quality falls below a predefined level, the first phase of ISHO is initiated and measurements are taken on the GSM system. If a second receiver is available, it could be tuned to GSM frequencies to take measurements while the first receiver ensures continuous operation in WCDMA. However, this mode of operation requires an additional radio frequency front-end on the UE, which increases the cost of the equipment.

To overcome this issue, UMTS introduced *Compressed Mode* (CM) operation [4,5]. In CM, gaps are created in the WCDMA frame structure, during which the UE measures the other system (see Section 6.3.3). Once the measurements are completed, the UE sends a report to the network identifying the best candidate for the ISHO, and the second phase is initiated. The handover is finally executed when the network sends the UE a handover from UTRAN command (see Section 6.3.2).

The GSM procedure is similar, with the UE measuring the WCDMA cells when the connection quality becomes too low, or always, depending on the parameter setting [6]. Because GSM systems receive information discontinuously based on the TDM frame, there is no need to create gaps in which to take the measurements. When the Measurement Report indicating the target cell is received at a sufficient level, the network initiates the ISHO execution.

6.3.1.2 Inter-System Transitions for Packet Switched Services

PS services are used mainly for data applications that are not subject to strict time constraints. Though some interactive applications (such as gaming) and streaming services (such as video streaming) may require limited delays, the majority of PS traffic is generated by file transfers, e-mail, and web surfing, for which high throughput is the prime concern.

ISHO for PS services consists of measurement and execution phases, which differ somewhat from the corresponding CS phases. In WCDMA, a UE in Connected Mode registered to the PS domain can assume different states depending on the amount of traffic, the application, and user activity. If a large amount of data is being transferred, the UE is in the CELL_DCH state and the inter-system measurement phase is similar to the one used for voice calls. However, after the measurements are completed and the decision is made to handover to GPRS, the procedure is initiated with the CCO from UTRAN command, which orders the UE to move to the target GPRS cell by means of an ISCR (see Figure 6.5). The UE enters the new system in GPRS Packet Idle Mode and the data transfer is resumed (if data are pending) with a transition to GPRS Packet Transfer Mode. On the other hand, if the UE has an active PDP context but is either not transferring data or is requesting a low throughput, it probably is in the CELL_FACH or CELL_PCH/URA_PCH state, depending on the network implementation and settings. In this case, the ISHO to GPRS is accomplished by cell reselection: the UE autonomously measures and decides which cell to camp on according to the broadcast information. The CELL_FACH state supports both CCO and cell reselection, but generally only one is implemented, depending on the architecture selected by the vendor.

For the transition from GPRS to WCDMA, either in GPRS Packet Idle Mode or in GPRS Packet Transfer Mode, no handover procedure has been defined for GPRS connections[1]. Instead, the UE performs an ISCR autonomously, if network-controlled cell reselection is not requested[2]. When selecting a new cell, the UE leaves Packet Transfer Mode and enters Packet Idle Mode. In Packet Idle Mode, the UE acquires a new WCDMA cell, reads the system information, and may then resume to UTRAN RRC Connected Mode (see Section 6.3.2).

When concurrent services are established (in other words, the UE is connected to the CS and PS domains), the CS handover procedures control inter-system changes. In GSM/GPRS mode, PS data calls are interrupted while a voice call is established. Similarly, during a WCDMA-to-GSM transition, the CS call hands over to GSM and the PS call is interrupted.

In contrast to voice calls, it is desirable that PS data calls remain in the WCDMA system as long as possible. Measurements have shown that WCDMA data throughput remains relatively high, even at very low E_c/N_o[3]. The need to quickly switch to WCDMA will be even greater for HSDPA because it provides significantly improved capacity, throughput, and latency compared to GPRS, as discussed in Chapter 7.

6.3.1.3 Blind Inter-System Handover

For blind ISHO, the strength of the new system is not measured before the transition. The transition is assisted by database information, which determines a target cell based on the cells in the current Active Set and eventually their reported quality. The handover is limited to the execution phase, which is usually triggered by measurements taken by the UE on the ongoing connection (for example, quality on its own frequency). Removing the GSM measurement phase simplifies the procedure and implementation, and speeds up the inter-system transition. Blind ISHOs are appropriate mostly for a WCDMA system that has a one-to-one overlay with a GSM/GPRS network, because there is only one GSM target cell for each WCDMA cell. For other types of overlays, blind handovers should be weighed carefully, because they pose a significant risk of handover failure. For example, if the WCDMA deployment covers an office building from outside while the GSM system is supported by multiple indoor microcells, it is not possible to clearly assign a GSM target cell to a serving WCDMA cell.

[1] Most recent versions of the standard (3GPP Release 6; see [7]) provide seamless inter-system handover from GPRS to WCDMA to support delay-sensitive applications such as video streaming. This is particularly interesting for areas (mainly rural) where the operator may decide to promote new services by updating the existing network to E-GPRS (e.g., EDGE) before completing the WCDMA coverage.

[2] If the *NETWORK_CONTROL_ORDER* parameter is broadcast on BCCH or PBCCH (and supported in the UE and the BSS), additional parameters not listed here apply and the network takes more control of the transition from GPRS to WCDMA [6]. This configuration is not detailed here because it is rarely implemented in current networks and UEs.

[3] In some implementations, Compressed Mode and inter-system handover are not configured for best effort PS services to save capacity and maximize WCDMA coverage. When moving out of the WCDMA area, the call simply drops and the UE must recover by using a cell update procedure with a cause of radio link failure. At this time, the UE selects the best-ranked cell. If parameters are appropriately set, the UE quickly reselects to GPRS, thus changing the system at the border of WCDMA coverage. The disadvantages of this approach are longer inter-system change delay and service interruption with a slightly increased risk of data stall and broken transfer.

For voice calls, triggering a blind ISHO may be too risky because of the increased probability of dropped calls. The PS domain is less sensitive to the risks associated with blind handover because the UE enters GPRS operation in Idle Mode and QoS is less of an issue. For packet data calls, data-assisted handover is an alternative. This handover activates UE measurements to monitor the WCDMA connection. When the network receives a Measurement Report, it does not activate CM and measure the GSM/GPRS system; instead, it sends a (blind) "CCO from UTRAN" command to the UE with a target cell provided by a database. The UE reselects to the indicated GPRS cell and, if this fails, returns to WCDMA as specified by the standard [8]. This avoids the CM disadvantages (throughput reduction, increased transmit power and interference, and reduced capacity) at the expense of a slightly higher probability of handover failure.

6.3.2 Message Flows and Delays

This section provides more detailed information on the ISHO procedure and the signaling message flow for both PS and CS services. It also discusses measurements, execution triggers, and expected handover delays.

Figure 6.6 shows a simplified generic message flow for WCDMA-to-GSM/GPRS ISHO[4].

As shown in the figure, when the WCDMA quality (continuously monitored) falls below a certain level, CM is activated and GSM/GPRS measurements are taken to identify the

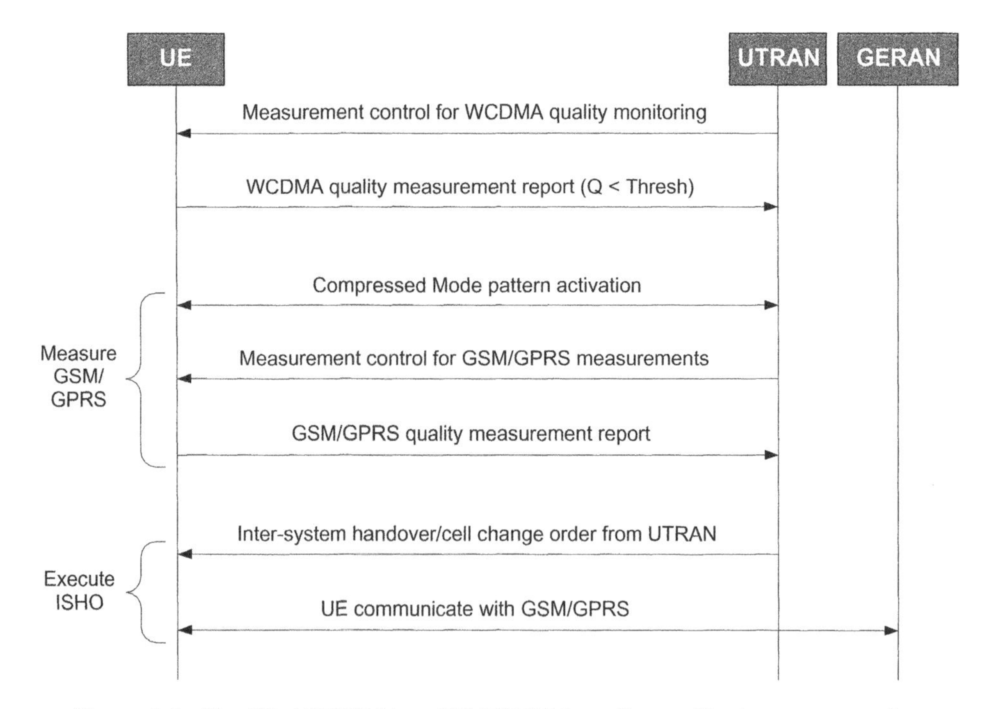

Figure 6.6 Simplified WCDMA-to-GSM/GPRS Inter-System Handover message flow

[4] For inter-system cell reselection triggered for PS services in CELL_FACH, CELL_PCH, and URA_PCH, see Section 6.4.

best target cell and estimate its quality. If specified conditions are met, UTRAN sends a handover (or cell change) order to the UE, which moves to the GSM/GPRS system.

This generic process is common to all message flows, with some variations. In particular, both the WCDMA quality monitoring and the GSM/GPRS measurements can be implemented with periodic reporting and event-triggered reporting:

- **Periodic reporting.** The UTRAN instructs the UE to send measurements periodically and the RNC decides how to use them (this is vendor-specific).
- **Event-triggered reporting.** The UTRAN instructs the UE to send measurements when defined events occur. The network usually reacts to a single event, or a combination of events, with a predefined action such as activating/deactivating CM, or sending a handover command.

Existing infrastructures use various combinations of the two reporting types.

6.3.2.1 WCDMA-to-GSM Handover Message Flow

Figure 6.7 shows the message flow for ISHO from WCDMA to GSM (CS calls). The figure identifies the main components of the handover: (a) setup quality measurements in UTRAN, (b) perform GSM measurements during CM, and (c) execute the ISHO.

When the UE requests a connection and gets a dedicated channel, it receives Measurement Control Messages (MCMs) to configure the reporting of measurements, which

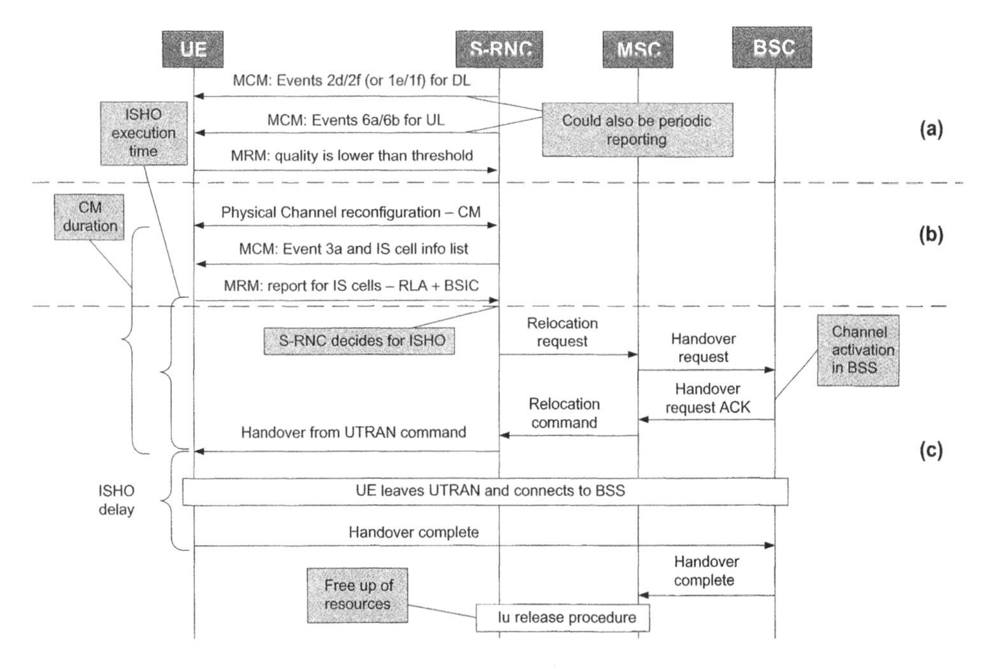

Figure 6.7 WCDMA-to-GSM Inter-System Handover message flow (CS calls)

it eventually sends to the UTRAN (a). Infrastructure vendors use different Measurement Reports to trigger CM and ISHO:

- When Downlink quality is monitored, CM can be activated by Measurement Report 2d (or 1f) and deactivated by Measurement Report 2f (or 1e). In all cases, the measurement quantity can be configured to either E_c/N_o or RSCP; multiple events can be set to monitor both quantities, depending on the implementation. Events 2d/2f relate to a connection quality measurement representing the entire Active Set. Events 1f/1e are triggered per cell, and the UTRAN must combine reports from each cell (usually belonging to the Active Set) to determine if and when to activate and deactivate CM.
- Uplink channel quality in terms of UE transmit power can be monitored with Measurement Reports 6a and 6b. Depending on the vendor, Uplink quality monitoring can be combined with Downlink quality monitoring to determine activation/deactivation of CM.
- Alternatively, vendors can use periodic reporting of WCDMA channel quality instead of event-trigger reporting. The UTRAN processes the measurement results to decide when to start and stop measuring the GSM system.

Once CM has been activated (b), another MCM tells the UE how to report GSM cell quality. Again, either event reporting (typically for Event 3a) or periodic reporting can be used, depending on the vendor implementation. When a good candidate cell is reported, the serving RNC may decide to initiate ISHO by forwarding a request for relocation to the Core Network (c). The request propagates to the BSS for resource allocation and, once accepted, the handover from UTRAN command is sent to the UE, which disconnects from the UTRAN and connects to the BSS. If successful, the UE completes the procedure by sending a Handover Complete message back to the network, and any resources still allocated on the Iu interface can be released.

As shown in Figure 6.7, the time between activating CM and receiving the Handover from UTRAN command corresponds to the duration of the CM session–usually several seconds, depending on signal quality, and the size of the Neighbor List of cells to be measured (see Ref [9] for requirements)[5], and the measurement to perform.

Another interesting quantity is the ISHO execution time, also shown in Figure 6.7. The execution time–usually several hundred milliseconds–depends on the infrastructure and on the interconnections between the UTRAN, Core Network, and GERAN.

Finally, the ISHO delay between the handover command and handover completion is also indicated. The delay time is too short to limit the interruption, which affects voice quality. The standard [9] provides the requirements for limiting ISHO delay.

6.3.2.2 WCDMA-to-GPRS Cell Change Message Flow

The measurements used for Inter-System Cell Change (ISCC) are identical for PS calls and CS calls; however, the handover execution procedure is different.[6] Figure 6.8 shows the

[5] If the quality of the ongoing connection becomes good again, Compressed Mode is deactivated, often leading to a shorter duration.

[6] The Cell Change Order from UTRAN command is used for UEs in dedicated channels or in the FACH Channel, while the URA_PCH or CELL_PCH states use the same cell reselection procedures as Idle Mode (see Section 6.3.2.1).

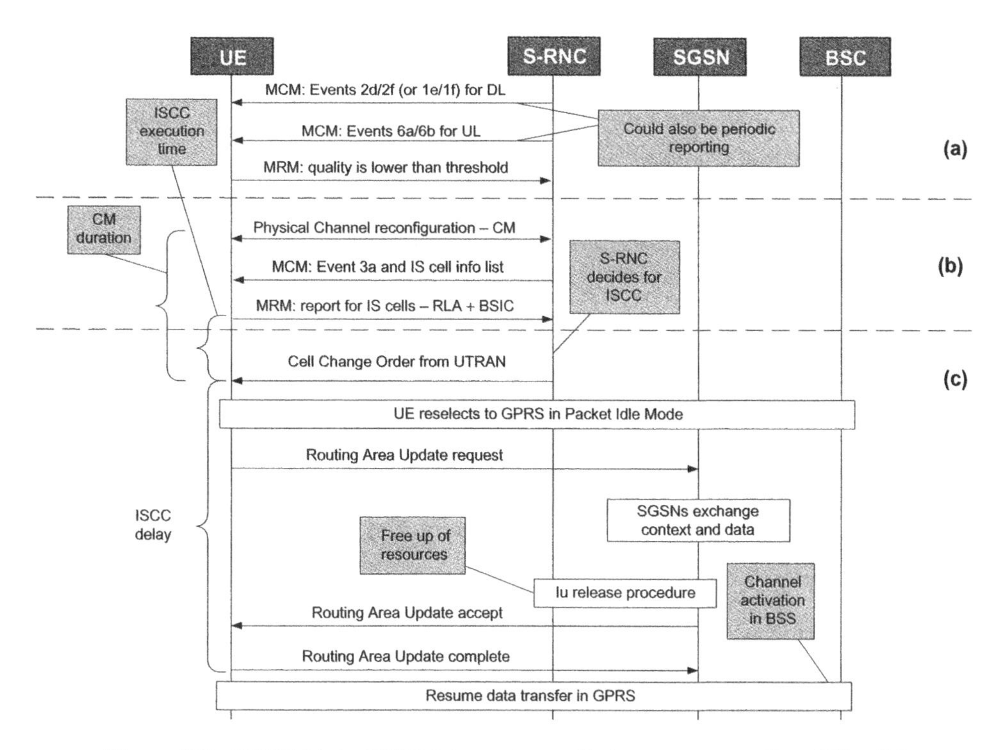

Figure 6.8 WCDMA-to-GPRS inter-system cell change message flow (PS calls)

message flow and identifies the main components of the cell change: (a) set up quality measurements in UTRAN, (b) take measurements on GSM/GPRS, and (c) execute the inter-system cell change. Up to the arrival of the CCO from the UTRAN, the message flow is the same as for the CS call ISHO flow described in Section 6.3.2.1. Therefore, the following discussion focuses on the remaining part of the procedure (c).

Upon receiving the order to move to the GPRS system, the UE reselects the indicated target cell with a cell reselection procedure ending in GPRS Packet Idle Mode. Next, the UE requests a Routing Area Update (RAU) to the new SGSN[7], which retrieves the context information and the buffered data from the old SGSN before releasing the Iu interface resources. After the update is completed, the UE resumes the PS data connection in GPRS as soon as resources are available (GPRS utilizes the remaining time slots after CS traffic has been accommodated in the GSM system).

The time to complete the transition to the other system is influenced by the CM duration and the inter-system cell change execution time, particularly in CS calls. But the transition time is strongly influenced also by the inter-system change delay–the interval between the cell change command and the completion of the RAU. The delay can be as long as 10 to 15 sec, during which no data is transferred between the UE and the network.

[7] If the UE is also attached to the CS domain, a location area update is performed as well, to allow the UE to receive paging for incoming CS calls. In this case, the SGSN passes UE location information to the MSC/VLR before completing the routing area update procedure. Systems are usually configured to update both routing and location area information. The UE leaves the PS domain, attaches to the CS domain to perform the location update, then returns to the PS domain.

Therefore, when optimizing inter-system change for PS domain services, this duration must be analyzed carefully.

6.3.2.3 GSM-to-WCDMA Handover Message Flow

The handover from GSM to WCDMA is not commonly implemented (or activated) for CS services because the GSM system provides excellent coverage for good-quality voice calls. When users move from a GSM-only coverage area into an area covered by WCDMA as well, handing over from GSM to WCDMA in Connected Mode is important only to provide access to advanced services or concurrent services (CS and PS simultaneously) that are supported only by WCDMA. It is also important for relieving capacity limitations in GSM.

Figure 6.9 depicts the typical message flow for the GSM-to-WCDMA handover. The process is similar to handover from WCDMA to GSM, except that CM is not needed in GSM. GSM employs Time Division Multiple Access (TDMA), which enables the UE to measure WCDMA neighbor cells during idle periods. Figure 6.9 also shows the execution time and ISHO delay.

6.3.2.4 GPRS-to-WCDMA Cell Change Message Flow

The inter-system transition from GPRS to WCDMA is actually a cell reselection procedure, since there is no support for handover to provide mobility when connected to GPRS. Figure 6.10 illustrates the message flow for the GPRS-to-WCDMA inter-system change.

After completing measurements on the WCDMA neighbor cells, the UE reselects to WCDMA if conditions are fulfilled, camps on the target WCDMA cell, and sends an RRC Connection Request to UTRAN to perform location updates and RAUs. Once the

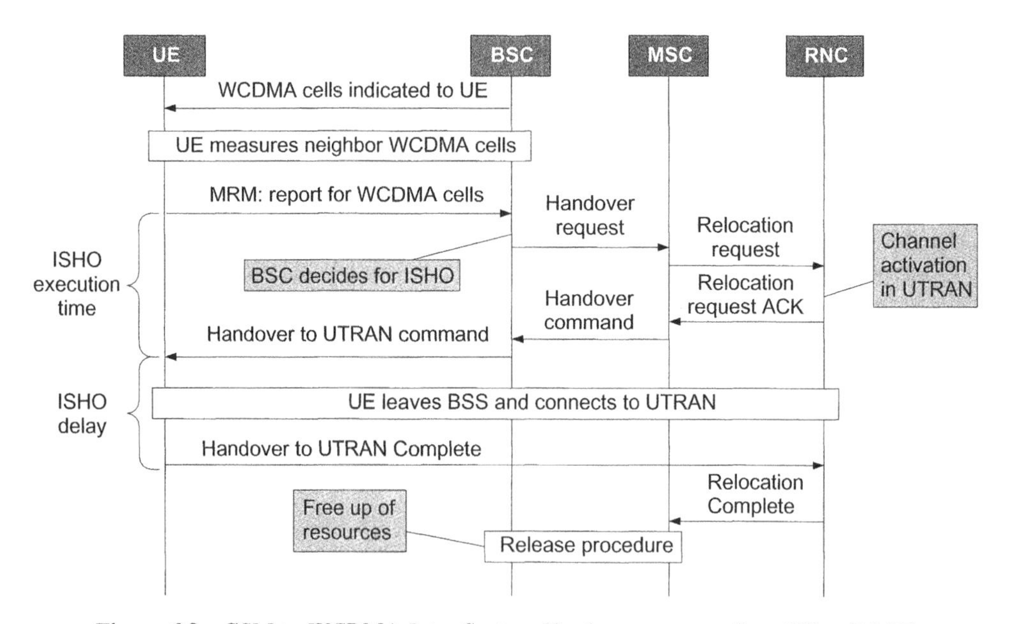

Figure 6.9 GSM-to-WCDMA Inter-System Handover message flow (CS calls) [3]

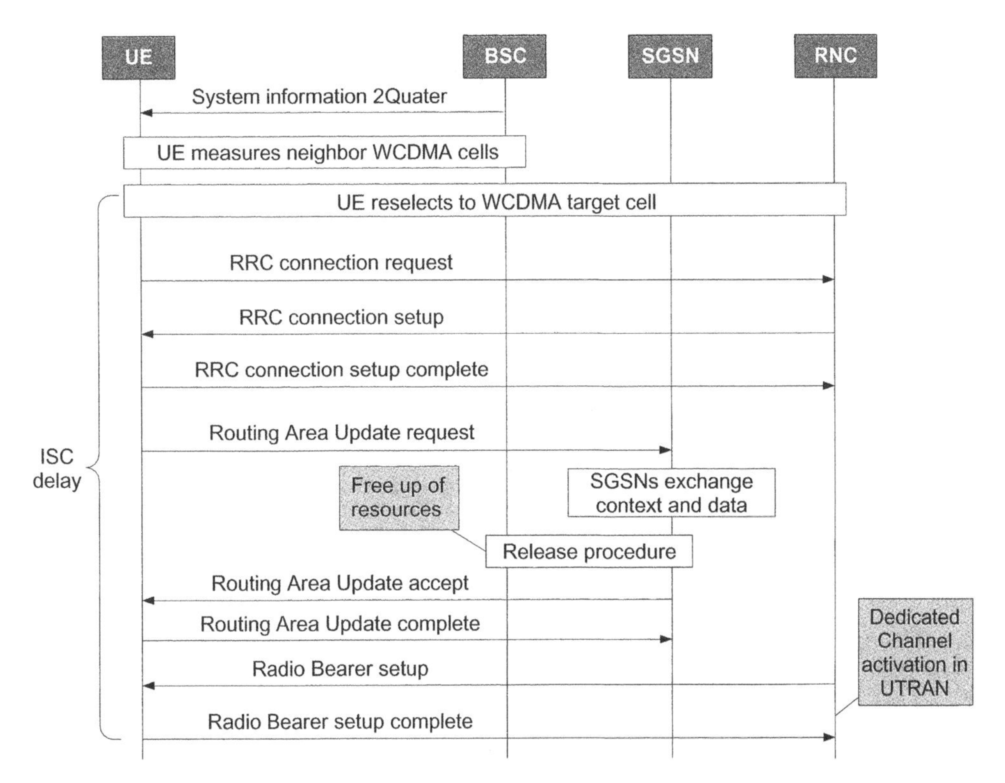

Figure 6.10 GPRS-to-WCDMA inter-system change message flow (PS calls)

Table 6.1 Delay associated with inter-system changes in Connected Mode

Procedure	Delay
WCDMA to GSM – ISHO delay	Several hundred milliseconds
WCDMA to GPRS – ISC delay	10–15 sec (value largely dependent on CN configuration)
GSM to WCDMA – ISHO delay	Several hundred milliseconds
GPRS to WCDMA – ISC delay	5–10 sec (value largely dependent on CN configuration)

registration is complete, the new SGSN retrieves the context information and the buffered data from the old SGSN before releasing the Core Network resources within the BSS. Finally, a Radio Bearer (RB) is set up for the PS data connection in WCDMA, either using the same RRC connection or a new one, depending on the configuration. An inter-system change delay of 5 to 10 sec can be expected before data transfer can continue, as summarized in Table 6.1.

6.3.3 Compressed Mode Issues

WCDMA introduced CM to allow the UE to suspend transmit-and-receive activity on the current frequency and to take measurements on other frequencies and/or systems.

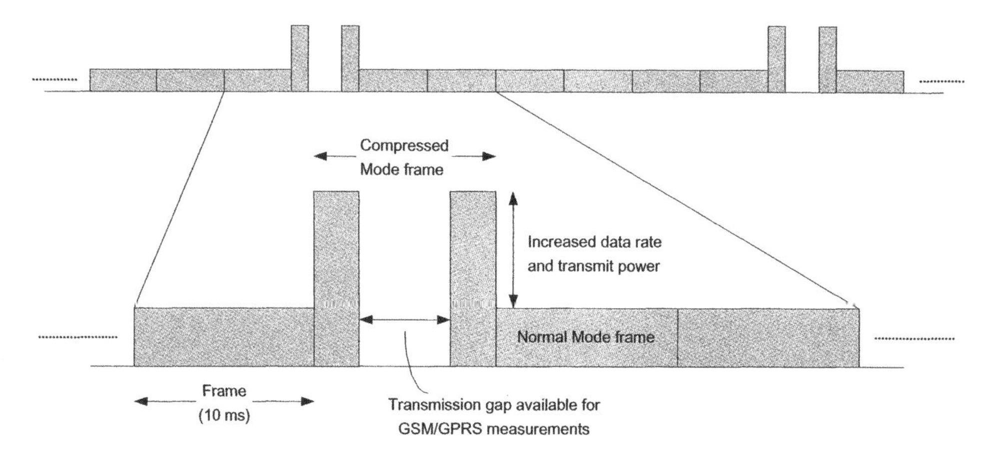

Figure 6.11 Compressed Mode frame characteristics (SF/2 method)

When CM is activated, silent gaps are created in the WCDMA frames. During the gaps, the receiver is tuned to other bands such as GSM/GPRS, in which measurements are collected.

In WCDMA, different methods were evaluated for CM [5]. Two are included in the current 3GPP standards: *Spreading Factor Reduction by 2* (commonly called SF/2), and *Higher Layer Scheduling* (HLS)[8]. The SF/2 method compensates for the lack of data transfer during the silent slots of the pattern gaps by doubling the data rate during the remaining slots of the compressed frames, as shown in Figure 6.11.

The spreading factor reduction method is particularly useful for delay-sensitive applications such as CS services, but has the disadvantage of consuming more network resources to compensate for the reduced spreading gain; this can be challenging at the edge of coverage. HLS, on the other hand, provides gaps by adapting the data rate from higher layers of the protocol stack, which delays some data transmissions to accommodate silent slots. This method does not increase the consumption of network resources, and is more suitable for PS services in which delay is less critical.

Compressed Mode gaps are organized in one or more CM patterns, which the UTRAN specifies. Typically, the UTRAN sends early signaling messages to the UE and the Node B to configure the CM patterns according to the scope and the defined parameters. In another step, when measurements of another system (or of another frequency) are required, the configured patterns are activated at a specified time. The standard is flexible on the supported pattern configurations [4], measurement requirements, and the density of the silent gaps [9]. The requirements are intended to ensure that the measurements are taken within a certain time, and that the gap density and the number of silent slots per frame are limited. Because no power control command is available during the silent slots, too-frequent gaps or too many silent slots per frame can dramatically worsen the connection quality.

As Figure 6.12 shows, gaps occur in Transmission Gap Patterns (TGP), which are repeated within the Transmission Gap Pattern Sequence (TGPS). The Transmission Gap Pattern Repetition Count (TGPRC) parameter specifies the sequence length. Two alternating patterns

[8] A third method, puncturing, was included in the initial release of the standard, but was later removed. In puncturing, measurement gaps are created by removing redundancy bits during rate matching.

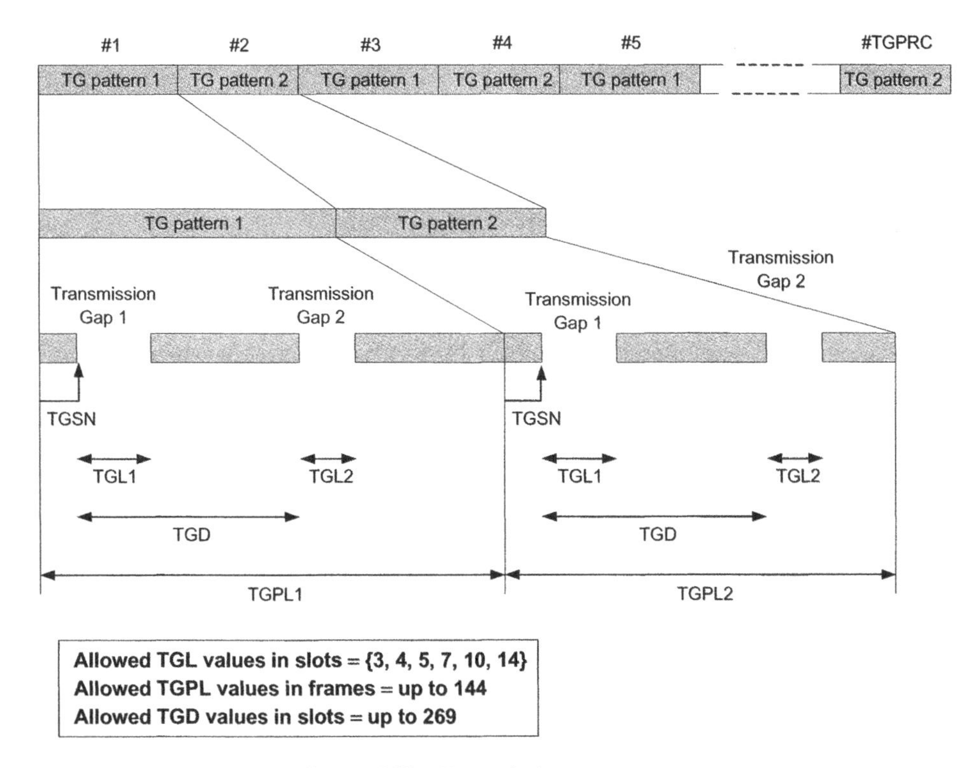

Figure 6.12 Transmission gap patterns

of lengths TGPL1 and TGPL2 frames (Transmission Gap Pattern Lengths 1 and 2) create the pattern sequence; each of them has (up to) two gaps of length TGL1 and TGL2 slots (Transmission Gap Lengths 1 and 2), separated by TGD (Transmission Gap Distance) slots. Each pattern begins in a radio frame called *first radio frame*, which contains at least one transmission gap slot. The Transmission Gap Slot Number (TGSN) parameter indicates the slot number of the first silent slot within the first radio frame.

Depending on the measurement strategy and the purpose of the measurement gaps, different TGPSs can be activated either sequentially or simultaneously, as long as they do not violate the requirement constraints [9]. A trade-off for CM parameters must be made: more aggressive patterns complete the requested measurements more quickly, but at the expense of higher resource consumption. Because the measurement gaps lack power control, an increased SIR target is expected for UEs in CM, increasing Uplink and Downlink transmitted power and hence the air-interface load. This effect is more evident for UEs in channel conditions in which inner loop power control works most effectively, such as slowly moving UEs that have no line-of-sight [10] connection to the Node B.

For inter-system measurements and measurements on the GSM/GPRS system, different pattern sequences are defined for signal strength measurements, initial BSIC (Base Station Identification Code) identification and BSIC verification[9] [9]. Table 6.2 shows possible

[9] The term Reconfirmation is sometimes used instead of BSIC verification. Some vendors skip BSIC verification to shorten the duration of Compressed Mode. BSIC identification synchronizes with the GSM channel to

Table 6.2 Selected Compressed Mode Pattern parameters for GSM/GPRS measurements (N is the number of cells in the measurement list)

IE	GSM RSSI measurement	Initial BSIC identification	BSIC verification
TGPSI	1	2	3
TGMP	2	3	4
TGPRC	6 (if N = <30, else 8)	Min(192, 24 × N)	24 + (300/TGPL1) × min(N, 8)
TGSN	8	8	8
TGL1	14	14	14
TGPL1	8 (if N = <30, else 6)	12	12
TGCFN	X	X + 48	X + 54

settings for the corresponding parameters, where gaps of 14 slots distributed over two consecutive frames are used for GSM/GPRS measurements with the SF/2 method.

The parameters controlling CM cannot be selected independently; they must be set jointly. In addition, the minimum UE performance or observed user performance is evaluated to ensure that all expected measurements can be taken during CM without impairing service to the user. The most critical interactions are highlighted below:

- The Transmit Gap Pattern Sequence Identifier (TGPSI) identifying the sequence is tied to the Transmission Gap Measurement Purpose (TGMP). The TGPSI must be unique for a given TGMP.
- The Transmit Gap Pattern Repetition Count (TGPRC) and Transmit Gap Length (TGL) are set high enough to allow the completion of the intended measurement. In Table 6.2, TGPRC is set as a function of the number of neighbors that the UE should evaluate, while TGL1 is set to the maximum allowed value, 14. With TGL1 = 14, at least 15 GSM RSSI measurement samples per gap can be collected [9], as shown in Table 6.3.
- Both the Transmission Gap Pattern Length (TGPL) and the TGL affect the total time the UE spends in CM (i.e., the time available to make measurements) as well as the throughput reduction when HLS is used in the PS domain. The ratio of TGL (in units of slots) over TGPL, multiplied by 15 to express it in slots, can estimate the throughput reduction. Using the values in Table 6.2, out of 120 slots (8 frames), 14 are used for CM. Therefore, a throughput reduction of about 11% is expected.
- The Transmission Gap Connection Frame Number (TGCFN) is set for different simultaneous patterns to avoid conflict in CM. Looking at the example in Table 6.2, if TGPSI 1 starts at CFN X, the BSIC identification starts 48 frames later, at X + 48. The patterns will not overlap because TGPSI 1 lasts for only 48 frames: TGPRC times TGLP1. The third pattern, BSIC verification, starts 54 frames after RSSI measurement, which is 6 frames after BSIC identification. At that time, the second pattern is still running, but no conflict occurs because the repetition patterns differ, as shown in Figure 6.13.

establish time tracking on the synchronization channel. If BSIC verification is skipped, a loss of synchronization–which can happen under bad RF conditions–may go undetected and the handover may fail. Currently, most vendors do perform BSIC verification.

Table 6.3 UE performance versus TGL1 [9]

TGL	Number of GSM carrier RSSI samples in each gap	Maximum time difference for BSIC verification [ms]	Maximum time for BSIC verification [sec]
3	1	Not specified	Not specified
4	2	Not specified	Not specified
5	3	±500	Not specified
7	6	±1200	10.2
10	10	±2200	8.1
14	15	±3500	5

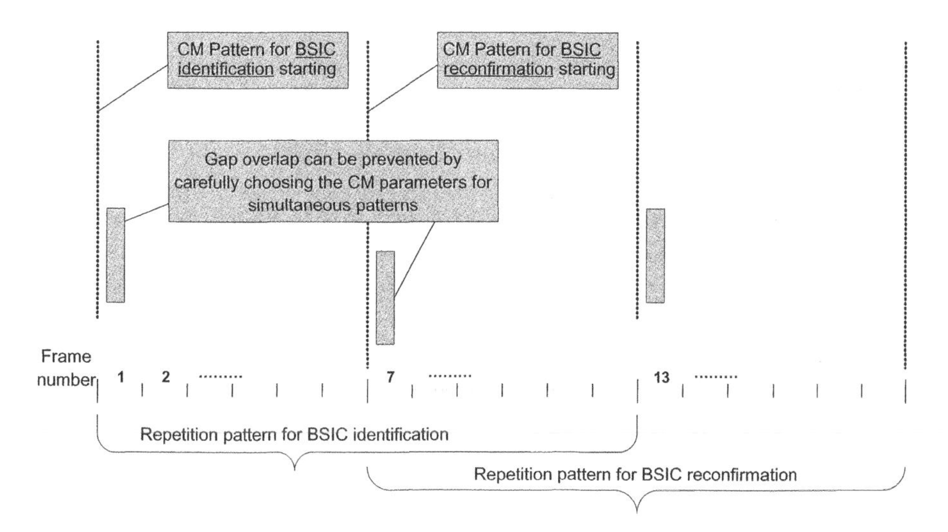

Figure 6.13 Using parameter settings to avoid conflict between patterns

6.3.4 Compressed Mode Performance Metrics

Compared to the metrics for *intra-frequency* handover discussed in Chapters 4 and 5, additional metrics are required for ISHO to characterize CM activation and deactivation, and ISHO triggering.

The performance metrics described in this section are compatible with the inter-system test setup described in Section 6.5.

6.3.4.1 Compressed Mode Metrics

Ideally, CM should start only when measurements on GSM/GPRS cells are really necessary, allowing enough time to take measurements and execute the ISHO before signal degradation disrupts service. Because channels and velocities vary in actual network implementations, it is not always possible to trigger CM precisely at the ideal time. For CS services, it is critical that ISHOs do not occur too late. However, activating

CM too early unnecessarily affects network resources. What indicates that CM has been triggered unnecessarily? We can suspect this if the WCDMA signal improves shortly after CM activation, which leads to an early deactivation of the measurements without triggering ISHO.

Section 6.3.3 suggested parameters for CM patterns, which the following impacts:

- Increased Uplink and Downlink transmit power required for CM based on SF/2, which reduces the air-interface capacity.
- Additional usage of code resources for SF/2.
- Throughput reduction for PS services when CM is based on HLS.
- Minimum duration of CM measurements.

The following metrics relate to CM triggers (for additional details, see Section 6.3.5):

- **Number of CM sessions per executed Inter-System Handover.** If CM parameters are optimally set, very few CM sessions (ideally only one) should be activated for each ISHO.
- **Time spent in CM.** If CM parameters are optimally set, the time spent in CM should be close to the minimum required to measure the other system and provide a narrow statistical distribution.
- **Channel quality during CM, as indicated by E_c/N_0, RSCP, BLER, and Call Drop statistics.** No additional dropped calls should be observed during CM.
- **Signaling.** Because of the necessary triggers for activating/deactivating CM, additional signaling in the form of Measurement Control and, especially, Measurement Report Messages are needed to activate/deactivate CM. CM measurement reporting is considered less important than intra-frequency (i.e., soft-handover) reporting because only a small fraction of users are simultaneously in CM.
- **Location of the CM activations.** If CM is activated well within the WCDMA network, coverage and capacity issues or suboptimal intra-frequency handover parameter settings may be the cause. This metric also depends on the measurement quantity used to trigger CM (CPICH E_c/N_0 or RSCP). If RSCP is the trigger, CM is activated independently of system load.
- **Increase in resource consumption at cell borders (e.g., Downlink and uplink power, active set size[10]).** As seen in Section 6.3.3, resource consumption at cell borders should not vary more than a few percent. A much larger increase could indicate a need for parameter tuning or revision of inter-system boundaries.

6.3.4.2 Inter-System Transition Metrics

ISHO should happen as late as possible without risking a call drop, to maximize WCDMA coverage while guaranteeing service continuity. Since GSM/GPRS networks do not support (or only partially support) advanced services such as video streaming or videotelephony, it is desirable to remain in WCDMA whenever possible.

[10] Since signal strength drops logarithmically with the distance from the transmitting node, signals from several cells may be equally strong at the coverage boundary of the network, thus increasing the Active Set size. As indicated in Section 6.2, boundaries should be planned to ensure that most UEs see one dominant cell before transitioning.

The following metrics are relevant for CS services:

- **Call Drops, BLER, E_c/N_0, and RSCP statistics.** CS services handover seamlessly from WCDMA to GSM/GPRS, to support low latency services such as voice. Therefore, it is important to transition to GSM/GPRS early enough to avoid call drops.
- **Coverage.** CS services typically do not return to WCDMA in Connected Mode[11] (see Section 6.3.1). To avoid limiting coverage for more advanced services such as video-telephony, the transition to GSM/GPRS should not take place too early.[12]
- **Inter-System Handover delay.**

The following metrics are relevant for PS services:

- **Coverage.** To maximize coverage for advanced services such as video streaming, and to avoid the ping-pong effect, the transition to GSM/GPRS for PS services should take place later than for CS services. Ideally, transitions to GSM/GPRS should occur only if staying in bad channel conditions on WCDMA would significantly affect the throughput.
- **Throughput interruptions due to:**
  - WCDMA-to-GSM/GPRS inter-system change delay
  - WCDMA-to-GSM/GPRS reselection delay
  - WCDMA-to-GSM/GPRS transition delay after the call dropped in WCDMA
  - GSM/GPRS-to-WCDMA reselection delay
  - GSM/GPRS-to-GSM/GPRS reselection delay

For PS services, call drops are less important than for CS services, because the UE usually maintains the PDP context and reselects to another cell upon Radio Link Failure (i.e., the UE recovers with a cell update procedure that has a Radio Link Failure cause). The UE resumes the data transfer from higher layers, either connecting to WCDMA or to GPRS. Although this process is defined in the CS domain as well, it has a noticeable impact on the user: namely, voice muting for the duration of the Cell Update procedure.

6.3.5 Compressed Mode Triggering and Inter-System Handover Parameters

This section summarizes the parameters that govern CM triggering and ISHO. Infrastructure vendors use different sets of events to activate and deactivate CM and to trigger ISHO. Section 6.3.1.2 discussed inter-system transitions for PS calls, which use the cell reselection procedure to return from GSM/GPRS and move to GSM/GPRS in all states except for CELL_DCH. The cell reselection parameters are the same as those used in Idle Mode; they will be explained in Section 6.4.4.

Periodic or event-triggered measurement reporting on the ongoing connection is performed to determine when to trigger CM. Periodic reporting is controlled by a parameter that specifies the measurement period. The UTRAN implementation (which is vendor-specific) processes the measurements and triggers ISHO. However, event-triggered measurement reporting is more common. Accordingly, the rest of this section concentrates on parameters related to event-triggered reporting [8].

[11] When GSM-to-WCDMA handover is activated, ping-pong effect between the two systems becomes an important metric because it increases the amount of signaling and also the risk of dropped calls.

[12] For new operators who do not own an underlying GSM network, the limited WCDMA coverage leading to inter-system handover to GSM incurs roaming fees, which reduces revenues.

Most vendor implementations use either Events 1f and 1e or Events 2d and 2f to activate and deactivate CM, based on the Downlink quality. Events 6a and 6b can be configured to assess the Uplink quality. Events 1f and 1e apply to single cells, whereas Events 2d and 2f apply to the complete Active Set. For example, the triggering conditions for Event 2d are outlined below [8]:

$$\text{Triggering condition:} \quad Q_{Used} \leq T_{Used-2d} - H_{2d}/2$$

$$\text{Leaving triggered state condition:} \quad Q_{Used} > T_{Used-2d} + H_{2d}/2$$

The variables in the formula are defined as follows:

- $\boldsymbol{Q_{Used}}$**.** The quality estimate of the used frequency.
- $\boldsymbol{T_{Used-2d}}$**.** The absolute threshold that applies to the used frequency and Event 2d.
- $\boldsymbol{H_{2d}}$**.** The hysteresis parameter for the Event 2d.

The connection quality estimate used in Events 2d and 2f is defined as:

$$Q_{Used\_frequency} = W_{Used} \times 10 \times Log\left(\sum_{i=1}^{N_{AS}} M_{i\_Used}\right) + (1 - W_{Used}) \times 10 \times Log M_{Best\_Used} \quad (6.1)$$

The variables in the equation are defined as follows [8]:

- $\boldsymbol{Q_{Used\_frequency}}$**.** The estimated quality of the current Active Set on the used frequency.
- $\boldsymbol{M_{i\_Used}}$**.** The measurement result of cell i in the current Active Set.
  — $\boldsymbol{N_{AS}}$**.** The number of cells in the current Active Set.
- $\boldsymbol{M_{Best\_Used}}$**.** The measurement result of the cell in the current Active Set with the highest measurement result.
- $\boldsymbol{W_{Used}}$**.** A parameter sent from the UTRAN to UE for weighting the used frequency.

For measurement results in CPICH E_c/N_0, M_{i_Used} and M_{Best_Used} are expressed as ratios. Measurement results in CPICH RSCP are expressed in mW.

Table 6.4 summarizes the triggering parameters for CM. This table and the sections that follow it refer to parameters, or Information Elements (IE), by their standard [8] name.

For ISHO, the UE performs measurements on GSM/GPRS candidate cells while in CM. Both periodic and event-triggered reporting are supported, independent of how CM was triggered, thus allowing different combinations and implementations. For periodic reporting, the UE continuously measures and reports GSM/GPRS cells at set intervals, and the UTRAN orders an ISHO (or cell change) when a suitable target is identified. Event-triggered reporting is based on events such as Event 3a, which weighs both WCDMA and GSM/GPRS quality to determine when to perform the handover; or Event 3c, which only considers the GSM/GPRS connection quality [8]. Table 6.5 summarizes the parameters for the most common events that trigger ISHO.

Table 6.6 lists the parameters related to GSM-to-WCDMA ISHO (CS services [6]). While in GSM, the UE can take measurements on WCDMA cells during the silent time slots without any need for CM (in GSM, as in any TDMA system, the transmission is not

Table 6.4 Compressed Mode triggering parameters [8]

Parameter	Range [units]	Explanation
Triggering conditions (Events 1f, 1e)	Active Set cells, Monitored Set cells, both	Types of cells to trigger Events 1f, 1e
Measurement quantity (Events 1f, 1e)	CPICH RSCP, CPICH E_c/N_o	Measurement quantity of Events 1f, 1e
Threshold (Events 1f, 1e)	CPICH RSCP: −115 to −25 [dBm] CPICH E_c/N_o: −24 to 0 [dB]	Absolute thresholds to report Events 1f, 1e
Hysteresis (Events 1f, 1e)	0–7.5 [dB], in 0.5 [dB] increments	Hysteresis for Events 1f, 1e reporting
Time-to-Trigger (Events 1f, 1e)	0, 10, 20, 40, 60, 80, 100, 120, 160, 200, 240, 320, 640, 1280, 2560, 5000 [ms]	Period of time for which the triggering conditions for Events 1f, 1e must be satisfied before transmission of the corresponding Measurement Report Message can occur.
Filter coefficient (Events 2x)	0, 1, 2, 3, 4, 5, 6, 7, 8, 9, 11, 13, 15, 17, 19	Inter-frequency measurement filter coefficient. The effective time constant of the low-pass filter depends on the measurement period [9] according to the calculation in [8]. The filter coefficient for Events 1x is discussed in Chapter 4 as part of the intra-frequency handover parameters.
W (Events 2d, 2f)	0–2, in 0.1 increments	Weighting factor (between sums of the Active Set versus best cell in the Active Set) in the inequality criterion that must be satisfied for Events 2d, 2f to occur.
Measurement quantity (Events 2d, 2f)	CPICH RSCP, CPICH E_c/N_o,	Measurement quantity of Events 2d, 2f
Threshold (Events 2d, 2f)	CPICH RSCP: −115 to −25 [dBm] CPICH E_c/N_o: −24 to 0 [dB]	Absolute threshold to report Events 2d, 2f
Hysteresis (Events 2d, 2f)	0–7.5 [dB], in 0.5 [dB] increments	Hysteresis for Events 2d, 2f reporting
Time-to-Trigger (Events 2d, 2f)	0, 10, 20, 40, 60, 80, 100, 120, 160, 200, 240, 320, 640, 1280, 2560, 5000 [ms]	Period of time for which the triggering conditions for Events 2d, 2f must be satisfied before transmission of the corresponding Measurement Report Message can occur.

(continued overleaf)

Table 6.4 (*continued*)

Parameter	Range [units]	Explanation
Measurement quantity (Events 6a, 6b)	UE Transmitted power	Measurement quantity of Events 6a, 6b
Threshold (Events 6a, 6b)	UE Tx Power: −50 to +33 [dBm]	Absolute threshold set by UTRAN for Events 6a, 6b reporting (Active Set quality crossing the thresholds is reported).
Time-to-Trigger (Events 6a, 6b)	0, 10, 20, 40, 60, 80, 100, 120, 160, 200, 240, 320, 640, 1280, 2560, 5000 [ms]	Period of time for which the triggering conditions for Events 6a, 6b must be satisfied before transmission of the corresponding Measurement Report Message can occur.
Filter coefficient (Events 6x)	0, 1, 2, 3, 4, 5, 6, 7, 8, 9, 11, 13, 15, 17, 19	Filter coefficient for measurement filter. The effective time constant of the low-pass filter depends on the measurement period [9], according to the calculation in [8].

Table 6.5 WCDMA-to-GSM/GPRS Inter-System Handover parameters [8]

Parameters	Range, units	Definitions
Measurement quantity (Event 3a)	CPICH E_c/N_o, CPICH RSCP	The inter-system measurement quantity the UE shall measure in a UTRAN cell.
Threshold own system (Event 3a)	CPICH RSCP: −115 to −25 [dBm] CPICH E_c/N_o: −24 to 0 [dB]	Absolute threshold set by UTRAN for Event 3a reporting. Primary CPICHs crossing the threshold are reported.
Threshold other system (Events 3a, 3c)	−115 to 0 [dBm]	This IE affects one of the inequality criteria that must be satisfied for an Event 3a to occur. It represents the minimum required GSM RSSI for reliable handover to GSM.
Hysteresis (Events 3a, 3c)	0–7.5 [dB], in 0.5 [dB] increments	Hysteresis between the conditions to activate and deactivate Event 3a reporting.
Time-to-Trigger (Events 3a, 3c)	0, 10, 20, 40, 60, 80, 100, 120, 160, 200, 240, 320, 640, 1280, 2560, 5000 [ms]	Period of time for which the event-triggering condition must be satisfied before transmission of the Measurement Report Message can occur.

Table 6.5 *(continued)*

Parameters	Range, units	Definitions
W (Event 3a)	0–2, in 0.1 increments	Weighting factor applied to the cell quality in the inequality criterion that must be satisfied for an Event 3a to occur.
Filter coefficient (Events 3x)	0, 1, 2, 3, 4, 5, 6, 7, 8, 9, 11, 13, 15, 17, 19	Filter coefficient for UTRAN inter-system measurement filter. The effective time constant of the low-pass filter depends on the measurement period [9], according to the calculation in [8].
GSM Filter coefficient (Events 3a, 3c)	0, 1, 2, 3, 4, 5, 6, 7, 8, 9, 11, 13, 15, 17, 19	Filter coefficient for GSM RSSI measurement filter. The effective time constant of the low-pass filter depends on the measurement period [9], according to the calculation in [8]. Time constant of the GSM RSSI measurement filter.
BSIC verification required	Required, not required	Defines whether the UE must verify the BSIC of measured GSM cells.

Table 6.6 GSM-to-WCDMA Inter-System Handover parameters [6]

Parameters	Range, units	Definitions
FDD_REP_QUANT	CPICH E_c/N_o, CPICH RSCP	The reporting quantity for UTRAN FDD cells that the UE shall measure in GSM.
FDD_MULTI-RATE_REPORTING	0–3	Specifies the number of WCDMA cells that the UE shall include in the list of strongest cells or in the Measurement Report.
Qsearch_C	−98, −94 to −74 [dBm], always −78, −74 to −54 [dBm], never	The signal level threshold above (upper line) or below (bottom line) which the UE shall search for WCDMA cells.

continuous). WCDMA cells are measured when the GSM signal level exceeds a defined threshold (above or below, depending on the parameter value). The selected reporting quantity is reported to the network for ISHO decisions.

6.4 Inter-System Transitions in Idle Mode

In Idle Mode, ISCR from WCDMA to GSM/GPRS enables the UE to choose a new cell in another system to camp on, thus ensuring service availability when WCDMA coverage deteriorates. Cell reselection from GSM/GPRS to WCDMA allows the UE to return to WCDMA, where it can access more advanced services.

PS services use ISCR to move from WCDMA to GPRS in Connected Mode, except when in the CELL_DCH state. ISCR is always used to return to WCDMA (see Section 6.3.1).

6.4.1 Overview of the Inter-System Cell Reselection Procedure

The ISCR procedure consists of the following three steps:

1. Measuring the neighbor cells on the current system and the other system.
2. Ranking the measured cells.
3. Determining reselection.

The measurements on neighbor cells can be continuous, or triggered when the quality of the camping cell falls below a defined threshold.

6.4.1.1 WCDMA-to-GSM/GPRS Cell Reselection

As specified in [11], the UE operates in discontinuous reception (DRX) mode during Idle Mode to improve its standby time. At the beginning of each DRX cycle, the UE wakes up, reacquires the camping cell, and reads its Paging Indicator Channel (PICH). Depending on the measured camping cell quality and on parameter settings, the UE may begin to make intra-frequency, inter-frequency, or inter-system measurements, and evaluate the respective cell reselection criteria. When the UE is camping on a WCDMA cell, inter-system measurements are triggered by one of the following two situations:

1. Measurements are triggered by the *measurement rules*, when

$$\mathrm{Q_{qualmeas}} < \mathrm{Qqualmin} + \mathrm{S_{searchRAT}}$$

2. Measurements could also be triggered if the serving cell does not fulfill the cell selection suitability criterion for consecutive Nserv DRX cycles [9]. According to the standard [11], a WCDMA cell is suitable if it fulfills the suitability criteria:

$$Srxlev>0 \text{ and } Squal>0$$

where:

$$Srxlev = Q_{rxlevmeas} - Qrxlevmin - Pcompensation[dB]$$

$$Squal = Q_{qualmeas} - Qqualmin[dB]$$

In the above equations, $Q_{qualmeas}$ represents the measured quality of the serving cell (typically the CPICH E_c/N_o), while $Q_{rxlevmeas}$ is the measured received signal CPICH RSCP. *Qqualmin*, $S_{searchRAT}$, *Qrxlevmin*, and *Pcompensation* are parameters broadcast by the UTRAN as system information (see Chapter 4 and Section 6.4.4). Both measurements for quality and received signal values must be filtered using at least two samples spaced by at least $T_{measFDD}/2$ sec (i.e., by half of the time specified for FDD measurements) [9].

To estimate the quality of the GSM/GPRS cells, GSM Broadcast Control Channel (BCCH) carrier measurements are collected at least every $T_{measGSM}$ sec [9], in terms of Received Signal Level Averaged (RLA). Measurements must be filtered for each suitable GSM BCCH carrier by using a running average of four samples, uniformly distributed over the averaging period. If the measurements are triggered by the measurement rules, the UE attempts to verify the BSIC at least every 30 sec for each of the four strongest GSM BCCH carriers. The UE also ranks the verified cells. To maintain control of the reselection procedure, cells are considered for cell reselection only if they are indicated by measurement control system information for which the BSIC can be decoded.

Once GSM cells are measured, the ISCR algorithm compares the CPICH RSCP of the WCDMA camping cell with the RLA of the measured GSM cells [9], according to the following ranking criteria:

$$\text{Serving cell: } R_s = Q_{meas,s} + Qhyst1_s[\text{dB}]$$

$$\text{Neighbor cell: } R_n = Q_{meas,n} - Qoffset1_{s,n}[\text{dB}]$$

The UE performs the cell ranking for all GSM/GPRS cells indicated in the measurement control system information of the serving cell. The GSM/GPRS inter-system cell ranking is always based on a RSCP versus RSSI comparison, whereas for WCDMA intra-frequency or inter-frequency cell ranking, the measurement quantity can be either E_c/N_o or RSCP. Most commonly, E_c/N_o is chosen because the RSCP is already included in the suitability criterion for the received signal level.

If a GSM/GPRS cell ranks higher than any WCDMA cell for Treselection seconds, the UE reselects to that GSM/GPRS cell. This implies it being at least $(Qhyst1_s + Qoffset1_{s,n})$ [dB] better than the WCDMA serving cell, and better than any other WCDMA measured cell for Treselection seconds:

$$R_{GSM} > R_s \text{ and } R_{GSM} > R_n \quad (\text{for } Treselection \text{ seconds})$$

6.4.1.2 GSM/GPRS-to-WCDMA Cell Reselection

Cell reselection from GSM/GPRS to WCDMA is supported by a mechanism similar to that for WCDMA-to-GSM/GPRS reselection, as specified in the GSM/GPRS standards [6,12,13]. During Idle Mode operations in the GSM/GPRS system, WCDMA cells included in the 3G reselection list are measured when the RLA of the serving cell differs (is lower or higher, depending on the setting) from the parameter *Qsearch_I* (or *Qsearch_P* for packet services) [6][13].

To estimate the quality of GSM/GPRS cells, measurements are filtered for each BCCH carrier using a running average of five unweighted samples, uniformly distributed over a period of at least 5 seconds (the value could be larger, depending on the paging block and on the number of nonserving cells). For WCDMA cells, the UE measures both CPICH E_c/N_o and CPICH RSCP and there is no specification about filtering and measurement time. However, the UE must be able to identify and reselect to a new WCDMA cell belonging to the Neighbor List within 30 seconds.

[13] If PBCCH (Packet BCCH) is used for GPRS, the cell reselection parameters can be assigned different values for CS and PS services; otherwise, the UE can receive only one common value on the BCCH.

After measuring the serving GSM/GPRS cell and the signaled neighbor cells (both in GSM/GPRS and WCDMA), the UE ranks the cells on the basis of RLA and RSCP measurements, for the serving cell and for the six strongest nonserving cells. The UE then reselects a suitable[14] UTRAN FDD cell if all the following conditions are true for at least 5 seconds:

- The measured CPICH E_c/N_0 value of the candidate WCDMA cell is equal to or greater than the parameter *FDD_Qmin*.
- The measured CPICH RSCP value of the candidate WCDMA cell is at least *FDD_Qoffset* [dB] better than the RLA of the serving cell and all suitable nonserving GSM/GPRS cells.
- The measured CPICH RSCP value of the candidate WCDMA cell is equal to or greater than the optional parameter *FDD_RSCP_threshold*, if supported by the UE.[15]

In the above conditions, *FDD_Qmin* and *FDD_Qoffset* (or *FDD_GPRS_Qoffset* for PS connections) are broadcast on the BCCH (or PBCCH) of the serving cell and *FDD_RSCP_threshold* is determined by the following equation:

$$FDD\_RSCP\_threshold = Qrxlevmin + Pcompensation + 10 \text{ [dB]} \tag{6.2}$$

If these parameters are not available, then *FDD_RSCP_threshold* is set to negative infinity, meaning that the criterion is not effective. If an inter-system WCDMA-to-GSM/GPRS cell reselection occurred within the previous 15 seconds, *FDD_Qoffset* is increased by 5 dB in this period.

Even if the above conditions are fulfilled, the WCDMA cell might not meet the suitability criterion if the CPICH RSCP is too weak, thus causing the UE to return to GSM (G2W-reject). In this case, access is disrupted while the UE attempts the reselection to WCDMA. A loss in service availability will occur because the UE will not receive a page from the system during this time.

If the above conditions are fulfilled *AND* a suitable WCDMA cell is found, the UE may return (ping-pong effect) to GSM after a short time, which also affects service availability. In fact, every time an ISCR takes place, the UE must perform location area updates and RAUs. These procedures can last several seconds, during which no service is available. The UE cannot be reached on the system it just left, nor on the system it is attempting to camp on, because it has not yet signaled to the network (through LA or RA update) that it has changed serving networks.

6.4.2 Message Flow and Delays

This section presents more detailed information on the ISCR procedure and on the signaling message flow. In Idle Mode, the procedure is the same for both the PS and CS domains

[14] A UE may start an inter-system cell reselection towards a WCDMA cell before decoding the BCCH of the WCDMA cell, leading to a short interruption of service if the WCDMA cell is not suitable. If this happens, the UE returns to GSM/GPRS and a loss in service availability can occur. This case is referred to as GSM-to-WCDMA cell reselection reject, or G2W-reject.

[15] A fixed RSCP-based threshold for GSM-to-WCDMA cell reselection has been added to the standard as an optional feature for the UE, to mitigate some of the parameter setting constraints (see Section 6.6.3).

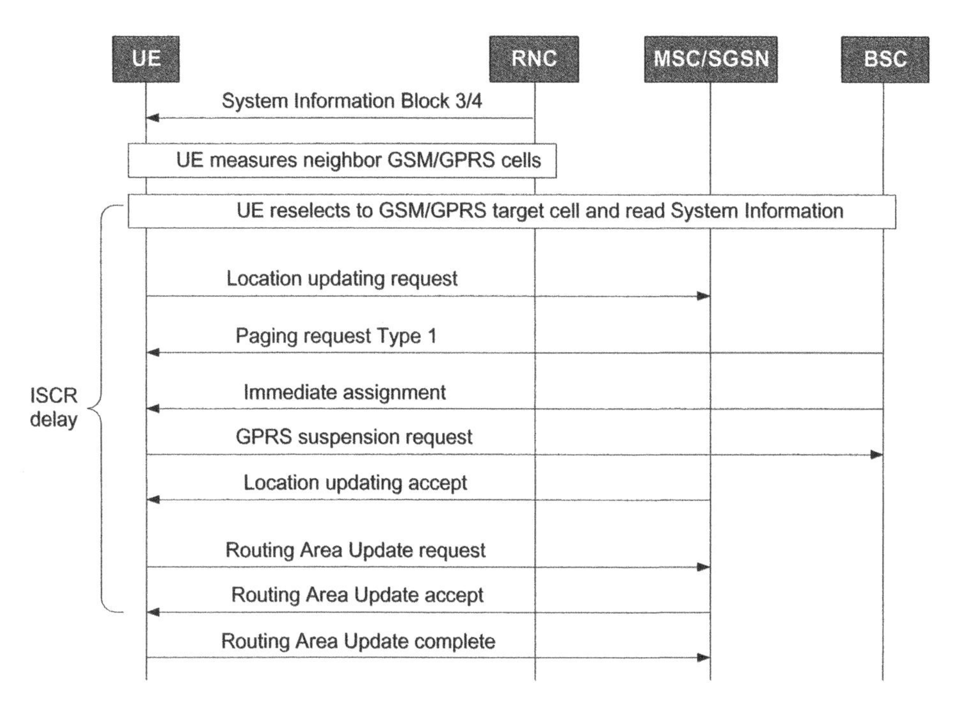

Figure 6.14 WCDMA-to-GSM/GPRS inter-system cell reselection message flow

and usually the UE initiates the reselection autonomously. Network-controlled [6] reselections are not included in this framework because they are not widely implemented in existing systems.

6.4.2.1 WCDMA-to-GSM/GPRS Cell Reselection Message Flow

Figure 6.14 illustrates a simplified WCDMA-to-GSM/GPRS cell reselection message flow.

First, the UE reads the system information to find which measurements shall be performed on the GSM/GPRS cells. Next, the UE applies the ranking procedure described in Section 6.4.1.1, which may initiate reselection to a target GSM/GPRS cell. Finally, the UE registers location area updates and RAUs, during which the UE cannot be paged nor can it access any service. This delay is normally several seconds, but could last longer depending on the Core Network implementation.

6.4.2.2 GSM/GPRS-to-WCDMA Cell Reselection Message Flow

Figure 6.15 illustrates the signaling for a GSM/GPRS-to-WCDMA cell reselection message flow.

The UE first reads the system information about the measurements to be performed on WCDMA cells and related parameters. System Information 2Quater [12] carries the WCDMA Neighbor List information. The UE then ranks the cells, evaluates the reselection criteria (see Section 6.4.1.2), and eventually reselects to the target WCDMA cell. Finally, the UE registers location area update and RAU, during which the UE cannot be paged nor can it access any service.

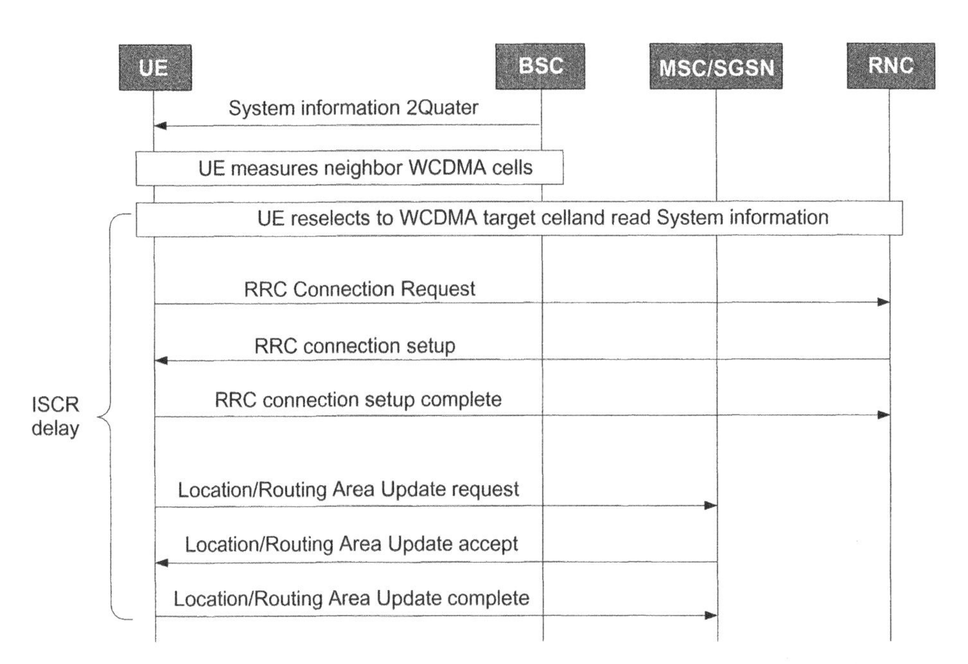

Figure 6.15 GSM/GPRS-to-WCDMA inter-system cell reselection message flow

The ISCR procedure–whether from GSM/GPRS to WCDMA or from WCDMA to GSM/GPRS–has a delay of several seconds. The call flow in Figure 6.15 applies to UEs registered in both the PS and CS domains. For a UE in GPRS Packet Transfer Mode, the first step (not shown in Figure 6.15) would be a transition from Packet Transfer Mode to Packet Idle Mode.

6.4.3 Idle Mode Performance Metrics

Compared to the intra-frequency cell reselection metrics discussed in Chapter 4, additional metrics are needed for ISCR, to address WCDMA coverage and service availability.

The two primary metrics are the WCDMA Idle Mode coverage and the WCDMA service availability:

- **WCDMA idle mode coverage.** Defined as the percentage of the route during which the UE is camped on WCDMA cells.
- **WCDMA service availability.** Defined through the following indirect metrics:
  - — Cumulative distribution of the WCDMA camping cell E_c/N_0 and RSCP collected over the metric route. Low E_c/N_0 and RSCP may lead to missed paging occasions and access failures, respectively.
  - — Number of ISCR and G2W-rejects. Both successful and attempted (but rejected) ISCR affect the service availability, because neither paging nor access is possible during the procedures. For every reselection, the UE performs a registration (location area update and RAU) and is unavailable for several seconds. In Idle Mode, call setup occasions (access and paging) are missed in both domains and, for PS connections, ongoing data transfer is interrupted.
  - — Out-of-Service (OOS) occurrences.

It is also useful to distinguish the WCDMA-to-GSM reselection trigger causes, namely, unsuitability and/or measurement rules; they indicate which parameters play a significant role under different loading and channel conditions.

For ISCR, standby time is not as important as it is for intra-frequency cell reselection, because inter-system reselections occur less frequently. However, standby time can be seriously affected if inter-system reselection occurs frequently in a stationary environment (ping-pong effect between the two systems).

6.4.4 Inter-System Cell Reselection Parameters

This section summarizes the ISCR parameters. Section 6.6.3 explains how to optimize the parameter settings for different scenarios. Tables 6.7 and 6.8 list the parameters for WCDMA-to-GSM/GPRS and GSM/GPRS-to-WCDMA cell reselection, respectively. The lists cover most of the commonly observed parameters in commercial networks.

Most infrastructure vendors allow cell reselection parameters to be set at the cell level, which provides better support for performance optimization when inter-system boundaries are present.

Table 6.7 WCDMA-to-GSM/GPRS Inter-System Cell Reselection parameters [8]

Parameters	Range [units]	Definitions
$T_{reselection}$	0–31 [sec]	Cell reselection timer value.
Qrxlevmin	−115 to −25 [dBm], in 2 [dB] increments	Minimum quality level in the cell, expressed in terms of CPICH RSCP.
Qqualmin	−24 to 0 [dB]	Minimum quality level in the cell, expressed in terms of CPICH E_c/N_o.
$S_{searchRAT}$	−32 to 20 [dB], in 2 [dB] increments	System-specific threshold provided by the serving cell and applied to the inter-system measurement rules.
$Qhyst1_s$	0–40 [dB], in 2 [dB] increments	Hysteresis. Used for TDD and GSM cells, and for FDD cells if the quality measure for cell selection and reselection is set to CPICH RSCP.
$Qoffset1_{s,n}$	−50 to 50 [dB]	Ranking offset between two cells. Used for TDD and GSM cells, and for FDD cells if the quality measure for cell selection and reselection is set to CPICH RSCP.
Pcompensation	Not signaled [dBm]	Max(UE_TXPWR_MAX_RACH - P_MAX, 0) *Pcompensation* is not signaled to the UE; it is calculated by the UE from internal or signaled parameters.
UE_TXPWR_ MAX_RACH	−50 to +33 [dBm]	Maximum transmit power level the UE may use when accessing the cell on RACH.
P_MAX	−50 to +33 [dBm]	Maximum RF output power of the UE.

Table 6.8 GSM/GPRS-to-WCDMA Inter-System Cell Reselection parameters [6]

Parameters	Range [units]	Definitions
FDD_Qmin	0 = −20 [dB], 1 = −6 [dB], 2 = −18 [dB], 3 = −8 [dB], 4 = −16 [dB], 5 = −10 [dB], 6 = −14 [dB], 7 = −12 [dB] (Default value = −12 [dB])	Minimum CPICH E_c/N_o threshold for WCDMA cell reselection
Qsearch_I (Qsearch_P)	Search for 3G cells if signal level is: below (0–7) threshold: 0 = −98 [dBm], 1 = −94 [dBm] to 6 = −74 [dBm], 7 = ∞ (always) above (8–15) threshold: 8 = −78 [dBm], 9 = −74 [dBm] to 14 = −54 [dBm], 15 = ∞ (never) Default value = 15 (never)	Qsearch_I (or Qsearch_P for PS services) indicates whether inter-system measurements shall be performed when RLA of the serving cell is below or above the threshold
FDD_Qoffset	0 = −∞ (always select a cell if acceptable), 1–15 = −28 to 28 [dB], in 4 [dB] increments Default value = 0 [dB]	Applies an offset to RLA for WCDMA cell reselection

6.5 Test Setup for Inter-System Handover and Cell Reselection Performance Assessment

To characterize the ISHO and cell reselection performance, we perform drive tests on selected routes where clear boundaries between WCDMA and GSM/GPRS coverage exist.

For the WCDMA-to-GSM/GPRS direction, the metrics discussed in Sections 6.3.4 and 6.4.3, starting from a point with good WCDMA coverage and finishing well inside the GSM/GPRS-only region are collected. In Connected Mode, a call is set up at the starting point and remains up while we drive across the boundary, collecting data and measurements from the test phone. If ISHO from GSM/GPRS to WCDMA is also supported, we perform drive testing in both directions for Connected Mode. For cell reselection, an identical test is performed, keeping the UE in Idle Mode during the entire route and driving in both directions.

The tests are repeated enough times to ensure that the measured metrics are statistically valid. The scope of the test and the metrics being measured may require a different amount of testing. For example, to test cell quality metrics, many measurements can be collected in Connected Mode with limited testing. In Idle Mode, however, a larger test effort would be needed because Idle Mode measurements occur only during the DRX cycle. For metrics such as call drop statistics, ISHO failures, and resource consumption, including the network counters in addition to collecting UE data can help increase the statistical relevance. These can be recorded for cells close to the boundary. To detect Neighbor List issues, we collect and evaluate scanner data as well. Figure 6.16 shows an example test setup for inter-system transitions.

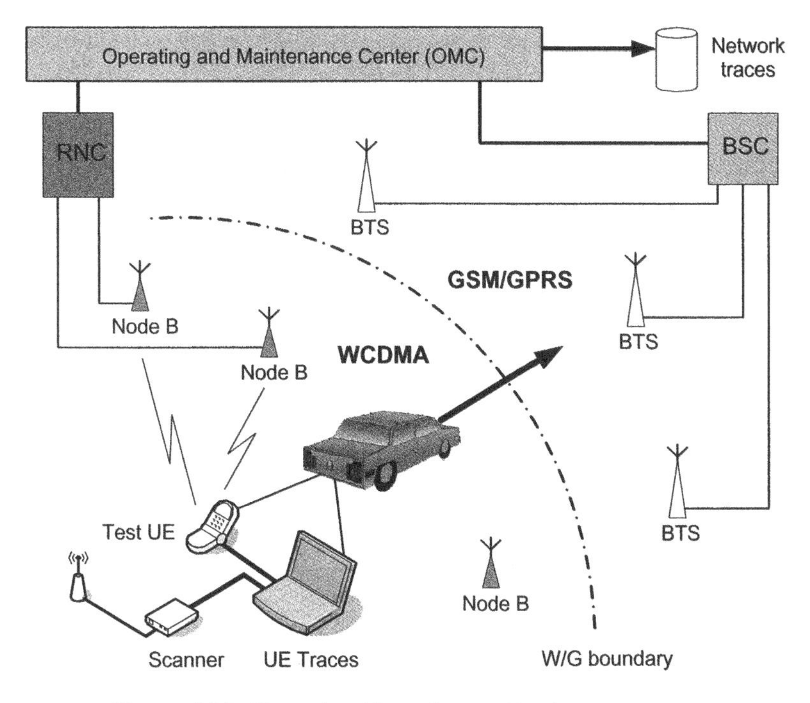

Figure 6.16 Example of Inter-System Handover test setup

6.6 Optimizing Inter-System Parameters

Sections 6.6.2 and 6.6.3 provide examples that explore how to optimize ISHO and cell reselection parameters. But first it is important to understand the relationship between both sets of parameters. Section 6.6.1 discusses the interplay between them.

6.6.1 Interplay between Inter-System Handover and Cell Reselection Parameters

The metrics used for WCDMA cell reselection characterization imply that maximizing WCDMA coverage and quality are the primary goals. However, maximizing Idle Mode WCDMA coverage should not be an optimization goal per se; instead, it should be evaluated in conjunction with Connected Mode coverage. In fact, there is no reason for having more coverage in Idle Mode than in Connected Mode, because the UE immediately activates Compressed Mode (CM) and initiates ISHO after setting up a call across the border region. If the user moves into a coverage hole and the channel deteriorates during CM, the call may even drop. Therefore, Idle Mode coverage should correspond to Connected Mode coverage.

By itself, Connected Mode coverage areas may be quite different for the CS and PS domains, according to the required data rates and the particular implementation:

- If PS services do not use CM and simply wait for the call to drop before handing over to GSM, the PS coverage in WCDMA may be significantly larger than the CS coverage.

- If CS call retention has the highest priority, Idle Mode coverage should not be much larger than CS coverage.
- If service availability has the highest priority, Idle Mode coverage should match PS coverage.

In the PS domain, state transitions (i.e., CELL_DCH to CELL_FACH to CELL_PCH to URA_PCH) will change the coverage definitions. Weigh these state changes carefully when defining the threshold for inter-system changes. Consider, for example, a case in which reselection occurs earlier than handover. In the PS domain, the coverage in CELL_FACH, governed by reselection, would be smaller than in CELL_DCH, governed by handover. As a result, a UE in CELL_FACH would perform an inter-system reselection to GSM/GPRS and then eventually return to WCDMA on CELL_DCH. In other words, the UE would be in a ping-pong scenario.

6.6.2 Optimizing Inter-System Handover Parameters

Several infrastructure vendors use Event 2d to activate and Event 2f to deactivate CM, and Event 3a to trigger ISHO. This section focuses on optimizing parameter settings related to Events 2d, 2f, and 3a. Most of these parameters can be set only RNC-wide, so this section does not cover cell-based parameter optimization.

6.6.2.1 Optimization Based on Events 2d, 2f, and 3a

ISHO performance does not depend on ISHO parameters alone; it also relies on intra-frequency handover parameters. For instance, in a scenario characterized by Pilot pollution and quickly changing best servers (that is, any scenario that leads to cells rapidly dropping from and re-adding to the Active Set), CM may be activated unnecessarily because it takes approximately 300 to 600 ms to add a cell after removing it.

Following the recommendation outlined in Chapter 4 for intra-frequency, handover parameters will lead to a reasonably long time-to-trigger for Event 1b (e.g., 640 ms) and a very short time-to-trigger for Event 1a (e.g., 0 ms), to ensure timely inclusion of fast-rising Pilots into the Active Set. Cell Individual Offset (CIO) could be used to mitigate difficult situations such as corner effects or narrow, urban canyons. Compared to a less dynamic setting, these recommended time-to-trigger settings for Events 1b and 1a significantly reduce the number of unnecessary CM activations. These settings also reduce intra-frequency measurement reporting, and, most importantly, improve call retention. This comes at the expense of a slight increase in mean Active Set size[16] and slightly reduced air-interface capacity.

Parameter optimization for CM and ISHO triggering is based on this key concept: activate CM only when an ISHO is necessary, to avoid having to deactivate it. "Necessary" in this context means that CM must be initiated early enough to fulfill the call drop probability requirement. Ideally, if CM is activated, it should rarely be deactivated, and ISHO should occur immediately after completing the measurements. This minimizes time in CM and avoids unnecessary network resource usage. At the same time, call retention

[16] This assumes that changes in inter-system parameters do not affect Connected Mode WCDMA coverage. Otherwise, mean Active Set size is influenced by two factors (the Active Set size tends to be higher at the cell edge).

performance also improves. Why? Because deactivations may delay ISHO and result in a call drop. (If CM must be reactivated, inter-system measurements restart from the beginning.) Using the CM optimization concept, you can derive the thresholds and time-to-trigger settings for the events listed above, and tune the hysteresis parameters of each single event to reduce CM measurement reporting.

The measurement quantity of Events 2d, 2f, and 3a can be set either to E_c/N_0 or to RSCP. Each configuration has advantages and disadvantages. Since the Downlink is usually limited by capacity and the Uplink by coverage, E_c/N_0 is better for tracking the Downlink, and RSCP is better for tracking the Uplink (alternatively, Uplink transmit power measurements could be used). Early deployments usually are coverage-limited, which makes RSCP a good-quality measurement. However, early deployments often lack radio frequency optimization and also exhibit Pilot pollution. For such deployments, it is more useful to measure E_c/N_0. In fact, E_c/N_0 is the recommended measurement quantity for both early and more mature deployments because E_c/N_0 can track weak coverage, whereas RSCP is not a good choice for tracking Pilot pollution.

While it is widely believed that GSM provides good coverage outdoors and indoors, that is not always the case. Therefore, the GSM threshold of Event 3a should not be set too high, because the UE may remain in CM for a long time without triggering the ISHO. Call retention performance would suffer.

6.6.2.2 Combining Coverage and Capacity Triggering

For the configuration described in Section 6.6.2.1, the E_c/N_0 thresholds must be set relatively high for coverage-limited situations. This unnecessarily reduces the WCDMA coverage. When infrastructure vendors support it, a combination of Events 2d and 2f, and Events 6a and 6b is used. Events 6a and 6b track the UE transmit power and are ideal for tracking the Uplink channel. CM is activated as soon as the UE transmit power exceeds a defined threshold for a specified amount of time. This reserves activation of the E_c/N_0 threshold setting only for capacity-limited and Pilot-polluted situations.

6.6.3 Optimizing Inter-System Cell Reselection Parameters

As described in Section 6.4, ISCR enables the UE in Idle Mode to choose a new cell of another system to camp on, thus providing service availability when WCDMA coverage deteriorates. Contrary to ISHO, the UE autonomously determines cell reselection, on the basis of measurements and parameters distributed in the system information.

This section discusses how to tune these parameters, based on examples that reflect boundary scenarios frequently observed in deployed networks. The examples are taken directly from Ref [14] and cover the most typical ISCR scenarios: WCDMA Network Boundary at the end of the WCDMA region, a coverage hole (i.e., a coverage imperfection) within the WCDMA region, and indoor situations (i.e., entering a building) where the WCDMA coverage is limited and comes from outdoor cells.

Section 6.6.3.1 discusses the scenarios and the collected field measurements. Section 6.6.3.2 introduces the optimization methodology and tools. Section 6.6.3.3 presents the optimization results for each analysis case, along with appropriate recommended parameter settings.

6.6.3.1 Scenarios and Field Measurements

The first step in optimizing ISCR parameters is to select representative scenarios and collect field measurements to evaluate their radio channel and mobility characteristics. The following examples cover three different ISHO scenarios in commercial UMTS networks in Europe [14]:

- **Network boundary.** Characterized by the loss of WCDMA coverage when leaving the WCDMA region.
- **Coverage hole.** Represents areas inside a WCDMA region that are subject to bad outdoor WCDMA coverage.
- **Entering a building.** Implies limited indoor WCDMA coverage within a WCDMA area.

Table 6.9 classifies the scenarios in terms of radio frequency environment and mobility.

6.6.3.2 Optimization Methodology

For each scenario, measurements are collected on several runs along a representative metric route. Because of the relatively short duration of the measurements across the border

Table 6.9 Classification of measured data[17]

Scenario	RF environment and mobility
Network Boundary	WCDMA coverage with high E_c/N_o for a given RSCP (fewer Node Bs at boundary to cause interference; thus less interference); Gradual fading of WCDMA signal strength when moving into GSM-only coverage; Suburban area, vehicular channel, average speed of 20 km/hr; Dominant Pilot in WCDMA coverage; Measurements for the GSM-to-WCDMA direction are obtained by reversing the time axis (see Section 6.6.3.2)
Coverage Hole	WCDMA coverage with low E_c/N_o for a given RSCP (Node Bs all around; thus more interference); Urban environment, vehicular channel, average speed of 25 km/hr with large speed variations between runs; Measurements are collected while crossing the coverage-hole area in both forward and reverse directions
Entering a Building	WCDMA coverage with medium E_c/N_o for a given RSCP (Node Bs all around, but UE is closer to one; medium level of interference); Entering and walking down a large indoor corridor of a large commercial building; Abrupt decrease in WCDMA signal strength when moving from WCDMA into GSM-only coverage; Dominant Pilot in WCDMA coverage, pedestrian channel, average speed of 3 km/hr; Measurements from indoor (GSM) to outdoor (WCDMA) are obtained by reversing the time axis (see Section 6.6.3.2)

[17] The networks were lightly loaded and characterized by a low Node B density. Both factors lead to higher E_c/N_o measurements for any given RSCP, and vice versa

region, and of the DRX operation in Idle Mode, comparing parameters using only a drive test may be too time-consuming for the requested statistical relevance. As an alternative, different parameter settings could be compared by collecting fewer field measurements and complementing that analysis with dynamic computer simulations that emulate the ISCR procedures, as detailed in Ref [14]. This method collects channel measurement data in Connected Mode while significantly reducing the amount of drive testing required to gather enough statistics (conventionally, performance metrics are collected once every DRX cycle directly from the UE). Advantages of this method include the following:

- Channel measurements are logged at a much higher rate than one sample per DRX cycle. By shifting the starting point by a few milliseconds (i.e., longer than the channel coherence time) and sampling once per DRX cycle, many instances of the measured multipath channel can be gained from a single measurement run.
- The same field data can be used to evaluate many parameter settings.
- The time axis of the field data can be reversed to emulate an identical measurement run in the opposite direction. This facilitates a side-by-side comparison between the two cell reselection schemes: GSM/GPRS to WCDMA and WCDMA to GSM/GPRS.

6.6.3.3 Optimization Results

Since the measurement data are collected in low-site-density networks under light loading conditions, WCDMA-to-GSM/GPRS reselections are more frequently triggered by unsuitability (low RSCP) rather than by the measurement rules. Accordingly, the *Qrxlevmin* parameter has the greatest influence on WCDMA-to-GSM/GPRS reselection, while *Qqualmin* plays only a secondary role. Favoring WCDMA, the *Qsearch_I* and *FDD_Qoffset* parameters are set so that WCDMA cells are always measured and reselected as soon as they meet the quality criteria. The GSM/GPRS to WCDMA reselection is controlled mainly by the *FDD_Qmin* parameter. As Table 6.10 shows, *Qrxlevmin* settings of −115, −113, and −111 dBm are tested together with *FDD_Qmin* settings of −12, −10, and −8 dB. The quantities defined in Section 6.3.4.1 and 6.3.4.2 are used for performance metrics.

6.6.3.3.1 Network Boundary Scenario

In the Network Boundary scenario (see results in Figures 6.17, 6.18, and 6.19), the route starts in good WCDMA coverage and ends in poor WCDMA coverage, while the GSM/GPRS coverage remains strong throughout. "WG→G" indicates the direction in which WCDMA coverage deteriorates, while "G→WG" indicates the reverse direction. Figure 6.17 shows that WCDMA quality and strength are low in the WG→G direction.

In the G→WG direction, WCDMA quality is better because of the high *FDD_Qmin* settings, which must be exceeded for 5 sec before reselecting to WCDMA. The downside is that the UE stays in GSM/GPRS longer, as shown in Figure 6.18, with most reselections in the WG→G direction triggered by unsuitability due to low RSCP.

Figure 6.19 shows that a high *FDD_Qmin* is the most effective way to reduce inter-system reselections and prevent G2W-rejects, thus improving service availability at the expense of a slight reduction in WCDMA Idle Mode coverage. A decrease in *Qrxlevmin* increases the number of ISCRs in the G→WG direction because of fewer G2W-rejects, and decreases the serving cell strength. Therefore, *Qrxlevmin* = −111 dBm and *FDD_Qmin* = −8 dB are recommended for this scenario.

Table 6.10 Parameter settings for inter-system cell reselection

Parameter	Setting	Comments
Qrxlevmin	−115, −113, −111 [dBm]	In this RF, main driver for WCDMA-to-GSM/GPRS reselection
$S_{searchRAT} + Qqualmin$	−14 [dB]	To avoid "ping-pong effect": $FDD_Qmin > S_{searchRAT} + Qqualmin$
Qqualmin	−18 [dB]	In this RF, secondary driver for WCDMA-to-GSM/GPRS reselection
$Qoffset1_n + Qhyst1_s$	3 [dB]	Ranking generally fulfilled
Treselection	1 [sec]	Common to intra-frequency cell reselection
FDD_Qmin	−12, −10, −8 [dB]	Drives GSM/GPRS-to-WCDMA cell reselection
Qsearch_I	Always measure *W*	To prioritize WCDMA
FDD_Qoffset	$-\infty$	To prioritize WCDMA

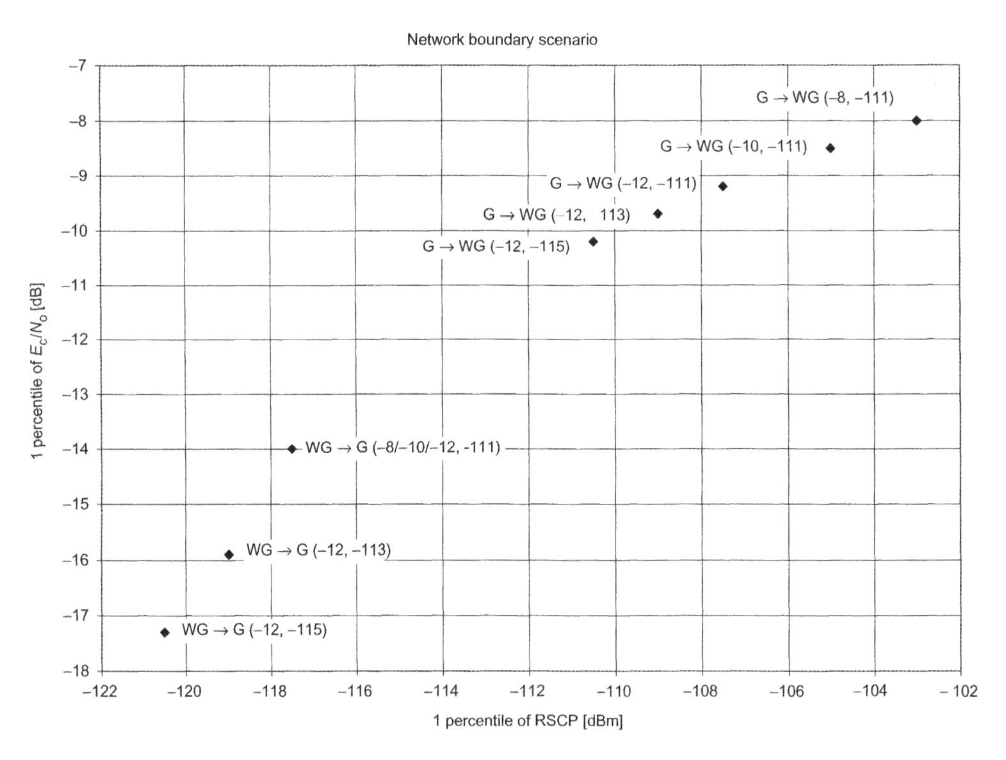

Figure 6.17 WCDMA serving cell quality in dB versus strength in dBm

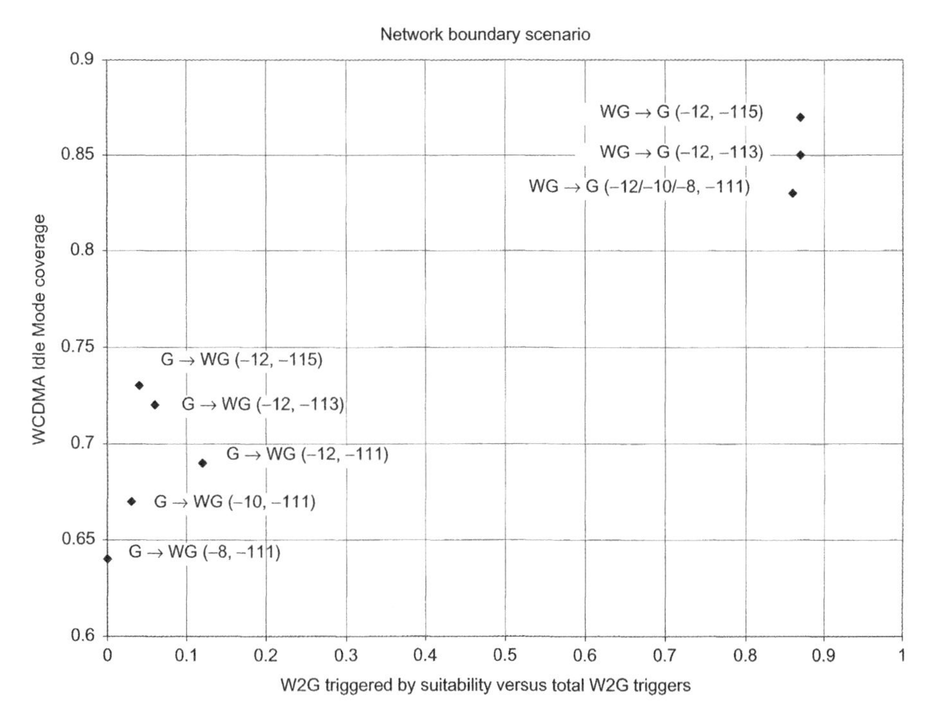

Figure 6.18 WCDMA Idle Mode coverage versus WCDMA to GSM/GPRS triggered by unsuitability relative to all WCDMA to GSM/GPRS triggers

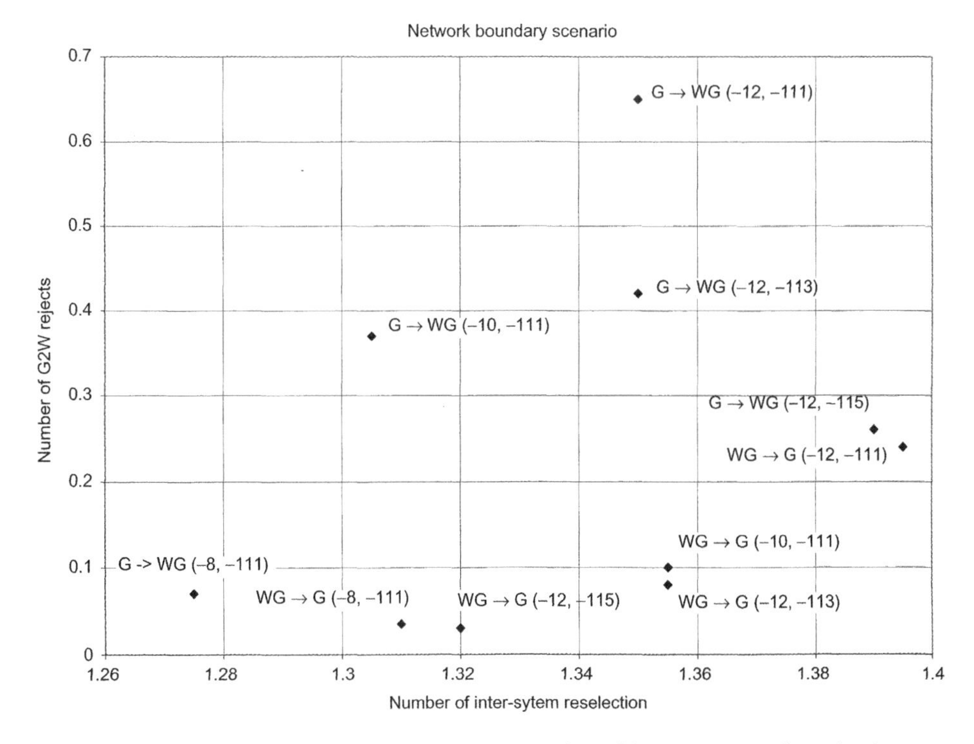

Figure 6.19 Number of G2W rejects versus number of inter-system cell reselections

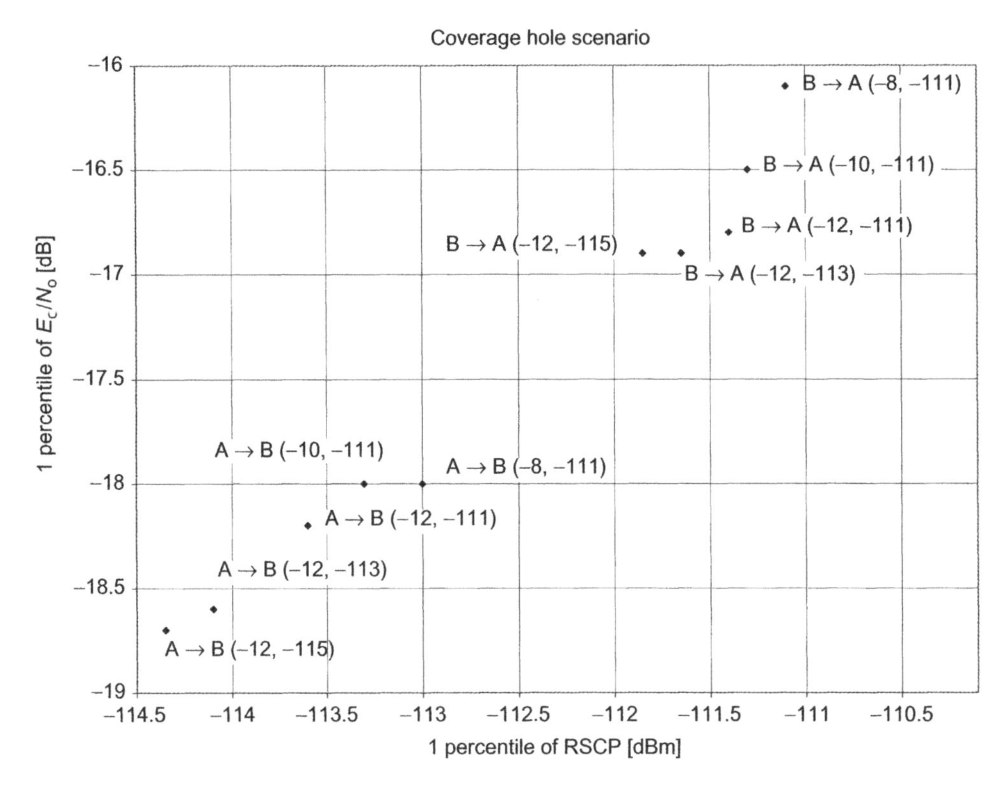

Figure 6.20 WCDMA serving cell quality in dB versus strength in dBm

6.6.3.3.2 Coverage Hole Scenario

In the Coverage Hole scenario (shown in Figures 6.20, 6.21, and 6.22), the route starts and ends in areas with good WCDMA coverage, denoted by A and B. Between these points, the route crosses an area with poor WCDMA coverage.

In this scenario, we use different channel measurements in the A→B and B→A directions (instead of reversing the time axis). Results are provided for both directions. As indicated in Table 6.9, the high interference results in low E_c/N_o for a given RSCP. Accordingly, WCDMA-to-GSM/GPRS reselections triggered by the measurement rules are more significant, as shown in Figure 6.22.

Because of the relatively high RSCP levels corresponding to a specific E_c/N_o (see Figure 6.20), a low *Qrxlevmin* setting can be used to reduce inter-system reselections and prevent G2W-rejects, while increasing WCDMA Idle Mode coverage. Therefore, a *Qrxlevmin* setting of −115 dBm is recommended for this scenario. We also recommend a low setting of *FDD_Qmin* (e.g., −12 dB) to increase WCDMA Idle Mode coverage at the expense of a tolerable reduction in serving cell quality.

6.6.3.3.3 Entering-a-Building Scenario

In the Entering-a-Building scenario, the results are given for the WG→G (entering the building) and G→WG (exiting the building) directions along the same route (see Figures 6.23, 6.24, and 6.25). The scenario is similar to the Network Boundary case and the inter-system reselection is unavoidable. Because of the channel conditions, the

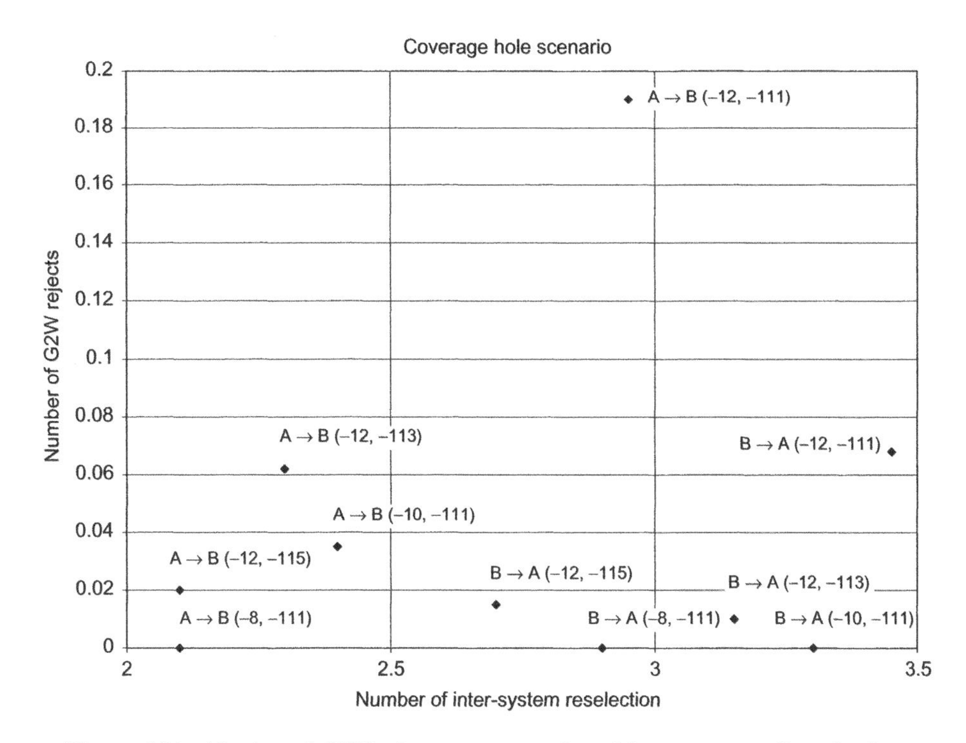

Figure 6.21 Number of G2W-rejects versus number of inter-system cell reselections

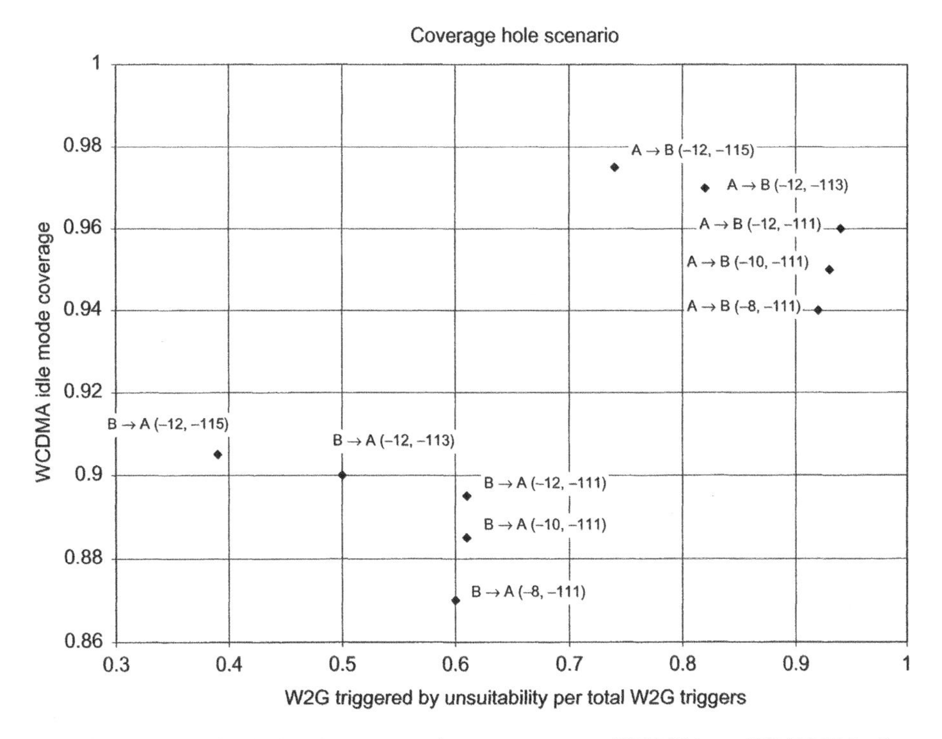

Figure 6.22 WCDMA Idle Mode coverage in percent versus WCDMA-to-GSM/GPRS triggered by unsuitability relative to all WCDMA to GSM/GPRS triggers

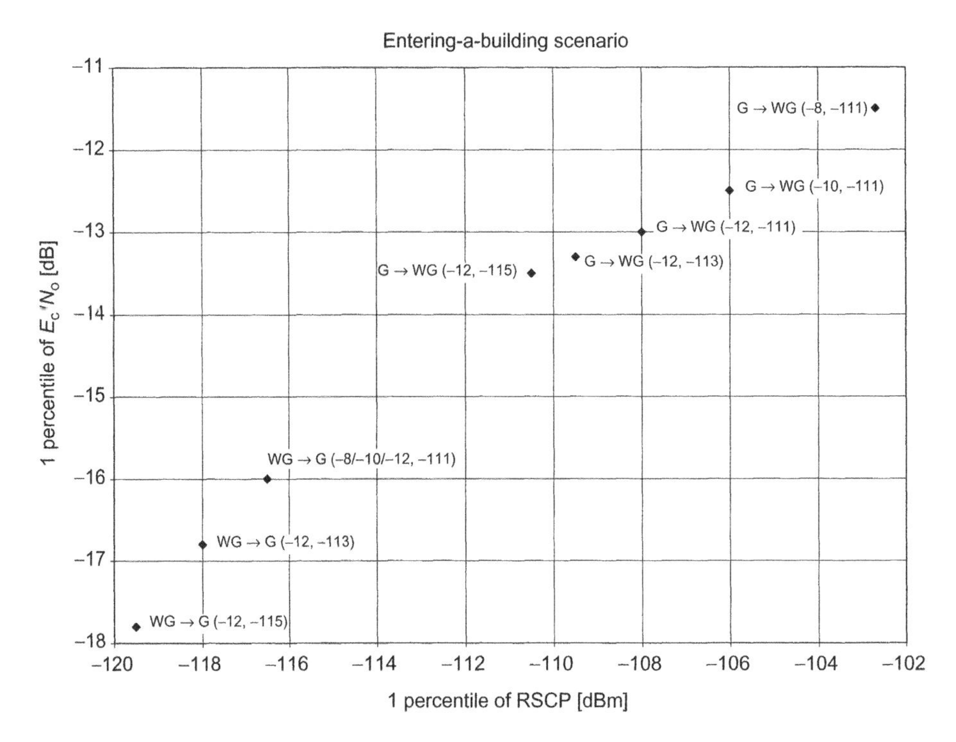

Figure 6.23 WCDMA serving cell quality in dB versus strength in dBm

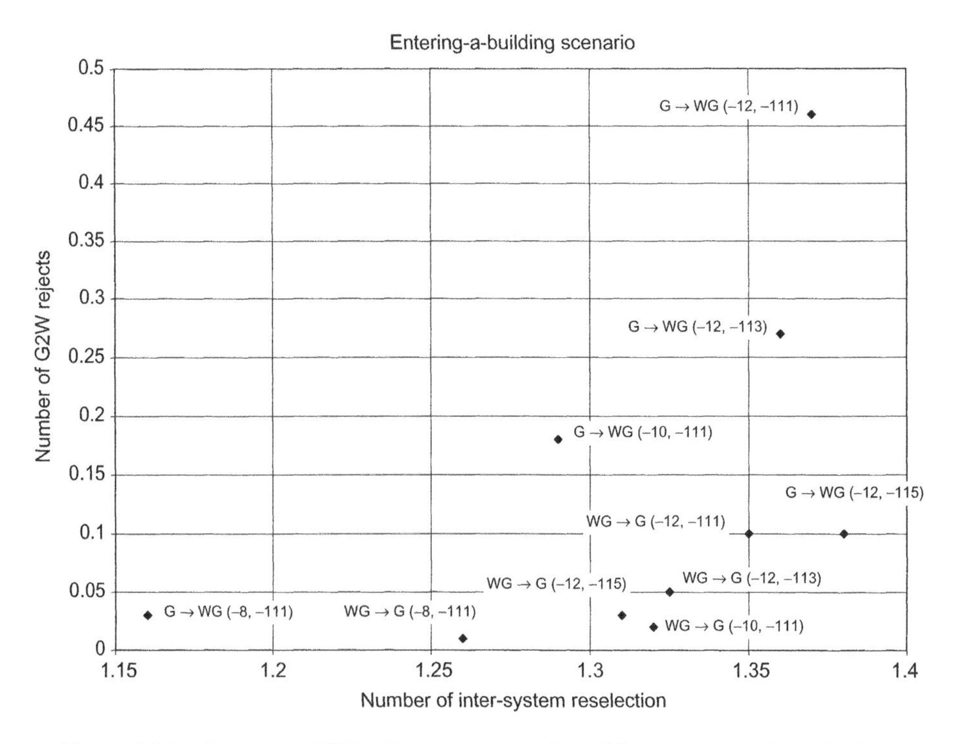

Figure 6.24 Number of G2W-rejects versus number of inter-system cell reselections

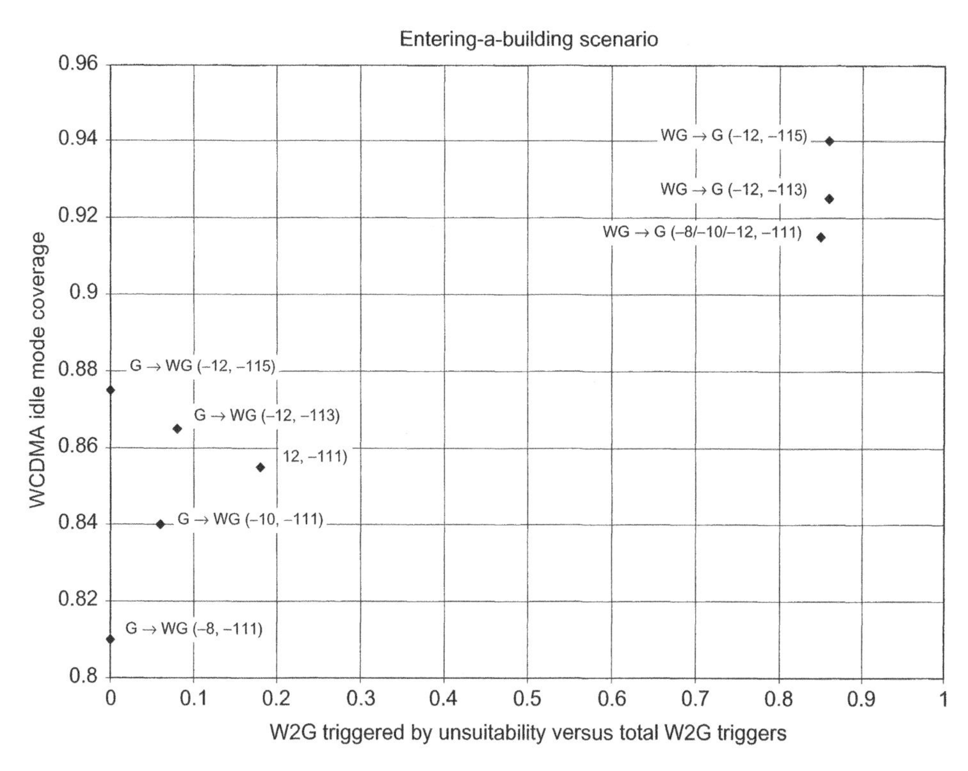

Figure 6.25 WCDMA Idle Mode coverage in percent versus W2G triggered by unsuitability relative to all W2G triggers

WCDMA signal deteriorates more abruptly in this scenario, but the low speed of a pedestrian user compensates for the quick drop.

As illustrated in Figure 6.23, a medium interference level and E_c/N_o values can be observed, owing to one dominant Pilot cell and several nearby cells. The tail of the RSCP distribution is relatively low in the WG→G direction, especially for low *Qrxlevmin* settings. Therefore, as in the Network Boundary scenario, a *Qrxlevmin* of −111 dBm is recommended.

Figure 6.24 indicates that a high *FDD_Qmin* is most effective in reducing inter-system reselections and preventing G2W-rejects while maximizing service availability. In addition, a high $FDD\_Qmin = -8$ dB improves the tail of the RSCP distribution at the expense of a slight reduction in WCDMA Idle Mode coverage, especially in indoor areas.

As in the Network Boundary scenario, most reselections in the WG→G direction are triggered by unsuitability criteria, as Figure 6.25 shows.

A stationary, indoor UE near a cell may experience high E_c/N_o values for low RSCP and may require an even higher *FDD_Qmin* to avoid frequent G2W-rejects[18]. This situation would severely degrade standby time performance, due to ISCR.

[18] This optimization study [14] assumes that the FDD_RSCP_threshold parameter (explained in Section 6.4.1.2) is not used. Although this parameter and the corresponding functionality have been added to the standard, most currently available UEs do not use this functionality.

6.6.3.3.4 Inter-System Cell Reselection Optimization Summary

On the basis of the results for ISCR optimization in Idle Mode, the following parameter combinations seem to perform well at the loading under which the tests were performed:

- **Coverage hole.** $Qrxlevmin = -115$ dBm; $FDD\_Qmin = -12$ dB
- **Network boundary.** $Qrxlevmin = -111$ dBm; $FDD\_Qmin = -8$ dB
- **Entering a building.** $Qrxlevmin = -111$ dBm; $FDD\_Qmin = -8$ dB

Infrastructure vendors usually allow cell reselection parameters to be set at the per-cell level. However, operators who want to simplify the task of deciding which scenarios apply to each cell can simply apply the most convenient settings for the entire area. In other words, the operator could determine which scenarios and environments would be the most representative and strategic for the area and tune the parameters' settings accordingly. When load increases, the tuning must be repeated to optimize performance at higher interference levels.

6.7 Additional Inter-System Planning and Optimization Issues

In addition to the concepts we have covered so far in this chapter, some other issues pertain to inter-system boundary planning, ISHO, and ISCR. The following sections briefly touch on two of these issues:

- ISHO in the presence of other WCDMA carriers
- ISHO triggering mechanisms in response to capacity or load distribution goals

6.7.1 Inter-System Handover when more WCDMA Carriers are Present

As traffic increases in WCDMA areas, operators can take advantage of additional WCDMA carriers in their license agreements to expand coverage for hotspots, or simply to distribute traffic better and maximize resource utilization. To do this, operators must activate inter-frequency cell reselections and handover procedures. This may require resolving compatibility issues with inter-system transition settings.

The procedure for inter-frequency handover is similar to the procedure for ISHO, and consists of a hard handover execution triggered by UE Measurement Reports. To measure the quality of other carriers, the UE uses CM, which can be activated for coverage or capacity reasons. Figure 6.26 represents a realistic scenario in which a hotspot with multiple WCDMA carriers is deployed on top of single-carrier WCDMA area, and GSM/GPRS is typically present everywhere.

When a UE connected to Frequency 2 (F2) approaches the border of the hotspot, the measured quality decreases and CM is activated. The network determines the patterns to measure the quality of Frequency 1 (F1) and to enable a possible inter-frequency handover. This can be done by taking inter-frequency and inter-system measurements simultaneously, and then determining which handover to execute when the measurements are reported.

Parameter settings that were optimized only for ISHO scenarios may not be the best settings for this situation and may require the following modifications:

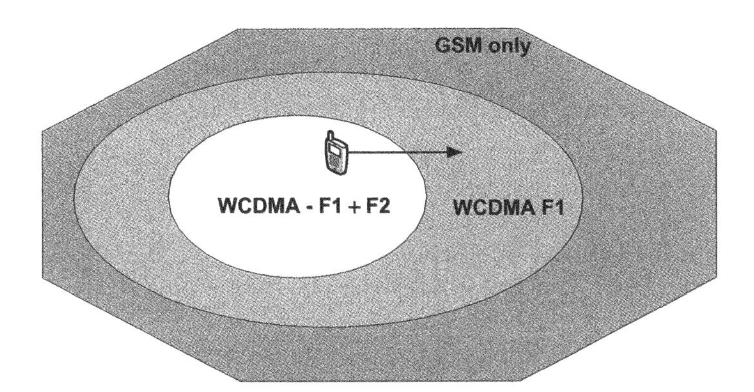

Figure 6.26 WCDMA hotspot scenario with multiple carriers

- **Compressed mode parameters.** These provide the correct triggers and configure appropriate patterns.
- **Inter-System Handover events/Inter-frequency handover events.** These events should be tuned together to prevent the UE from switching to GSM/GPRS when good coverage is available on another WCDMA carrier. The joint tuning also prevents call drops caused by late triggering of ISHO when WCDMA deteriorates on all carriers (similar to the case in which the user enters a building).
- **Additional parameters.** These minimize the ping-pong effects caused by WCDMA inter-frequency handovers.

Cell reselection is somewhat simpler than ISHOs because parameters such as $S_{intersearch}$ can independently trigger the inter-frequency measurements, and operators can determine the settings according to their desired strategy. For example, when moving from the area covered by Frequency 1 alone into the area where both WCDMA carriers are available, inter-frequency cell reselection could be used to achieve a first load distribution, while inter-frequency handover could be triggered for capacity reasons, to further adjust the network load [15].

6.7.2 Inter-System Triggered for Capacity Reasons

In addition to inter-system transitions triggered for coverage reasons, some scenarios may require inter-system changes for capacity reasons. With the increasing demand for new advanced services that can only be offered cost-effectively on WCDMA, operators may decide to accommodate voice and low-rate data services mainly on the GSM/GPRS network where good coverage is available.

The standard provides several mechanisms to support this, some of which involve inter-system transitions triggered for capacity reasons. This could be achieved by configuring Measurement Report events for CM and ISHO triggering, which can sense the loading of the WCDMA network. In this case, E_c/N_o is the suggested measurement quantity for CM triggering, and additional measurements can identify critical capacity situations.

ISCR can also redistribute users between WCDMA and GSM, in conjunction with other mechanisms such as directed retry or redirection of signaling connections. Parameters can be set to distribute users in Idle Mode evenly between the two systems, but to make the mechanism more effective a network intervention at call setup is probably necessary to ensure that every user gets access to the system that is offering the best quality for the requested service.

References

[1] TS 23.009. Handover procedures. 3GPP; 2003.
[2] TS 23.060. General packet radio service, service description, stage 2. 3GPP; 2004.
[3] TR 25.931. UTRAN functions, examples on signalling procedures. 3GPP; 2002.
[4] TS 25.212. Multiplexing and channel coding (FDD). 3GPP; 2002.
[5] Dahlman E, Gustafsson M, Jamal K. Compressed mode techniques for inter-frequency measurements in a wide-band DS-CDMA system. *Proceedings of PIMRC '97*; Helsinki, Finland; Sep. 1997; pp 23–35.
[6] TS 05.08. Radio subsystem link control. 3GPP; 2004.
[7] TS 43.129. Packet-switched handover for GERAN A/Gb mode. 3GPP - Release 6; 2006.
[8] TS 25.331. Radio Resource Control (RRC) protocol specification. 3GPP; 2004.
[9] TS 25.133. Requirements for support of radio resource management (FDD). 3GPP; 2004.
[10] Hamalainen S, Henttonen T, Numminen J, Vikstedt J. Network effects of WCDMA compressed mode. *Proceedings of VTC Spring 2003*; Jeju, Korea; Apr. 2003.
[11] TS 25.304. User Equipment (UE) procedures in idle mode and procedures for cell reselection in connected mode. 3GPP; 2004.
[12] TS 04.18. Mobile radio interface layer 3 specification, radio resource control protocol. 3GPP; 2004.
[13] TS 03.22. Functions related to Mobile Station (MS) in idle mode and group receive mode. 3GPP; 2002.
[14] Garavaglia A, Brunner C, Flore D, Yang M, Pica F. Inter-system cell reselection parameter optimization in UMTS. *Proceedings of PIMRC 2005*; Berlin, Germany; Sept. 2005.
[15] Fiorini A, De Bernardi R. Load sharing methods in a WCDMA macro multi-carrier scenario. *Proceedings of VTC*; Orlando, FL, USA; 2003-Fall.

7

HSDPA

Kevin P. Murray and Sunil Patil

7.1 Motivations for High Speed Downlink Packet Access (HSDPA)

Release 99 originally defined three techniques to enable Downlink (DL) packet data. Most commonly, data transmission is supported using either the Dedicated Channel (DCH) or the Forward Access Channel (FACH). The Downlink Shared Channel (DSCH) was also defined, but was not widely adopted or implemented for FDD and was eventually removed from the specifications [1].

CELL_DCH (DCH) is considered the primary means of supporting any significant data transmission. Each DCH is transmitted on a Dedicated Physical Channel (DPCH). Individual users are assigned a unique Orthogonal Variable Spreading Factor (OVSF) code, depending on the required data rate; the lower the spreading factor the higher the data rate that can be supported. This, however, comes at the expense of coverage and capacity, because the spreading gain is reduced (a coverage limitation) and the number of OVSF codes are limited (a capacity limitation). Fast closed loop power control is employed to ensure that the target Signal-to-Interference Ratio (SIR) is maintained, as needed, to sustain a required Block Error Rate (BLER). Macro diversity is also supported for the DPCH with soft handover.

Alternatively, packet data transfer can be supported in the CELL_FACH state, using the FACH channel on the DL and the Random Access Channel (RACH) on the Uplink (UL). The FACH is a common channel transmitted on the Secondary Common Control Physical Channel (SCCPCH), which employs a fixed length OVSF code with a spreading factor configurable from 4 to 256. Because it generally must be received by all UEs in a cell's coverage area, a relatively high spreading factor (128 or 256) is employed. As a result, only relatively low data rates are supported, 4 to 16 kbps. Macro diversity is not supported and the channel operates with a fixed (or slow changing) power allocation. Although multiple SCCPCH channels could be defined per cell to increase the capacity, the lack of any macro diversity or fast power control would render this inefficient.

WCDMA (UMTS) Deployment Handbook: Planning and Optimization Aspects QUALCOMM Incorporated

Table 7.1 Comparison summary among CELL_DCH, CELL_FACH, and HSDPA

Mode	CELL_DCH	CELL_FACH	HSDPA
Channel type	Dedicated	Common	Common
Power control	Closed inner loop: 1500 Hz Slow outer loop	None	Fixed power with link adaptation
Soft handover	Allowed	Not allowed	Not allowed
Suitability for bursty traffic	Poor	Good	Good
Data rate	Medium (limited to 384 kbps)	Low (limited to 4 to 16 kbps in practical deployment)	High (theoretically up to 14 Mbps depending on channel condition and UE category)

Release 99 defined a third technique for DL packet data for bursty, low-duty-cycle data: the DSCH. This common channel, transmitted on the Physical Downlink Shared Channel (PDSCH), is always associated with a DCH. Many users share a common, variable spreading factor, with assignments controlled by Physical Layer signaling. Multicode operation is also possible, enabling data rates higher than with the DCH alone. Although the DSCH has been implemented in commercial TDD networks, there have been no such implementations in Release 99 FDD systems and the technique was removed from the specifications completely in 2005 for Release 5. Table 7.1 summarizes the main Release 99 as well as HSDPA DL data transfer modes.

There are two motivations for deploying HSDPA, both associated with the limitations of using either CELL_DCH or CELL_FACH for packet data applications. The first is peak data rate. Although Release 99 allows for data rates of up to 2.0 Mbps (with DSCH), it has only been implemented with a maximum DL data rate of 384 kbps (with DCH), usually with limited coverage and capacity as highlighted in Chapters 2 and 3, respectively. Although this data rate is adequate for many existing applications, the substantial ongoing growth in data services implies an increasing demand for high-data-rate, content-rich, multimedia services. HSDPA addresses this need by offering significantly higher peak data rates than Release 99, theoretically up to 14 Mbps for a fully capable user equipment (UE) (category 10).

The second and most significant advantage of HSDPA compared to Release 99 relates to cell capacity. The previously stated inability to use CELL_FACH for any significant data transfer means that dedicated resources must be employed to meet the demands of majority of packet data applications. However, the use of dedicated resources is inherently inefficient for many applications (see Chapter 5) where the demand is of a rapidly changing, bursty nature.

For the DCH, first consider code tree management where each DCH uses a single OVSF code. Shorter codes are used for higher data rates and longer codes for lower data rates. When an OVSF code of a particular length is employed, all longer codes derived from it become unavailable [2]. Shorter codes above the assigned code on the same branch of the tree are also unavailable. This limits how many simultaneous users a given cell can support. The lack of codes could be addressed using a secondary scrambling code, but this introduces nonorthogonality between channels that limits capacity from a power point of view.

For example, consider a 384 kbps bearer that uses an OVSF code with a spreading factor of 8. As discussed in Chapter 3, this limits the maximum number of simultaneous users to seven after the minimum required DL common channels (CPICH, PCCPCH, SCCPCH, AICH, and PICH) are accounted for. Under this limitation, channel switching should be used to ease the OVSF code space limitation and thus the capacity limitation, but only if users do not require continuous data transmission. Typical deployments introduce further limitations, admission control, code space, UL interference, and DL power reduces the capacity to two to three simultaneous packet switched (PS) 384 Radio Bearers.

Even when used, the effectiveness of channel switching limits the overall achievable efficiency. Two factors affect this: the time taken to switch rate or type, and the event or measurement that triggers the change. The change itself takes between 80 and 160 ms, depending on the type of switching and the specific infrastructure. This is an intrinsic delay, caused by the message exchange and the 40 ms Transmission Time Interval (TTI) of the Signaling Radio Bearers (SRBs), apparent regardless of the sensitivity of the trigger used to initiate the switch. Single or multiple events or measurements can trigger the change, including UE measurement reports, DL data buffer occupancy, network inactivity timers or Node B code/power usage. For a packet data application with a predominantly bursty traffic profile, the reactive nature of these triggers–along with the time to complete the change–makes high utilization difficult to achieve without affecting user perception.

In contrast, HSDPA addresses the slow channel switching, rate or type, by using a common channel for data transfer along with Node B scheduling. In this implementation the Node B, not the Radio Network Controller (RNC), is responsible for scheduling a set of high-speed channels among multiple users. This allocation uses a 2 ms TTI, significantly smaller than those available in Release 99. This rapid scheduling of shared resources is well suited to the bursty nature of packet data.

Another Release 99 capacity limitation relates to the coding and modulation scheme employed with DCH dedicated resources. For modulation, a single Quadrature Phase Shift Keying (QPSK) option is available, where one modulation symbol represents two information bits. Forward Error Correction (FEC) using $R = 1/2$ or $R = 1/3$ convolution, or $R = 1/3$ turbo coding, is allowed, although the latter is always employed for higher rate data services because of its greater efficiency. Both these factors can be restrictive, especially in good radio conditions where employing higher order modulation or higher coding rates could significantly increase spectral efficiency. HSDPA addresses both these issues by allowing 16-QAM modulation (where one modulation symbol represents four information bits) and effective coding rates between $R = 1/3$ and close to the theoretical limit of $R = 1$. In HSDPA, modulation and coding are adapted to the channel condition. This requires accurate and recent knowledge of the channel condition, which is achieved with fast feedback from the UE to the Node B scheduler of a so-called Channel Quality Indicator (CQI). For HSDPA, Adaptive Modulation and Coding (AMC), often referred to as link adaptation, replaces fast power control.

The limitation of the power control mechanism employed for Release 99 DCHs relates to the relatively slow outer loop. The fast inner loop varies the Node B transmit power at a rate of 1500 Hz so that a DL target SIR is achieved. The outer loop then varies the SIR target based on the achieved BLER. Different radio conditions produce different SIR to BLER mappings. The frequency and manner in which the SIR target is changed can result in a waste of resources. The rate at which the SIR target is changed is not defined in the specifications but will clearly be a much slower rate than the inner loop.

7.2 HSDPA Concepts

Figure 7.1 illustrates a basic functional overview of HSDPA. For HSDPA, four new channels are defined [3]. The Data payload is carried on a transport channel called the High Speed Downlink Shared Channel (HS-DSCH), which operates on a fixed TTI of 2 ms. At the Physical Layer, the HS-DSCH is mapped onto the High Speed Physical Downlink Shared Channel (HS-PDSCH). This common channel, which is capable of multicode transmission, is shared among users by employing a combination of time and code division multiplexing. Scheduling and control information relating to each HS-PDSCH transmission is communicated to a UE on one of several possible High Speed Shared Control Channels (HS-SCCH). During each 2 ms TTI, a single HS-SCCH carries control information for one UE. Therefore, the number of necessary HS-SCCH channels is defined by the number of UEs that will have concurrent HS-PDSCH transmissions.

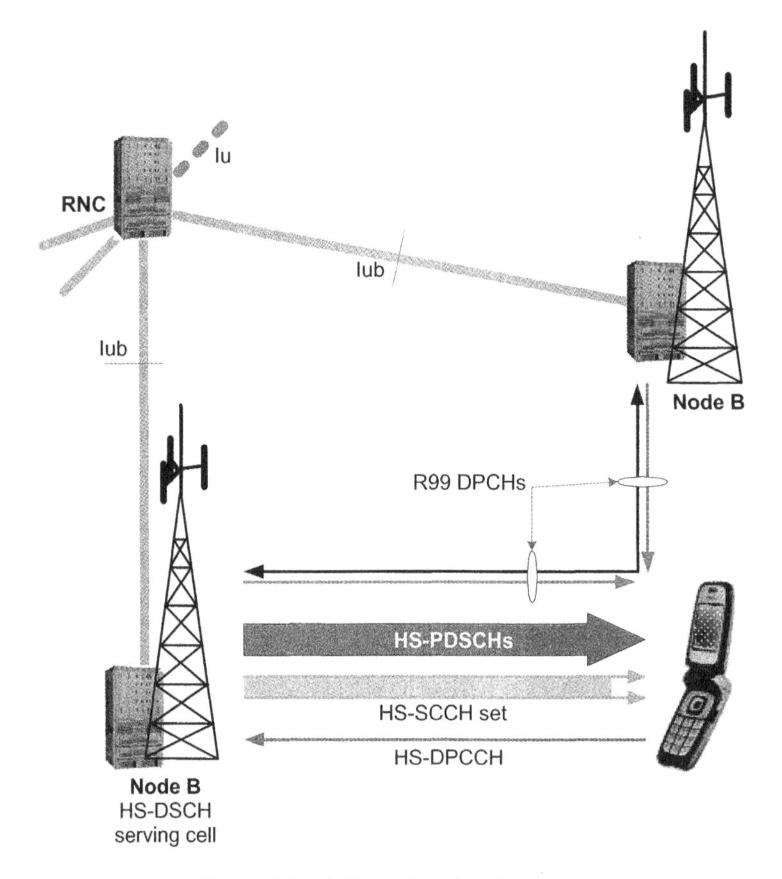

Figure 7.1 HSDPA functional overview

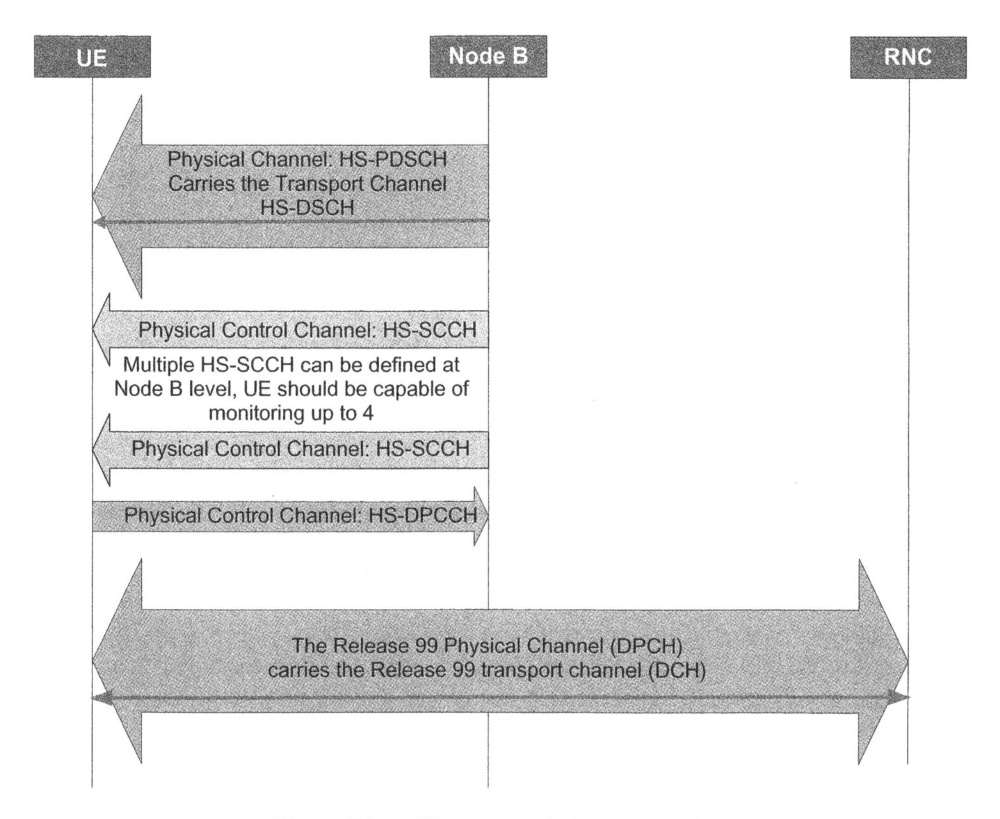

Figure 7.2 HSDPA-related channel mapping

On the UL, Release 5 introduces a new Physical Layer channel called the High Speed Dedicated Physical Control Channel (HS-DPCCH). Each UE operating in HSDPA mode has a single HS-DPCCH, which carries a positive or negative acknowledgment (ACK/NAK) of a transmitted HS-DSCH data transmission and also the CQI report that informs the Node B scheduler of the quality of the channel. Figure 7.2 shows the mapping of transport and physical channels.

The functional overview illustrates the fact that, in this example (Figure 7.1), the UE is in soft handover with two Release 99 DPCHs. For HSDPA to function, a UE must always be in CELL_DCH (associated DCH hereafter), notably to carry the UL user payload and to transfer the Layer 3 signaling. Furthermore, there is no soft handover for HSDPA, meaning that there can only be one serving HS-DSCH cell at any one time. Both of these factors represent important issues for HSDPA dimensioning and performance. Their impact is discussed in more detail in upcoming sections.

7.2.1 Common Channel with Multicode Operation

HSDPA achieves high data rates by allocating multiple codes to a single user. In Release 99 a single, dedicated OVSF code is allocated to each user with a spreading factor dependent on the required data rate. HSDPA allocates codes with a fixed spreading factor of 16 to either a single or multiple users. A code set, between 1 and 15 SF16 OVSF codes, can be reserved for HS-PDSCH transmissions. An additional feature of HSDPA is that both the HS-DSCH and the HS-PDSCH operate with a 2 ms TTI, compared to the minimum 10 ms TTI employed with Release 99.

7.2.2 Adaptive Modulation and Coding

In Release 5 for the HS-PDSCH, AMC replaces power control. For Dedicated Mode in Release 99, fast power control is used so that a target SIR is met, but with a relatively slow outer loop adjusting the target based on a required BLER. As a result of this mechanism, a progressively larger proportion of the total available PA power is allocated to a user when the distance from the serving cell increases, the interference increases, or the propagation conditions worsen. The total power available for one user is constrained, however, so the highest data rates are not offered over the complete cell coverage area for all geometry (c.f. Section 4.2.3). This prevents a single, high-data-rate user at the cell edge from utilizing all the available power and hence all DL capacity. Down-switching (384 kbps to 128 kbps to 64 kbps, for example) is generally employed to enable continuous PS data service.

In HSDPA, a relatively fixed power allocation is employed while both the modulation scheme and the effective coding rate are adapted to the prevailing RF conditions. In contrast to Release 99, this results in an achieved data rate that varies dynamically over the coverage area of the cell, depending on the condition of the radio channel. The advantage of this technique over power control is that it more effectively utilizes resources on the basis of the current channel conditions, overcoming the inherent slow nature with which the target SIR is adjusted in Release 99.

For modulation, Release 99 offers only QPSK, where two information bits are represented by a single modulation symbol. HSDPA offers both QPSK and 16-QAM modulations. 16-QAM doubles the data rate as compared to QPSK, by representing four information bits per modulation symbol. While advantageous, 16-QAM's application is limited to areas with good RF conditions, because accurate magnitude information and phase information are both required to correctly discern the position of the symbol in the resulting constellation. Figure 7.3 shows the difference between QPSK and 16-QAM modulations.

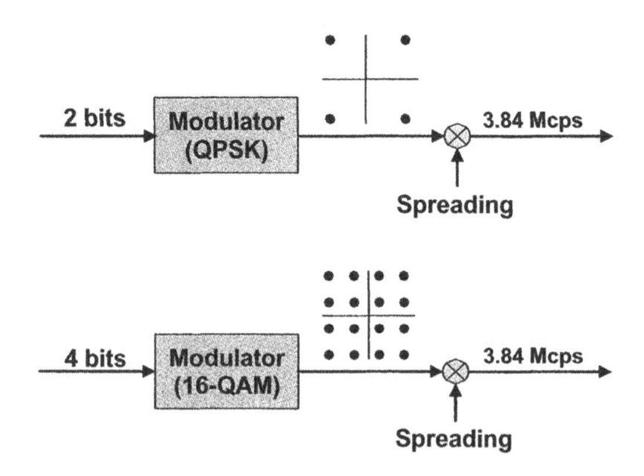

Figure 7.3 Conceptual representation of QPSK and 16-QAM modulations

For FEC, Release 99 employs $R = 1/3$ turbo coding for higher data rates. The output of the encoder consists of systematic bits (the original input data bits) and two sets of parity bits. Thus, each input information bit yields three output bits. In Release 5, $R = 1/3$ turbo coding remains the basis of FEC but Hybrid Automatic Repeat Request (HARQ) is also implemented. HARQ combines FEC with ARQ, resulting in a technique that allows a wider range of effective coding rates by enabling the transmission of only some of the coded information, if desired. A two-stage rate-matching procedure allows coding rates close to unity, this being necessary to achieve the maximum theoretical HSDPA data rate.

Another key aspect of HSDPA is that the UE saves information from previously failed decoding attempts, then combines it with retransmissions for future decoding. A first transmission could consist of predominantly systematic bits, with later retransmissions sending redundant information if the first transmission fails. As shown in Figure 7.4, two types of HARQ schemes are supported:

- **Chase.** Each transmission is self-decodable; therefore, the systematic bits are always prioritized. This makes it possible to achieve a time diversity gain whereby the same coded packet could be transmitted many times.
- **Incremental redundancy.** Parity bits are prioritized with retransmissions, such that additional redundant parity information is transmitted if decoding is unsuccessful on the previous attempt.

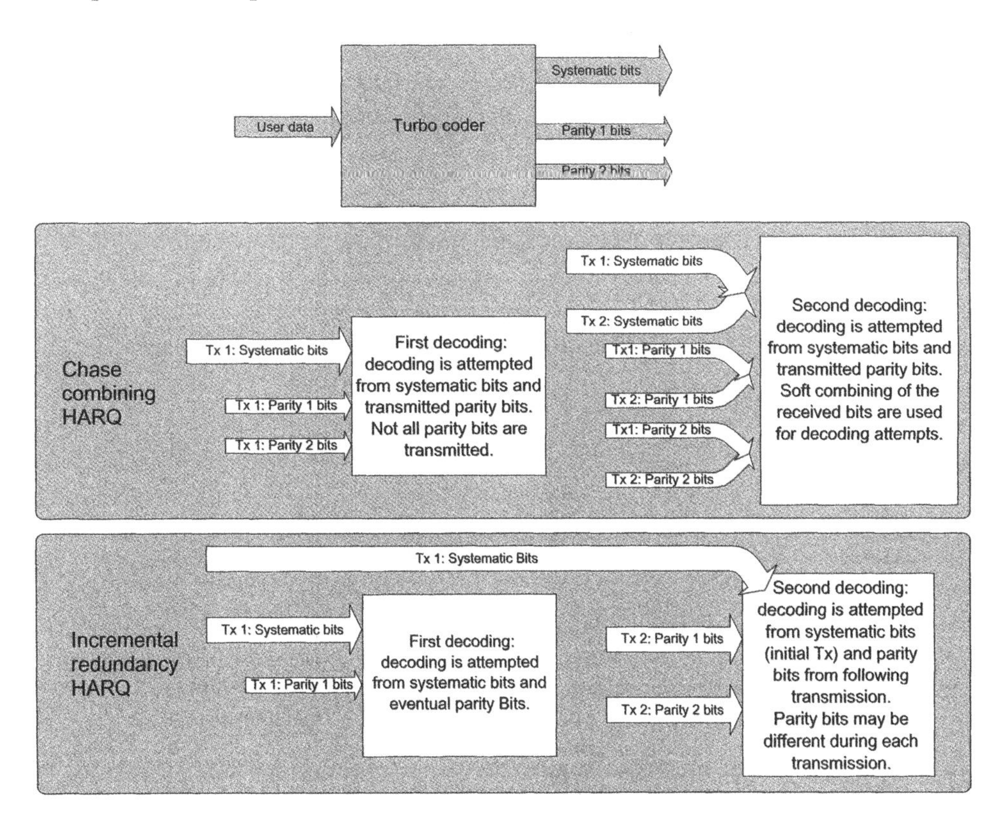

Figure 7.4 Conceptual description of chase and incremental redundancy HARQ

The proprietary scheduling algorithm of the Node B determines the type of HARQ scheme employed on each retransmission, which is signaled to the UE on the HS-SCCH by the Redundancy Version (RV) parameters s and r.

7.2.3 Fast Scheduling and Retransmissions

A key aspect of AMC described above is the ability of the scheduler to respond quickly to changes in radio channel conditions. To enable this, Release 5 makes a significant change to the protocol stack. In Release 99, Radio Resource Control (RRC), Radio Link Control (RLC), and Medium Access Control (MAC) all terminate at the RNC. The Release 5 specification [4] defines a new MAC sub-layer called MAC-hs, which terminates at the Node B. MAC-hs works in conjunction with the MAC-d entity at the RNC, the latter mapping user data from logical DTCH channels to MAC-d flows. These arrive at MAC-hs located in the Node B, which then routes the data to *priority queues*. Figure 7.5 shows the components of the MAC-hs entity.

The scheduler is responsible for taking data from each priority queue and assigning it to an individual HARQ process. There can be up to eight HARQ processes associated with a single UE. Multiple processes are necessary to ensure that a continuous stream of information can be transmitted to the UE. The scheduler determines the amount of data sent, on the basis of the CQI report along with the number of codes and power available for HS-PDSCH transmission. The choice of modulation scheme also affects the amount of data that can be sent. Figure 7.5 also shows the reception of the ACK/NAK by the HARQ processor and scheduler. Although each UE makes a synchronized acknowledgment of a HS-PDSCH transmission, the timing of retransmissions is not standardized; the scheduler decides if and when to retransmit any NAK'ed data.

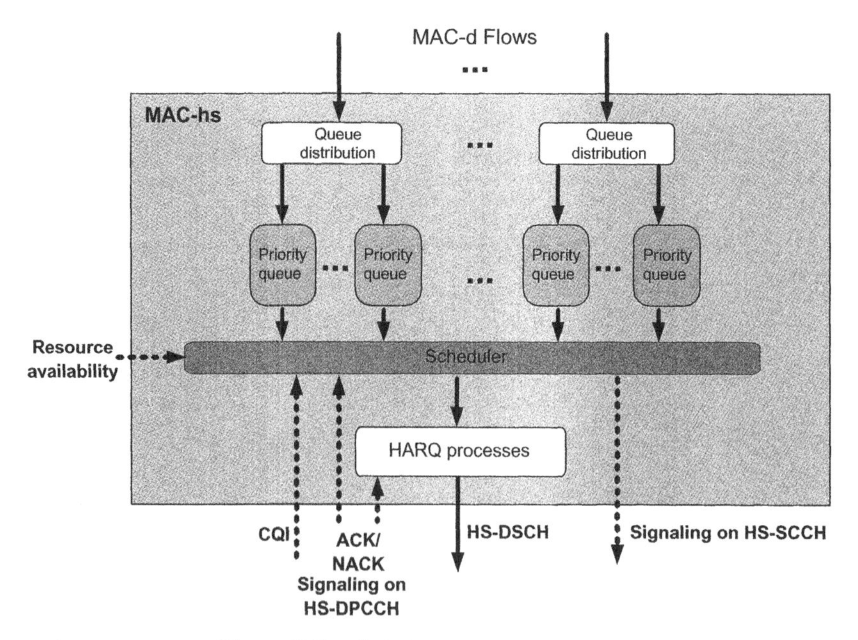

Figure 7.5 Node B MAC-hs architecture

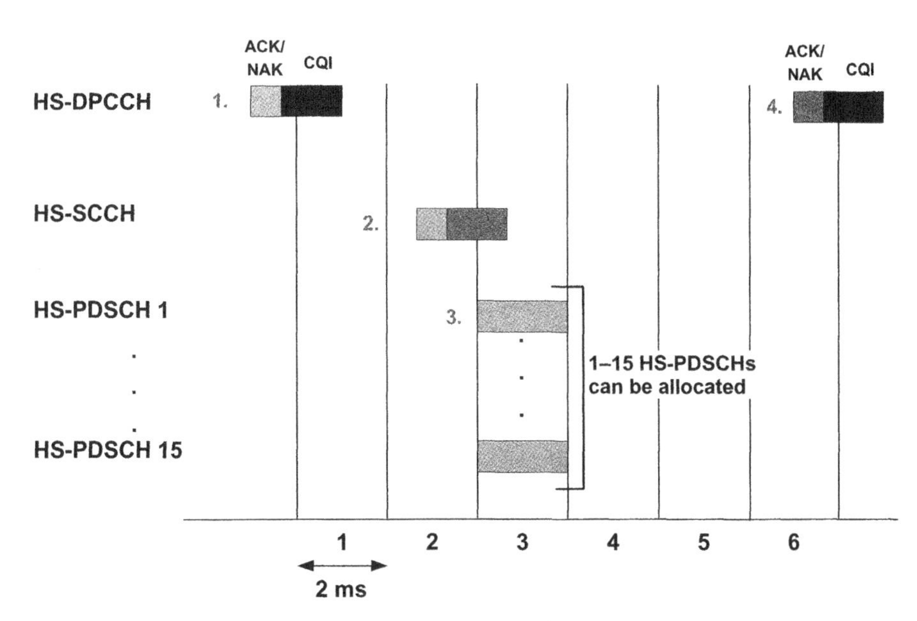

Figure 7.6 HSDPA channel timing diagram

To understand the impact of Node B scheduling, consider the HSDPA timing diagram shown in Figure 7.6, which illustrates the relationship between the UL and DL Physical Layer channels. The timing is based on a time interval of 2 ms, corresponding to 3 slots.

Figure 7.6 illustrates the following steps:

Step 1. When the UE enters the HS state, it starts transmitting a CQI report based on the measured channel quality. This is transmitted on the HS-DPCCH with any required ACK/NAK from a previous transmission, if applicable. In most implementations, when the UE first enters the HS state, the Node B will not schedule the UE until a CQI is received.

Step 2. Also, as the UE enters the HS state, it starts decoding the HS-SCCH. The Node B scheduler decides to send data to a UE and sends control information on the HS-SCCH. This transmission tells the UE that it has been scheduled and also provides the information necessary for the UE to demodulate and attempt to decode the data transmission. The timing position of the HS-SCCH transmission shown is the fastest that data could be scheduled while taking into account the CQI report shown in Step 1.

Step 3. During subframe 3, the data payload itself is transmitted from 1 through 15 HS-PDSCH in accordance with the scheduling information transmitted during Step 2.

Step 4. An ACK or NAK for the transmitted data is generated and sent (along with a CQI report) on the HS-DPCCH 7.5 slots (5 ms) at the end of the data transmission.

Many of the timing relationships shown are set by the standard. First, the HS-PDSCH must be transmitted two slots after the start of its control information transmitted on the HS-SCCH, with the illustrated one-slot overlap. The overlap is possible because the HS-SCCH has two distinct parts. The first (single slot) part tells a single UE which

OVSF codes to monitor, along with the type of modulation scheme employed. As will be discussed later, the addressing inherent to part 1 is critical to the optimal performance of the system. Part 2, which is not received in its entirety until one slot after the start of the HS-PDSCH transmission, contains the information necessary for the UE to decode the data, primarily Transport Block Size (TBS), HARQ processes identification, redundancy information, and identification of new or retransmitted data.

Second, there is always a synchronized response to a transmitted block of data, with a UE generating an ACK or NAK 7.5 slots (5 ms) after the end of the data transmission. However, retransmissions need not be synchronized. If an initial transmission is NAK'ed, according to the timing diagram shown in Figure 7.6, a retransmission would be possible 12 ms after the original transmission, at the earliest, if the scheduler waits to receive the entire HS-DPCCH transmission. It is up to the scheduler, though, to decide on the timing of retransmissions. This must be considered along with the role of HARQ. Each HS-PDSCH transmission is associated with a single HARQ process. To support consecutive assignments of data, multiple processes are required–a minimum of six to achieve the theoretical maximum data rate, assuming the 12 ms round trip time described above. However, the specification [4] defines that up to eight HARQ processes can run simultaneously for a single UE. Therefore, when a UE sends a NAK, this allows up to 16 ms for the reception of a NAK for retransmission after the original transmission. In addition, multiple HARQ processes can manage multiple data flows with differing priorities and still achieve the maximum theoretical data rate.

The Node B defines the frequency of the CQI report. The report itself is based on a three-slot measurement period that ends one slot before the start of the report transmission. This introduces a nominal offset of 6 ms between the beginning of the measurement and its reporting. The report is a five-bit number taking values 0 through 30, which is an index for a table in which each row corresponds to a requested Transport Format Resource Combination (TFRC). The TFRC includes a TBS defining the data rate that the UE can support. The CQI is further defined as the TBS that can be supported with a BLER no greater than 10% [5]. Because of the limited number of reporting bits, different CQI index tables are defined in Ref [5] for the different UE capabilities. Table 7.2 summarizes these for UE categories 1 through 10. With the maximum TBS and an assumed inter-TTI, the supported data rate can be calculated from the following equation:

$$Supported\_data\_rate\ [bps] = \frac{TBS\ [bits]}{TTI\ [sec]} \tag{7.1}$$

This calculation does not consider the number of HS-PDSCH channels, since the entire transport block is transmitted over the code set, not per code.

Table 7.2 and equation 7.2 suggest that the maximum bit rate that a category 10 UE (the highest capability) would support is 12.8 Mbps, significantly less than the theoretical maximum of 14.4 Mbps. However, the actual TBS sent would ultimately be left to the scheduler, up to a maximum of 27952 or 14.0 Mbps [4]. The differences among the 12 different types of UE categories will be examined in Table 7.7.

Another important aspect of CQI reporting is that it is implementation-dependent because the specifications do not define how the Node B processes the reports. Consequently, the Node B scheduler does not have to react to an individual report and may use a filtering algorithm that considers their variation. For example, the scheduler could be more or less

Table 7.2 CQI index for UE categories 1 through 10

CQI value	Transport block size, per UE category				Number of codes (HS-PDSCH)	Modulation
	Cat. 1-6	Cat. 7-8	Cat. 9-8	Cat. 10		
0	NA	NA	NA	NA	HSDPA not supported	
1	137	137	137	137	1	QPSK
2	173	173	173	173	1	QPSK
3	233	233	233	233	1	QPSK
4	317	317	317	317	1	QPSK
5	377	377	377	377	1	QPSK
6	461	461	461	461	1	QPSK
7	650	650	650	650	2	QPSK
8	792	792	792	792	2	QPSK
9	931	931	931	931	2	QPSK
10	1262	1262	1262	1262	3	QPSK
11	1483	1483	1483	1483	3	QPSK
12	1742	1742	1742	1742	3	QPSK
13	2279	2279	2279	2279	4	QPSK
14	2583	2583	2583	2583	4	QPSK
15	3319	3319	3319	3319	5	QPSK
16	3565	3565	3565	3565	5	16-QAM
17	4189	4189	4189	4189	5	16-QAM
18	4664	4664	4664	4664	5	16-QAM
19	5287	5287	5287	5287	5	16-QAM
20	5887	5887	5887	5887	5	16-QAM
21	6554	6554	6554	6554	5	16-QAM
22	7168	7168	7168	7168	5	16-QAM
23	7168	9719	9719	9719	5-Cat. 1 to 6; 7-All other Cat.	16-QAM
24	7168	11418	11418	11418	5-Cat. 1 to 6; 8-All other Cat.	16-QAM
25	7168	14411	14411	14411	5-Cat. 1 to 6; 10-All other Cat.	16-QAM
26	7168	14411	17300	17300	5-Cat. 1 to 6; 10-Cat. 7 or 8; 12-Cat. 9 or 10	16-QAM
27	7168	14411	17300	21754	5-Cat. 1 to 6; 10-Cat. 7 or 8; 12-Cat. 9; 15-Cat. 10	16-QAM
28	7168	14411	17300	23370	5-Cat. 1 to 6; 10-Cat. 7 or 8; 12-Cat. 9; 15-Cat. 10	16-QAM
29	7168	14411	17300	24222	5-Cat. 1 to 6; 10-Cat. 7 or 8; 12-Cat. 9; 15-Cat. 10	16-QAM
30	7168	14411	17300	25558	5-Cat. 1 to 6; 10-Cat. 7 or 8; 12-Cat. 9; 15-Cat. 10	16-QAM

aggressive in the TFRC that it chooses for a particular transmission, depending both on the most recent value and on the recent rate of change of the report.

7.3 HSDPA Planning

7.3.1 HSDPA Deployment Scenarios

Deployment strategy is a key consideration for implementing HSDPA. This section examines three distinct deployment scenarios:

- One-to-one overlay
- Single carrier shared between Release 99 and HSDPA
- Deployment in hotspots

7.3.1.1 Scenario 1: One-to-One Overlay

When an operator has an existing operational Release 99 network, HSDPA technology will likely be deployed initially as a one-to-one overlay to existing sites, either in the entire network or in a selected area (Figure 7.7).

This approach makes an incrementally efficient use of spectrum for PS services, as HSDPA-capable devices are gradually introduced into the market and existing Release 99

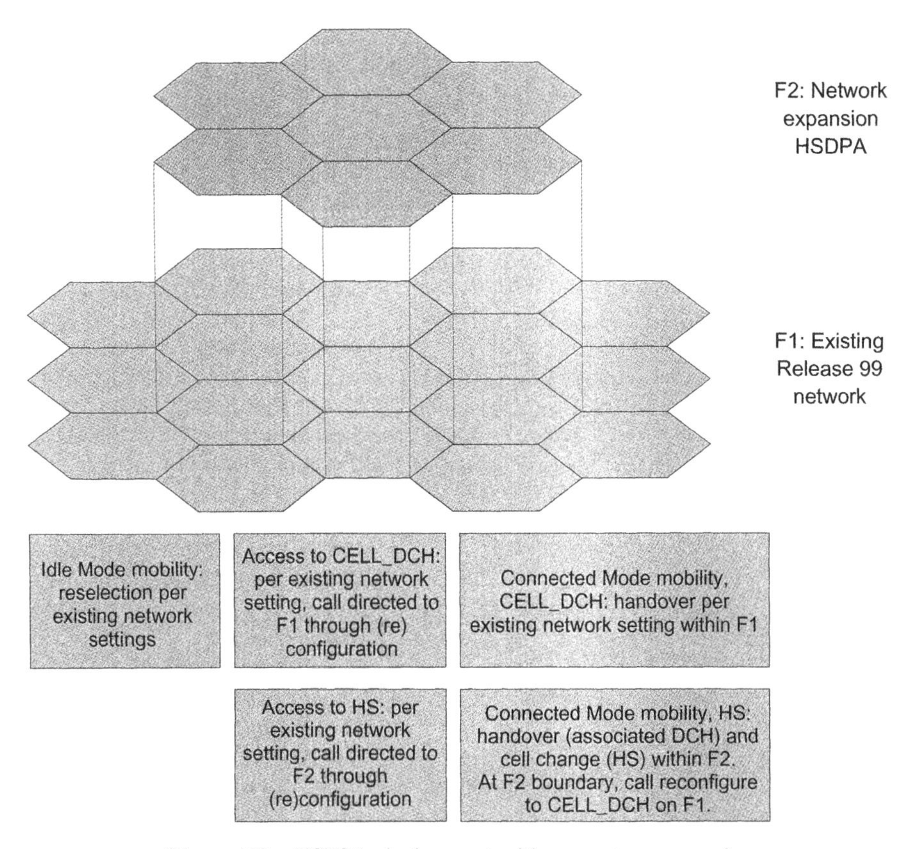

Figure 7.7 HSDPA deployment with a one-to-one overlay

PS users migrate to HSDPA. Ultimately, all PS traffic–using HS–would move to a single carrier, with the initial carrier being used only for circuit switched (CS) traffic. At first, this type of deployment makes limited use of the HSDPA carrier, but does not cause any significant impact on Release 99 users. Such a deployment also significantly reduces the need to manage mobility in both Idle and Connected Modes, because users, regardless of their capabilities, can use the schemes already in place on the initial carrier. The hand-down to the HS carrier can be done easily by reconfiguring the Physical Channel when the HSDPA-capable users are moved to the HS state, with no impact on the legacy users.

7.3.1.2 Scenario 2: Single Carrier Shared Release 99 and HSDPA

In the early stages of a one-to-one overlay, spectrum is used relatively inefficiently. To avoid this, a second deployment scenario shares the initially deployed single carrier among all types of traffic (CS and PS, CELL_DCH, and HS) (Figure 7.8). In this scenario, it is important to estimate the impact on both the new HS-capable and legacy UE and the traffic domain. Also, management of the OVSF code tree and HPA power allocations becomes important. Not only does HSDPA use more of these resources but also implementation-dependent issues may exist.

For example, the specifications do not define the question of how HPA power is allocated to HSDPA. One infrastructure may require that HSDPA power be fixed as a percentage of the total, while another may dynamically allocate power rapidly on the basis of usage by other services. The fixed power allocation benefits HSDPA users, because a minimum cell edge throughput can be planned, as Section 7.3.3 will discuss. Because available power is arbitrarily limited, however, this setting may limit the capacity for

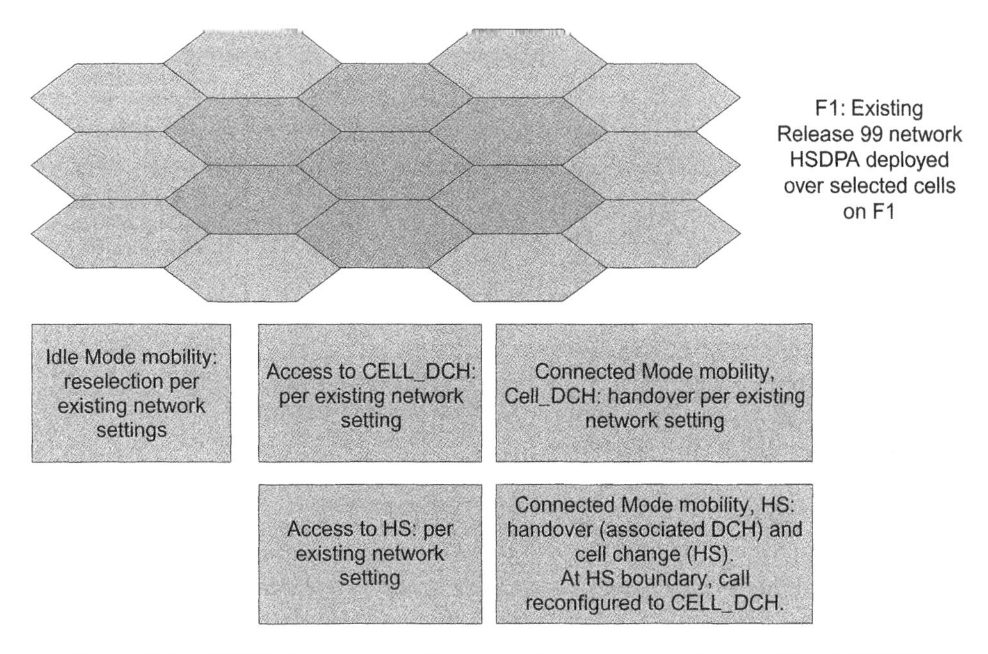

Figure 7.8 HSDPA deployment with carrier sharing between Release 99 and HSDPA

legacy traffic. On the other hand, a dynamic power allocation may leave little power available for HS users, limiting performance and thus impeding the commercial uptake of the technology. In either case, the OVSF code tree management will likely limit the maximum data rate available to an HS user. Mobility management here is simple, because it relies exclusively on the existing Release 99 network settings. The only additional issue is reconfiguring the HS users in CELL_DCH when they reach the boundary of the HS-enabled cells. However, this applies to all deployment scenarios.

7.3.1.3 Scenario 3: Deployment with Hotspots

Another deployment alternative to the one-to-one overlay is to deploy HSDPA only in *hotspot* configurations, where smaller, localized high-demand areas are served by microcells or picocells (to a greater extent than in a typical macro network), or by dedicated in-building systems (Figure 7.9).

The hotspot scenario offers the following advantage: its radio environment, with generally higher cell geometry and lower multipaths, enables increased 16-QAM modulation and higher coding gains. These factors produce significantly higher available peak data rates and improved spectral efficiency for HSDPA compared to a macro environment deployment. The disadvantage of this approach centers on controlling the transition of the HSDPA-capable users between macro- and micro-layers.

As mentioned in Chapter 2, using Hierarchical Cell Structure (HCS) in WCDMA is impractical when a single carrier is used, which limits deployment options for microcells and picocells. Accordingly, the most appropriate hotspot deployments would be over a dedicated frequency, with either *HS only* or *shared HS and Release 99* selected. Sharing the frequency ensures similar coverage for both CS and PS traffic, but this comes at

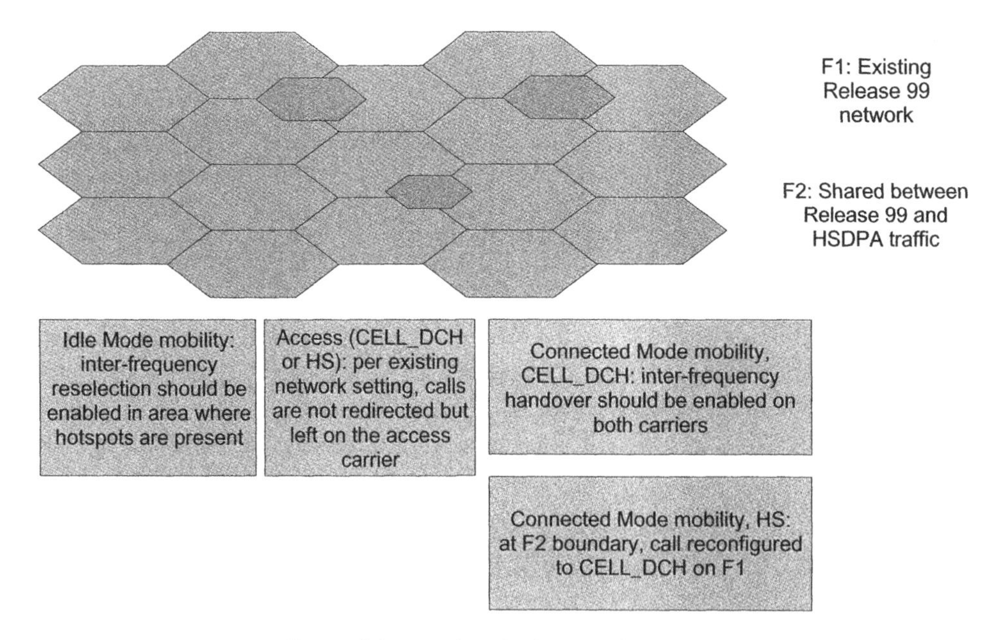

Figure 7.9 HSDPA deployment in hotspots

the expense of throughput for the HSDPA-capable PS users. Capacity management is less critical in this case because the improved channel conditions also improve overall capacity. Mobility management can be achieved relatively easily through inter-frequency reselection or handover, as discussed in Chapter 4. On the other hand, if a dedicated HS carrier is deployed in the hotspots, mobility management is significantly affected. The UE would need to camp on the macro network and be directed to the HS carrier only when activating HS service. The main drawback is that, in indoor environments notably, HS could only be enabled for users that have service on the macro network.

When a dedicated carrier is used for HSDPA operation, the associated DCH will always be carried on the HSDPA carrier. The DCH channel carried by the Release 99 carrier would be the one for CS or legacy PS traffic.

7.3.1.4 Deploying HSDPA

Clearly, both technical and business factors influence how an operator chooses to deploy HSDPA. The overlay and hotspot configurations described above offer advantages for specific situations, but these come with associated performance issues. From a technical perspective, consider the following issues when beginning to plan for an HSDPA deployment:

- **Existing Release 99 network.** What is the extent of coverage? How deep is it in terms of in-building penetration?
- **Existing Release 99 service mix.** What is the breakdown of the current usage between the offered CS and PS services?
- **Available spectrum.** How many carriers are available? What costs and issues are associated with upgrading to a multicarrier network?
- **Planned HSDPA services.** What PS services will be offered once HSDPA is deployed? How do these differ from those offered with Release 99? What average HSDPA rates will be available to the users? What percentage of PS traffic will be supported on Release 99 and HSDPA?
- **Expected HSDPA user mobility.** What is the predominant mode in which the HSDPA users will operate? Will it be stationary in-building (with laptop data cards), or mobile (pedestrian or vehicular)?

Initial analysis should focus on the existing network deployment. The network's current utilization and any planned future expansion both determine how to deploy HSDPA. Initially, deploying HSDPA on the same carrier as existing Release 99 services is cost-effective. As HSDPA usage increases, however, the quality of existing services may eventually degrade, as the HSDPA users demand more of the available power. To assess the impact of sharing HSDPA with current users, the dimensioning should consider both the current network and the planned PS traffic that HSDPA will support. In this process, it is important to keep in mind that many deployed Release 99 networks support significant circuit switched services, in the form of 12.2 kbps AMR voice and 64 kbps circuit switched video-telephony. Demand and usage for these will likely be unaffected with the introduction of HSDPA; therefore, both current and predicted CS loading should be considered. Another PS traffic issue is the impact of existing PS users who migrate to

HSDPA. The increased available data rates, plus any change in billing rate could affect demand patterns and must be considered.

User mobility and distribution also play a key role in HSDPA deployment and in assessing its relative benefit. HSDPA offers 16-QAM modulation, which doubles throughput as compared to QPSK. As discussed previously, the stronger modulation and higher coding gains possible with HSDPA are best suited to scenarios where the UE has line of sight to the Node B and the propagation conditions are characterized by a strong dominant path. As a result, HSDPA offers significantly higher spectral efficiencies if deployed in a microcell environment. Therefore, when planning the deployment, examine the predominant mode of operation and location of the HSDPA users as this may influence the scenario choice.

7.3.2 HSDPA Link Budget

After quickly reviewing the CELL_DCH Link Budget concept detailed in Chapter 2, this section discusses HSDPA Link Budgets, their detailed parameters, and an example of such a Link Budget. For HSDPA planning, the CELL_DCH Link Budget should always be considered first. This is because HS operation requires the simultaneous assignment of DCH channels, notably to carry signaling and UL user payload. As a result, the CELL_DCH Link Budget defines the extent of the coverage, while the HSDPA Link Budget defines the maximum user throughput at this defined edge of coverage. More specifically, the coverage will be limited by the UL Link Budget for PS 64. On the DL, a low data rate would be required on the associated Release 99 DCH because only signaling has to flow through this Radio Bearer: SF256 would be sufficient to support the SRB at 3.4 kbps.

7.3.2.1 Downlink Budget Key Concepts

One of the key initial steps in planning any radio network is generating a Link Budget for the UL and DL. As shown in Chapter 2, this generally involves accounting for all the gains and losses in the system, then determining a Maximum Allowable Path Loss (MAPL) for each service. In turn, the coverage area associated with each service can be determined, followed by the number of Node Bs necessary to meet a predefined coverage objective. Two inputs to a Release 99 DL Link Budget are also vital when defining an HSDPA Link Budget:

- **Maximum DPCH E_c/I_{or}.** E_c/I_{or} represents the fraction of the total transmitted HPA power (I_{or}) used for a given service (E_c for the code or channel or code set). Its maximum value is the upper limit of power that can be allocated to a single Traffic Channel for a particular service, when the HPA is transmitting at its rated maximum power. As this maximum is increased, the MAPL and, hence, the coverage area for a particular service also increases. E_c/I_{or} could be set at any value within the limits of the HPA but is normally set so that higher data rate services have a smaller coverage area than lower rate services. This permits acceptable capacity to be obtained across a cell so that, for example, a single high-rate user at the cell edge does not exhaust all available power.
- **Cell geometry: $\hat{I}_{or}/I_{oc}$.** In terms of the cell geometry, I_{oc} is the spectral density of all channels of all cells that a UE is not connected to. Effectively, it represents interference.

$\hat{I}_{or}$ is the spectral density of the signal from the target cell but also includes cells that the UE is in soft handover with. Conceptually, therefore, cell geometry is the defined SIR at the UE.

Following Ref [6] and as shown in equation 2.7 (Chapter 2), both of these parameters define the achieved DPCH E_b/N_t at the cell edge and hence whether the required E_b/N_t is met at the chosen operating point. In equation 2.7, the only other variables are β, the combining gain due to soft handover, and a_i, which represents the weighting factors of each path of the chosen channel model, such that:

$$\sum_{i=1}^{N} a_i^2 = 1 \tag{7.2}$$

Once the DPCH power allocation and cell geometry are fixed and an adequate E_b/N_t has been achieved, the effective receiver sensitivity (RX_{sens}) can be calculated. Initially, this involves calculating the sensitivity without considering interference. The median maximum path loss can then be calculated by taking the difference between this value and the maximum DPCH EIRP. The total power received from the serving cell is calculated by subtracting this path loss from the total cell EIRP. Dividing this calculated absolute power by the spreading bandwidth of 3.84 MHz yields $\hat{I}_{or}$. The absolute other-cell interference I_{oc} can thus be determined from the defined ratio of $\hat{I}_{or}/I_{oc}$. Finally, the effective receiver sensitivity RX_{sens} can be calculated such that:

$$RX_{sens} = (\hat{I}_{or} + I_{oc}) \times (E_b/N_t)_{achieved} \times R \tag{7.3}$$

where, R is the information rate.

Once RX_{sens} is calculated, the MAPL can be determined by subtracting this value from the maximum DPCH EIRP and factoring in the propagation components associated with the transmit and receive chains.

The role of the maximum DPCH E_c/I_{or} and cell geometry in the Link Budget definition for Release 99 is significant. For an HSDPA Link Budget, these two variables remain equally important. However, compared to Release 99, a number of substantial differences exist, as the following section explains. The basic principle of DL modeling is the same, though, with interference from adjacent cells and power allocation controlling the available coverage area. An important consideration for HSDPA is the difference in a cell geometry compared to Release 99. In Release 99, $\hat{I}_{or}$ includes the power from all cells in the Active Set, whereas, in HSDPA, no soft handover is allowed, thus leading to different geometry between these two cases. This results in noticeably lower cell geometries for HSDPA than for Release 99, especially in the transition region between adjacent cells where the absolute coverage (in terms of RSSI) of each cell is similar. As an example, referring to Table 2.9 in Chapter 2, the geometry degradation between Release 99 channels and HSDPA could be up to 4.7 dB.

7.3.2.2 HSDPA Link Budget Parameters

HSDPA uses a relatively constant transmission power with fast link adaptation, so the effective coding rate and modulation scheme are adjusted to account for changes in radio conditions. This is fundamentally different from the fast power control with fixed coding and modulation employed in Release 99, where the transmission power is adjusted to achieve a target SIR. For HSDPA, data rates are distributed throughout a cell coverage area, with their distribution relating to both the distance from the antenna and the propagation conditions associated with each user. Therefore, in contrast to Release 99 where a DL Budget defines a MAPL for a specific service, an HSDPA Link Budget defines the achievable data rate either at the cell edge or at any other position throughout the cell.

In addition to the general propagation parameters necessary for compiling a Release 99 DL Budget as shown in Section 2.3, this section introduces several additional concepts and parameters that facilitate HSDPA analysis:

- HSDPA Power Allocation
- E_c/N_t to Data Rate Mapping
- Operating Margin
- Scheduling Gain

7.3.2.2.1 HSDPA Power Allocation

For HSDPA Link Budget, the power allocation algorithm influences the achieved data rate. The Release 5 specifications do not define how power should be allocated for HSDPA. However, in general terms, either a static or dynamic methodology, as shown in Figure 7.10, could be applied.

Because HSDPA operates in conjunction with Release 99, both schemes must allocate sufficient power from the total available cell power for the mandatory common channels (CPICH, PCCPCH, SCCPCH, PICH, and AICH). In the static scheme, a fixed portion of total PA power is reserved for the HSDPA, HS-PDSCH, and HS-SCCH channels, with the remainder available for any of the Release 99 power-controlled DCHs. By contrast, with dynamic power allocation, HSDPA can utilize the net remainder of HPA power

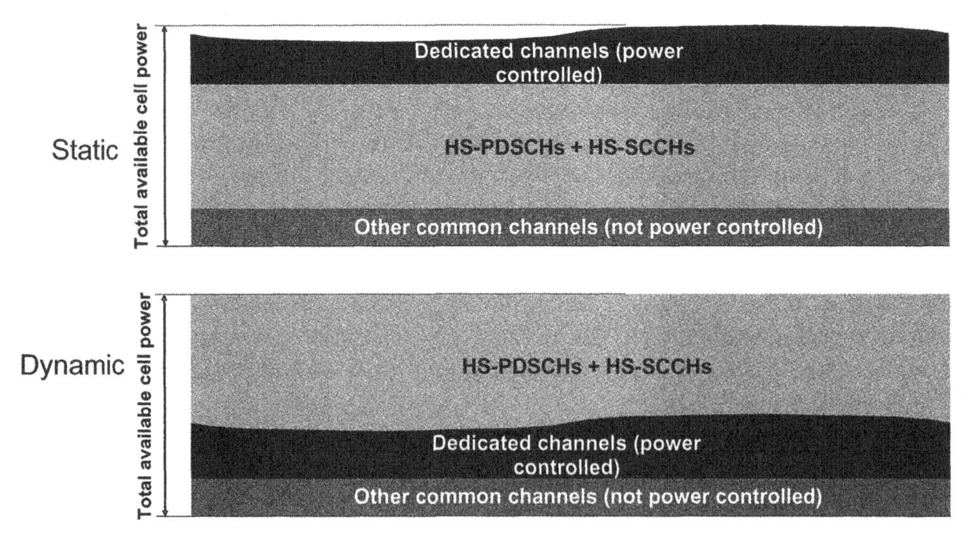

Figure 7.10 Possible HSDPA power allocation schemes

after all necessary power has been allocated to the required common and any Release 99 power-controlled DCHs.

The chosen power allocation scheme is managed within the limits imposed by the specific UTRAN infrastructure. Different vendors have proposed and implemented different methods. For example, in Ref [7], the power available for HSDPA is dynamically adjusted so that the entire power is always utilized. This allows the highest utilization of PA power but with the disadvantage of a more complex scheduler and admission/congestion control algorithm that must rapidly manage both Release 99 and HSDPA power allocations. Simpler schemes involving a fixed allocation of power are possible and would be equally suitable, especially in a situation where HSDPA is deployed alone on a single carrier. Another factor in a power allocation scheme, in a shared carrier configuration, is the relative priority of HSDPA services over any existing Release 99 services.

7.3.2.2.2 E_c/N_t to Data Rate Mapping

The next HSDPA-specific Link Budget item is the E_c/N_t to data rate mapping. As described in Section 7.3.2.1, a Release 99 DL Budget allows calculation of the MAPL, and hence the cell coverage area for a specific service. The DL operating point is set so that the combination of maximum DPCH E_c/I_{or} and cell geometry achieves a target E_b/N_t. This is the value required to maintain a specific BLER for a particular service in a specific set of propagation conditions. With HSDPA, a similar situation exists: a specific allocation of PA power is reserved for HSDPA with the UE in a specific location within a cell described by the cell geometry. In the HSDPA Link Budget, an achieved E_c/N_t (chip-to-total noise and interference ratio) is determined, in contrast to determining the achieved E_b/N_t (information bit-to-noise and interference ratio) in Release 99. This difference illustrates the fact that the actual information rate is not yet defined.

The achieved E_c/N_t relates to an achievable data rate based on the TBS, the modulation scheme, and the coding that could be supported successfully in the defined interference environment. Obviously, this depends on the implemented receiver architecture. Figure 7.11 shows an example of achievable data rate with the achieved E_c/N_t determined through a link-level simulator. These values were derived from a receiver performance model that assumed AWGN propagation conditions with a full capability UE. Any HSDPA Link Budget requires a set of such curves or lookup tables, corresponding to each UE (both manufacturer and capability) that a network will support.

As stated above, the curve shown in Figure 7.11 assumes an AWGN environment. In an actual deployment, the radio environment varies dramatically throughout the coverage area of a cell. The mapping of achievable data rate to achieved E_c/N_t depends on both the multipath delay profile and the speed of the UE. To account for this in the Link Budget developed here, the concept of Sub-Packet Error Rate (SPER) is introduced. This is the achieved BLER of HS-PDSCH transmissions when the modulation and coding inferred from the curve in Figure 7.11 are applied. The curve shows the Physical Layer data rate with the SPER defining how many retransmissions would be necessary.

7.3.2.2.3 Operating Margin

AWGN generally represents the most favorable propagation condition. To account for less favorable conditions, an *operating margin* is defined. This offsets the operating point of the E_c/N_t by an amount designed to optimize a fixed number of retransmissions. The

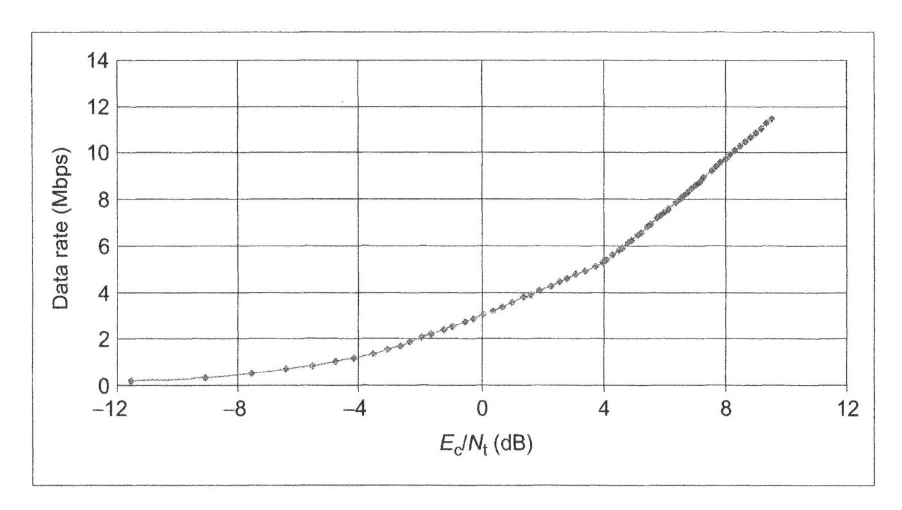

Figure 7.11 Sample E_c/N_t to data rate mapping

operating margin is subtracted from the achieved E_c/N_t, and the resultant effective E_c/N_t value is used to select a TBS that maps to a Physical Layer data rate. With this technique, different operating margins can be adopted to reflect either conservative or aggressive strategies. An aggressive strategy, with a small or negative margin, would allocate less power per information bit, thereby achieving a relatively high Physical Layer data rate but with many retransmissions. A conservative approach with a large or positive margin would allocate more power per information bit, resulting in a lower Physical Layer data rate but with fewer retransmissions.

Table 7.3 shows an example of how SPER varies with both operating margin and channel model. The channel models [8] shown in Table 7.4 are the standard ITU models: PA3 and PB3 correspond to Pedestrian A and B multipath profiles, respectively, at 3 km/hr; VA30 and VA120 correspond to Vehicular A multipath profile at 30 km/hr and 120 km/hr, respectively. The operating margins shown are optimum values that enable the decoding of a packet in either one or two transmissions for a low velocity channel, and correspond to 1.3 dB for one transmission (operating margin 1) and −2.2 dB for two transmissions (operating margin 2).

Table 7.3 Variation of SPER with channel model and operating margin

Channel model	Sub-Packet Error Rate (SPER)	
	Operating margin 1 (%)	Operating margin 2 (%)
PA3	8	40
PB3	30	60
VA30	30	60
VA120	30	60

Table 7.4 Channel models considered for HSDPA evaluation

Multipath profile	ITU Pedestrian A speed 3 km/hr (PA3)		ITU Pedestrian B speed 3 km/hr (PB3)		ITU Vehicular A speed 30 km/hr (VA30)		ITU Vehicular A speed 120 km/hr (VA120)	
	Relative delay [ns]	Relative mean power [dB]	Relative delay [ns]	Relative mean power [dB]	Relative delay [ns]	Relative mean power [dB]	Relative delay [ns]	Relative mean power [dB]
First (reference) path	0	0	0	0	0	0	0	0
Second path	110	−9.7	200	−0.9	310	−1.0	310	−1.0
Third path	190	−19.2	800	−4.9	710	−9.0	710	−9.0
Fourth path	410	−22.8	1200	−8.0	1090	−10.0	1090	−10.0
Fifth path	NA	NA	2300	−7.8	1730	−15.0	1730	−15.0
Sixth path	NA	NA	3700	−23.9	2510	−20.0	2510	−20.0

7.3.2.2.4 Scheduling Gain

An additional factor in an HSDPA Link Budget is the effect of the scheduler. When allocating resources, the scheduling algorithm likely accounts for the propagation conditions in which the UE is experiencing. If a proportionally fair or CQI-based scheduling algorithm is employed (as will be described in Section 7.4.3.6), the uncorrelated demand profile and radio conditions of the UEs being served are exploited to increase overall cell throughput–by scheduling a UE for data transfer when conditions are more, rather than less, favorable. Depending on which algorithm is employed, a scheduling gain could be added to the Link Budget to account for this effect; this would offset the achieved E_c/N_t, defining a higher instantaneous data rate.

Such a scheduling gain in the Link Budget could be useful, given the knowledge of the scheduler employed and the traffic profile of the users. In the analysis shown in the following section, the absolute data rate of a single user is determined without regard for the scheduler; thus this factor is set to zero.

7.3.2.3 Sample HSDPA Link Budget

This section uses the concepts described above to develop a sample HSDPA Link Budget. Because HSDPA always requires the presence of Release 99 UL and DL DCH channels, the HSDPA cell edge is defined by the cell edge of the service with which HSDPA will coexist. This analysis assumes that the UL is the limiting link and that the UL PS bearer is 64 kbps. The Link Budget could, however, be adjusted to assume any limiting link with any associated Release 99 service. Specifically, for HSDPA, it is likely that a 384 kbps UL bearer will be implemented where full capability UEs are employed, in situations where higher sustained peak data rates are probable. The throughput achieved with a 384 kbps UL bearer might be necessary to support the size and rate of TCP layer acknowledgments, depending on the configuration, particularly the maximum TCP segment size and acknowledgement frequency. A higher UL data rate may also be required to address the asymmetry of data rates that HSDPA will introduce between the UL and DL.

Table 7.5 Sample Release 99 Uplink budget for HSDPA[1]

Reference	Description	Value	Units
a	Maximum transmit power	19.0	dBm
b	UE cable and connector losses	0.0	dB
c	UE antenna gain	0.0	dBi
d	EIRP	19.0	dBm
e	Thermal noise density	−174.0	dBm/Hz
f	Information rate	48.1	dBHz
g	Thermal noise floor	−125.9	dBm
h	Receive noise figure	5.0	dB
I	Uplink load	50	%
j	Noise rise over thermal	3.0	dB
k	Required E_b/N_t	3.4	dB
l	Receiver sensitivity	−114.5	
m	Node B antenna gain	17.0	dBi
n	Node B cable connector and combiner losses	3.0	dB
o1	Cell-edge confidence	90	%
o2	Standard deviation	8.0	dB
o	Log normal fade argin	10.3	dB
p	Soft handover gain	4.1	dB
q	Building penetration loss	20.0	dB
r	Body loss	3.0	dB
s	Gain of propagation components	−29.2	dB
t	Maximum Allowable Path Loss (MAPL)	118.4	dB

[1] Details of calculations are available in Chapter 2, Section 2.3.1

Table 7.5 shows the UL Budget for this analysis. It is similar to the Link Budget described in Chapter 2, with two important differences.

First, row a shows a reduced UE transmit power of 19 dBm, compared to 21 dBm used in the Release 99 Link Budget (see Chapter 2). As before, a Class 4 UE is assumed. However, the 3GPP specification [9] allows a reduction in maximum transmit power to mitigate the increased Peak-to-Average Ratio (PAR) because now three OVSF codes are employed in the spreading for the UL. In Release 99, the control and data portions of the UL DCH are transmitted on the DPDCH and DPCCH, respectively. These are I/Q multiplexed, with the DPDCH mapped to the I branch and the DPCCH mapped to the Q branch. To permit cost-effective and efficient amplifier design, Hybrid Phase Shift Keying (HPSK) modulation using complex scrambling is employed on the UL. The benefit is twofold: reducing the number of zero-crossings, and distributing the power more evenly across the constellation. Amplifier operation becomes more efficient, resulting in higher possible maximum transmitted power while maintaining an adequate Adjacent Channel Leakage Ratio (ACLR). In Release 5, the HS-DPCCH is multiplexed onto the Q branch with the DPCCH. This results in a three-code UL with the power of each of the three channels specified by the signaled scaling factors β_d, β_c, and β_{hs} for data, control, and

HS respectively. The effectiveness of HPSK is subsequently reduced, resulting in a likely higher PAR when HSDPA is active. The allowed reduction in maximum UE transmit power depends on the ratio of the control and data parts of the Release 99 DPCH, with the largest reduction of 2 dB allowed when $15/7 \leq \beta_c/\beta_d \leq 15/0$.

The second difference between the Release 99 Link Budget illustrated in Chapter 2 and the present one is the information rate. As stated above, a 64 kbps PS UL is assumed, to model the following possible implemented scenario: an application using HSDPA for DL data transfer, and a Release 99 64 kbps bearer for the UL.

Once the UL Budget has been defined, the resulting MAPL can be used as an input to the HSDPA DL Budget, which is shown in Table 7.6.

Notice that, in the HSDPA Link Budget, the calculation of achievable data rate has four distinct parts. Rows a and b define the operating environment by specifying the channel model, in this case Vehicular A at 30 km/hr, and the UL MAPL that was defined as 118.4 dB in Table 7.5.

Rows c through j define HSDPA-specific parameters:

Table 7.6 Sample HSDPA Link Budget

Reference	Description	Value	Units	Formula
a	Channel model	VA30		
b	Uplink MAPL	118.4	dB	Input
c	Power allocated to Release 99 overhead channels (CPICH, PCCPCH, SCCPCH, AICH)	20	%	Input
d	Percentage of Remaining power allocated to HSDPA	100	%	Input
e	Operating margin	1.3	dB	Input
f	$I_{oc}/\hat{I}_{or}$ (1/geometry)	1.0	dB	Input
g	Required HS-SCCH E_c/N_t	−14.4	dB	Input
h	Orthogonality factor	0.5		Input
I	Scheduling gain	0.0	dB	Input
j	Average SPER	30	%	Dependent on row a
k	BTS antenna gain	17.0	dBi	Input
l	BTS cable losses	3.0	dB	Input
m	Body loss	3.0	dB	Input
n	BTS Tx power (dBm)	43.0	dBm	Input
o	UE noise figure	8.0	dB	Input
p	UE antenna gain	0.0	dBi	Input
q	Total traffic E_c/I_{or}	−1.0	DB	$= 10\log(1 - c)$
r	Control Channel E_c/I_{or}	−11.9	DB	N_t/I_{or}(dB) + HS-SCCH E_c/N_t(dB)
s	Available E_c/I_{or} for HS-DSCH	−1.3	DB	$= 10\log((10^{q/10}) - (10^{r/10}))$
t	Achieved E_c/N_t for HS-DSCH	−3.8	DB	see equation 7.4
u	Max PHY data rate	800	kbps	Lookup of E_c/N_t using (for example) Figure 7.11
v	Max MAC data rate	615	kbps	= u/(1 + j)

- **Power allocations.** Rows c and d show that 20% of power is reserved for the Release 99 mandatory common channels with all the remaining HPA power allocated for HSDPA.
- **Required HS-SCCH E_c/N_t.** This is necessary to calculate how much of the HSDPA allocated PA power must ensure that the DL HSDPA control information is successfully received by the UE.
- **$I_{oc}/\hat{I}_{or}$ (1/geometry).** The assumed value of 1 dB shown in row f is somewhat higher than that which would be expected in an embedded cell within a typical macro network. A higher value is specified here because, in this configuration with a 64 kbps UL bearer, the cell edge is assumed to be closer to the Node B than the associated CPICH cell edge.
- **Operating margin and SPER.** A conservative operating margin of 1.3 dB is assumed. This results in a SPER of 30% for VA30 channel conditions, as shown in Table 7.3.
- **Orthogonality factor.** Set at 0.5 (50%) in this analysis, this factor accounts for the loss of orthogonality due to multipaths, as described in Chapter 2.
- **Scheduling gain.** Set to 0 dB in the analysis; the effect of any scheduling algorithm is not included in calculating the achievable data rate, so would be representative of round robin scheduling, or a single user, full capability UE.

The third section of the Link Budget (rows k to p) defines the propagation components and UE characteristics that are unaffected by HSDPA. For the Node B transmit chain, a 43 dBm PA is assumed with a 17 dBi gain antenna and 3 dB feeder loss. At the UE, an 8 dB noise figure is assumed with 3 dB body loss and a 0 dBi gain antenna.

The final section of the Link Budget focuses on determining the achievable data rate using the parameters defined in the first three sections. This is calculated as follows:

- The available E_c/I_{or} for HSDPA is calculated on the basis of the defined power allocations for the Release 99 common channels and the percentage of the available remaining power that will be used for HSDPA (row q). The stated value (100% of remaining power) would correspond to a carrier dedicated to HSDPA, as no power would be available to support CS or legacy PS traffic.
- The required power for the HS-SCCH is calculated in terms of an E_c/I_{or} (row r). Only a single HS-SCCH is specified in this analysis but the number could be changed to reflect multiple HS-SCCH channels, to enable HS-PDSCH transmission to multiple UEs concurrently.
- Available E_c/I_{or} for HSDPA traffic (HS-DSCH) is calculated (row s) on the basis of the powers defined in the previous two points.
- Achieved E_c/N_t is calculated (row t) using the following equation:

$$E_c/N_t = \left(\frac{E_c}{I_{or}}\right)_{HSDPA\ Traffic} \times \frac{1}{\left(\frac{kTWN_F}{(P_{BTS}MAPL)} + \frac{I_{oc}}{\hat{I}_{or}} + (1-\alpha)\right)}. \quad (7.4)$$

Equation 7.4 takes into account the power allocated to the HS-DSCH E_c/I_{or}, the path loss between the Node B and the UE (*MAPL*), the interference condition as specified by the cell geometry, and the assumed orthogonality factor (α). In the equation, k is Boltzmann constant, T is temperature in kelvin, W is the WCDMA spreading bandwidth, and N_F is the UE noise factor.

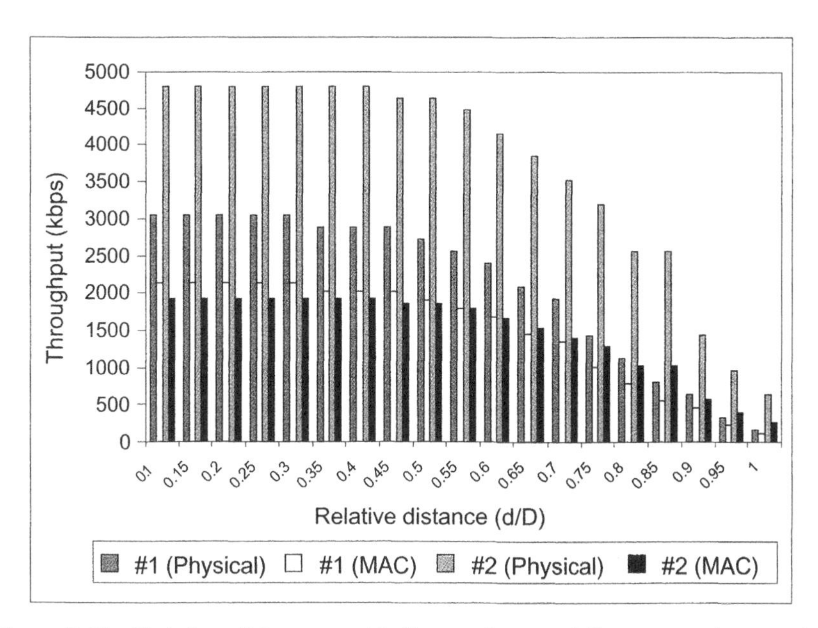

Figure 7.12 Variation of data rate with distance for two different operating margins

Once the achieved E_c/N_t has been determined, this value minus the operating margin can be referenced in a lookup table, such as that shown graphically in Figure 7.11, to determine the Physical Layer data rate (row u).

The MAC Layer data rate is determined from the Physical Layer data rate, taking into account the retransmissions that are necessary due to the defined SPER.

As previously stated, the data rate defined by this Link Budget yields the data rate at the cell edge, with the cell edge being defined by the UL MAPL of the associated Release 99 DPCH. To illustrate the variation of peak data rate throughout a cell's coverage area, Figure 7.12 shows how Physical Layer and MAC Layer data rates vary with relative distance from the center of the cell. The distance relates to a cell geometry defined, following Ref [10], by assuming a hexagonal grid of three-sectored Node Bs with a propagation constant of $n = 4$. The cell edge ($d/D = 1$) in this example is for the DL CPICH, whereas the cell edge used in the Link Budget above corresponds to that defined for the UL 64 kbps bearer at $d/D = 0.87$. The two values of operating margin shown correspond to offsets of +1.3 dB and −2.2 dB for the achieved E_c/N_t.

As expected, there is a clear reduction in data rate with increasing distance from the cell center. The Physical Layer and MAC Layer data rates achieved with the two different operating margins can be noted. The Physical Layer data rate is significantly higher when a negative margin is used, as opposed to a positive margin. Significantly, the resultant MAC Layer data rate is relatively similar in both cases, with the positive operating margin (case 1 in Figure 7.12) showing a small advantage for most distances.

7.3.3 HSDPA Capacity and Performance

This section discusses HSDPA capacity and performance. A number of factors influence the achieved HSDPA performance, including the capability and configuration of the UE and the network, along with the mobility and propagation conditions of the users.

7.3.3.1 UE Capability

As mentioned in Section 7.2, the maximum theoretical data rate for HSDPA is 14.4 Mbps. However, in reality, twelve different categories of HSDPA-capable terminals exist [11], offering maximum data rates ranging from 0.9 to 14.0 Mbps. The differences are due to the following factors:

- **HS-PDSCH codes.** The maximum number of HS-PDSCH channels that a UE can simultaneously decode. This can be 5, 10, or 15. A lower number of codes than the maximum can be supported and are signaled over the HS-SCCH.
- **Inter-TTI interval.** The minimum interval, in terms of 2-ms TTIs, between successive HS-PDSCH assignments. Values of 1, 2, or 3 are possible, with 1 corresponding to an ability to decode consecutive HS-PDSCH channels. Values other than 1 reduce the maximum data rate, as observed by comparing categories 11 and 12, for example.
- **Transport block size.** The maximum HS-DSCH TBS that can be supported by the UE in a single TTI.
- **Incremental redundancy buffer size.** The number of soft bits that can be buffered by a UE across all of the active HARQ processes.
- **Modulation.** The ability of a UE to support both QPSK and 16-QAM or solely QPSK.

Table 7.7 shows the characteristics of the 12 UE categories in terms of HS-PDSCH codes, Inter-TTI interval, Maximum TBS, and Incremental Redundancy (IR) Buffer Size. All categories support both QPSK and 16-QAM modulations, with the exception of categories 11 and 12 that support only QPSK. Table 7.7 also shows the maximum supported data rate, which is a function of only the Maximum TBS and the Inter-TTI Interval.

The IR buffer size, while not affecting peak theoretical data rate for the defined configuration, does influence the effective throughput, because of the number of soft bits available to support either Chase or IR combining. For example, consider categories 7 and 8, which support the same data rate but with different IR buffer sizes. For a typical

Table 7.7 HSDPA UE categories

HS-DSCH category	HS-PDSCH codes	Modulation supported	Min. inter-TTI	Max. TBS [bits]	UE IR buffer size [SML]	Peak data rate [Mbps]
1	5	QPSK and 16-QAM	3	7298	19200	1.2
2	5	QPSK and 16-QAM	3	7298	28800	1.2
3	5	QPSK and 16-QAM	2	7298	28800	1.8
4	5	QPSK and 16-QAM	2	7298	38400	1.8
5	5	QPSK and 16-QAM	1	7298	57600	3.6
6	5	QPSK and 16-QAM	1	7298	67200	3.6
7	10	QPSK and 16-QAM	1	14411	115200	7.2
8	10	QPSK and 16-QAM	1	14411	134400	7.2
9	15	QPSK and 16-QAM	1	20251	172800	10.1
10	15	QPSK and 16-QAM	1	27952	172800	14.0
11	5	QPSK	2	3630	14400	0.9
12	5	QPSK	1	3630	14400	1.8

HSDPA session where six HARQ processes may be operating with equal division of soft bits between processes, 19200 and 22400 bits are available per process for categories 7 and 8, respectively. This corresponds to nominally 1.33 and 1.55 times the maximum TBS, respectively. For peak data rate operation, Chase combining would be more suitable for the category 7 UE, while the category 8 UE would be best suited to support both Chase and IR because it can buffer more redundant information.

7.3.3.2 OVSF Code Tree Utilization

A key issue when considering HSDPA capacity, especially for a network with concurrent CS and PS users, is utilization of the OVSF code tree. Each HS-PDSCH uses a SF16 code with a maximum of 15 allocated to HSDPA, out of the 16 available in the Primary Scrambling Code OVSF tree. For example, consider the code tree diagram shown in Figure 7.13, showing 15 SF16 codes allocated for HS-PDSCHs. On the tree, once a code with a specific spreading factor has been utilized, the codes above and below on the same branch become unavailable. In this example, this leaves a single SF16 branch available for the mandatory Release 99 common channels, one or more HS-SCCH channels, and the required Release 99 DCHs for each HSDPA active user.

As further illustrated in Figure 7.13, the PICH, AICH, CPICH, and PCCPCH each require an SF256 code. The figure also shows the utilization of a SF128 code for a single SCCPCH (this channel can be configured with any code from SF4 to SF256) and a SF128 code for a single HS-SCCH. As a result, a single SF32 branch is available for any Release 99 DCH. A further limitation in this example is that only one HSDPA user can be allocated an HS-PDSCH set in any 2-ms TTI, because only one HS-SCCH is defined and an HS-SCCH is required for each concurrent user. If a SF256 code, the minimum possible, is assumed for the associated Release 99 DCH of each HSDPA user, eight active

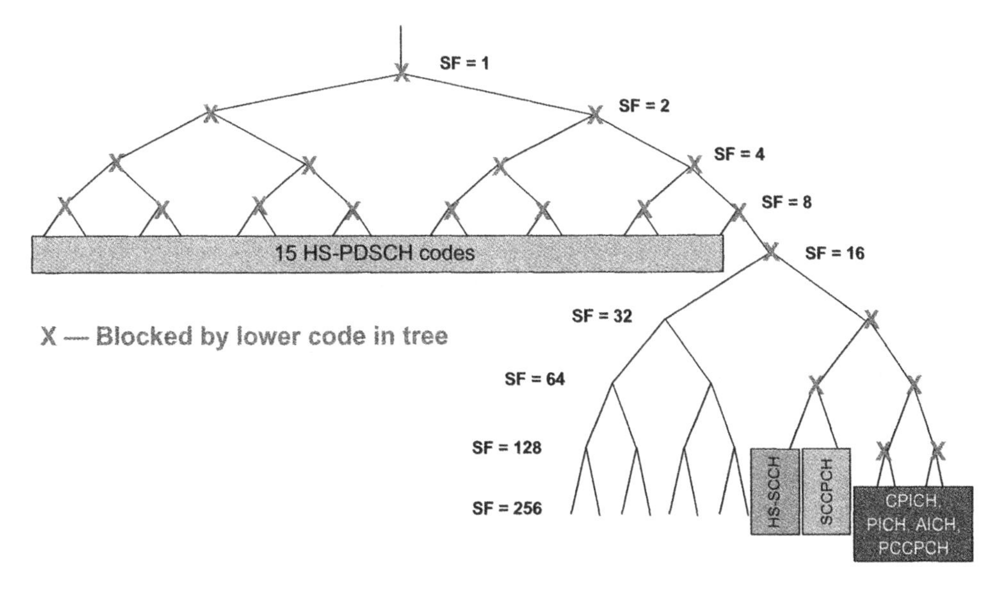

Figure 7.13 Example utilization of the PSC OVSF code tree with HSDPA

Table 7.8 Number of supported active HS users as a function of number of codes and HS-SCCH channels

	Number of SF16 OVSF reserved for HS-PDSCH	Number of HS-SCCH				
		1	2	3	4	5
Maximum number of supported HS users assuming SF256 for associated DCH	5	168	166	164	162	160
	10	88	86	84	82	80
	15	8	6	4	2	NA

HSDPA users could be supported with this configuration but, as described above, only one user would be served in any 2-ms TTI.

Allowing HSDPA operation using 15 codes would be unrealistic in most cases, while 5 or 10 code allocations offer a reasonable trade-off between individual user throughput and cell capacity, measured in terms of concurrent active HS users. Table 7.8 illustrates this trade-off, that is, the number of possible active HS users when the number of HS-SCCH channels varies from one to five. In this estimation, a SF256 is used in the DL for the associated DCH. This is sufficient to carry the RRC signaling. Consequently, the results shown in this table represent the absolute maximum number of users, because some users may require a shorter OVSF.

7.3.3.3 Impact of HSDPA on a Release 99 Network

To illustrate how deploying HSDPA affects an existing network, this section explores how the capacity available for existing Release 99 services is affected when more power is allocated to HSDPA.

Figure 7.14 shows the DL PS coverage for a typical macro network with a site spacing of nominally 1 to 1.5 km. No HSDPA is present in this prediction and the coverage areas for 384 kbps and 64 kbps are shown on the basis of predominantly vehicular traffic, according to a relatively large number of Monte-Carlo simulations.

Table 7.9 shows the capacity that would be supported for 384 kbps and 64 kbps PS service and voice (AMR 12.2 kbps) and video-telephony (64 kbps) CS service. For the PS services, this is expressed as a mean throughput and mean number of users per cell, while for CS services only the mean number of users per cell is shown. The table also lists the maximum DPCH power allocation available for each service.

Table 7.10 shows the effect of implementing HSDPA with a fixed allocation of total PA power. For three cases, corresponding to 20, 50, and 80% of total power, the reduction in available Release 99 DL capacity is clearly demonstrated and expected. This capacity reduction, however, must be considered in light of the resulting gain in HSDPA capacity. Figure 7.15 demonstrates this point, by showing the DCH and HSDPA cell capacity for a range of HSDPA power allocations, both individually and as a total cell capacity. In this simulation, the Release 99 DCH throughput was derived from a combination of 384 kbps and 64 kbps users.

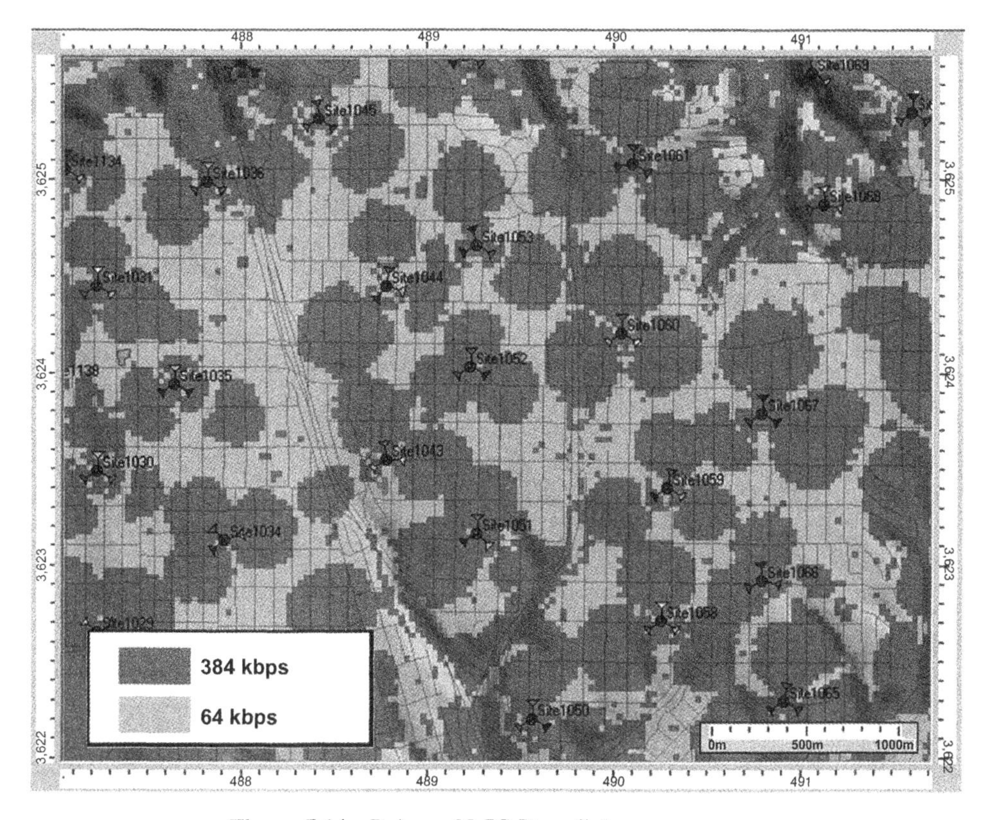

Figure 7.14 Release 99 PS Downlink coverage areas

Table 7.9 Release 99 available cell capacity, based on Monte Carlo simulation

Service	Maximum DPCH power allocation (dBm)	Mean throughput per cell (kbps)	Standard deviation (kbps)	Mean users per cell	Standard deviation	Mean DPCH power (dBm)
384 kbps PS	37.0	742.8	182.7	2.3	0.7	34.0
64 kbps PS	36.5	715.0	161.3	11.0	2.3	29.9
64 kbps CS (Video-telephony)	36.0			7.1	1.7	31.8
12.2 kbps CS (AMR voice)	31.5			108.7	17.5	22.85

Figure 7.15 illustrates two important points. At the higher HSDPA power allocations, the improved spectral efficiency achieved with HSDPA is quite clear, with mean cell throughputs twice over those attainable with Release 99 alone. Equally important, however, are the composite throughputs obtained with relatively small HSDPA power allocations. With only a small impact on the Release 99 network in terms of Release 99 capacity, HSDPA can significantly increase the overall PS cell capacity. An additional benefit of HSDPA is clearly its increased peak data rates, compared to Release 99. These allow users to enjoy bandwidth-hungry, content-rich applications.

Table 7.10 Release 99 capacity with different HSDPA power allocations

Service	Mean throughput per cell (kbps)	Standard deviation (kbps)	Mean users per cell	Standard deviation
20% of total power allocated for HSDPA				
384 kbps PS	594.0	129.4	1.7	0.6
64 kbps PS	388.9	60.2	5.6	1.0
64 kbps CS (Video-telephony)			3.7	0.7
12.2 kbps CS (AMR voice)			63.0	8.5
50% of total power allocated for HSDPA				
384 kbps PS	200.83	127.3	0.7	0.6
64 kbps PS	169.12	26.4	2.1	0.5
64 kbps CS (Video-telephony)			0.8	0.4
12.2 kbps CS (AMR voice)			29.9	4.6
80% of total power allocated for HSDPA				
384 kbps PS	0.00	0.0	0.0	0.0
64 kbps PS	0.00	0.0	0.0	0.0
64 kbps CS (Video-telephony)			0.0	0.0
12.2 kbps CS (AMR voice)			0.7	0.6

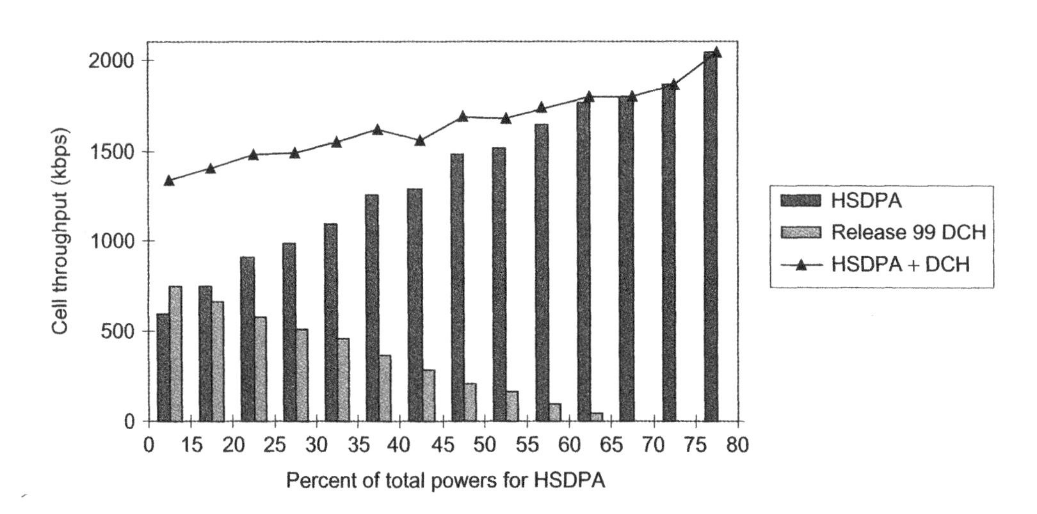

Figure 7.15 Concurrent DCH and HSDPA cell capacity

Figure 7.16 shows the peak achievable data rates available with a 50% power allocation for HSDPA.

From this analysis, it is obvious that sharing a single carrier between Release 99 and HSDPA offers good improvement, in terms of capacity, over a Release 99-only network. The improvement becomes more significant as the portion of PS data traffic carried over HSDPA increases. A secondary benefit, even for limited HSDPA power allocations, is the increased user data rate at the cell edge: from 64 kbps in Release 99-only to 320 kbps

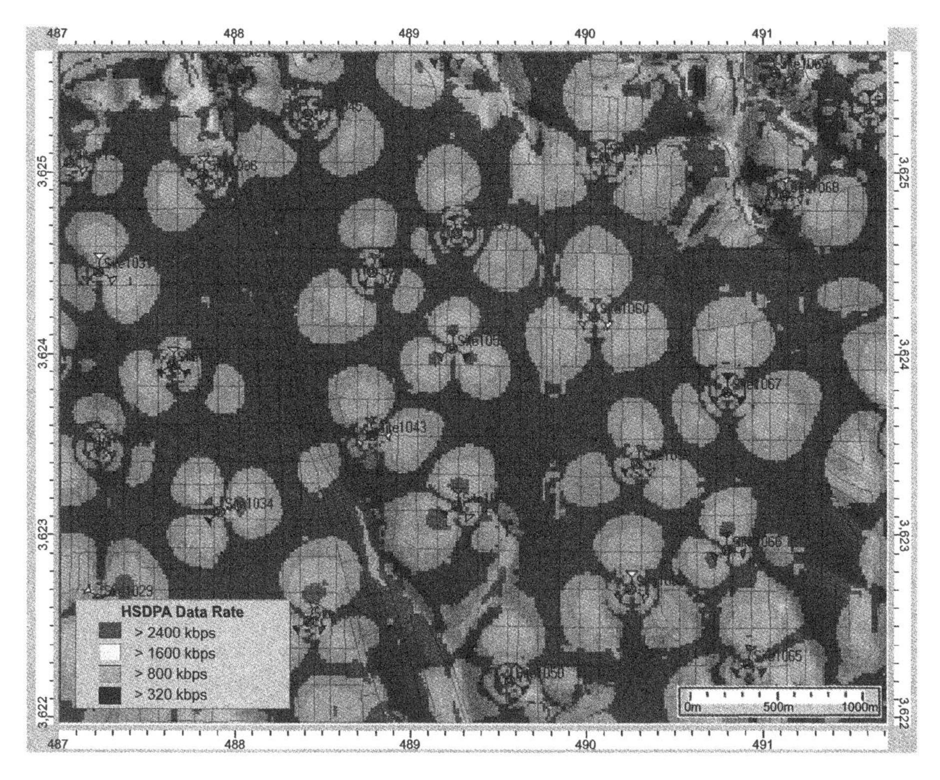

Figure 7.16 HSDPA data rate coverage with 50% power allocation

with HSDPA, at 50% power allocation. For lower HSDPA power allocations, the cell edge user data rate decreases to a point where HSDPA is not available over the entire area (CQI = 0 would be reported). This effect would be more pronounced for relatively smaller power allocations.

7.3.3.4 HSDPA Cell Capacity

This section describes the results of several link-level simulations that show the effect on cell capacity of several factors relating to UE capability and network scheduling configuration. The simulations assume a suburban environment, with users distributed according to three scenarios, as shown in Table 7.11. Each scenario assumes a different mix of users in PA3, PB3, and VA30 channel conditions:

- In Scenarios 1 and 2, all users are distributed in an environment with significant multipaths.
- In Scenario 3, 25% of all users are in a PA3, that is, with a strong dominant path, representing in-building or microcell circumstances, among others.

For all of the simulations presented, a full traffic buffer is assumed such that all users always have data to send. For the CQI measurements, delay, measurement, and

Table 7.11 HSDPA performance simulation scenarios

Channel model	Scenario 1 (%)	Scenario 2 (%)	Scenario 3 (%)
PA3	0	0	25
PB3	75	50	50
VA30	25	50	25

quantization errors are accounted for. Transmission errors are not modeled, so a transmitted CQI report is always correctly received. For the HS-SCCH, only the decoding is simulated; false alarms, whereby a UE incorrectly assumes that it has been scheduled, are not modeled. Per cell, each simulation assigns 30 users, each having a UE capable of supporting an Inter-TTI Interval = 1 with six active HARQ processes.

Table 7.12 shows the cell throughput results for each scenario under two conditions: first, 16-QAM modulation becomes available; second, the number of supported HS-PDSCH codes increases from 5 to 15. The Relative Impact column expresses the throughput as a percentage of the throughput for the basic UE (5 codes and no 16-QAM). A Proportional Fair Scheduler is assumed for all simulations.

In Table 7.12, first consider the number of HS-PDSCH codes. When 15-code UEs are employed instead of 5-code UEs, cell throughput increases by approximately 12 to 15%. Next, consider the introduction of 16-QAM modulation. This increases throughput only negligibly compared to the QPSK-only UEs for Scenarios 1 and 2. Only for Scenario 3 is the impact of 16-QAM noticeable, increasing throughput approximately 6 to 9% compared

Table 7.12 Simulation results for each scenario with different supported HS-PDSCH codes and modulation

Scenario	Number of codes/ HS-PDSCH	Availability of 16-QAM	Cell throughput [Mbps]	Relative impact (same scenario) [%]
1	5	No	1.780	Reference
	5	Yes	1.783	0.2
	15	No	2.017	13.3
	15	Yes	2.023	13.7
2	5	No	1.519	Reference
	5	Yes	1.523	0.3
	15	No	1.750	15.2
	15	Yes	1.754	15.5
3	5	No	2.171	Reference
	5	Yes	2.301	6.0
	15	No	2.404	10.8
	15	Yes	2.573	18.5

to the QPSK-only UEs. Overall, the maximum obtainable cell throughput–obtained by increasing the number of codes the UEs can support from 5 to 15 and introducing 16-QAM modulation–is nominally 18%.

Although the UE's capability can significantly affect cell throughput, the Node B's scheduling algorithm can have a much greater effect. Section 7.4.3.6 describes possible HSDPA scheduling techniques in detail. Here, Table 7.13 shows the results of the same simulations shown in Table 7.12, but using both a Proportional Fair and Round Robin scheduling scheme. In Table 7.13, the Relative Improvement column shows the relative throughput increase of Proportional Fair over Round Robin, for each illustrated UE capability. Depending on the scenario, the cell throughput obtained with the Proportional Fair scheduler is 32 to 76% greater than that obtained using a simple Round Robin approach.

Clearly, the scheduling algorithm has a greater effect on cell throughput than UE capability differences. From a network–and thus an operator's–perspective, the Proportional Fair scheduler has an undeniable advantage. From the user's perspective, though, only those users in good RF conditions see the advantage. Such users would be scheduled with

Table 7.13 Simulation results for each scenario with both a Round Robin and Proportional Fair scheduler

Scenario	Number of codes	16-QAM available	Scheduler	Cell throughput (Mbps)	Relative improvement (%)
1	5	No	Round Robin	1.196	Reference
	5	No	Proportional Fair	1.780	48.8
	5	Yes	Round Robin	1.197	Reference
	5	Yes	Proportional Fair	1.783	48.9
	15	No	Round Robin	1.280	Reference
	15	No	Proportional Fair	2.017	57.6
	15	Yes	Round Robin	1.280	Reference
	15	Yes	Proportional Fair	2.023	58.1
2	5	No	Round Robin	1.146	Reference
	5	No	Proportional Fair	1.519	32.5
	5	Yes	Round Robin	1.147	Reference
	5	Yes	Proportional Fair	1.523	32.8
	15	No	Round Robin	1.254	Reference
	15	No	Proportional Fair	1.750	39.5
	15	Yes	Round Robin	1.254	Reference
	15	Yes	Proportional Fair	1.754	39.9
3	5	No	Round Robin	1.368	Reference
	5	No	Proportional Fair	2.171	58.6
	5	Yes	Round Robin	1.402	Reference
	5	Yes	Proportional Fair	2.301	64.2
	15	No	Round Robin	1.434	Reference
	15	No	Proportional Fair	2.404	67.7
	15	Yes	Round Robin	1.462	Reference
	15	Yes	Proportional Fair	2.573	76.0

higher priority than users experiencing less favorable conditions. In less favorable conditions, users could observe a degradation of service: either reduced application throughput, or increased end-to-end latency, which would be even more pronounced. Unfortunately, from a user or operator point of view, little can be done on this subject as schedulers are typically vendor implementation-dependent with no control to fine-tune their behavior.

To summarize the effects of the scheduler, the number of UE codes, and the UE modulation, Figure 7.17 shows their cumulative effect on cell throughput for each of the three scenarios. The impact of each factor is clearly visible:

- **For Scenarios 1 and 2.** Only the type of scheduler and number of codes increase the cell throughput.
- **For Scenario 3.** All three factors yield an improvement.

The figure also shows a nominal range of cell throughput that would be expected for typical PS service in a Release 99-only macro network for a range of typical fading environments.

Figure 7.17 illustrates what is probably the main justification for HSDPA. This is true even though the achieved cell or user throughput, in a realistic deployment situation, is far below the theoretical limits. For such realistic user profiles, the throughput improvement over Release 99 CELL_DCH is still two to three times better, thus reducing the cost per bits in the same ratio.

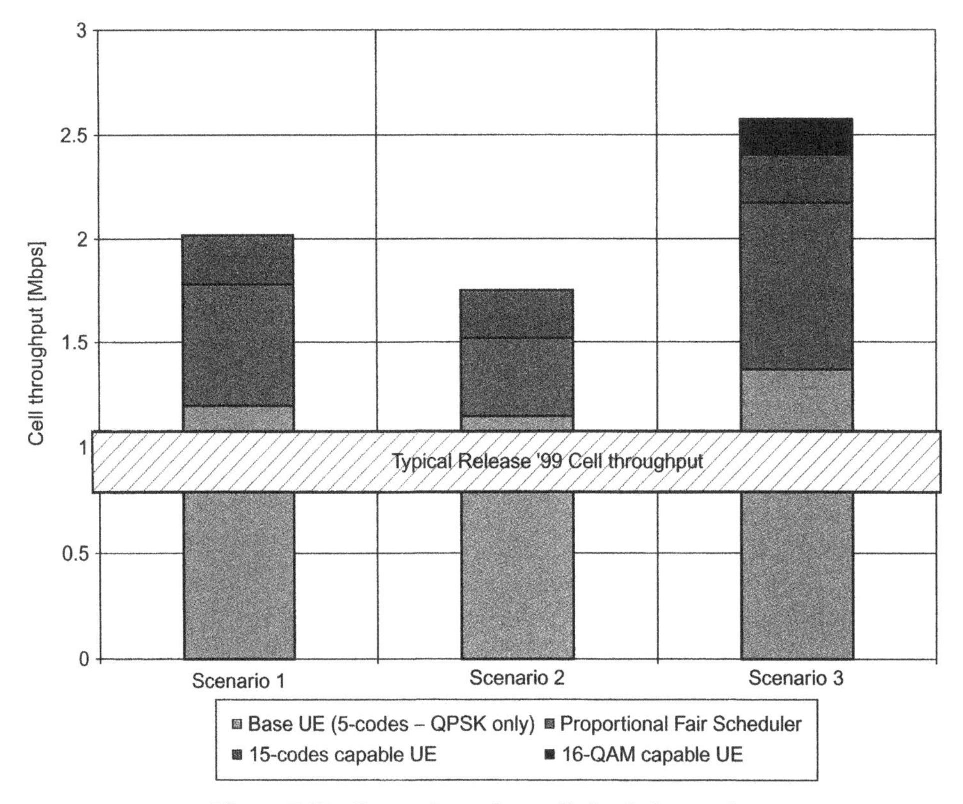

Figure 7.17 Comparison of overall simulation results

7.4 HSDPA Operation and Optimization

Previous sections have dealt with only HSDPA concepts and performance. This section discusses the signaling required to enable and operate HSDPA. The purpose of this signaling is mainly to start, stop, or reconfigure the HS bearers. Optimization engineers should take a keen interest in signaling, because it can explain why HSDPA performance might be suboptimal in a deployed network.

7.4.1 HSDPA Configuration

7.4.1.1 Establishing and Stopping HSDPA

The RRC is responsible for setting up and tearing down HSDPA operations, as well as for the signaling of all necessary Layer 1 and Layer 2 parameters. It is a requirement that the UE be in CELL_DCH for HSDPA to function, such that the serving HSDPA cell will always have an associated Release 99 DPCH. One function of this channel is the transmission of RRC information on the DCCH logical channel, which cannot be mapped to the HS-DSCH in Release 5.

Figure 7.18 shows a sample message flow to enable the establishment of HSDPA.

The sequence of messages is as follows:

Step 1. With the UE already in CELL_DCH, the RNC decides to start HSDPA operations. This could be based on availability of DL data from the PS Core Network, and

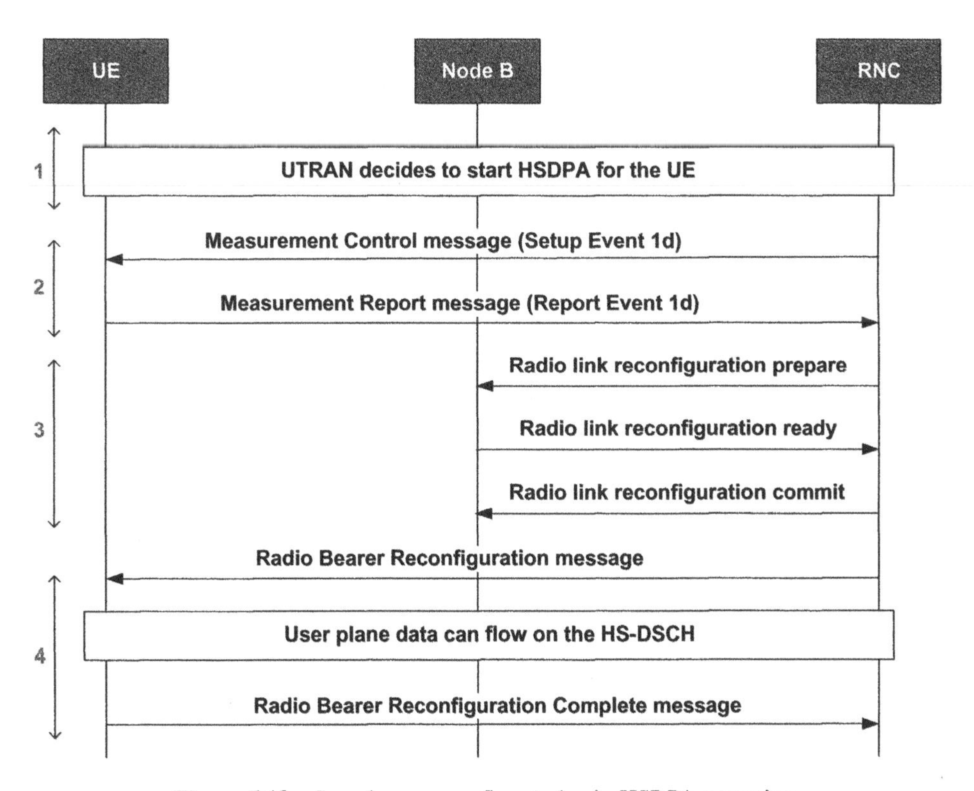

Figure 7.18 Sample message flow to begin HSDPA operations

QoS negotiations that occurred at the NAS layer when the data call was initially established. UTRAN is notified of a UE's HSDPA capability during initial RRC Connection Setup.

Step 2. The RNC sends a Measurement Control message requesting that the UE perform measurements to determine the best cell in its Active Set. Event 1d is used for this purpose, causing the UE to send a Measurement Report whenever there is a change in the best cell. Release 5 modified this event to require the UE to immediately report the initial best cell when the measurement is set up if there are two or more members of the Active Set, to always start HSDPA service on the best serving cell within the Active Set.

Step 3. The RNC notifies the Node B that controls the best cell to set up an HS-DSCH radio link on that cell. Practically, in the Reconfiguration message, the Serving HSDPA link would be identified by its PSC. An activation time in terms of Connection Frame Number (CFN) will be included in this message exchange.

Step 4. The RNC sends a Reconfiguration message to notify the UE that the HS-DSCH is available, and to signal all the parameters associated with HSDPA operations. After that point, as mentioned before, the scheduling of the UE is independent from RNC signaling; the Node B performs this task using the HS-SCCH.

This sequence of messages is one example of how HSDPA could be initialized using the Radio Bearer Reconfiguration message when a UE is already in Connected Mode. However, all required Information Elements (IEs) are also contained in the Radio Bearer Setup, Radio Bearer Release, and Cell Update messages. The option to start HSDPA operation using the Radio Bearer Setup message is of particular interest as it would allow an HSDPA-capable UE to start operating on HSDPA whenever a PS data call is originated. In this case, the RNC should configure both the associated DCH and the HS channels. Additionally, Layer 1 and Layer 2 HSDPA configurations could also be changed using the Physical Channel Reconfiguration and Transport Channel Reconfiguration messages, respectively. Section 7.4.1.2 describes the specific IE contained in each of these messages.

Stopping HSDPA service involves either removing the HS-DSCH transport channel while the UE remains in CELL_DCH, or removing the HS-DSCH with a transition to CELL_FACH. This is useful when a large number of codes are assigned to HS-PDSCH because code tree limitations can be mitigated by switching low activity or low priority users to CELL_FACH when resources are limited. Figure 7.19 shows a sample message flow with the following message sequence:

Step 1. With the UE in CELL_DCH state, the RNC decides to stop HSDPA operations. Possible reasons include a lack of DL activity, a change in RF conditions, high mobility detected, or resource management or when a UE can no longer be served by an HSDPA-capable Node B.

Step 2. The RNC notifies the Node B that controls the Serving HS-DSCH Radio link to stop transmitting to the UE on the HS-DSCH.

Step 3. The RNC sends a Reconfiguration message to notify the UE that the HS-DSCH is not available. This is accomplished by removing the Radio Bearer mappings to the HS-DSCH, removing MAC-d flows, and/or indicating that no radio link is a Serving HS-DSCH Radio link (i.e., Serving HS-DSCH set to none). When the UE

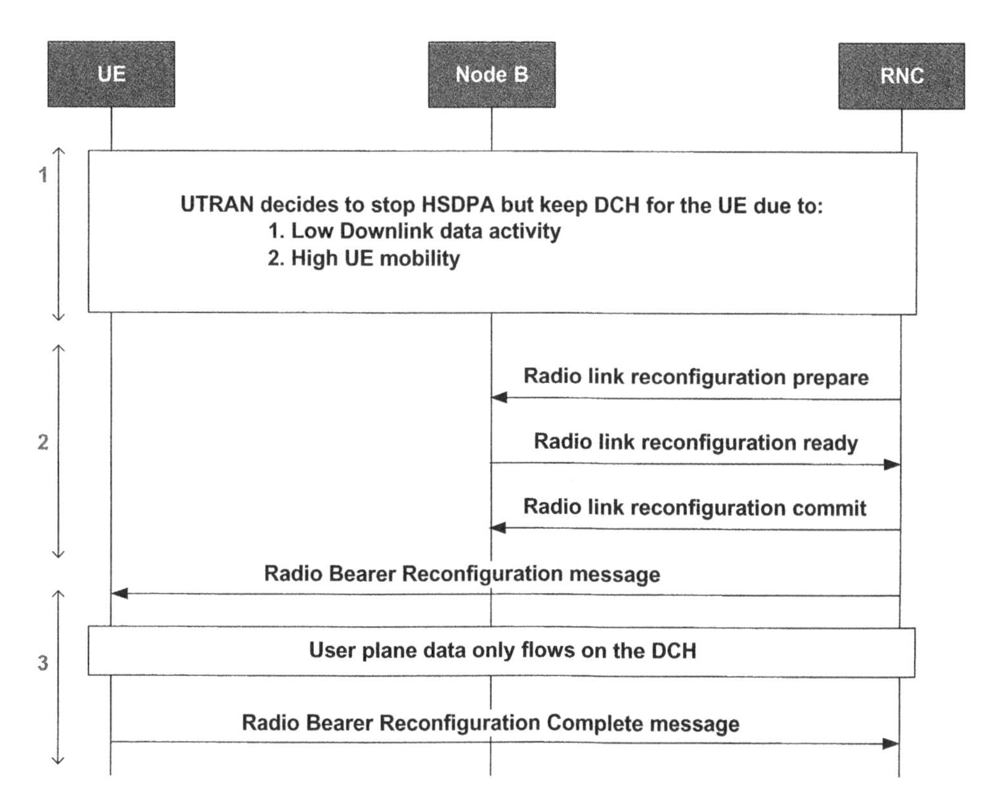

Figure 7.19 Sample message flow to stop HSDPA operations and remain in CELL_DCH

remains in CELL-DCH (as shown here), the message may also alter the data rate of the DCH. If the UE is to transition to the CELL_FACH state, DCCH/DTCH logical channels would also be remapped to RACH and FACH.

As in the previous example, the Radio Bearer Reconfiguration message was used here to stop HSDPA services. As with the initialization of HSDPA, all necessary IEs are contained in a number of other messages. Specifically, to stop HSDPA while remaining in CELL_DCH, the following messages could be used: Radio Bearer Reconfiguration, Radio Bearer Setup, Radio Bearer Release, Physical Channel Reconfiguration, and Transport Channel Reconfiguration. For a transition to CELL_FACH, however, logical channels can only be remapped by using the Radio Bearer Setup, Reconfiguration, and Release messages. An Active Set update would also stop HSDPA if the serving HS cell were removed from the Active Set.

7.4.1.2 HSDPA Information Elements

This section describes the IEs that are signaled to the UE to configure HSDPA. As stated above, all the IEs necessary to configure or reconfigure HSDPA service are contained in the Radio Bearer Setup, Reconfiguration, and Release messages as well as in the Cell Update Confirm message. Signaling of Layer 1 or Layer 2 IEs alone can be achieved using the Physical Channel Reconfiguration and Transport Channel Reconfiguration messages, respectively.

For the Physical Layer, the following IEs are signaled and are necessary for HSDPA operation:

- **Downlink scrambling code.** This enables possible use of secondary scrambling codes for the HS-PDSCHs as a solution to OVSF code tree exhaustion; it defaults to the Primary Scrambling Code if not signaled.
- **HS-SCCH channelization codes.** Each UE must be able to monitor at least four HS-SCCH channels. The OVSF codes applied to each HS-SCCH are defined by the UTRAN and must be signaled to the UE. UTRAN may however signal less than four codes to monitor.
- **Serving HS-DSCH radio link indicator.** Because the UE can be in a multiway handover, a signal must be sent to indicate which scrambling code of the Active Set is used as the serving HS-DSCH radio link.
- **ACK, NAK, and CQI power offsets.** The power allocated to each portion of the HS-DPCCH is defined by the offsets Δ_{ACK}, Δ_{NACK}, and Δ_{CQI}, which are defined in relation to the associated Release 99 DPCCH.
- **CQI feedback cycle and repetition factor.** Each CQI report can be configured to be repeated up to three times. The frequency at which the report is transmitted is also configurable in 2-ms increments ranging from 2 to 160 ms (1 to 80 subframes). If no CQI report is scheduled, DTX bits are transmitted. A CQI repetition factor greater than 1 will reduce the maximum throughput by a factor equal to the number of repetitions.
- **ACK/NAK repetition factor.** Each positive or negative acknowledgment can be configured to be repeated up to three times. Because of the synchronous nature of the ACK/NAK with the HS-PDSCH, this reduces the maximum achievable data rate by a factor equal to the number of retransmissions.
- **HS-PDSCH power offset from CPICH (Γ).** In calculating the CQI, the UE assumes that the power of the HS-PDSCH is greater than the Primary CPICH by $\Gamma/2$ dB.

The signaled HSDPA Layer 2 IEs relate to the MAC-d and MAC-hs entities. DTCH data arrives at MAC-hs on MAC-d flows. There can be up to eight MAC-d flows defined per UE. Radio bearer mapping information signaled to the UE denotes the mapping of each DTCH to either HS-DSCH or DCH. For HSDPA channels, the UE is further signaled an identifier (0–7) that associates each with a MAC-d flow. A single MAC-d flow can also contain multiple logical channels if C/T multiplexing is applied. Regarding the logical DCCH channels, their mapping to HS-DSCH is not allowed in Release 5 and all Layer 3 messaging will therefore always be sent to the UE on the existing Release 99 DCH.

The remaining Layer 2 IEs are associated with the configuration of MAC-hs and the HARQ mechanism. Each MAC-d flow arrives at MAC-hs and is distributed to one or more of up to eight priority queues. Only one MAC-d flow can be associated with each queue. It is then the responsibility of the scheduler within MAC-hs to assign data from each priority queue to an individual HARQ process. To enable this functionality, the following IEs are signaled to the UE:

- **Number of HARQ processes.** A maximum of eight processes can be defined for each UE.

- **MAC-d to priority queue mapping.** The mapping of each MAC-d flow to a priority queue is configured before HSDPA operations start; mappings can be altered only by Layer 3 signaling.
- **Division of soft buffer bits between HARQ processes.** The total available UE HARQ buffer can be divided equally or with an explicit division between processes.
- **Size index identifier.** MAC-hs concatenates MAC-d RLC PDUs and adds a header to form the HS-DCSH transport blocks. The header contains the Size Index Identifier (SID), a three-bit number that represents up to eight valid MAC-d PDU sizes. The mapping table of PDU size to index number must be signaled to the UE before HSDPA can begin operation.

Two additional Layer 2 IEs relate to HARQ. While multiple HARQ processes are operating simultaneously, data can arrive at the UE out of order. Because RLC requires that its PDUs be delivered in the correct sequence, a reordering queue is necessary. Here, data can be buffered while missing NAK'ed data is retransmitted. MAC-hs has a limited sequence number space of 6 bits limiting how long the protocol can wait for a missing block of data before either stalling or experiencing sequence number wrap. Two mechanisms, one window-based and one timer-based, are employed to overcome this problem with the following IEs:

- **Reordering queue timer.** Each reordering queue has an associated timer, which flushes the queue of data after the timer expires. This is used to avoid stalling if the UE misses the last in a sequence of transport blocks or the Node B mistakes a NAK for an ACK. The timer starts when a hole is detected.
- **Reordering queue transmit window.** Each reordering queue has an associated window, which is used to avoid sequence number ambiguity due to the limited sequence number space. The upper edge of the window advances whenever the Node B transmits a new transport block; the lower edge of the window advances whenever the Node B receives an acknowledgment from the oldest transmitted transport block. Any previously received blocks of data that move outside the window to a more recent transmission are delivered to RLC regardless of any holes.

If a PDU is discarded in either case, it would be up to higher layers, typically RLC if data is send in Acknowledge Mode (AM), or the application layer to either conceal the missing data or request further retransmissions. Such retransmissions originate at the RNC or beyond, rather than at the Node B level as is the case for MAC-hs retransmissions. Clearly, additional retransmission delays would be observed.

7.4.2 HSDPA Serving Cell Change

An important distinction between HSDPA and Release 99 DCH is that HSDPA lacks soft handover. In Release 99 for the DPCH, the UE may have an Active Set consisting of multiple cells from multiple Node Bs. As the radio conditions change, radio links are added and deleted to and from the Active Set, depending on the strength of the associated CPICHs and the defined event triggers. However, HSDPA can have only one serving HS cell at a time. If the radio conditions change, the serving cell must change; thus HS data transfer is discontinued on the current serving cell and reestablished in the new target cell.

Why is this significant? At the Node B, data may be buffered in MAC-hs. No mechanism is defined to recover untransmitted data from a Node B once a cell change has been completed. As a result, any data remaining in the MAC-hs buffer of a previously serving cell is flushed and discarded. In terms of recovery, the defined transmission mode (AM or UM) then determines whether this data is retransmitted by RLC; higher layers may also cope with the missing data.

Cell change is performed independently of the Active Set update procedure and can be completed with any one of the six reconfiguration messages described in Section 7.4.1. Dependencies between cell change and Active Set still exist, because the serving HS cell must be within the Active Set. The exclusion of the Active Set Update message to complete the cell change introduces a number of potential performance issues. For example, a cell that is increasing in strength but not yet in the Active Set cannot become the serving HSDPA cell without first being added to the Active Set by an Active Set Update message, which would then have to be followed by the cell change procedure. Furthermore, if an Active Set Update message removes the serving HSDPA cell from the Active Set, HSDPA operations would stop. The network must manage these transitions and their associated delays. In a rapidly changing RF environment, this becomes increasingly complex, leading to the possibility of race conditions between Active Set management and cell changes.

7.4.2.1 Cell Change Procedure

The HSDPA Serving Cell change procedure itself may be either synchronized or unsynchronized. The difference between the two techniques is whether an Activation Time is included in the messages used to carry out the cell change.

The unsynchronized procedure does not include the action time. This produces a much faster cell change than the synchronized procedure would, but it flushes any untransmitted data in the MAC-hs buffer of the previous Node B. For RLC AM, this produces significant retransmissions; for RLC UM, permanent loss occurs.

In the synchronized case, mandating an action time allows the Node B to attempt to transmit any remaining data before the cell change occurs. Hence, data loss (for UM services) or retransmissions (for AM services) is minimized. However, if the activation time is set far enough ahead to ensure that all data from the MAC-hs buffer has been transmitted, a significant delay can be introduced. One possible result would be a suboptimal or even unsuitable serving HSDPA cell. The impact of a cell change delay would be most apparent in a rapidly changing RF environment, where the time duration for which a cell is the best one may be less than the time taken to switch HSDPA to that cell. Optimizing the MAC-hs buffer size and Activation Time, as well as the Event 1d-associated parameters would alleviate but not eliminate this issue, for given RF and mobility conditions.

To illustrate the concept of cell change, Figure 7.20 shows a sample message flow for a synchronized HSDPA cell change. An existing source Node B is assumed to be already serving the HSDPA cell.

The figure shows the following steps:

Step 1. In response to a previous Measurement Report (not shown), a new Node B (the target) is added to the Active Set. This target Node B is first instructed to set up a radio link, and then the RNC sends the Active Set Update message to the UE.

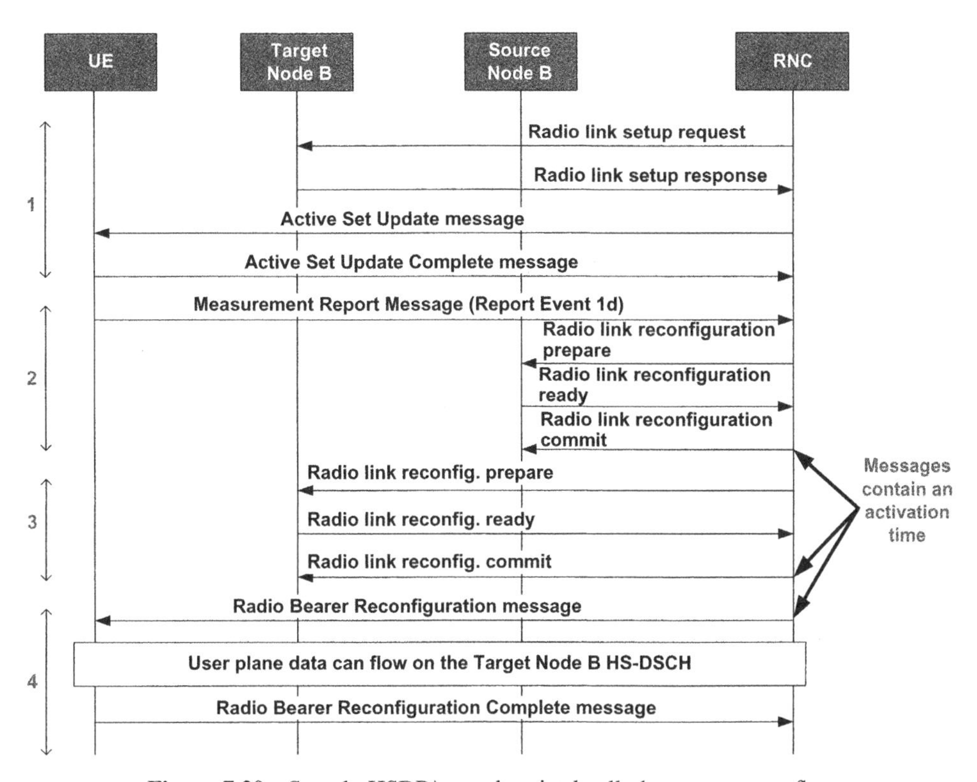

Figure 7.20 Sample HSDPA synchronized cell change message flow

For the associated DCH channel, the UE is now in soft handover and has a DPCH on both target and source Node Bs, and an HS-DSCH on the source Node B.

Step 2. Eventually Event 1d is triggered, in which the UE reports that the best cell is the one just added: the target Node B. This triggers the beginning of the Serving HS-DSCH Cell change procedure. The RNC sends a message to the current Serving HS-DSCH Node B to prepare the Release of the HS-DSCH. This gives the Node B a chance to finish transmitting data already in its MAC-hs buffers. The Commit message contains an action time that specifies the time after which this Node B will stop sending HS-DSCH data.

Step 3. The RNC instructs the target Node B to establish HSDPA channels for the UE. The Commit message contains the same action time, indicating the time at which HS-DSCH operations begin for this UE.

Step 4. The RNC sends a Radio Bearer Reconfiguration message to change the Serving HS-DSCH Radio link. This message contains the same action time, at which time the UE begins receiving HS-DSCH data from the new cell.

For comparison, Figure 7.21 shows a sample message flow for an unsynchronized cell change.

The figure shows the following steps:

Step 1. In response to a previous Measurement Report (not shown), a new Node B (the target) is added to the Active Set. This target Node B is first instructed to set

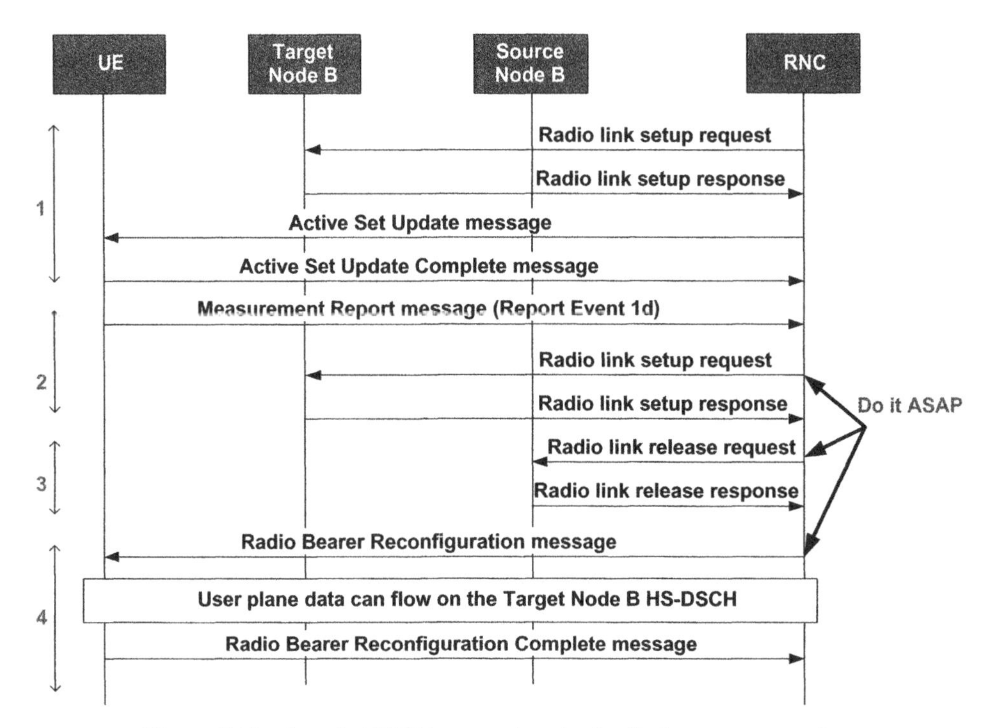

Figure 7.21 Sample HSDPA unsynchronized cell change message flow

up a radio link, and then the RNC sends the Active Set Update message to the UE. The UE is now in soft handover and has a DPCH on both target and source Node Bs, and an HS-DSCH on the source Node B. This process is identical to the Synchronized process.

Step 2. Event 1d is triggered and the UE reports that the best cell is the one just added; the target Node B. This triggers the beginning of the Serving HS-DSCH Cell change procedure. The RNC sends a message to the target Node B instructing it to establish HSDPA channels for the UE, with no Action Time specified in this case.

Step 3. The RNC sends a message to the Serving HS-DSCH Node B to Release the HS-DSCH with no Action Time specified.

Step 4. The RNC sends a Radio Bearer Reconfiguration message to notify the UE that the HS-DSCH is available on the new Serving HS-DSCH Radio link (no Action Time is specified).

Looking at the two message flows shown in Figures 7.20 and 7.21, the most significant difference is the inclusion, in the synchronized case, of an Action Time, which is sent in the Radio Link Reconfiguration Commit messages–both to the source and to the target Node B–and also in the Radio Bearer Reconfiguration message to the UE. In the synchronized case, setting the Action would have to take two things into account: the amount of data likely buffered at the Node B, and the time likely taken to complete the flow of messages. In the unsynchronized case, without the Action Time, some data may be lost, mainly due to the data in the source Node B being flushed at the completion of the cell change.

7.4.2.2 Measurement Event for Triggering Cell Change

With the specification that there can be only one serving HSDPA cell, a key aspect of performance is focused on triggering the cell change procedure. As described above, this is achieved by the definition of Event 1d, which is triggered when the best cell changes. Figure 7.22 shows an example of such a triggering.

The figure shows CPICH E_c/N_o varying in two cells. Initially Cell 1 is providing HSDPA. When Cell 2 becomes stronger than Cell 1 plus a hysteresis (point A), a timer starts. At point B the timer equals the Time-to-Trigger (TTT) of the event, which generates a Measurement Report. This initiates the cell change procedure, which is completed at point C when the HS-DSCH data transfer commences on Cell 2.

In the example described above, the Pilot measurement quality for each Pilot is E_c/N_o, although RSCP or path loss could also be used. The specific parameters associated with this event are the hysteresis, the TTT, and the filter coefficient. All of these, along with their setting trade-offs, are described in the following section.

7.4.3 HSDPA Parameter Tuning

HSDPA parameter tuning is vital to ensuring adequate performance. This section focuses on the HSDPA-specific parameters that can be set and optimized to ensure that the high-speed services possible with HSDPA can be realized. This section does not cover the RF and Release 99 parameters, but they are of the utmost importance for achieving good HSDPA performance. The RF configuration affects HSDPA by maximizing the geometry,

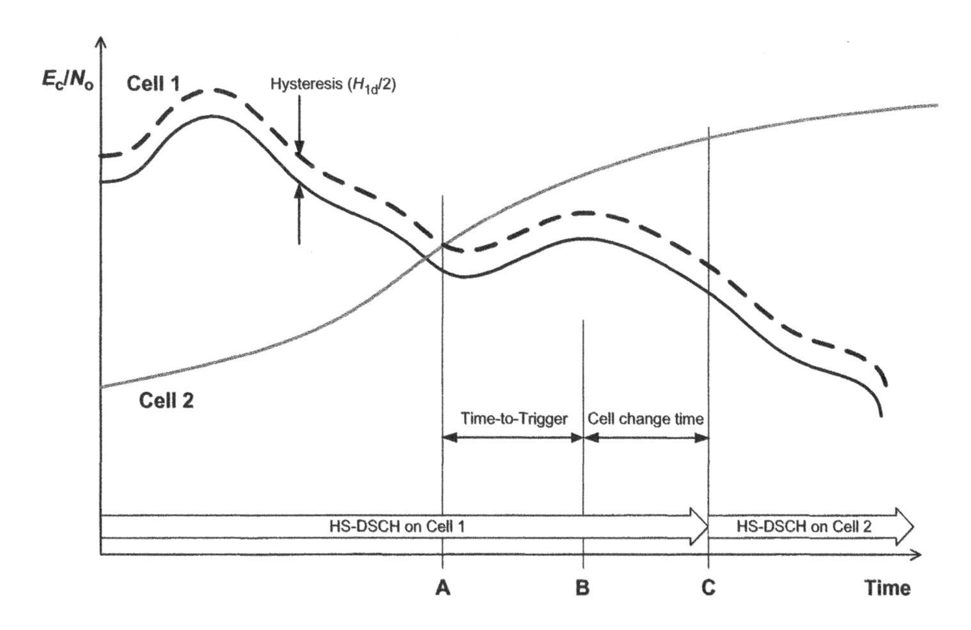

Figure 7.22 Cell change triggering with Event 1d

which is one of the main drivers of HSDPA throughput, as discussed in Section 7.3.2. Release 99 parameters for cell reselection and handover also play a role in HSDPA performance. Both sets of parameters ensure that the most appropriate cell is in the Active Set when HSDPA operation starts (for details, see Chapter 4).

7.4.3.1 HS-SCCH Parameters

The HS-SCCH channel tells the UE that it has been scheduled. Each HSDPA UE may continuously monitor up to four HS-SCCH channels. The information transmitted on the channel enables the UE to decode packets scheduled on the HS-PDSCH channels. It is critical, therefore, that the UE can successfully decode the HS-SCCH. Any misdetection by the UE while monitoring the HS-SCCH channels results in the UE not receiving the packets intended for it, thereby increasing the overall number of HARQ retransmissions and reducing throughput.

The following parameters are associated with the HS-SCCH:

- **Number of codes.** The number of codes assigned for the HS-SCCH channels determines how many users can be concurrently scheduled for HS-PDSCH transmission in a single subframe. Assigning fewer codes impact the number of concurrent HS users; assigning more codes impacts the resource (code and power) available for both HS and non-HS users.
- **HS-SCCH power allocation.** The power allocated to the HS-SCCH not only determines its BLER but also affects the decoding performance of the UE. An HS-SCCH detection error is defined in Ref [9] as the event when the UE is signaled on the HS-SCCH, but a DTX is observed in the corresponding HS-DPCCH ACK/NAK field. This failure is observed when the CRC check fails in decoding Part 2 of the HS-SCCH. Though the CRC is attached only to the Part 2 bits, it is calculated jointly over both Part 1 and Part 2 of a HS-SCCH transmission. A consistency check is also carried out by the UE. The Node B conformance specification [12] defines an acceptable error rate of 1 to 5%, depending on the channel condition, for HS-SCCH power assignment of -10 dB, E_c/I_{or}. This results in a significant amount of power being utilized for the HS-SCCH, reducing power that could be allocated for the data payload on the HS-PDSCH, and reducing overall cell throughput. This effect increases in severity as more HS-SCCHs are added to support concurrent assignments to multiple UEs in the same subframe. However, the above HS-SCCH power assignment would be necessary only in poor RF conditions at the cell edge. Significantly less power would be required in all other scenarios. This implies that a power control mechanism is necessary for the HS-SCCH to use HPA power effectively. The Release 5 specifications neither specify nor prevent HS-SCCH power control; the decision to implement such a scheme is left to the infrastructure vendor's discretion. Two schemes likely to be implemented could be based either on the DL DPCH power control feedback provided by the UE, or on the CQI feedback report, as described below:
  - **DPCH-based HS-SCCH power control.** In this case, the power allocated to the HS-SCCH is a fixed offset from the power allocated to the serving DL DPCH. Because soft handover is not supported for the HS-SCCH, different offsets will be necessary depending on the handover state of the DPCH. In the presence of multiple

soft DPCH links, each Active Set member transmits lower power. This necessitates a higher HS-SCCH to DPCH offset to ensure the same HS-SCCH BLER. If HS-SCCH power offset cannot be changed according to the handover state, then the fixed offset should be calculated for the worst-case handover state. Dynamically changing the offset allows an optimum HS-SCCH power setting that most effectively utilizes available power but at the expense of additional processing at the Node B.

- **CQI-based HS-SCCH power control.** It can be shown [13] that there is a linear relationship between the reported CQI and power allocation required to ensure that the HS-SCCH BLER meets defined performance requirements. Thus, HS-SCCH power can be adjusted as a function of the reported CQI on a per-UE basis. An advantage of this approach is that the actual HS-SCCH transmitted power is a direct function of the link quality associated with the serving cell and there is no dependency on the handover state.

7.4.3.2 CQI Reporting

The Release 5 specification [5] mandates that the CQI be based on the perceived performance during a three-slot period that ends one slot prior to the two-slot transmission of the actual CQI value, as shown in Figure 7.23. The reported value corresponds to the highest HS-DSCH TBS that could have been supported with a BLER less than 10%.

Figure 7.24 illustrates the CQI reporting parameters that govern how often CQI is reported. Each block represents a two-slot CQI report; the space between reports represents the single-slot ACK or NAK. The best HSDPA performance is obtained when the Node B scheduler has the most recent measurement of channel conditions, enabling an appropriate TBS to be selected for the HS-SCCH.

The CQI feedback cycle defines how often a CQI report is transmitted to the Node B. The CQI Repetition Factor defines how many times the same CQI is reported during the same feedback cycle. In Figure 7.24, the feedback cycle is 8 ms for all three cases; the repetition factor is 1 for Cases A and B, and 2 for Case C. The repetition factor is

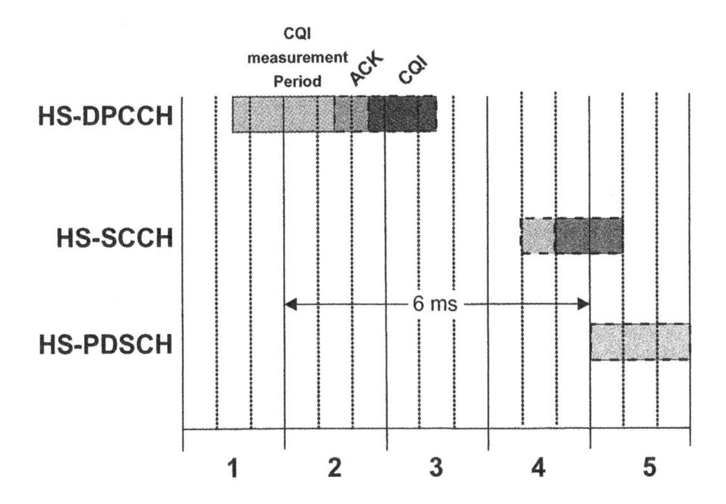

Figure 7.23 CQI measurement period

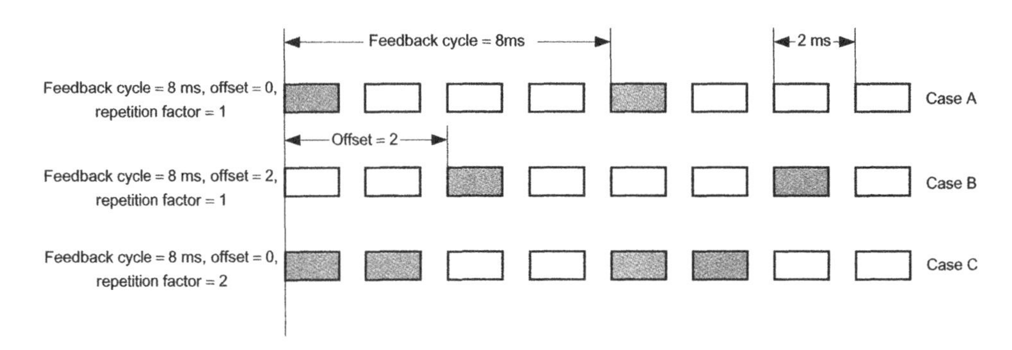

Figure 7.24 CQI reporting parameters

up-bounded by the feedback cycle, such that a CQI report cannot be repeated more times than would be allowed by the cycle length.

The CQI Reporting Offset is a number of subframes that a CQI report may be delayed, with respect to the earliest time it could have reported during a reporting cycle. This could be used to reduce the UL interference by distributing the CQI reports of different UEs over a number of subframes. Case B shows an offset of 2 in this example.

The CQI Reporting parameters must be set in conjunction with one another. To properly schedule an appropriate TBS, the Node B requires an accurate estimate of the channel condition. If the CQI feedback cycle is set too large, the CQI report may be substantially out of date by the time it is received by the Node B. However, this must be considered in conjunction with the increased required power and UL interference associated with a low setting. For the Repetition Factor, repeating the CQI report relatively few times will reduce any delay in reception but may require higher power. Conversely, a relatively high repetition rate allows a lower power (which may be beneficial in cell-edge scenarios) but may delay accurate reception of the report. In any case, CQI feedback cycle, offset, and repetition factor setting should also consider any CQI filtering done at the Node B. Node B CQI filtering, a vendor implementation process, will affect how CQI is used and would impact the validity of the following discussion.

Simulations [14] have shown how the CQI feedback cycle affects the achieved throughput in a single-path Rayleigh fading environment for different UE velocities. When using the CQI report alone to decide on TBS, there is little difference between the throughputs obtained using a CQI feedback cycle of either 1 or 3 subframes at 3 km/hr. At 10 km/hr, however, the 1-subframe feedback cycle produces the highest throughput; all other combinations appreciably reduce the throughput. At 30 and 120 km/hr, there is little difference in the throughputs obtained using any feedback cycle between 1 and 30 subframes. Because of the high mobility, the CQI report is substantially out of date by the time a transport block can be scheduled on the basis of the report. A feedback cycle beyond 30 does appreciably degrade the throughput. These results suggest that in static or low velocity conditions, a CQI feedback cycle greater than 2 ms performs adequately. As velocity increases, a CQI report is necessary every 2 ms, although beyond 30 km/hr a relatively high CQI feedback cycle affects the achievable throughput only marginally.

Another aspect of the channel quality reporting mechanism is the CQI Power Offset; Δ_{CQI}. This defines an offset between the CQI transmitted on the HS-DPCCH and the associated Release 99 DPCCH. This offset must be set in conjunction with the feedback cycle parameter described above, to ensure that the CQI report is successfully received within the defined performance range:

- Setting it too low affects the DL HSDPA throughput if the CQI report is not detected often enough.
- Setting it too high unnecessarily increases UL interference and the PAR, and also negatively affects UE battery standby time.

Another consideration for setting CQI Power Offset is the UE handover state. To ensure that UL capacity is not adversely affected by HSDPA operation, the HS-DPCCH is power-controlled on the basis of the associated DCH, offset by Δ_{CQI}. Because of macro diversity gain, the transmit power of the UE may be reduced while in handover. The HS-DPCCH, however, is detected only at the serving HSDPA cell. This situation is further complicated by the fact that serving cell change may occur at a lower rate than Active Set updates, potentially resulting in a UE, sending the CQI (and associated ACK/NAK) to a weak member of the Active Set. To mitigate this problem, different CQI Power Offsets can be set according to the handover state so that in handover, Δ_{CQI} could be increased by the typical handover gain, 3 dB for 2-way, and 4.7 dB for 3-way. The ability to configure a network in this manner is not standardized and is infrastructure-dependent.

7.4.3.3 ACK-NAK Reporting

The positive or negative acknowledgment of a HS-PDSCH transmission is always sent 7.5 slots (5 ms) after its transmission; it may be configured to repeat up to four times. As with the CQI report, both the ACK and NAK have individual configurable power offsets compared to the associated Release 99 DPCCH. Trade-offs for setting these offsets are similar to those for the CQI power offset: that is, the higher misdetection rate of a smaller offset must be weighed against the increased UL interference and PAR, and reduced battery life, of a higher offset. The soft handover issue is also similar: the DPCH on which the offset is based may be transmitted at a lower level if multiple radio links exist.

In setting the ACK/NAK offsets, the relative impact of an ACK misinterpretation versus a NAK misinterpretation should be considered:

- **ACK misdetected as either a DTX or a NAK.** This causes the HARQ process at the Node B to retransmit that data packet. If correctly decoded by the UE, this duplicate transmission is simply disregarded. Although this second packet could have been used for a new data transmission, the overall impact is relatively small with just one extra transmission required by a single HARQ process.
- **NAK misdetected as an ACK.** This error has a more significant impact. Here in this case, the Node B assumes that the UE correctly decoded the data; thus a new data block is sent associated with that same HARQ process. When the UE detects that new data has been sent, it flushes any previously stored transmissions of the original data block. RLC retransmissions would recover this discarded data, assuming AM. Clearly, the effect is more significant than a single retransmission of a HARQ as in the previous case.

The other item associated with ACK/NAK reporting is the Repetition Factor. Increasing the number of times that an acknowledgment is transmitted improves the reliability of detection. However, because of the synchronized relationship of HS-PDSCH and ACK/NAK transmissions, repeating an acknowledgment disallows certain subframes for HS-PDSCH data transmission. As a result, the achievable maximum data rate decreases by the number of times that ACK/NAK is repeated. To maximize data throughput, the repetition factor is usually set to 1 for initial deployments, except when the infrastructure allows different settings for UEs in soft handover. In that case, allowing multiple ACK/NAK transmissions is acceptable; it reduces achievable throughput, but maintains the integrity of the HARQ processes, compared to significant RLC retransmissions.

7.4.3.4 MAC-hs Reordering Queue Parameters

As described in Section 7.4.1.2, a reordering mechanism must buffer and sort the packets from multiple HARQ processes before PDUs can be delivered to RLC, in order. The MAC-hs reordering entity employs both window and timer mechanisms to flush holes, avoid sequence number ambiguity, and prevent stalling.

First, the MAC-hs Transmit/Receive Window size should be set to allow the maximum desired number of transmissions associated with each HARQ process. If the window size value is too small, significant RLC retransmissions–as substantially incomplete blocks of data–would be delivered to RLC, even though the UE Physical Layer successfully received them. This could possibly cause stalls in environments with rapidly changing RF. Conversely, if the window size value is too large, latency increases. Data delivery could be delayed while the mechanism waits for delivery of a single HARQ transmission, which has been unsuccessfully transmitted multiple times. If IR combining is being used instead of Chase combining, this also increases the required UE memory utilization.

In terms of what is an appropriate number of retransmissions, Ref [15] demonstrates that, usually, little gain is obtained beyond the second transmission for both Chase and IR combining. Assuming that a UE could have the maximum number of 8 HARQ processes assigned to it, this implies an appropriate window size of 16.

The MAC-hs Reordering Release (T1) Timer is used for stall avoidance. The timer is started on the successful reception of a later packet after a previous packet has been NAK'ed, except when the timer is already running. On expiration of the timer, all PDUs are delivered, regardless of whether the hole has been filled or not. The timer should be set in such a way that the maximum desired retransmissions from all HARQ processes are allowed to complete. Setting the timer too high, particularly in relation to the Status Prohibit Timer at the RLC layer, results in unnecessary RLC retransmissions. Also, this could introduce significant delays in delivering data to higher layers.

The Release 5 specifications allow different receiver window sizes and timer settings for different reordering queues. While initial deployments will likely be implemented with a single queue, future implementations could set characteristics for queues depending on their applications. For example, for queues serving delay-sensitive applications, relatively low values would achieve better performance. For queues serving delay-tolerant applications, relatively high values would be more appropriate. Both parameters, though, must be set in conjunction with the RLC parameters of the respective logical channels that are being served.

7.4.3.5 Serving Cell Change Parameters

As mentioned in Section 7.4.2, HSDPA does not allow soft handover for the serving HS cell. If the UE has multiple Release 99 radio links in its Active Set, the cell with the strongest radio link should be the HSDPA server. The UE uses Event 1d to report the strongest cell to the RNC, reporting whenever the strongest cell changes. Also, when set up with a Measurement Control Message, the event immediately reports the best cell regardless of whether the strongest cell has recently changed.

The following parameters define how an Event 1d is reported:

- **Measurement quantity.** The measurement quantity can be path loss, CPICH RSCP, or CPICH E_c/N_o. The most appropriate measure for triggering the event is CPICH E_c/N_o because it provides the closest estimate of the radio link quality.
- **Event 1d hysteresis (H_{1d}).** Using CPICH E_c/N_o as the measurement quantity, an Event 1d is triggered when:

$$E_c/N_{o(\text{NotBest})} \geq E_c/N_{o(\text{Best})} + H_{1d}/2 \tag{7.5}$$

where:

- $E_c/N_{o(\text{NotBest})}$ is the E_c/N_o of any cell that is currently not the best cell[2]
- $E_c/N_{o(\text{Best})}$ is the E_c/N_o of any cell that is currently the best cell[2]
- H_{1d} is the hysteresis for Event 1d
- A cell individual offset (CIO) could also be applied but is not considered here.

Because of the inherent delay associated with cell change, depending on whether synchronized or unsynchronized cell change is implemented, an appropriate hysteresis must be chosen that avoids ping-pong effect between two cells of similar strength, yet does not delay cell change to the extent that the serving HS cell is significantly weaker than another available cell. In the synchronized cell change case, frequent Event 1d triggering increases signaling load and also causes frequent flushing of the MAC-hs buffers of the cells involved in the cell change. This triggers RLC retransmissions that significantly affect the achievable user data rate and thus capacity. In the unsynchronized cell change case, using a large hysteresis causes the serving HSDPA cell to be of a relatively poor or even unsuitable quality compared to others. This could lead to possible data transfer interruption and throughput degradation.

- **Event 1d Time-to-Trigger (TTT).** The condition shown in equation 7.5 must exist for a time equal to the TTT before a Measurement Report is generated. If this parameter is set to a relatively high value, the necessary triggering of the event due to a significantly stronger cell is delayed or may not take place at all. If set to a relatively low value, small fluctuations in CPICH E_c/N_o of nonbest server can trigger Event 1d, leading to frequent cell changes. TTT must also be set in conjunction with the filter coefficient, which was described in Chapter 4, because they have similar effects.

  As handover parameters, TTT and hysteresis have opposite effects: larger TTT and small H_{1d} is almost equivalent to low TTT and large H_{1d}. The selection of a suitable set should then be based on the overall number of cell changes and on throughput. But it depends on the mobility observed: in low mobility, long TTT and low H_{1d} work best; in high mobility, short TTT and high H_{1d} work best.

---

[2] Either best or not best cell can be the currently serving cell.

Another factor affecting the cell change parameters is how much time the specific infrastructure typically takes to complete the cell change. A network capable of rapid cell change is more suitable for UEs in RF environments where the best cell changes frequently; the hysteresis and TTT can then be set accordingly. At the time of this writing, typical cell change times can range from 300 to 800 ms depending on infrastructure vendor.

7.4.3.6 Scheduler

The throughput and latency performance of an HSDPA system is a function of available resources. This includes power allocated to the HS-PDSCH and HS-SCCH channels, as well as the number of HS-SCCH defined and the number of OVSF codes that are available for the HS-PDSCH. If only a single user accesses HSDPA services from a cell, the cell can allocate all available resources to that user. In the real world, however, there will always be multiple users–each with an individual demand profile–attempting to utilize the available resources. The HSDPA cell is expected to manage the available limited resources to serve all the users accessing the system, as effectively as possible. The function of allocating resources to various users is performed by the scheduler, which, as described previously, resides at the Node B for HSDPA. To decide when, and with what resources, a user is served, the scheduler requires the following inputs:

- Available HS-PDSCH power
- Available HS-SCCH power
- Number of HS-SCCH
- Available OVSF codes for HS-PDSCH
- User CQI reports
- User data availability (buffer occupancy)
- Buffer data arrival rate
- Number of users to be served

Additionally, the scheduler may also have to satisfy users' different QoS requirements [16], such as the following:

- User priority
- Guaranteed data rate
- Latency requirement
- User traffic class

On the basis of these factors, the scheduler must decide when to schedule a user and which TBS to assign. No criteria specify how the scheduler should make this decision; thus scheduling can be designed to achieve one of several goals. The following are the main scheduling algorithms for HSDPA:

- **Round robin.** This is the simplest of all scheduling algorithms. Resources are allocated to all users cyclically and for the same duration. Depending on implementation, the size of the transport block allocated to each user can be based both on the reported CQI and the availability of power and codes, or the reported CQI can be ignored such that the TBS allocated is blind to RF conditions.

- **Maximum CQI.** Scheduling is based on the CQI report of each user with the highest priority given to whoever reports the greatest value. This approach is substantially biased towards users in good RF conditions, and it produces very different results than Round Robin. While the overall cell throughput is significantly increased, users at the edge of coverage rarely get service. From a fairness perspective, therefore, this technique likely produces an unacceptable variation in service quality that would negatively affect overall user satisfaction.
- **Proportional fair.** This type of scheduler takes advantage of the fact that the fading experienced by the signals received by different UEs is uncorrelated. It attempts to serve a user when the experienced channel conditions are relatively good, while still maintaining equitable access to the available resources for all users. Figure 7.25 shows this concept schematically. The figure shows the variation of the Signal-to-Interference and Noise Ratio (SINR) with time for three different UEs. It also illustrates *Opportune Service Times* where the SINR is relatively high within the time interval shown. The result is the scheduling order shown in this example. In this type of scheduler, the evolution of the CQI of a given user is more important than its absolute value (as in the Maximum CQI scheme).

To decide which user to schedule, the Proportional Fair algorithm defines a priority for each user. Assuming N active HSDPA users, the priority of each user P_i at subframe t is:

$$P_i = \frac{CQI_i(t)}{\lambda_i(t)} \tag{7.6}$$

such that:

$$\lambda_i(t+1) = (1 - 1/\alpha)\lambda_i(t) + 1/\alpha \times \textit{Current data rate} \tag{7.7}$$

where α is a time specified in terms of a number of slots.

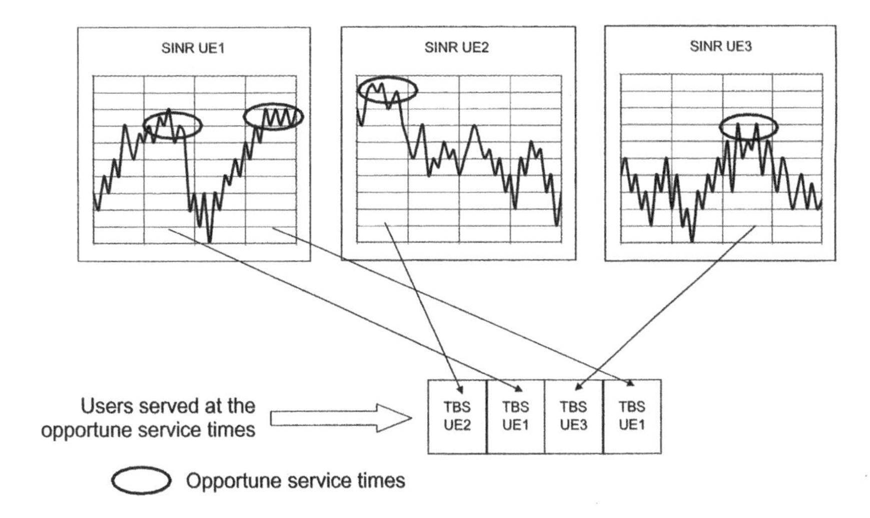

Figure 7.25 Schematic diagram of a Proportional Fair Scheduler

The two equations above indicate that if a user's channel condition improves, the priority increases; if it worsens, the priority decreases. In steady channel conditions, the priority stays the same. The parameter α sets the sensitivity and can be changed to adjust the rate at which a user's priority recovers as channel conditions improve:

- If parameter α is set too low, the user's priority increases too quickly after only a brief recovery in channel quality.
- If parameter α is set too high, it takes too long for the user's priority to increase, resulting in possible missed opportunities where that user could have been scheduled for data transmission.

An inherent property of the Proportional Fair algorithm is that it gives higher priority to delay-sensitive applications such as HTTP, as opposed to delay-tolerant applications [17].

In addition to choosing which algorithm to employ, the scheduler also decides the TBS to be transmitted to a UE. The scheduler bases this not only on CQI reports, but also on the available OVSF codes and power, allocating these resources as required. For example, if a cell is code-limited, the scheduler may decide to use a relatively low number of codes but with a relatively high power allocation. If a cell is power-limited, the scheduler may decide to use a relatively high number of codes but with a relatively low power allocation.

The scheduler also has one more dimension that can be changed in terms of the modulation scheme. The Node B can choose between QPSK and 16-QAM (except for UE categories 11 and 12 which support only QPSK) based on its perception of RF conditions. Finally, scheduling decisions also may be influenced by the relative priority of the users and the traffic class.

7.4.4 RLC Parameters and HSDPA

The Release 5 specifications do not define any new RLC parameters for HSDPA. However, the smaller 2-ms TTI and the higher data rates associated with HSDPA mean that typical Round Trip Times (RTT) are lower than those in a Release 99 system. Hence, for HSDPA, the RLC parameter setting recommendations that are described in Chapter 5 must be reevaluated and new settings recomputed.

A specific RLC problem arises because of the higher data rates of HSDPA and the transmit window associated with RLC. In Acknowledged Mode (AM), this window defines the maximum number of RLC PDUs that can be transmitted before an acknowledgment is required. RLC also normally uses the Transmit Status Prohibit (TSP) timer to keep the receiver from sending status reports more often than one RTT. For a specific RLC PDU size, RTT, and TSP, [18] the maximum data rate would be defined as:

$$Maximum\ data\ rate = \frac{RX\_window\_size \times PDU\ size}{RTT + TSP}. \tag{7.8}$$

The window size has a maximum value of 4096 PDUs, but a number of error conditions would arise if this value were used. This leads to an actual maximum utilizable window size of 2048 PDUs. In Release 99, if RTT is assumed to be 200 ms and the maximum PDU size is 320 bits, this would limit the maximum HSDPA data rate to 1.6 Mbps, assuming that the setting recommended in Chapter 5 is followed for TSP. This is well below the maximum data rate that many types of UE support.

To address this issue, one partial solution is to double the RLC PDU size to 640 bits, thereby doubling the maximum supported data rate to 3.2 Mbps. Another factor is that the RTT achieved with HSDPA is smaller than that of Release 99. If an RTT of 100 ms is assumed, this limits the maximum supported data rate to 3.2 and 6.4 Mbps for 320 and 640 bit PDUs, respectively.

7.5 HSDPA Key Performance Indicators (KPI)

HSDPA is designed to provide high-speed PS data service. PS data service is used for many diverse applications such as Web browsing, e-mail access, and so on. On the basis of the performance requirements for each application, and as discussed in Chapter 5, they can be classified into four traffic class categories: conversational, streaming, interactive, and background. Each class has different requirements that can be specified in terms of two key indicators for PS data categorization, namely, throughput and latency. To determine how well HSDPA can support different traffic classes, these two indicators should be used to qualify the network. Chapter 5 described how they could be measured. The metrics described here focus more closely on HSDPA performance measurements.

7.5.1 Physical Layer Metrics

The following Physical Layer metrics allow a UE's performance to be categorized throughout a typical stationary or mobile test:

- **Reported Channel Quality Indicator (CQI).** Both the instantaneous and median CQI reported by a UE are important, because they indicate the quality of the network coverage area. Different UEs in the same radio conditions may report different CQIs, but the Node B scheduler uses the absolute value and rate of variation to decide what TBS, and hence data rate, to allocate.
- **Transport Block Size (TBS).** The TBS that the cell schedules for the UE under test. The Physical Layer rate can be computed on the basis of the TBS information according to equation 7.2.
- **Percentage of time scheduled.** The percentage of total time the tested UE is scheduled by the cell, during the test. This is an important metric to evaluate when performing tests that measure achievable UE throughput, assuming a full buffer. In a multiple UE test, this metric can also be used to track the scheduler performances based on the PS traffic profile.
- **Mean time between schedules.** The mean time between consecutive schedules for the same user. Clearly, the scheduling algorithm significantly affects this metric.

7.5.2 MAC Layer Metrics

For the MAC layer, performance metrics evaluate the achieved BLER and how this relates to HARQ. One such metric is *first transmission BLER*. This is the BLER calculated for only the first transmission of a given HARQ process. This can simply be evaluated by counting the number of NACKs received after every first transmission (New Data Indicator(NDI) set), as follows:

$$\text{First transmission BLER} = \frac{\text{Total number of NACKs after first transmission}}{\text{Total number of new transmissions}}. \tag{7.9}$$

In static conditions, the BLER is expected to be 10%, based on the definition of CQI, assuming that the scheduler reacts and schedules resources as indicated by the CQI rcport.

Residual BLER is the BLER at the end of the last retransmission for a given HARQ process. This is the BLER that is presented to the RLC layer, thus governing the number of RLC transmissions. A large Residual BLER leads to significant RLC retransmissions and also increases latency. It is calculated as follows:

$$\text{Residual transmission BLER} = \frac{\text{Total number of NACKs after last retransmission}}{\text{Total number of new transmissions} + \text{Total number of ACKs}}. \tag{7.10}$$

An additional concern for the MAC Layer performance is accurate reception of any acknowledgments sent on the HS-DPCCH. Of specific interest are the following:

- **Percentage of ACK to NAK/DTX interpretations.** The percentage of the total number of ACKs transmitted by the UE that are detected either as NACKs or DTXs. This results in a retransmission of the transport block by the associated HARQ process but does not involve RLC. Any retransmitted transport blocks that are successfully decoded by the UE are discarded. Throughput degrades when many of these misdetections occur.
- **Percentage of NAK to ACK interpretations.** The percentage of the total number of NACKs transmitted by the UE that are detected as ACKs. In this case, the Node B assumes that the transmission has been successfully received by the UE. As a result, the Node B never retransmits the transport block associated with that particular NAK. Hence, every NAK to ACK conversion result in RLC retransmissions and an associated increase in latency.

To avoid Transmission Sequence Number (TSN) wrap and stalling of the reordering queue, two mechanisms are used to flush the MAC-hs buffer: a transmit window that aligns its leading edge with the most recent transmission, and a timer (T1) that starts when a hole is detected in the reordering queue. A measurable performance metric is the number of T1 expirations, because each expiration results in an RLC retransmission.

7.5.3 Serving Cell Change Metrics

As discussed in Section 7.4.2, there are two techniques for performing a serving HSDPA cell change: synchronized or unsynchronized. The technique being used affects the time taken to complete the change and also the amount of data lost. The following metrics categorize the performance associated with the change:

- **Duration of data interruption.** A change in serving cell is a form of hard handover and the data flow is interrupted during its occurrence. This interruption causes lower throughput and increased latency during the change. It is computed as the time difference between the last packet received on the source cell and the first packet received on the target cell.
- **RLC bytes retransmitted.** The number of RLC bytes retransmitted during Serving Cell Change. In this case, RLC retransmissions are a result of a MAC-hs reset, which is inevitable with an unsynchronized change but can also occur during a synchronized change if the specified action time does not allow the Node B enough time to successfully transmit all remaining data. Other causes of a MAC-hs reset include nonresidual BLER after the maximum number of HARQ retransmissions, or NAK to ACK conversions due to a weak UL. Therefore, when computing RLC bytes Retransmitted to study the impact of the cell change, retransmissions must be appropriately categorized so as to exclude any retransmissions not related to cell change.
- **Activation time.** The duration of the Action Time for a synchronized cell change; this is implementation-dependent but generally based on proprietary buffer management schemes. A long activation time means that the UE is forced to use nonbest server for HSDPA for longer than may be necessary. A short activation time leaves insufficient time for the Node B to transmit all the data in the MAC-hs buffer, thus causing RLC retransmissions.

7.6 Test Setup

As described earlier, throughput and latency are important performance indicators for a packet data system. Throughput can be measured using applications such as FTP, perfmon, Ethereal, or Iperf. While FTP uses TCP as its transport protocol, iperf can use either TCP or UDP. Latency can be measured with ping or http. For HSDPA, the test methodology used to measure throughput and latency is similar to that for Release 99, as described in Chapter 5.

HSDPA allows priorities to be assigned at the MAC layer for the data flows carrying the payloads of different applications. These applications may have different QoS requirements. Figure 7.26 shows such a situation, with two MAC-d flows assigned to two different priority queues.

As an example of how applications fall into different traffic classes, e-mail fits into the background traffic class and is delay-tolerant. Video streaming, however, fits into the streaming traffic class and is more sensitive to delay variations. Hence, a MAC-d flow corresponding to an e-mail application would be assigned to a lower priority queue, whereas a MAC-d flow corresponding to video streaming application would be assigned a higher priority. The scheduler is expected to consider the priorities of the different flows while scheduling data on HS-PDSCH.

To determine the capabilities of an HSDPA system, tests must be carried out to evaluate the performance of applications with different QoS requirements when these applications are used simultaneously. Figure 7.26 illustrates one of the many possible combinations. To fully qualify the user's experience computing all of the following: throughput, latency, HSDPA-specific metrics, and application-specific metrics is required.

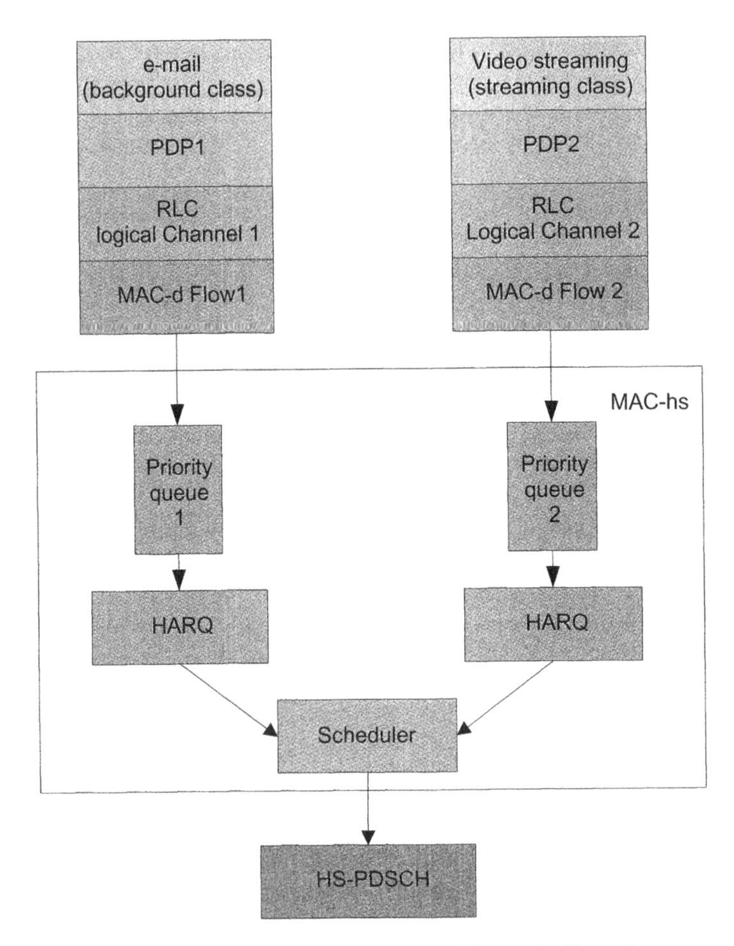

Figure 7.26 Multiple applications with different QoS requirements

References

[1] RP-050248. Removal of DSCH (FDD Mode). 3GPP. Jun. 2005.
[2] 25.213. Spreading and modulation (FDD). 3GPP. Jun. 2002.
[3] 25.211. Physical channels and mapping of transport channels onto physical channels. GPP. Jun. 2005.
[4] 25.321. Medium Access Control (MAC) protocol specification. Jun. 2005.
[5] 25.214. Physical layer procedures (FDD). 3GPP. Jun. 2005.
[6] 25.942. Radio Frequency (RF) system scenarios. 3GPP. Jun. 2004.
[7] WCDMA Evolved: The first Step–HSDPA. Ericsson White Paper. http://www.ericsson.com/technology/whitepapers/wcdma_evolved.pdf. May 2004.
[8] 34.121. Terminal conformance specification; radio transmission and reception (FDD). 3GPP. Apr. 2005.
[9] 25.101. User Equipment (UE) radio transmission and reception (FDD). 3GPP. Jun. 2005.
[10] Lee J, Miller L. *CDMA Systems Engineering Handbook*. Artech House; 1996.
[11] 25.306. UE radio access capabilities. 3GPP. Jun. 2005.
[12] 25.104. Base Station (BS) radio transmission and reception (FDD). 3GPP. Jun. 2005.
[13] HSDPA and Beyond. Nortel White Paper. Version 1.02. Feb. 2005.
[14] HSDPA performance and CQI reporting cycle. 3GPP TSG RAN WG1#27. Document R1-02-0920. Jul. 2002.

[15] Frenger P, Parkvall S, Dahlman E. Performance comparison of HARQ with chase combining and incremental redundancy for HSDPA. *Proceeding of the IEEE 54th; VTS Vehicular Technology Conference*, Rhodes, Greece, Vol 3. Oct. 2001; pp 1829–1833.
[16] 23.107. Quality of Service (QoS) concept and architecture. 3GPP. Dec. 2004.
[17] Mohanty B, Rezaiifer R, Pankaj R. Application layer capacity of the cdma2000 1xEV wireless internet access system. *Proceeding of the World Wireless Congress*, San Francisco, CA, USA, May 2002.
[18] General RLC requirements. 3GPP TSG-RAN WG2#44. Document R2-042029. Oct. 2004.
[19] 25.308. HSDPA overall description stage 2. 3GPP. Dec. 2004.

8

Indoor Coverage

Patrick Chan, Kenneth R. Baker and Christophe Chevallier

8.1 Introduction

It could be said that nobody plans indoor systems: people just walk into a building, look around, and figure out where to put coverage antennas. While the cellular industry is populated with experienced engineers who can do a pretty good job of providing coverage inside individual buildings, it is our contention that such an approach does not fully consider the best operation of the network as a whole. Is it more economically feasible to provide coverage inside buildings from the outside network? Can a cell, or perhaps a microcell, be placed adjacent to a building to provide coverage to the occupants or does it make more sense to place a cell in the building? What impact will that microcell have on the performance of surrounding cells or of the network as a whole? Would a repeater, perhaps in combination with a Distributed Antenna System (DAS), provide a more economical solution? Which will provide the best service for the lowest cost? This chapter addresses these questions and more.

8.2 Design Approach and Economic Considerations

8.2.1 Indoor Coverage: The Traditional Approach

Traditionally, indoor coverage has been provided from outdoors. This approach has its roots in the analog origins of cellular telephony in the late 1970s. Cellular telephones used to fit neatly under the seat of a car, or perhaps in the trunk. These phones were used outdoors, in a vehicle, often with an antenna mounted on the roof and power supplied from the vehicle's electrical system. Network planners concentrated on the obvious: providing coverage and capacity to the outdoor environment, with the emphasis on the streets and highways where customers used their phones.

As cellular telephones became more portable, coverage for indoor environments gained importance. The migration of cellular telephones indoors changed the way networks were planned. The natural and most straightforward tendency was to put cells closer together:

a stronger signal outside implies more signal inside. How much closer could cells be positioned? This was dictated by the Link Budget in concert with the maximum Signal-to-Noise Ratio (SNR) that the air interface technology could tolerate, because putting cells closer to one another increases the co-channel and adjacent channel noise. FDMA and TDMA air interface technologies could, to some extent, use the frequency plan to help mitigate interference–both co-channel and adjacent channel–as cell sites were spaced more closely together. It was also possible to reserve certain sets of frequencies for use indoors or in places where capacity or coverage dictated the use of a microcell.

Microcells and picocells were not initially part of network architecture. In other words, the vendors of cellular networking equipment were not quick to introduce smaller cells (lower output power, smaller equipment footprint) into their product lines. Even today, these components are the exception for WCDMA. Yet, according to some estimates, 60 to 70% of cellular traffic originates from indoors [1]. As data rates increase, the Link Budgets for user equipments (UEs) with finite output transmit power will define cells with smaller radii. It makes sense to put cells closer to where the customers are.

8.2.1.1 Placing Cells Closer Together

In most network plans, to make cells smaller and closer together, a Building Penetration Loss (BPL), also referred to as a building penetration margin, is inserted into the Link Budget. The Link Budget predicts the Maximum Allowable Path Loss (MAPL) that can be sustained between transmitter and receiver. The MAPL in turn defines the cell radius by casting the maximum sustainable path loss across some suitable propagation model. It is important to remember that the BPL is a loss margin assumed and inserted into the Link Budget to account for the anticipated reduction of power as signals propagate into buildings. Outside the building, this reduction in received power does not exist. The result is that cells are closer together and interference increases, on both the Uplink and Downlink. In WCDMA, there is no frequency plan to help hide the extra interference between cells. For any individual channel (UARFCN), all users utilize the same frequency spectrum resource. As cells are placed closer together, interference goes up and, as a collateral effect, per-cell capacity goes down.

It is important to specify that it is the capacity of an individual cell that decreases. If many more cells are placed, the increased number of cells compensates for the reduced capacity per cell and the overall network capacity increases. But at what cost? Every cell site has an associated capital cost as well as continuing operating expenses. There is also cost associated with the optimization effort, which tends to increase as the density of cells increases. It becomes clear that as the density of cells in a given area increases, the cost increases. At some point, the network operator encounters diminishing returns.

8.2.1.2 Analyzing Network Capacity as a Function of Cell Spacing

To illustrate this situation, we used a commercial network planning tool to simulate a 19-cell microcell system as shown in Figure 8.1. We analyzed the capacity for two propagation conditions: (1) line of sight (LOS) and (2) obstructed (OBS). Both propagation conditions were modeled using two different antenna heights, for a total of four cases:

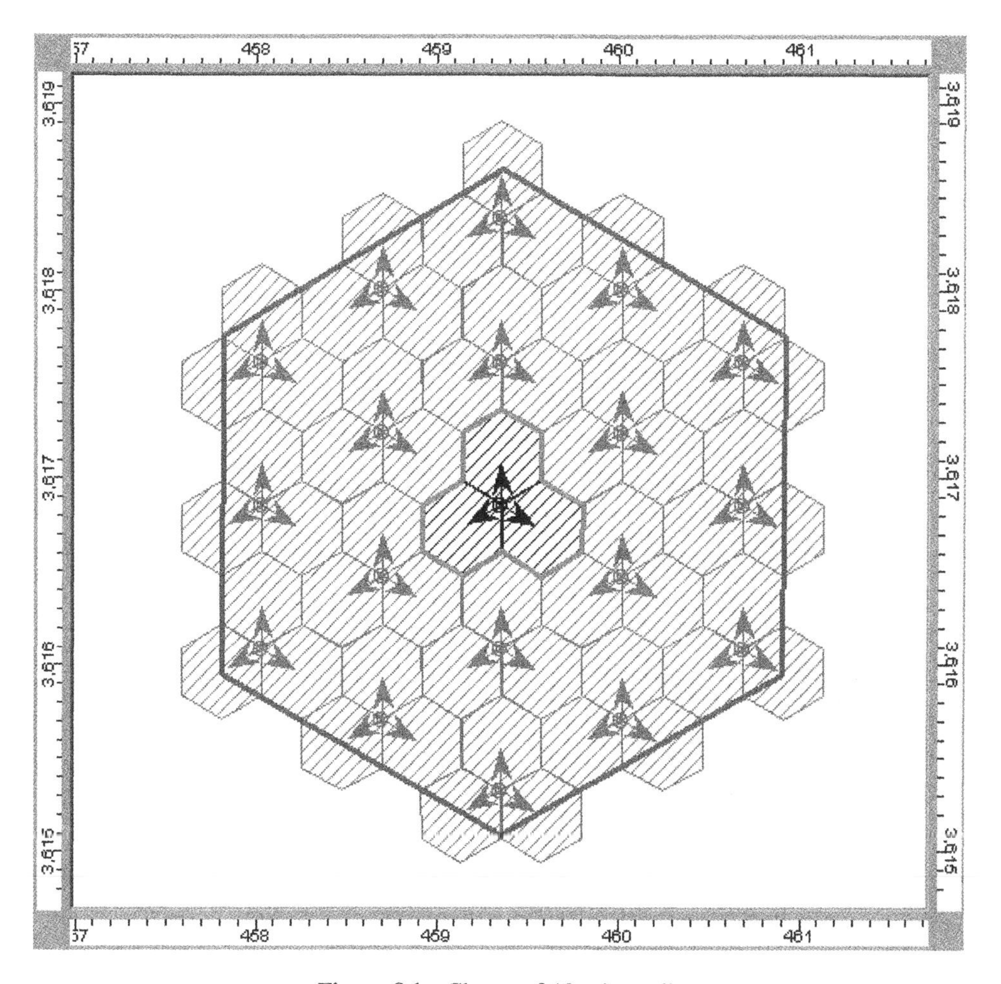

Figure 8.1 Cluster of 19 microcells

1. LOS propagation from low antennas (3.7 m)
2. LOS propagation from high antennas (13.3 m)
3. OBS propagation from low antennas (3.7 m)
4. OBS propagation from high antennas (13.3 m)

The simulation was repeated for different cell spacing distances; that is, the cell density, in terms of number of cells per square kilometer, became a variable in the simulations. In this way, total system capacity was calculated as a function of the distance between cell sites. System performance as a function of antenna height and propagation environment can also be compared.

The LOS cases used a Fresnel zone break point model as shown in Figure 8.2, while the OBS cases used a one slope path loss model as shown in Figure 8.3. For realistic results, the propagation models used were based on measurement studies in a dense

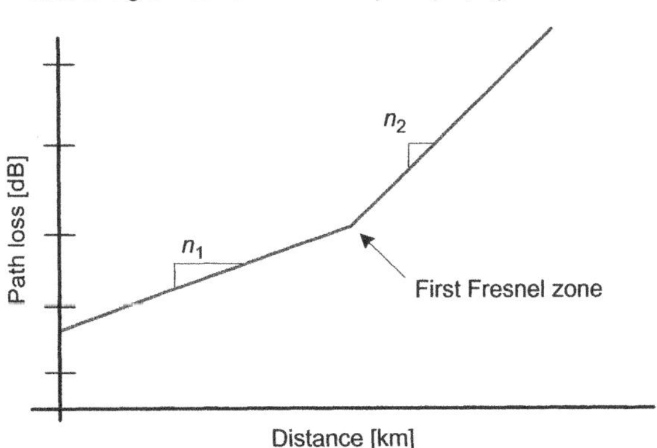

Figure 8.2 LOS propagation model

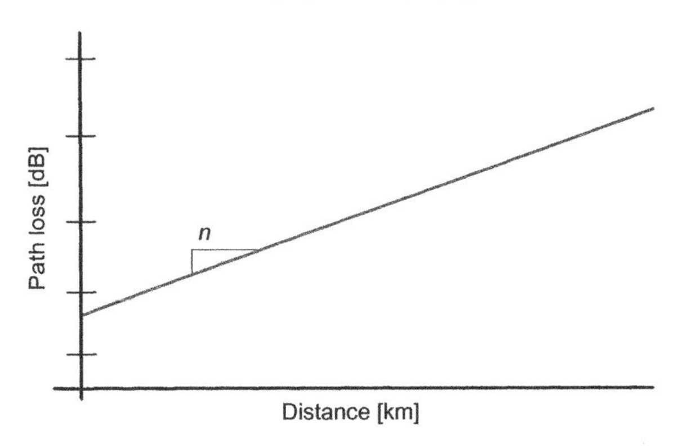

Figure 8.3 OBS propagation model

urban environment where transmitter heights were low and cell spacing was less than 1 km [2].

For the relevant topology, it can be shown that this model is accurate if the break point distance is selected carefully, as by the standard deviation indicated in Tables 8.1 and 8.2. In these tables, the measured standard deviation (σ) is around 8 dB, a typical value recognized by network planners as the limit of accuracy for a statistical model.

Table 8.1 LOS model accuracy and path loss exponents before (n_1) and after (n_2), the breakpoint

Antenna height	n_1	n_2	σ (dB)
Low (3.7 m)	2.18	3.29	8.76
High (13.3 m)	2.07	4.16	8.77

Table 8.2 OBS model accuracy and path loss exponents before (n), the breakpoint

Antenna height	n	σ (dB)
Low (3.7 m)	2.58	9.31
High (13.3 m)	2.69	7.94

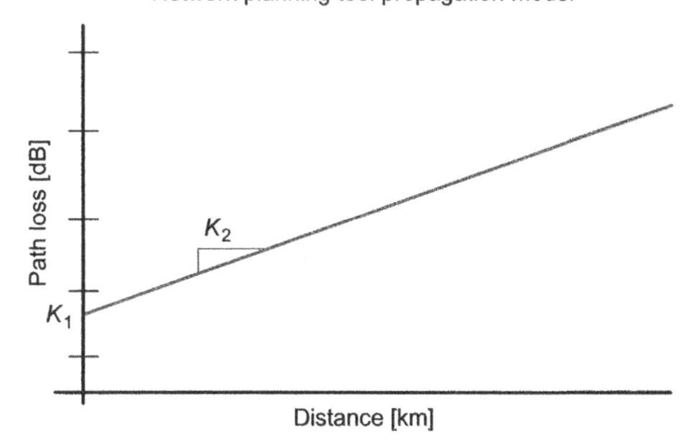

Figure 8.4 Network planning tool standard propagation model

In the network planning tool used for the simulation, a standard point slope model takes the form presented in Equation 8.1 and illustrated in Figure 8.4 [3]. This model must be adapted to the considered model, as presented in Table 8.3.

$$PL = K_1 + K_2 \log(D) \tag{8.1}$$

The model, originally developed for 1900 MHz, was adjusted for the IMT-2000 frequency band (2.1 GHz) by augmenting the intercepts by 0.91 dB, as in equation 8.2.

$$\Delta PL = 20 \times \log\left(\frac{2110}{1900}\right) \tag{8.2}$$

The slopes and standard deviations remained unchanged, taken as reasonable assumptions within a limited frequency change.

Table 8.3 Summary of propagation models for capacity analysis

Low antenna height, LOS	$K_1 = 36.2$ and $K_2 = 21.8$ for $d<159$ m
	$K_1 = 11.76$ and $K_2 = 32.9$ for $d>159$ m
High antenna height, LOS	$K_1 = 37.4$ and $K_2 = 20.7$ for $d<573$ m
	$K_1 = -20.245$ and $K_2 = 41.6$ for $d>573$ m
Low antenna heights, OBS	$K_1 = 39$ and $K_2 = 26.9$
High antenna heights, OBS	$K_1 = 39$ and $K_2 = 25.8$

Table 8.3 summarizes the four variants of this model used in this study.

To illustrate network performance, we analyzed the 19-sector system shown in Figure 8.1 using multiple Monte Carlo simulations to determine the maximum traffic that could be sustained by the central cells. Considering only the central cells capacity does not factor in the effects of network boundary (cell edge); thus, the simulation results are consistent with the expected results for an embedded cell.

Figure 8.5 shows how the capacity of an individual cell decreases as the distance between the cells decreases.

In Figure 8.5, the cells with lower antennas have more capacity than those with higher antennas. Also notice that the OBS propagation models show greater capacity than the LOS cases. Both these observations illustrate the effect of increased interference on the air link capacity. These observations are consistent with expected results, because either reducing the antenna height or increasing the isolation between cells (using the OBS model in our case) is consistent with optimization techniques.

While the capacity of a single cell decreases, it is important to understand how the overall network capacity evolves with reduced site-to-site distance. Figure 8.6 illustrates how the number of calls per unit area increases as the cell density increases. This occurs

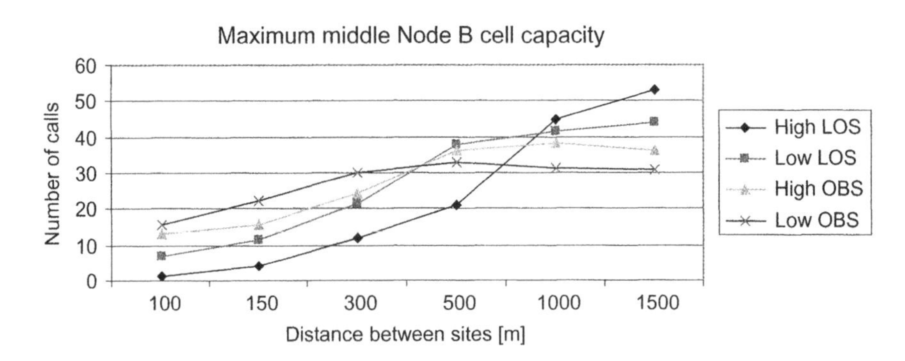

Figure 8.5 Node B capacity as a function of cell distance

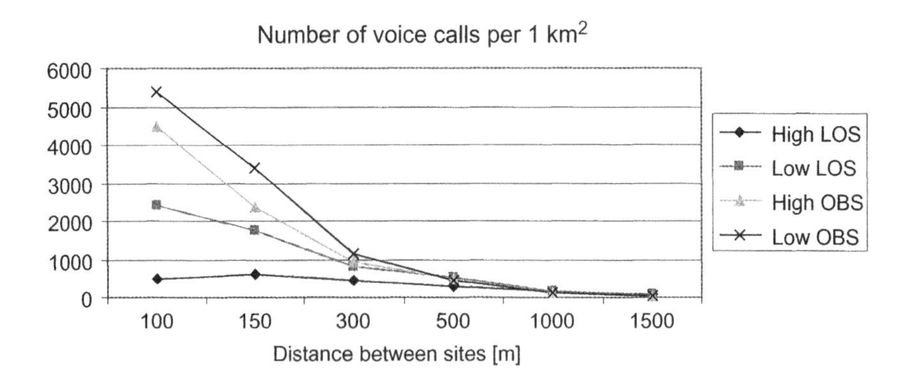

Figure 8.6 Number of calls per square kilometer

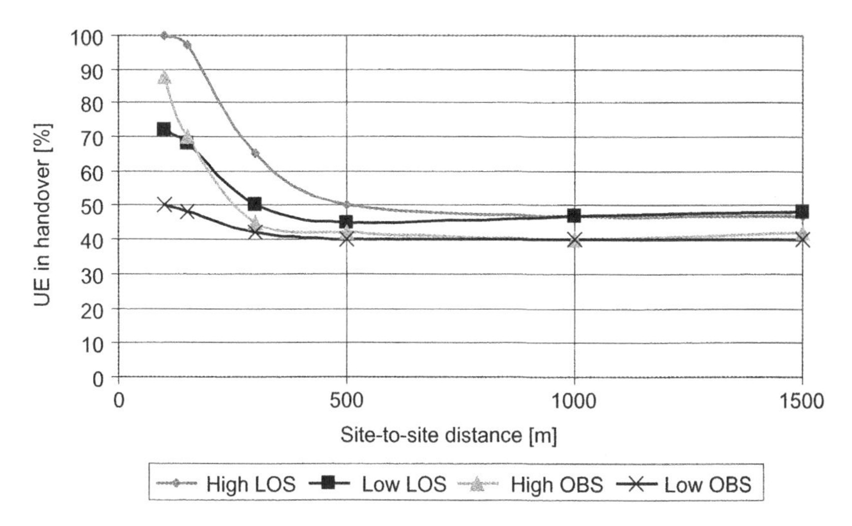

Figure 8.7 Handover status as a function of cell distance

despite the fact that the capacity per sector decreases. In other words, increasing the number of cells per square kilometer does increase the overall network capacity; however, the cost per square kilometer increases dramatically as well.

Figure 8.6 shows that for the high antenna height/LOS model, capacity decreased when the site spacing was less than 150 m. (This is approximately 50 cells per square kilometer.) This scenario is consistent with a network that is not perfectly optimized for RF configuration. It also represents what can be observed indoors at higher elevations (i.e., in tall buildings) on floors higher than the average clutter height in urban environments.

Figure 8.6 also shows that the number of calls per unit area increased as the density of cells increased, except in the High LOS case just mentioned.

In addition to the obvious capital costs of such a dense network of microcells, there is also a network resource cost, in the form of increased soft handover. Naturally, as the cells get closer together, the amount of soft handover increases as well, assuming that the handover parameters remain constant. Figure 8.7 illustrates the increase in soft handover as a function of the distance between cells. Once again, lower cells exhibit less soft handover than higher cells.

To summarize, a network can be deployed with site-to-site distances that could overcome building penetration of 40 dB. Such a network would provide a large, even if not optimal, capacity because of the large overlap between cells.

8.2.2 Indoor Coverage: A Hypothetical Approach

Section 8.2.1 described the traditional approach to covering the indoors from the outdoor network. But a key question might suggest another approach to indoor network planning: where are the network users, the customers? Think of a typical urban environment characterized by dense population, tall buildings, and often narrow streets. Sometimes the buildings are extremely tall, but not necessarily, and usually not over the entire urban area of interest to WCDMA operators. In such an environment, the users are generally at street level or inside the buildings. Indeed, if one believes the marketing research, 70% of the users are in the buildings.

As a counter-example of planning, what if all cell sites for this urban area are inside the buildings? Imagine a hypothetical world where there are no outdoor cell sites, only cells in the buildings. Suddenly 70% of the user population has excellent indoor coverage. As for users in the street, many of them will be covered by signals that leak out of the buildings. This hypothetical world illustrates a second approach to indoor network planning, the antithesis of the traditional approach. Here, the cells are placed inside the buildings and the leakage from the buildings is allowed to cover the streets. BPL is symmetric. Radio signals exit and provide coverage in the streets.

Some obvious questions when considering a network of indoor cells are what is the capacity of each individual cell and how will such a network operate as a whole? We can estimate that the cells, considered individually, have more capacity. This is because the UEs, most certainly those inside the buildings, would perceive less Downlink interference. Similarly, our indoor cells would receive less Uplink interference from UEs in other cells, consistent with the interference factor (α) introduced in the Uplink capacity equation in Chapter 3 (see equation 3.5). This occurs because the BPL would now be working in our favor. The cell receivers inside a building do not hear users in other cells, because most of them are in other buildings. On the Downlink, the BPL shields the UE from other Downlink signals. Experiments have been done that support these hypotheses, for High Speed Downlink Packet Access (HSDPA) performance indoors over an optical-fiber-based DAS [4].

As for the 30% of users in the street, they also receive less Downlink interference. This occurs because our cells are low and small in power. A user would receive signals from the building nearby, possibly also signals from the building across the street, but still in a limited amount.

Imagine other advantages of our hypothetical world. Users, both inside and out, would be in more intimate contact with the serving cell. Outdoor users would see a reduction in reselection errors and perhaps a reduction in reselection altogether when mobility is limited. Gone would be the urban canyon effect, in which a UE rounds the corner and is immediately inundated with a powerful new cell. That environment is prone to reselection errors and often requires significant optimization effort. Consider also the users inside the buildings, who are more likely to be using high-speed data services. In our hypothetical world of indoor cells, the users are now closer to the cell's transmitters and receivers, perhaps even in a LOS propagation environment. Such users find themselves enjoying a much better QoS. Naturally, this perfect world also has drawbacks, mainly for the outdoor users traveling at higher velocities. For them, fast reselection and handover, definitely become issues and highlights the need for some level of outdoor-to-outdoor coverage.

8.2.3 Indoor Coverage: The Hybrid Approach

So far, we have considered two radically different network planning approaches. The first is traditional: a network is planned and deployed from the outside with the intent of covering the inside areas with increased signal strength from outdoors. The second is hypothetical: the majority of cells are indoors and the streets are covered via leakage from the buildings. We have seen some advantages and disadvantages to both approaches.

We now consider a third approach to indoor network planning, which amounts to a hybrid of the two previous approaches. The reality is that, at least for the foreseeable future, network deployments will always begin from the outside. The reasons for this are based in tradition as well as economics. Network operators planning and deploying a new service must provide coverage to a large section of the population and over the widest area as quickly as possible. Capital budgets are always limited, but they are especially so when the business is not yet up and running, before it generates income to offset the investment in spectrum and equipment.

A hybrid approach is simple: the network is deployed primarily from the outside looking in, but migrates toward the hypothetical network of predominately indoor cells as the number of users grows and the network matures. The benefits of this hypothetical indoor-looking-out network take shape as more indoor cells are added while lowering and removing the outdoor cells. Let us be clear: outdoor cells will always be needed, for parks and outdoor gathering areas, as well as for the high-speed mobility user on the streets. There will always be buildings with no access for installing indoor coverage solutions, or where the limited traffic does not justify the cost. The methodology of this hybrid indoor design approach is to begin with the outdoor-to-indoor network, assuming a reasonable BPL, typically around 20 dB, but to keep the hypothetical world of indoor cells as the goal for network growth. The real key to this method is this: operators must recognize that putting more cells outdoors (increasing outdoor cell density) is no longer to their advantage. It is also important to keep in mind that indoor users will affect the performance and capacity of the outdoor network [5,6].

For each network operator, a different set of circumstances defines the point at which the next new cell is indoors, not outdoors. Operators must weigh many factors. Each operator has a different cost for capital purchases and associated labor. Carriers offer different services to different types of users. For example, a given operator's business might rely on providing high-speed data services to business users at their places of business. Other operators might primarily provide voice service to residential users when located in or around their own home. Competition among service providers strongly drives the amount of expected indoor coverage. Labor costs, which vary around the world, drive the economics of indoor solutions so that it is difficult to define a universal guideline. Until more economically viable picocells come on the market, many indoor networks will require a DAS; the associated labor costs of pulling cables can dominate such networks. Backhaul requirements for picocells also drive up costs, but with large variations that depend on the topology (daisy chain, star, and so on). In this decision-making process, one factor is consistent irrespective of the operator approach: prioritization of the building to be covered is paramount. Prioritization is necessary to deploy the indoor cells, and can also help in selecting the location of the outdoor macro sites. An appropriately placed macro cell can cover inside buildings with BPL up to 40 dB!

The rest of this chapter describes various methods for introducing picocell and microcell Node Bs into the network. It also discusses other indoor coverage solutions such as repeaters.

8.3 Coverage Planning and Impact on Capacity

This section begins with a review of the main techniques available for distributing signals indoors. Next the differences in coverage planning between indoor systems and the outdoor macronetwork are discussed, including estimates of how the indoor coverage is affected when provided by the outdoor network or by a dedicated indoor solution. Finally,

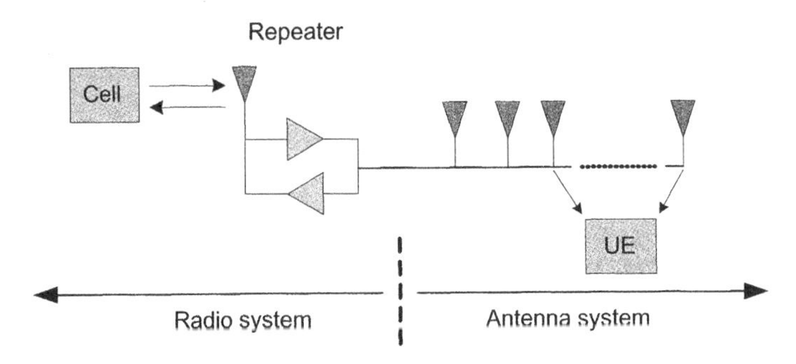

Figure 8.8 Indoor system illustrating radio system and antenna system

because the introduction of 3G mostly intends to provide improved PS data service, what is required in different scenarios to reach a data rate of 384 kbps is estimated, after looking at indoor capacity.

8.3.1 Indoor Coverage Systems

An indoor network solution can be divided into two parts: the radio system and the antenna system, as shown in Figure 8.8. The radio system determines how to facilitate the provision of radio bearers. For the radio system, this section discusses the practical aspects of applying either repeaters and/or a dedicated Node B. The antenna system defines how to distribute the radio bearers and radiate them to different indoor areas of the building. Antenna systems can be divided into two major categories: passive, where only passive components are used; and active, in which amplifiers and signal converters constitute all or part of the distribution network. For the purposes of discussion, the antenna system and the radio system are considered independent. This implies that any radio system can be connected to any antenna system.

8.3.1.1 Radio System Components: Repeaters

A repeater is one of the most common radio signal sources for in-building coverage extension. Repeaters come in a variety of types and sizes. In its most basic form, a repeater consists of linear amplifiers connected in a back-to-back fashion, as shown in the block diagram of Figure 8.9.

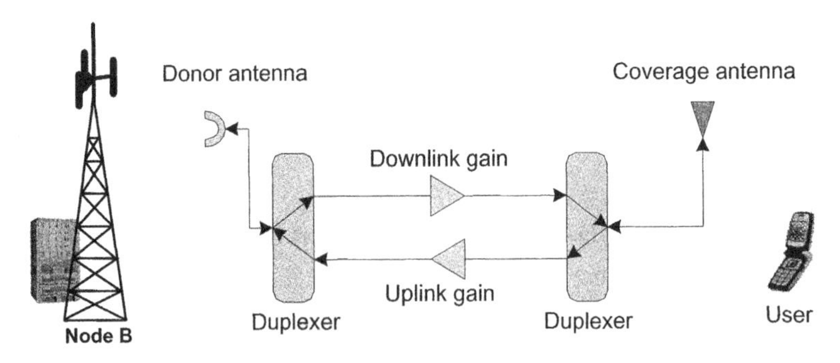

Figure 8.9 Repeater block diagram

In Figure 8.9, the duplexers separate the Uplink and Downlink signals to the appropriate amplifier chain. The gains of the Uplink and Downlink amplifiers are often controlled individually. On some smaller, lower-power repeaters, the gain is not controllable but instead is preset and determined by an automatic gain control (AGC) function. On other units, the Uplink and Downlink gain can be set, but not independently for each link. As discussed below, setting the repeater gain correctly is one of the main ways to optimize the use of a repeater in a WCDMA network.

Repeaters can be connected to a donor Node B in three ways: an over-the-air link (Figure 8.10 (a)), via a cable, either coaxial cable or an optical fiber link (Figure 8.10 (b)), or daisy chained using either cable or over-the-air links (Figure 8.10 (c)). In all cases, each repeater is treated as an individual radio system component that is connected to an antenna system to distribute signals to the desired coverage area.

When connected to the donor cell via a radio connection, the repeater for indoor application has two antennas; one for internal and one for external operation. The external antenna, or donor antenna, is located usually on the roof of the building being covered, and receives the Downlink signal coming from the donor cell site. Simultaneously, the donor antenna transmits the Uplink signals to the donor cell. The repeater receives and amplifies the Downlink signal and then retransmits the Downlink through the serving antenna or antenna system to cover the building. The serving antenna or antenna system receives the Uplink signals from UEs in the repeater coverage area.

Repeaters do not add capacity to the network. They only move the available traffic capacity of the donor cell to areas that the cell could not otherwise directly access. These areas can be indoors, as is the focus of this discussion. However, repeaters apply equally well to areas where network service is desired but direct coverage from the macrocellular

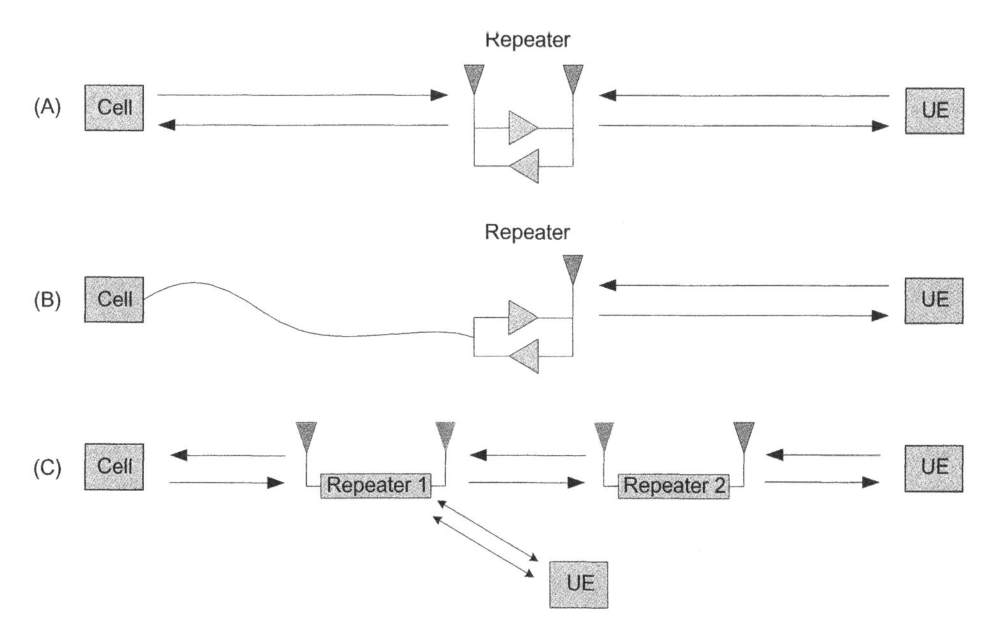

Figure 8.10 Various repeater connection options

Table 8.4 Main criteria for repeater selection

Parameter	Recommended value	Comments
Gain	Maximum 90 dB	Gain affects the isolation requirement
Isolation	Gain + 15 dB	Prevents oscillation
Donor side antenna	High gain and high directivity	Reduces repeater gain requirement and helps ensure that a single donor cell is repeated
Server side antenna	Depends on coverage requirements	Either a single antenna or a distributed antenna system is possible

network is not possible: rural areas, tunnels, canyons, urban underground garages, train platforms, airport terminals, shopping malls, and the like.

Table 8.4 summarizes the main criteria for planning and optimizing repeaters. The discussion following the table provides more details.

Noise figure and peak power handling are important specifications of a repeater. The power handling capability is probably the most obvious RF parameter. To achieve the desired coverage, the repeater unit must be capable of providing enough Downlink output power to drive the antenna system. Repeaters are available with output powers that range from 10 mW to 10 W.

While sufficient output power is easy to understand, network engineers should be aware of some nuances associated with the choice of output power. As output power increases on the Downlink, the power required at the repeater donor side input becomes greater. This is because, in practice, the gain available from input to output of a repeater is limited to values of approximately 90 dB. Above 90 dB or so, repeaters become difficult (and expensive) to build because of the practical limitations on achieving sufficient RF isolation within an enclosed box. Furthermore, as the gain goes up, the requirement to isolate the donor antenna from the server antenna increases, as discussed below. Consequently, for a repeater to have 10 W (40 dBm) of output power, assuming 90 dB of gain, the input signal to the repeater must be approximately 40–90 = −50 dBm. Of course, the donor antenna gain helps gather this energy. Nevertheless, the requirement of a −50 dBm donor side input signal limits how far such a large repeater can be from the donor cell.

We must also keep in mind that the Downlink signal feeding the repeater must be of high quality, that is, the SNR of the input signal must be sufficient for good quality communications at the output. In general, the repeater does very little filtering and, for cost reasons, does not regenerate the signal. The Downlink signal passes through the repeater and is broadcast at a higher power, but not with better SNR than was available at the input. If anything, the Downlink output SNR will be slightly worse, owing to the noise figure of the repeater on the Downlink. This makes the Downlink noise figure of the repeater an important parameter, along with the filtering and intermodulation characteristics of this transmitter device. In general, it is good practice to maintain at least a 20 dB SNR at the donor input of the repeater so that the repeated noise does not significantly impact the Downlink capacity in the repeater coverage area. At the same time, the repeated signal should be of high quality, characterized, for example, by a strong CPICH E_c/N_o, ideally better than −7 dB, in unloaded conditions.

It can be shown that high power repeaters will, by necessity, inject more Uplink signal into the cell so as to maintain Uplink and Downlink balance at the UE. This strong Uplink signal includes noise generated by the electronics of the repeater. Thus, the Uplink noise figure of the repeater also becomes an issue.

To prevent a repeater from oscillating, sufficient isolation is maintained between its donor and server antennas. An oscillating repeater broadcasts noise and interference into the donor cell site (perhaps into several cell sites) as well as into the UE coverage area. Such interference can severely disrupt communication. The amount of isolation required is a function of the gain of the repeater. In general, an antenna isolation that is at least 15 dB greater than the gain is established, as measured from input to output of the repeater antenna ports. (This includes the antenna cables.) In the extreme case, if the repeater has 90 dB of gain, 105 dB of antenna isolation would be required. Needless to say, it can be very challenging to obtain this level of isolation between two antennas that are physically connected to the same device.

The isolation requirements between antennas are also factors in the planning of a repeater. For example, an engineer may determine that a large repeater gain is required for a desired repeater location. If this location does not allow the donor and sector antennas to be physically mounted with sufficient isolation, then a new location must be sought. For indoor applications, the isolation requirements are aided by the building, which provides isolation between the donor antenna on the outside of the building and the system antenna inside the building.

Several types of donor antennas are possible. For urban applications, a low gain (6 to 8 dB) Yagi–Uda antenna placed at the corner of a building is a common choice. A parabolic dish is also common, especially when greater antenna gain is required. Both Yagi–Uda and parabolic dish antennas provide good front-to-back performance as well as directivity, which helps increase the isolation between the donor and server antennas. These directive antennas also assist in targeting a single cell as input to the repeater so that a single server dominates. Panel antennas can also be used in applications where low profiles are needed, but their patterns should be selected carefully; their wider beamwidth may cause several donor cells to be repeated.

For optimal performance, the donor antenna should be deployed so that propagation between the donor site and the repeater is LOS. This minimizes the multipath and fading impacts on this critical link. In addition, the donor antenna should have a view of only a single cell so as to rebroadcast only a single PSC. To do otherwise puts all users in the repeater coverage into soft handover. In this case, soft handover will not yield its natural advantages because no link diversity would be provided (the link through the repeater would be the same). However, the handover would adversely affect system resources (such as OVSF code tree, channel elements, and Downlink power).

In summary, when using repeaters in a WCDMA network, engineers must follow some basic procedures for installing, adjusting, and fine-tuning to minimize any impact on the capacity and coverage of the entire system. General recommendations for installing a repeater for in-building coverage and for using a macronetwork cell as a donor are given below. All these recommendations can be accomplished with a test phone that measures receive power, transmit power, CPICH_RSCP, and CPICH_E_c/N_o.

- On the Downlink, ensure that there is only one dominant server, characterized by a single dominant CPICH. The goal is to have only one cell in the Active Set of any UE in the repeater coverage area. This requires that no other PSCs are within the reporting range defined for Event 1a; otherwise every UE in the repeater coverage area ends up in soft handover. Verify the dominant server by establishing a call with a test phone and confirm that the Active Set size is limited to one within the repeater coverage area.
- When setting the Downlink gain, check the output from each antenna node in the building with a power meter or spectrum analyzer and make sure that the radiated power is at the designed power level. A less rigorous alternative is to check the UE operation within the building and make sure there is adequate coverage around each antenna node.
- If time permits, perform a complete in-building test to verify coverage within the intended areas.
- To allow for future growth, when setting the Downlink gain, check the repeater's WCDMA Downlink to ensure that there is head room for full power from the cell when fully loaded on all frequencies.
- The most important requirement for WCDMA is to ensure that there is a good balance between Uplink and Downlink coverage. Strive for link balance at the repeater server antenna that is consistent with the link balance of the outside network.
- Because conditions can change with time, periodically carry out a routine audit of the repeater and the antenna systems, both indoor and outdoor.
- When a new repeater is added, it is strongly recommended to monitor the donor cell, in particular the level of UL interference as broadcast in SIB 7.
- When the donor cell is modified (by down tilting or re-pointing antennas, relocating the cell site, or modifying output power), audit the repeater and adjust the repeater parameters as required.

The last recommendation is important for making sure that the donor cell has not changed and the monitored cell list contains the appropriate cells. This reflects an attribute of repeaters: they extend the coverage of a particular cell (the donor cell) to locations that were not originally part of the parameter data fill. As such, network optimization following the installation of a repeater often requires reevaluation of the Neighbor Lists for the donor and surrounding cells to include new donor cell coverage areas.

All these recommendations also naturally apply to cascaded repeaters. In this case, monitoring injected UL interference is even more critical, because noise rise would also cascade. Also, auditing the monitored cell list is especially important because each added repeater affects the coverage of the original donor cell.

8.3.1.2 Radio System Components: Picocells

This discussion deals with picocells, but it would hold true for microcells or even macrocells used indoors. While no formal definition exists, generally a picocell is defined as a Node B with low power (20 to 30 dBm) and low capacity (a single carrier and limited channel elements), which connects to the UTRAN through the Iub.

Compared to repeaters, picocells are generally more appropriate for providing in-building radio solutions when additional capacity is required. The main advantage of using a dedicated Node B for an indoor system is that it provides additional radio and network capacity. Higher output power is possible, depending on the equipment, without the risk of oscillation that a repeater can bring. Furthermore, Node B equipment has an inherent Operations, Administration, and Maintenance (OA&M) reporting capability that repeaters usually lack. The greater output power permits large buildings to be served from a central point, rather than from multiple repeaters and multiple antenna systems. High data rate services can benefit from the increased capacity and power. This requirement alone can drive the need for a picocell over a repeater, because many indoor venues demand high-rate data services.

The primary disadvantage of a dedicated indoor Node B is that it can be more expensive than repeaters. Even if picocell electronics became equivalent to repeaters in cost, providing adequate backhaul always adds to the cost of indoor cells. Repeaters are popular because they reduce the backhaul requirements. Also, cells often require a shelter space and corresponding support facilities (such as air-conditioning and power), which contribute to the cost.

Introducing a picocell into the network is similar, in many respects, to introducing any new cell into the network: all the usual commissioning and optimization procedures apply. The unique effort centers on the fact that this cell exists within a building. In particular, drive testing becomes walk testing. While this change may sound humorous, the reality is that much of the world's test equipment and post-processing software assumes outdoor drive paths. Most of the test equipment available today are intended for mounting on a vehicle. Often, the post-processing software assumes that GPS data provides location and positioning of the measurements. As a result, adopting standard drive test equipment to the indoor space is generally anything but routine.

For a newly added picocell, just as for any new cells introduced into an established network, the Neighbor List must contain the appropriate cells. The fact that this new cell exists in a building, perhaps with antennas on multiple floors, compounds the problem three-dimensionally, that is, Neighbor Lists must be accurate on all levels of the building. If the building is in a cluster of tall buildings, it is important to verify that service in surrounding buildings has not been altered by the introduction of this new cell. Naturally, the street-level performance must also be part of any commissioning or optimization effort.

Finally, capacity requirements within the building dictate whether to select a picocell, microcell, or a macrocell. If more than one cell is needed to accommodate expected traffic, a macrocell with a DAS may be a better option than several picocells/microcells. For this option, both the interaction with the macro layer and the interactions among the different indoor cells must be planned and optimized. In particular, to maximize indoor capacity, plan and optimize indoor cells so that they overlap in spaces where traffic is low.

8.3.1.3 Antenna System Components

CDMA works well with DASs in general [7]. The options are diverse for antenna systems, as outlined in Table 8.5 and given in further detail below.

Table 8.5 Summary of antenna system components

Type	Distribution methods	Main advantage
Passive	Coaxial feed to antenna nodes	Cheap compared to active system
	Coaxial feed to leaky coax	Most appropriate for linear coverage: tunnels, walkways, and so on
Active	Optical distribution	Allows distribution over long distance
	Bidirectional amplifiers	Often part of leaky coax or large DAS systems
	Cat 5 or other digital cable	A recent alternative to optical distribution, presenting similar advantages but with a lower installation cost and shorter distances

8.3.1.4 Indoor Antennas

The obvious component of an indoor antenna system is the antenna itself. Currently, several manufacturers produce indoor antennas, in a large selection of types and sizes. Electrically, most indoor antennas have lower gain than outdoor antennas. Indoor antennas fall into two categories: omnidirectional antennas, generally for ceiling installation; and directional microwave-patch-type antennas, generally for wall installation. The most common directional antenna for IMT-2000 band frequencies is a patch structure with a typical beam width of approximately 90 degrees in both horizontal and vertical dimensions, linearly polarized, and with directivity that provides approximately 5 dB of gain. In addition, these antennas can have built-in electrical downtilt that can help manage overlap when multiple cells are deployed in an undivided enclosure. Usually, indoor antennas can be designed to cover multiple operating bands, such as GSM900 and DCS1800, as well as the IMT-2000 frequency band (2.1 GHz), so that separate antennas are not required.

As noted above, indoor antennas can be mounted either on the ceiling, on the wall, or in some cases above the ceiling tiles for a truly invisible deployment. Aesthetics play a part in all antenna deployments, including indoors. Building owners and occupants are very sensitive to the aesthetics of their building and usually do not want indoor antennas to be conspicuous. For optimal performance, the antennas must be mounted higher rather than lower. Placing them above head level–just below or on the ceiling–gives the greatest coverage range, while ensuring that building occupants are at a safe distance from Electromagnetic Radiation (EMR).

Tests have shown that propagation polarization is not a significant issue for the indoor environment [8]. This is largely due to two factors. First, the indoor environment is full of reflecting objects, so that polarization is not maintained except in the shortest of LOS propagation conditions. Second, today's handheld devices are always held at random angles, often next to the body of the user. The overall result is that the UE exhibits poor polarization discrimination, both on the Uplink and Downlink. Nonetheless, most indoor applications are vertically polarized.

Antenna mounting should also help meet the coverage goals. For example, if the goal is to capture the indoor traffic on the indoor system and minimize the amount of traffic captured by the outdoor network, the indoor antennas should be placed at the external walls and windows pointing into the building core. In this configuration, power control keeps the indoor UE from transmitting at higher powers near the windows, thus minimizing the interfering signals that radiate outside and affect the outdoor network capacity. The trade-off is that it can be more costly to provide antenna nodes to the exterior walls and windows of a building because more cables must be run from some central location. If cost is more important than capacity, mounting indoor antennas well within the building would be appropriate. In this case, users tend to operate on the indoor system when well inside the building and on the exterior macronetwork when near the walls and windows; however, in many areas of the building, users would likely be in soft handover between the indoor and outdoor system. In this last case, having similar link balancing for the indoor and outdoor cells would be critical to avoid large fluctuations in UE transmit power.

When placing antennas indoors, engineers must comply with electromagnetic exposure limits for people in the vicinity. Most countries legally mandate a maximum permitted exposure limit for radio frequency energy. These legal requirements limit the maximum permitted EIRP, as well as how close antennas can be mounted to people. In other words, a minimum separation distance is often mandated between the public and the antenna. Specific local requirements are outside the scope of this section, but optimization/deployment engineers should learn the applicable regulations.

Before considering the distribution systems commonly used indoors, a special case should be made for leaky coaxial cable, as it can replace both the antennas and the distribution system. Leaky coax has been used in mines and tunnels for many years [9], and is widely deployed in mobile systems [10]. The cable itself appears similar to any coaxial cable except slots or holes have been applied to the outer conductor allowing radio signals to propagate (or leak) outward. The slots produce a distributed radiating source along the length of the cable. This works well in tunnels and can also be used in other indoor areas with a long coverage.

In all cables, power dissipates as the distance from the source increases. As a result, coverage should be designed on the basis of the power level needed at the end of the radiating cable. Despite the extra loss in a leaky coaxial cable due to radio signal leakage, the loss per unit distance is not much greater than for nonradiating coax of similar size and construction. The exact amount of extra loss per unit length is a function of the construction and frequency of operation. Leaky coax may have 15 to 20% greater loss per unit length over that of similar nonradiating coaxial cable. Apart from longitudinal loss, the main parameter for leaky coaxial system design is the coupling loss. The coupling loss is generally defined as the signal loss between the leaky coaxial cable and a standard dipole receiver antenna at a distance of 2 m. Definitions of coupling loss can vary from manufacturer to manufacturer, so it is important to follow their guidelines when designing and deploying a leaky coax DAS. Most leaky coaxial cable products can cover a wide frequency band (450 MHz to 2.6 GHz is possible) although the coupling loss may vary significantly over such a wide frequency range. The wide bandwidth permits the antenna system to be easily shared by multiple technologies/systems.

8.3.1.5 Passive Distributed Antenna Systems

Various indoor antenna solutions can meet different deployment requirements. A DAS solution distributes many antenna nodes throughout the coverage area. Each antenna node can transmit and receive WCDMA signals. The Uplink signals from all the antenna nodes are collected to a central point, the point of attachment to the radio system. Simultaneously, the Downlink signals from the radio system are divided and distributed to each antenna node for transmission. In general, DASs are either passive or active.

A DAS consists of a multicarrier combiner, coaxial cables, jumpers, power splitters, directional couplers, attenuators, terminators, connectors/adaptors, and indoor antennas. The multicarrier combiner is normally used in a shared DAS, which is an indoor distribution system that is shared among operators or different radio access technologies. A shared DAS is sometimes called a *neutral host system*.

Deployment and optimization become more complicated with a neutral host system, because different radio technologies have different Link Budgets. Consequently, they could benefit from different output powers and, perhaps, different antenna locations. Optimizing one service may affect the performance of another service that is also using the system. Furthermore, maintaining suitable intermodulation performance among the different technologies can increase the cost of the DAS. On the other hand, sharing the DAS among operators can reduce the cost of capital equipment and of installation and maintenance.

If a multicarrier combiner is used, it usually is the first major component to be connected to the radio system. It combines and divides signals in both directions (Downlink and Uplink) in different frequency bands while keeping the required isolation high (typically >70 dB) and the intermodulation low (typically < -140 dBc) between different systems connected to different ports. Insertion loss is low, about 6 to 8 dB, depending on the number of system ports available on the multicarrier combiner.

Coaxial cables interconnect the radio system and the indoor antennas distributed over multiple locations in the target building. In a passive system, these coaxial cables traverse the building to reach the indoor antennas at different locations. They should have fire-resistant jackets and, if exposed to sunlight, UV-resistant jackets as well. They should also comply with the building regulations. It is important to choose a coaxial cable size that is the correct diameter to minimize loss and meet the DAS loss design. However, we must ensure that the building conduits are large enough to accommodate the selected feeder type.

Passive power-splitting devices permit the radio signal to arrive at antenna nodes in different locations. Many passive power-splitting devices exist, with specifications for different uses. Equal-power-splitting devices (two-way: 3 dB; three-way: 4.8 dB; four-way: 6 dB; and so on) are commonly called power splitters or dividers. In reality, these devices have slightly higher insertion loss figures (an additional 0.1 to 0.5 dB) than the perfect splitting loss, due to losses within the device. In some cases, unequal power splitting must balance the input power levels at different antennas that are connected through feeders of varying lengths. This can be accomplished using directional couplers, which couple only a portion of the power and only in one particular direction of propagation. Nonsymmetric power dividers are also available, which send unequal amounts of power to different parts of the building. For example, a 60 : 30 power splitter sends 1/3 of the energy to one output port and 2/3 of the power to the second output port. By combining splitters and couplers, system losses can be managed so that all antennas have nearly equal Downlink output power levels. This then becomes the design criterion for a passive indoor DAS: to specify cable losses and coupler/splitter losses to distribute energy to each antenna node at the desired level.

If this is achieved, antennas can be located at any distance from the Node B equipment. To begin optimization, it must be confirmed that each indoor antenna node is generating the specified DL output power, not only at the antenna input but also over the intended coverage area, while the UE transmit power remains within the valid range. This dynamic range is greater than 70 dB, according to the standard [11]. For a typical Class 3 UE, the transmit power will be from +24 dBm to −46 dBm. If the RF conditions require the UE to operate below the latter condition, the UE will not be properly power-controlled and will generate UL interference.

8.3.1.6 Active Distributed Antenna Systems

An active DAS contains the same components found in a passive DAS, with the addition of active devices such as amplifiers, repeaters, or bidirectional amplifiers (BDAs) to amplify and/or transport radio signals. This enables an active DAS to cover an area much larger than a passive DAS. Active distribution systems, on the other hand, are more complicated and have more problems with distortion and intermodulation. They may also oscillate if excessive gain is used. Link balance remains critical, although the use of active components with variable gain may make it easer to adjust link balance and output power.

An active DAS system may also have a fiber optic distribution system. Fiber distribution networks not only apply to indoor systems but have also been incorporated into repeaters and directly into cells. They work well as a DAS for large indoor coverage areas and for remote distribution. RF signal on a fiber could be transported over tens of kilometers by converting the RF to light over the fiber and back to RF for radiation using a local antenna. A long feed results in long propagation delays that may need to be accounted for. A large delay could affect timing-related processing functions and power control.

The fiber optic system has an interface that converts RF signal to light. The light is transported over an optical cable through a number of optical connectors and splitters, finally reaching the remote unit with a photo detector. The light wave is then converted back into an RF signal and radiated by the indoor antenna nodes. The system usually has an AGC to compensate for any loss in optical transmission, resulting in virtually no loss in RF signal strength on the Downlink and Uplink at both ends.

The installation of optical fiber transmission lines and their supporting hardware is generally more expensive than a coaxial cable or other transmission lines. Usually, optical distribution systems are bought in a package that includes distribution hubs and antenna nodes from the same vendor to ensure interoperability. Fiber hubs combine multiple Node B signals, use band filtering, and covert RF to and from optical signals before and after distribution.

The most important specifications for fiber are noise and dynamic range. The laser diode and the photodiode receivers contribute to noise and limit the dynamic range. Optical transmission usually operates at wavelengths in the 1550 nm range and uses single-mode fiber cables. Single-mode fiber cables have lower loss per meter and larger bandwidth, and can transport signals over many kilometers. Multimode fibers are used for shorter runs of approximately 100 m.

Table 8.6 Indoor (UE) to Outdoor (Node B) Uplink Link Budget

		WCDMA 12.2 speech		WCDMA 64 CS data		WCDMA 64 PS data	
a	UE Maximum transmit power	21	dBm	21	dBm	21	dBm
b	Cable, connector, combiner losses	0	dB	0	dB	0	dB
c	UE transmit antenna gain	0	dBi	0	dBi	0	dBi
d	**Mobile ERP**	**21**	**dBm**	**21**	**dBm**	**21**	**dBm**
e	Thermal noise density = kT	−174.0	dBm/Hz	−174.0	dBm/Hz	−174.0	dBm/Hz
f	Information full rate	40.9	dB-Hz	48.1	dB-Hz	48.1	dB-Hz
g	Thermal noise floor	−133.1	dBm	−125.9	dBm	−125.9	dBm
h	Receiver noise figure	5	dB	5	dB	5	dB
I	Load	50%		50%		50%	
j	Rise over thermal aka – interference margin aka – load margin	3.0	dB	3.0	dB	3.0	dB
k	Required E_b/N_t	5.1	dB	1.7	dB	1.5	dB
l	**Node B Rx sensitivity**	**−120.0**	**dBm**	**−116.2**	**dBm**	**−116.4**	**dBm**
m	Receive antenna gain	17	dBi	17	dBi	17	dBi
n	Cable, connector, combiner losses	−3	dB	−3	dB	−3	dB
o	**Rx antenna gain−cable losses**	**14**	**dB**	**14**	**dB**	**14**	**dB**
p	Lognormal fading	−10.3	dB	−10.3	dB	−10.3	dB
q	Handover gain	4.1	dB	4.1	dB	4.1	dB
r	Diversity gain	0	dB	0	dB	0	dB
s	Car penetration losses	0	dB	0	dB	0	dB
t	Building penetration losses	−20	dB	−20	dB	−20	dB
u	Body loss	−3	dB	0	dB	0	dB
v	**Σ Propagation components**	**−29.2**	**dB**	**−26.2**	**dB**	**−26.2**	**dB**
w	**Maximum Allowable Path Loss**	**125.8**	**dB**	**125.0**	**dB**	**125.2**	**dB**

8.3.2 Service Indoors from the Outdoors

As we have seen, there are various ways to provide coverage indoors. This section discusses covering an indoor environment from an outdoor network. We can begin by examining a Link Budget for the Downlink and Uplink. Table 8.6 shows the Uplink Link Budget.

The WCDMA Uplink for outdoor networks is modeled by examining cell loading from UEs within the cell. This assumes that UEs in the cell are the dominant source of interference. Some key assumptions in the Uplink Link Budget, which may differ from the Link Budget shown in Chapter 2, are listed below:

- In WCDMA systems, cell loading produces noise Rise Over Thermal (ROT). WCDMA Uplink load is typically assumed to be 50% for the purposes of planning, which results in an ROT of 3 dB. Compared to outdoor systems serving only outdoor users, the ROT for a system carrying both indoor and outdoor traffic is affected by the different traffic profiles of the two user groups.

- Required E_b/N_t depends on channel conditions and vendor demodulator implementation. In addition, compared to outdoor systems serving outdoor users, the E_b/N_t requirement might be reduced, especially at higher floors under LOS conditions where multipaths become negligible compared to the main path. User velocity is also lower. In some situations, the difference in propagation may limit the effectiveness of diversity.
- Handover gain is included in most outdoor Link Budget models. Compared to outdoor systems serving outdoor users, the handover gain outdoors to indoors may be reduced, because buildings tend to provide attenuation and thus limit the number of detected Pilots, particularly at lower floors. At higher stories, the attenuation is counterbalanced by the advantageous propagation condition leading to the detection of multiple distant servers.
- Regarding fade margin, compared to outdoor systems serving only outdoor users, the indoor cell edge probability may be selected differently. The standard deviation, which defines the fade margin, will depend on the position of the serving cell relative to the indoor user. For a building located in LOS, the standard deviation would be most affected by the propagation and shadowing within the building. For a building located in non-line-of-sight (NLOS), the standard deviation is affected by the propagation condition first outside and then inside the building for a resulting decrease in standard deviation [12].

Table 8.7 shows the Downlink Link Budget.

The Downlink for outdoor to indoor users is modeled by determining the required power received (receiver sensitivity) at the UE to maintain a desired Signal-to-Interference Ratio. Several assumptions specific to indoor users must be made regarding interference from other cells and handover conditions:

- The WCDMA Downlink Link Budget uses a ratio of other-cell interference to same-cell received power density ($I_{oc}/\hat{I}_{or}$) to specify the Downlink interference from other cells. $I_{oc}/\hat{I}_{or}$ is the inverse of the quantity known as the cell geometry. It can be recalled that the cell geometry defines the DL interference for the UE at a location relative to other cells in the network.
- Required E_b/N_t is determined when setting the minimum traffic channel power E_c/I_{or} to achieve a required Block Error Rate (BLER) for a specified interference condition, $I_{oc}/\hat{I}_{or}$.
- Downlink capacity is determined from the fraction of transmit power allocated to the traffic channel.
- In WCDMA networks, voice calls and data links are modeled in soft handover (combining gain $\beta = 3$ dB).

Our Downlink Link Budget assumes no antenna diversity at the UE receiver terminal. Some newer devices may have receive diversity, effectively reducing the E_b/N_t requirements.

Comparing these Link Budgets to the outdoor-to-outdoor Link Budget discussed in Chapter 2, the primary difference is an additional loss associated with radio transmission into and out of the building. This is commonly called BPL.

Table 8.7 Outdoor to Indoor Downlink Link Budget

		WCDMA 12.2 speech		WCDMA 64 CS data		WCDMA 384 PS data	
a	HPA maximum transmit power	43.0	dBm	43.0	dBm	43.0	dBm
b	**Maximum traffic channel fraction of total power (E_c/I_{or})**	**−17.0**	**dB**	**−8.0**	**dB**	**−6.0**	**dB**
c	Maximum traffic channel transmit power	26.0	dBm	35.0	dBm	37.0	dBm
d	Tower cable, connector, combiner losses	−3.0	dB	−3.0	dB	−3.0	dB
e	Transmit antenna gain	17.0	dBi	17.0	dBi	17.0	dBi
f	**Maximum per traffic channel ERP**	**40.0**	**dBm**	**49.0**	**dBm**	**51.0**	**dBm**
h	Maximum ERP	57.0	dBm	57.0	dBm	57.0	dBm
k	Thermal noise density = kT	−174.0	dBm/Hz	−174.0	dBm/Hz	−174.0	dBm/Hz
l	Information full rate	40.9	dB-Hz	48.1	dB-Hz	55.8	dB-Hz
m	Receiver noise figure	7.0	dB	7.0	dB	7.0	dB
p	**$I_{oc}/\hat{I}_{or}$**	**3.0**	**dB**	**3.0**	**dB**	**3.0**	**dB**
q	β-combining gain	3.0	dB	3.0	dB	3.0	dB
s	Predicted E_b/N_t–with 2-way soft handover	5.5	dB	7.3	dB	1.6	dB
t	**Required E_b/N_t**	**5.6**	**dB**	**2.5**	**dB**	**2.1**	**dB**
u	Sensitivity, discounting interference, and propagation components	−120.5	dBm	−116.4	dBm	−109.3	dBm
v	Maximum path loss, without interference	160.5	dB	165.4	dB	160.3	dB
w	Received power from target cell	−103.5	dBm	−108.4	dBm	−103.3	dBm
x	Other-cell interference power density, I_{oc}	−163.4	dBm/Hz	−168.3	dBm/Hz	−162.9	dBm/Hz
z	**Sensitivity**	**−115.3**	**dBm**	**−114.0**	**dBm**	**−103.5**	**dBm**
ak	Receive antenna gain	0.0	dBi	0.0	dBi	0.0	dBi
al	Cable, connector, combiner losses	0.0	dB	0.0	dB	0.0	dB
am	**UE Rx antenna gain–cable loss**	**0.0**	**dB**	**0.0**	**dB**	**0.0**	**dB**
ap	Lognormal fading	−10.3	dB	−10.3	dB	−10.3	dB
aq	Handoff gain	4.1	dB	4.1	dB	4.1	dB
ar	Diversity gain	0.0	dB	0.0	dB	0.0	dB
as	Car penetration losses	0.0	dB	0.0	dB	0.0	dB
at	Building penetration losses	−20.0	dB	−20.0	dB	−20.0	dB
au	Body loss	−3.0	dB	0.0	dB	0.0	dB
av	**Σ Propagation components**	**−29.2**	**dB**	**−26.2**	**dB**	**−26.2**	**dB**
aw	**Maximum Path Loss**	**126.1**	**dB**	**136.8**	**dB**	**128.3**	**dB**

The BPL varies in different buildings, depending on many factors. The BPL also varies in different regions of the world, because construction techniques vary widely. Nevertheless, it is possible (perhaps necessary) to generalize the BPL for a given area to produce a network plan. We must keep in mind that planning a BPL that is too large has a detrimental effect on the overall network performance, and the optimization effort grows exponentially.

When engineering coverage into a specific building from a specific nearby cell site, the BPL should be estimated differently than when it is planning coverage into buildings from a macronetwork. In the latter case, a statistical BPL is sufficient even if its accuracy is limited. For a specific building, applying a statistical BPL limits the accuracy even more. Measuring actual loss for the local materials and construction is well worth the effort. We must be aware that for higher angles of incidence (more than 45 degrees), the loss measurement should be scaled. Table 8.8 shows the scaling factors [13]. As an example, a material presenting an attenuation of 10 dB would have an effective attenuation of about 15 dB when illuminated by an antenna with an incidence angle of 45 to 70 dB.

Any energy not transferred into the building is reflected and scattered. Such scattering may help provide coverage to the surrounding areas, particularly in an urban setting; however, it can also be detrimental if it scatters beyond desired cell boundaries, because it increases other-cell interference. This phenomenon would be more noticeable at higher elevations than at street level because statistical BPL tends to change with elevation, as shown in Figure 8.11.

Several researchers [12,14], have observed and documented the change in statistical BPL as a function of floor height. Most studies suggest that BPL decreases with height. Figure 8.11 charts the measured losses on the floors of a modern, multiple-story building. Floor level 0 corresponds to the ground floor. In the experiment, received signal strength was measured inside with a transmitter feeding a collinear, omnidirectional antenna mounted on the outside of another building. The transmitting antenna was raised clear of any local obstructions. The chart shows the BPL at three frequencies from a receiver located on various floors. The general trend of the data is for penetration loss to decrease with increasing frequency, even if local variation can be observed on the basis of the construction type. Also, the penetration loss decreases at higher floor levels, although there is much variation in the transition from ground level to the first floor. Finally, the loss trend is approximately 2 dB per floor. A line at a slope of 2 dB per floor illustrates the trend.

The decrease of BPL with height can be attributed to the fact that the receiver obtains a more direct (more LOS) link to the transmitter. In other words, the UE eventually clears the clutter or reaches the same elevation as the Node B antennas. When this occurs,

Table 8.8 Material loss scaling factor for different angles of incidence

Angle of incidence [degrees]	Scaling factor
0 to 45	0
45 to 70	1.5
70 to 90	2

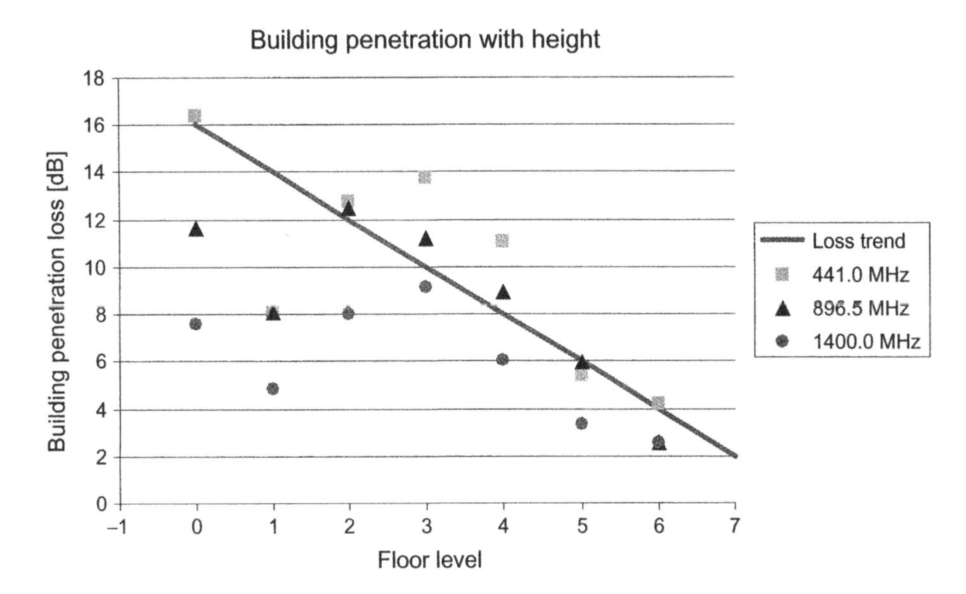

Figure 8.11 Building penetration as a function of height [14]

the path is more likely LOS and the UE may have a shorter distance from the Node B antennas. This is confirmed by measuring the changes in BPL for different floors in a building that stands significantly higher than the average clutter. As the UE is positioned above the average clutter height, a reduction of 1.5 to 2 dB in BPL can be measured, as shown in Figure 8.11. Although not illustrated in Figure 8.11, the supposition is that if the building is much taller than the transmitter, the BPL eventually increases on floors that are above the height of the transmitter.

The variation of BPL with height is important for several reasons. It affects not only coverage but also performance. Within a building, as a UE moves upward floor by floor, it experiences a changing RF environment. One of two things can happen. If the outdoor cells are low in elevation, and/or if the outdoor antennas have large downtilt, the UE gets less coverage from outdoors as it moves up, resulting in no or limited service. On the other hand, if the outdoor cells are not low and the antenna downtilt is not large, the UE sees more cells as it moves higher above the surrounding clutter. This results in a nondominant Pilot or Pilot pollution problems.

Pilot pollution is often responsible for poor UE performance indoors, including dropped calls and failure to reach Connected Mode from Idle Mode. Pilot pollution often corresponds to a condition of strong Downlink interference, meaning that E_c/N_o at the UE is low, even if the Received Signal Code Power (RSCP) is strong. In this case, as a UE moves up in a building, it is more likely to see signals from many cells. Having several Pilots to choose from–assuming that all the unwanted servers are in the Monitored Cell List–the UE performs many handovers. In the Idle state, this may make accessing the network unreliable, because reselection might be required while the UE is attempting to perform access-related communication.

Many performance issues can be observed in such Pilot polluted environments. Calls can drop because of the high number of handovers required when signals from many cells are present. In CELL_DCH, the UE and the network will be very active with signaling operations related to handovers. If the infrastructure is slow to respond to measurement messages, a call can be lost. The monitored set at the 10th floor of a building can be very different than that found at ground level. As we have seen, this is because the cells visible from a height may be different from those in the surrounding neighborhood. Close cells may be OBS, while distant cells may be in a LOS condition. These distant cells may not be neighbors for UEs on the ground. If a Pilot is not in the Monitored Set, it is unlikely to be added to the Active Set and thus becomes a strong interferer. This causes poor call quality and possibly dropped calls. Finally, a UE trying to enter Connected Mode from Idle Mode in a tall building may also have trouble as a result of constantly changing Pilot conditions. In such a heavy interference environment with many Downlink signals, it may not be possible to complete all of the required signaling messages for service connection before the call is lost.

The solution is to limit the strength and the number of Pilots that enter the building, particularly for the upper floors. At the same time, a dominant cell must be provided to communicate with the UE. Pilot numbers can be reduced by downtilting antennas and/or changing antenna orientation, perhaps in combination with reducing CPICH output power as necessary. Of course, any such changes will likely change the coverage on the ground or in other areas of the network, requiring additional optimization. For those cells that appear in the upper floors of a building, the Monitored Set may also need to be changed. If the signals cannot be removed from the building, they must at least be identified as neighbors for the UE. This often means that cells several tiers distant must be added to the Monitored Set. If network propagation prediction tools consider only receivers at ground level, they may not predict this.

Providing a dominant Pilot may be possible by reorienting the antenna of a desired outdoor cell (or cells) to promote coverage within the building. Other methods are to bring coverage into the building via a DAS, perhaps fed from a repeater or picocell. Regardless of the method, if the CPICH E_c/N_o of a single cell can be made dominant, indoor performance is enhanced.

8.3.3 Service Indoors from the Indoors

Sometimes it is appropriate to place a picocell or a repeater inside a building for coverage or capacity reasons. This section describes how the general WCDMA Link Budget would change in that situation, including an MAPL for indoor design. We also present a method for predicting indoor transmission loss, based on measured output Pilot power combined with knowledge of the other-cell interference component. This particularly applies to repeater installations where the percentage of CPICH power is not under direct control, as it is when deploying a cell. Either of these methods defines a MAPL, which can then be used to predict indoor coverage. To translate the maximum transmission loss into a propagation distance and indoor coverage area, we can apply an appropriate indoor propagation model, as discussed later in this section.

8.3.3.1 Indoor Link Budget Review

Table 8.9 shows a representative Downlink Link Budget.

Signal fluctuations can be represented by large-scale and small-scale effects similar to the model commonly used for outdoor propagation. Indoor shadowing results in large-scale signal variations, which can be represented by a lognormal distribution in a manner analogous to macrocell propagation models. However, the standard deviation can be much higher in the indoor environment, and some measurement results show standard deviation of 16.0, 16.5, and 17.8 dB at 900, 1800, and 2300 MHz respectively [15]. The standard deviation of lognormal fading can easily be translated into a fade margin to be included in the Link Budget calculation, based on the required coverage probability (reliability), the corresponding edge probability, and the standard normal distribution curve [16]. These are similar to the well-established techniques used for outdoor reliability. The motion of

Table 8.9 Indoor to Indoor Downlink Link Budget

	Item		Downlink	Downlink	Downlink
	Test environment		Indoor repeater	Indoor repeater	Indoor repeater
	Test service		WCDMA 12.2 kbps Speech	WCDMA 64 kbps CS data	WCDMA 384 kbps PS data
(a0)	**Maximum total repeater transmitter power**	**(dBm)**	**23.0**	**23.0**	**23.0**
(a1)	**Fraction of total repeater Tx power for the Pilot = 10 log(CPICH_E_c/I_{or})**	**(dB)**	**−10.0**	**−10.0**	**−10.0**
(b)	Repeater cable, connector, and combiner losses	(dB)	17.0	17.0	17.0
(c)	Repeater coverage antenna gain	(dBi)	5.0	5.0	5.0
(e)	UE Receive antenna gain	(dBi)	0.0	0.0	0.0
(f)	Cable and connector + body losses	(dB)	3.0	0.0	0.0
(g)	UE receiver noise figure, NF	(dB)	8.0	8.0	8.0
(h)	Thermal noise density, N_{th}	(dBm/Hz)	−174.0	−174.0	−174.0
(i)	Receiver thermal noise power = $N_{th}W + NF$	(dBm)	−100.2	−100.2	−100.2
(j)	**Ratio of other cells (not in HO) to target cell RX power densities, I_{oc}/I_{tc}**	**(dB)**	**2.5**	**2.5**	**2.5**
(l)	**Required receive CPICH E_c/I_o**	**(dB)**	**−16.0**	**−16.0**	**−16.0**
(m)	Transmission loss to achieve required CPICH E_c/I_o	(dB)	124.0	124.0	124.0
(p)	Lognormal fade margin (90% edge coverage, standard deviation = 6 dB)	(dB)	7.7	7.7	7.7
	Maximum Allowable Path Loss = (l − b + c + e − f − p)	(dB)	101	104	104

objects surrounding the antennas can cause small-scale signal variations due to multipaths. This must also be taken into account. The small-scale multipath process can be considered the same within each room or contiguous areas. Rayleigh fading is usually used when no dominant multipath component exists, while Rician distribution is a more accurate model when both a dominant path and a number of weaker multipaths are present. Other empirical fading models for indoor propagation are the Suzuki [17], Weibull [18], and Nakagami [19] distributions. Because indoor traffic is typically slow moving, multipath fading is not easily averaged out. In contrast to macrocell traffic, multipath fading indoors has a greater impact on performance.

8.3.3.2 Defining Microcell or Repeater Range on the Basis of Pilot Power

Particularly for indoor applications, it is often desirable to predict the coverage of a repeater or microcell on the basis of CPICH power. During repeater installations and commissioning, CPICH power can be determined by measuring the output (or input) E_c/N_o and the received signal strength. The goal is to estimate the range of the repeater or, equivalently, the microcell, given a known Pilot power in dBm. While this technique can apply to any WCDMA source, it especially applies to repeater installations because repeater gain can be adjusted to provide a desired amount of output power; however, the individual channel powers cannot be adjusted as is possible for a picocell. Relative channel powers are determined at the donor cell. The only signal held constant at the donor cell is the CPICH power. If repeater coverage can be predicted on the basis of the amount of CPICH power coming from the repeater, then the repeater gain setting will be more accurate than trying to predict coverage on the basis of composite WCDMA power.

To this end, Equation 8.11 is derived below. The inputs to this equation are the minimum required CPICH E_c/N_o at the UE and an estimate, or assumption, of the other-cell interference component that the UE experiences at cell edge. The output of this equation is the maximum transmission loss that can be sustained between the indoor source (repeater or picocell) and the UE.

8.3.3.3 Downlink (DL) Boundary

The DL boundary is the position in which a UE receives a specified Pilot energy relative to the total interference, E_c/N_o.

By definition, the CPICH E_c/N_o received from the radio system (microcell or repeater) is:

$$\frac{E_c}{N_o} = \frac{\dfrac{\left[\dfrac{\xi \times P \times G_m \times G_s}{L}\right]}{W}}{N_{th} + I_{oc} + I_{sc}} \tag{8.3}$$

where:

ξ = the proportion of Downlink power from the source that is the CPICH

P = total Downlink power of the source

G_m = UE antenna gain

G_s = antenna gain at the source
L = path loss from source to UE
W = bandwidth of the CDMA waveform
N_{th} = thermal noise density at the UE receiver
I_{oc} = other-cell interference at the UE receiver
I_{sc} = same-cell interference at the UE receiver from the source

Equation 8.3 can be simplified into equation 8.4:

$$\frac{E_c}{N_o} = \frac{\left[\dfrac{\xi \times P}{T}\right]}{N_{th} \times W + I_{oc} \times W + I_{sc} \times W} \tag{8.4}$$

where we have defined the transmission loss T, between the source and the UE as:

$$T = \frac{L}{G_m \times G_S} \tag{8.5}$$

It is commonly assumed that the same-cell interference is equal to the signal power coming from the cell, minus that portion which is the CPICH power. Thus:

$$I_{sc} W = \frac{P(1-\xi)}{T} \tag{8.6}$$

Substituting this expression and casting the equation into terms of normalized cell transmit power (normalized with respect to noise power) we can write:

$$\frac{E_c}{N_o} = \frac{\left[\dfrac{\xi \times P}{N_{th} \times W \times T}\right]}{1 + \dfrac{I_{oc}}{N_{th}} + \dfrac{\overline{P} \times (1-\xi)}{N_{th} \times W \times T}} \tag{8.7}$$

$$\frac{E_c}{N_o} = \frac{\left[\dfrac{\xi \times \overline{P}}{T}\right]}{1 + \dfrac{I_{oc}}{N_{th}} + \dfrac{\overline{P} \times (1-\xi)}{T}} \tag{8.8}$$

where: $\overline{P} = \dfrac{P}{N_{th} \times W}$ is the normalized Downlink power

$$\text{Then:} \quad \frac{E_c}{N_o}\left(1 + \frac{I_{oc}}{N_{th}}\right) = \frac{\overline{P}}{T}\left(\xi - \frac{E_c}{N_o}(1-\xi)\right) \tag{8.9}$$

$$\text{Solving for } T: \quad T = \overline{P}\frac{\left(\xi - \dfrac{E_c}{N_o}(1-\xi)\right)}{\dfrac{E_c}{N_o}\left(1 + \dfrac{I_{oc}}{N_{th}}\right)} \tag{8.10}$$

For a specified minimum CPICH E_c/N_o, equation 8.11 yields the maximum transmission loss:

$$T_{max} = \overline{P}\frac{\left(\xi - \frac{E_c}{N_o}\Big|_{min}(1-\xi)\right)}{\frac{E_c}{N_o}\Big|_{min}\left(1+\frac{I_{oc}}{N_{th}}\right)} \tag{8.11}$$

To use equation 8.11, we must estimate the I_{oc}/N_{th} in the environment of interest. Because this is the ratio of other-cell noise energy to thermal noise energy, we should be able to monitor the ambient signal strength with the repeater off, using a UE.

With an estimate of the I_{oc}/N_{th} value, we can apply Equation 8.11 for a given minimum E_c/N_o value, which defines the edge of coverage.

Figure 8.12 shows a plotting of Equation 8.11 over a range of I_{oc}/N_{th} for four different minimum CPICH E_c/N_o values.

Figure 8.12 is based on 10% CPICH power from a 23 dBm total power repeater/microcell transmitter and assumes a 7 dB receive noise figure at the UE.

To put the I_{oc}/N_{th} parameter in perspective, the noise power floor of a 3.84 MHz bandwidth UE with a 7 dB noise figure would be approximately −101 dBm. Thus, if the ambient signal power measured by that UE (with the repeater off) is −80 dBm, the ratio of I_{oc}/N_{th} would be 21 dB. Figure 8.12 shows that if the repeater or microcell is installed deep inside a building (perhaps in a basement) where there is no other-cell interference, the range of the Pilot increases significantly. For comparison, we can consider a low other-cell interference case such as $I_{oc}/N_{th} = -20$ dB, and estimate the path loss supported to reach a minimum CPICH $E_c/N_o = -16$ dB. From Figure 8.12, path loss would be approximately 129 dB, or the RSCP at the cell edge would be −116 dBm, assuming a

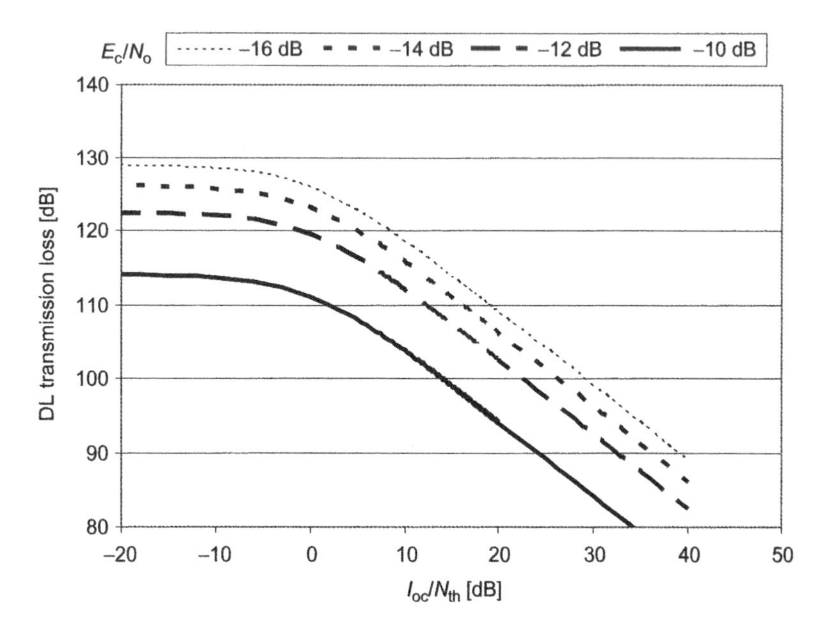

Figure 8.12 DL maximum allowable path Loss as a function of I_{oc}/N_{th}

CPICH power of 13 dBm. This value is consistent with the limit of sensitivity for the UE found in Ref [11].

8.3.4 Indoor RF Models

Whether we determine MAPL from a Link Budget or from the technique above, the next step in the indoor design process is to determine the range defined by this path loss and, ultimately, the amount of area covered by the antenna node. To make the leap from path loss to distance, we must apply a propagation model for the indoor environment.

Different buildings have very different indoor propagation losses. Path loss within a particular building depends on building construction and layout. In general, office buildings, high-rise buildings, manufacturing facilities, warehouses, shopping malls, stadiums, and concert halls have different propagation characteristics. This section summarizes the various propagation models available for predicting coverage within a building.

These models broadly fall into two categories: non-site-specific or site-specific. Non-site-specific models are generalized propagation models. For example, a non-site-specific model might provide an expression for path loss as a function of distance for a "home" or an "office environment." In other words, these models do not attempt to account for a specific set of walls or features and generally yield circular contours of equal power around a given indoor transmitter location. In contrast, site-specific models consider the details of a particular building when predicting propagation losses. Often these models count the number of walls or floors that the signal must traverse and produce an attenuation factor based on the number and type of obstacles through which the signal must pass. While this process can be performed manually, propagation predictions of this type can be automated by computers. This requires importing the building plans into a software tool, which then applies the appropriate propagation model for the building from a specific transmitter location, taking into account the antenna radiation pattern.

A subcategory of site-specific models is propagation prediction based on ray tracing techniques. Ray tracing is a form of computer automated propagation prediction in which rays are postulated to depart at different angles from a transmitter. Their paths are followed after accounting for reflections from walls and obstacles. A computer adeptly calculates reflection angles off various objects and calculates losses from these reflections, given the material properties. These techniques tend to be computationally intensive, often requiring sophisticated expertise to operate, and have not demonstrated significant improvement, in terms of accuracy, over the simpler site-specific models of the type presented here. For that reason, ray tracing techniques are not discussed further in this section. As computing power increases, ray tracing models will become more usable.

Currently there are no universally accepted path loss models for indoor propagation prediction [18,20]. For comparison purposes, we present four general indoor propagation models. Three of these are non-site-specific and one is site-specific, as shown in Table 8.10.

Table 8.10 Main indoor RF propagation models

Non-site-specific	Inverse exponent law [21–23]
	Distance dependent exponent [24]
	Linear loss per unit distance rule [25–27]
Site-specific	"Motley–Keenan model" [28]

The best place to begin the comparison is to consider free space radio propagation. Basic electromagnetic transmission theory shows that free space propagation loss follows the following formula [29]:

$$L_O = \left[\frac{4\pi d}{\lambda}\right]^2 \tag{8.12}$$

where:
d = distance from the transmitter, in meters
λ = wavelength of propagation, in meters

Converting this equation to decibel form and rewriting the wavelength term as the ratio of the speed of light, c, over the frequency, f, yields:

$$10\log(L_O) = 20\log\left[\frac{4\pi df}{c}\right] \tag{8.13}$$

which can be written as:

$$L_{O\,\mathrm{dB}} = -27.56 + 20\log(d) + 20\log(f) \tag{8.14}$$

where:
d = distance from the transmitter, in meters
f = frequency, in MHz

The formula for free space radio propagation loss, Equation 8.12 or Equation 8.14, is based on the physics of radio wave propagation. It assumes that no obstacles reflect or scatter the radio wave as it propagates, and applies the principle that radio wave energy fills the volume of a sphere, expanding as it propagates. Technically, free space loss is not a loss of energy as much as it is a spreading of the available energy across an ever-increasing sphere. Just as the area of a sphere is a function of the square of the radius (to the power of 2), it can be seen in Equation 8.12 that energy decrease is a function of the square of the distance. Thus, the energy of propagating radio waves spreads at a power of 2 as a function of d, the distance from the transmitter. The spread of energy is a function of the wavelength. The final formula converts the original formula to decibels, with the final constant (−27.56) a function of the units chosen: distance in meters (m) and frequency in megahertz (MHz). The most important thing is that, for a fixed frequency, the formula in equation 8.14 is simply the equation of a line in "log(d)" space. The slope of the line is 20 and the intercept of the line is 20log(f)−27.56.

Both indoor and outdoor propagation commonly refer to propagation constants that are larger than 20log(d). For example, UE propagation in the outdoor environment routinely follows a 40log(d) slope. Such an equation might be written as:

$$L_{O\,\mathrm{dB}} = -27.56 + n \times 10\log(d) + 20\log(f) \tag{8.15}$$

This is a modification of the free space propagation equation, allowing for an increase in the propagation constant from the free space value ($n = 2$) to values in the range of $3<n<5$ for outdoor UE environments, with a typical value of $n = 4$ in urban areas. This represents the increased scattering and absorption found in the UE channel as radio wave signals traverse from a Node B antenna to a UE antenna located amid the local clutter.

This same concept applies equally well to many indoor environments. As discussed below, several indoor non-site-specific radio propagation models use this same equation with values of n in the range of $3<n<6$. Other models modify n as a function of distance from the transmitter, or augment the above equation with additional loss terms as discussed below.

8.3.4.1 Inverse Exponential Law

Generalizing equation 8.15, we can write:

$$L(d)_{\mathrm{dB}} = P_{\mathrm{O}} + n \times 10\log(d) \tag{8.16}$$

where P_{O} = the power at $d = 1$ m from the transmitter.

P_{O} is a function of such elements as transmitter power, the frequency, antenna heights, and gains. This expression is a generalized restatement of the propagation equation derived above.

Various values for n have been reported in the literature:

- $1.2<n<6.5$ for 900 MHz for different indoor environments [21].
- $1<n\leq 2$ for LOS measurements in an office building [22].
- $n = 3.9$ for inside a commercial office building when the transmitter and receiver are on different floors [23].

How can n be less than 1 if the laws of physics say that $n = 2$ for free space propagation? The answer is that some indoor hallways or corridor-like environments seem to produce a "wave guide effect" in which the signal energy propagates down the hallway with a standing wave pattern, similar to how electromagnetic energy propagates along a wave guide. This focuses the energy along a particular path so that propagation losses are observed to diminish with less than a power of 2. This phenomenon is possible when considering the placement of, and interference between, indoor antennas.

8.3.4.2 Distance Dependent Exponent

A variation of the exponential law derived above is that of the distance dependent exponent. In this model, the path loss exponent n changes as a function of distance from the transmitter. The equation for this propagation loss model can be written as [24]:

$$L(d)_{\mathrm{dB}} = P_{\mathrm{O}d} + n_d \times 10\log(d) \tag{8.17}$$

where:

$n_d = 2, \quad 1\text{ m} < d < 10\text{ m}$

$n_d = 3, \quad 10\text{ m} < d < 20\text{ m}$

$n_d = 6, \quad 20\text{ m} < d < 40\text{ m}$

$n_d = 12, \quad d > 40\text{ m}$

Note that P_{O} is now written as $P_{\mathrm{O}d}$, indicating that the intercept point is also a function of d. The constant $P_{\mathrm{O}d}$ must be adjusted to produce a continuous function at each breakpoint; for simplicity, this is not shown here.

The larger values of n_d are postulated to be due to an increase in the number of walls and partitions between the transmitter and receiver, as d increases.

8.3.4.3 Linear Loss per Unit Distance

As a final category of indoor non-site-specific propagation models, we consider two formulas that insert a linear distance dependent term. The first inserts a linear distance dependent loss in addition to the log(d) free space loss component, as shown below [25]:

$$L(d)_{\mathrm{dB}} = P_{\mathrm{O}} + 20\log(d) + \kappa \times d \tag{8.18}$$

where $0.3 < \kappa < 0.6$ dB/m

This model has been shown to be valid over a wide frequency range, demonstrated at 850 MHz, 1.9 GHz, 4 GHz, and 5.8 GHz [25]. The conclusion is that the linear dependence of loss with distance corresponds to the indoor clutter environment.

Alternatively, there is a model for indoor propagation that eliminates the log(d) component altogether. The model for this type of relationship is written with the following equation [26,27]:

$$\begin{aligned} L(d)_{\mathrm{dB}} &= P_{\mathrm{O}} + \alpha \times d, \quad d < d_1 \\ &= (P_{\mathrm{O}} + \alpha \times d_1) + \beta \times d, \quad d_1 < d \end{aligned} \tag{8.19}$$

where:

$\alpha = 0.45$ dB/ft, $\beta = 0.22$ dB/ft, $d_1 = 150$ ft @ 150 MHz
$\alpha = 0.38$ dB/ft, $\beta = 0.24$ dB/ft, $d_1 = 130$ ft @ 450 MHz
$\alpha = 0.42$ dB/ft, $\beta = 0.27$ dB/ft, $d_1 - 70$ ft @ 850 MHz

8.3.4.4 Summary of Non-Site-Specific Models

We have surveyed a collection of models, but which one should be used? The answer depends on the application and the building in question. Large open areas such as concert halls, sports arenas, large hotel lobbies, and similar indoor areas with high ceilings and a predominance of open space would likely follow the free space path loss equation. Office buildings, hotels, and apartment living spaces tend to follow one of the models that exhibit larger attenuation per unit distance than free space loss. The linear loss models are convenient mathematically, while the 40 log(d) exponential model is also easily applied.

For comparison, we can consider the graph in Figure 8.13.

Most models, with the exception of the $-40\log(d)$ model, tend to produce about the same loss per unit distance. In other words, we have arrived at four or five equations that produce almost the same curve for a similar set of assumptions. It must be noted that, in the range of 50 m to approximately 120 m, all these models yield similar loss values. Because these models are not very precise without tuning, it is difficult to say that one is significantly better than another.

For tuning, as in all propagation predictions for indoors or outdoors, results are always better if the propagation over the area of interest can be measured using a test transmitter and receiver. Most of the constants in the models we presented can be adjusted for specific circumstances.

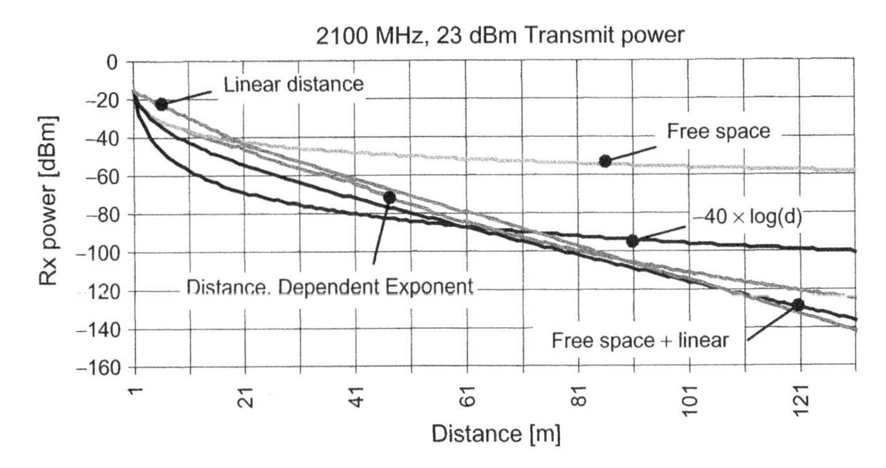

Figure 8.13 Comparison of non-site-specific indoor propagation models

With a proper Link Budget, all the non-site-specific propagation models produce a radius of coverage for a particular antenna gain. If the antenna is omnidirectional, we can draw a circle around each transmitter site. Given a floor plan of the building, we can draw circles of coverage. These coverage circles ignore the walls and obstructions in the building; such is the nature of non-site-specific propagation modeling. The models represent generalized loss for a building, not a prediction of specific losses for a given set of walls, doors, and windows. For that reason, we should keep in mind that reality will be quite different from any circles drawn on a floor plan. Nevertheless, using these simple circles to plan indoor coverage can quite effectively estimate the number of transmitting nodes required and determine the required (desired) output power for each node.

To predict propagation more precisely, we must use site-specific propagation modeling. The following section discusses site-specific propagation prediction.

8.3.4.5 Motley–Keenan Model [28]

Site-specific indoor propagation models, as the name implies, take the specific features of a particular building as their input. Often these models count the number of walls, floors, or other obstructions between the transmitter and receiver to produce an attenuation value for propagation along that path. If we repeat this procedure for many hypothetical receiver (UE) locations, we can draw contour lines of equivalent received signal strength. These propagation predictions are ideally suited for computer automation. At the time of this writing, several indoor propagation prediction tools in the market use variations of this technique.

One such model, often cited, has come to be known as the Motley–Keenan model after the researchers who first published it [28]. This model takes the form of a free space propagation model with attenuation added for each wall, floor, or obstruction encountered along the path. Mathematically, we may write the path loss L_O as:

$$L_O = P_O + 20\log(d) + p \times WAF + k \times FAF \tag{8.20}$$

where:

$$
\begin{aligned}
P_O &= 20\log\left[\frac{4\pi f}{c}\right]\\
d &= \text{distance between transmitter and receiver}\\
p &= \text{number of walls between transmitter and receiver}\\
WAF &= \text{Wall Attenuation Factor}\\
k &= \text{number of floors between transmitter and receiver}\\
FAF &= \text{Floor Attenuation Factor}
\end{aligned}
$$

The free space propagation model should be evident from the first two terms of the equation. The additional terms account for the loss through each obstruction. This site-specific model is general enough to accommodate a broad range of frequencies with the appropriate choice of attenuation factors. The literature is full of measured attenuation values for a wide variety of materials and at different frequencies, and more are published regularly [30–34]. In general, at frequencies used for personal communication (from 800 to 2100 MHz), interior walls of a modern office building have a WAF of approximately 3 dB. Interior brick structural walls and supports have a WAF of approximately 10 dB. Floors vary depending on the construction, but usually have an FAF between 13 and 18 dB.

8.3.5 Capacity Dimensioning

Chapter 3 presented a principle for macronetwork capacity dimensioning, which can also be applied to indoor systems, with some adaptation. For properly planned indoor systems, the main difference is limited to other-cell interference. To that end, setting the geometry as a design target and minimizing the Active Set size ensures that capacity is maximized, as shown in Figure 8.14. In this figure, the Link Budget previously defined is used to estimate the required DPCH power that meets the E_b/N_t requirements. In all cases, the combining gain is set to 0, corresponding to no handover. Also, a single path is used to reflect the AWGN channel condition. The result is the Downlink capacity for voice plotted against the geometry. On the basis of this, designing the indoor network with a geometry target of 2 dB ensures that OVSF and power are limiting the capacity at the same time.

This estimation ignores the cell dynamic, thus allowing the DPCH to decrease as much as possible. In a real system, the cell power dynamic is limited from 20 dB to 24 dB, thus causing the capacity to level off.

By extension, Table 8.11 summarizes the target geometry for OVSF and power to be limiting at the same time, for different services, in the same multipath and handover conditions. This shows that, on the Downlink OVSF, the absence of a secondary code tree would likely be limiting. If we compare these results to the macrolayer capacity estimated in Section 3.3.2, we can see that indoor cell capacity is 2 to 3 times larger than macrolayer capacity.

A similar calculation can be done for the Uplink. In contrast to the Downlink, the Uplink OVSF is not limiting, because the code tree is not shared on the Uplink.

Such capacity numbers cannot be scaled up by introducing multiple cells over an enclosed area. Either handover (if enabled) or reduced geometry diminishes the available capacity. When handover is enabled, the capacity decreases as the handover overhead factor increases. When handover is disabled, the geometry in the area where cells overlap

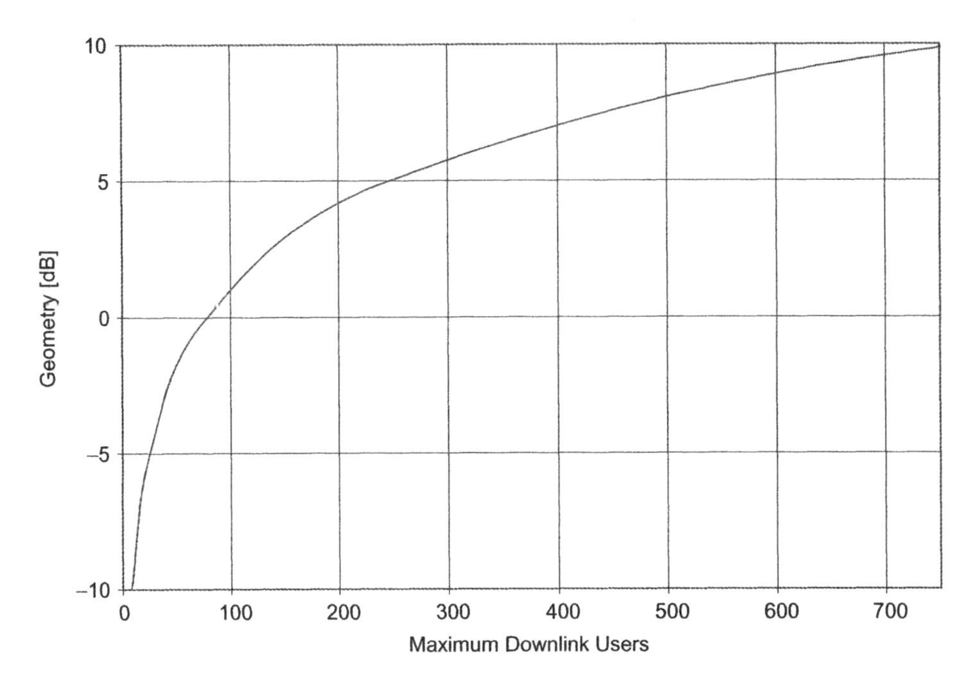

Figure 8.14 Impact of geometry on capacity, voice (AMR 12.2) services, and AWGN channel, considering power limitation only

Table 8.11 Geometry targets for different services, AWGN channel

	DPCH E_c/I_{or} [dB]	Geometry [dB]	Available OVSF
Voice	−21.8	2	121
CS 64 (VT)	−15.9	0.6	31
PS 64	−15.9	0.4	31
PS 128	−12.8	−0.7	15
PS 384	−9.5	−0.6	7

increases, thereby increasing the required DPCH power to meet the E_b/N_t target. Once the maximum DPCH is reached, the quality of the serviced call degrades: BLER increases to a point where the call cannot be sustained.

8.3.6 Achieving Higher Throughput Indoors

From the discussion in the previous section, it can be shown that 384 kbps can be achieved over the entire indoor cell area, if enough DPCH power is assigned. The limitation in this discussion is that the AWGN channel is used. An extreme channel condition for the indoor environment (case 3 of Ref [27], see also Table 2.3) is used here to verify where PS 384 can be achieved.

As discussed in Chapter 2, the limiting link for PS data is the Downlink if a high data rate (384 kbps) is assumed in the Downlink and only a medium data rate (64 kbps) is assumed on the Uplink. We can estimate where PS 384 can be achieved indoors by using the Link Budget defined in Chapter 2, for the following specific indoor conditions:

- Maximum transmit power depends on the type of cell, varying from 43 dBm for a macrocell to 30 dBm for a micro or picocell.
- Antenna gain and cable loss vary in different installations, depending on the type of antenna system used. To simplify the estimate, these values are included in the MAPL, thus trading cable attenuation and gains for propagation losses.
- Geometry during indoor planning should be a design criterion rather than an assumption. Such a criterion maximizes the capacity of the indoor system and ensures that high data rate service is not limited in terms of coverage area. A suggested geometry target of 1.5 dB, over 90% of high-traffic floor space, is reasonable.
- Combining gain indoors is typically set to 0, because indoor systems are designed to cover all areas with a single dominant server, to favor capacity.
- The target E_b/N_t value is selected to reflect the low amount of discernable multipath observed indoors, as well as the low velocity of the indoor user. If the UL Link Budget is used, the target E_b/N_t must be derated for the absence of diversity typical of indoor systems.

Before inserting these assumptions into the referenced Link Budget, MAPL that satisfies the geometry requirement must be estimated. From the indoor field measurements, we can easily estimate the other-cell interference: I_o, or RSSI, measured indoors from the outdoor macronetwork. Figure 8.15 shows such values, typical for a building located in a dense urban environment. For this example, the I_{or} target is I_{oc} + Geometry target = −68 dBm.

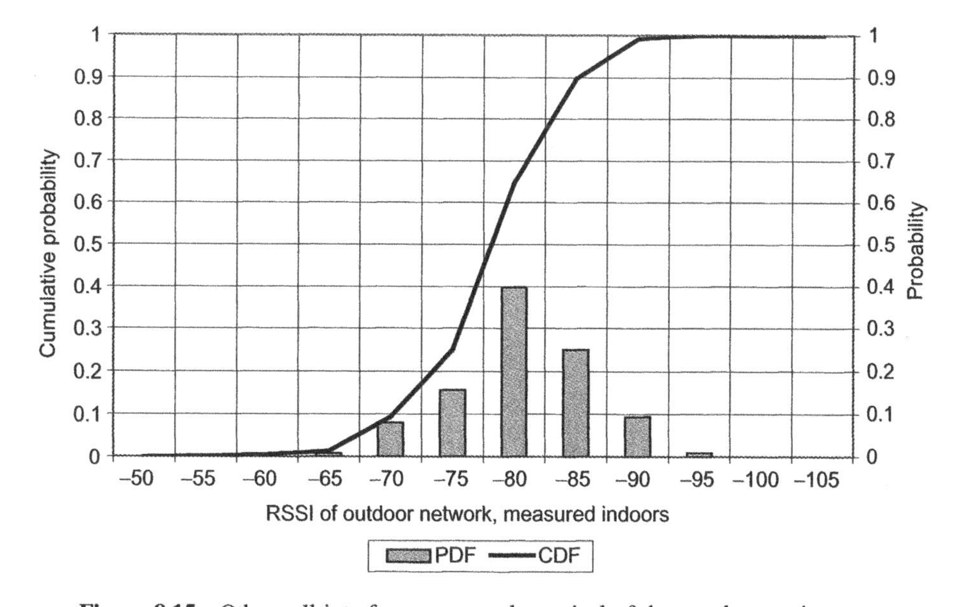

Figure 8.15 Other-cell interference example, typical of dense urban environment

The estimated maximum path loss would be 110 dB for a 43-dBm power amplifier, 80% loaded. Under such conditions, the required DPCH power for PS 384 would be 32.5 dBm, thus allowing a maximum of seven supported PS 384 users, identical to the OVSF code tree limitation.

As seen in the previous discussion, for a given system we can estimate the requirement on maximum transmit power and the MAPL simply by measuring the level of interference (other-cell interference) from the outdoor macrosystem.

8.4 Optimizing Indoor Systems

8.4.1 Practical Considerations for Indoor Deployments

An indoor system can be well optimized only when optimization is considered at the very beginning, that is, when the system is designed. After system deployment, it is often tremendously difficult to alter the installed indoor network components because of construction cost, time, network service delays, and inconvenience to the people who are using the building. After deployment, minor changes can be made to fine-tune the performance, for instance to account for discrepancies between projected and actual building details, or unforeseen changes to the technical dimensions and forecasted traffic. Just as for a macronetwork, it is almost impossible to design an indoor system that needs no changes after the installation. However, following a well structured planning and deployment process can minimize the amount of rework required after deployment.

Previous sections of this chapter covered the technical details of Link Budgets and other planning calculation. This section covers the end-to-end indoor deployment approach, from planning through optimization. Because many planning and optimization tasks are common to both macrocell and indoor network design, this section highlights only the elements that differ from an outdoor deployment.

Figure 8.16 illustrates an indoor project lifecycle from a high-level point of view.

8.4.1.1 Determining Indoor Network Requirements

Like outdoor network planning, indoor network planning starts with obtaining network requirements for coverage and capacity. As will be discussed in Section 8.4.1.2, the emphasis on coverage or capacity affects the type of system deployed. For example, a DAS fed by a dedicated indoor Node B would offer high capacity; however, if a dominant serving macrocell with sufficient capacity already exists, repeaters could be used. Therefore, the very first step is to identify what areas need coverage, and then evaluate the capacity requirements on the basis of the macrocell/microcell traffic split.

For a given building, the indoor coverage requirements should include how the coverage will be planned. Table 8.12 summarizes the types of coverage requirements.

Values for the three coverage requirements are planned differently:

- **Statistical building penetration.** Typical of indoor coverage from outdoor sites. It is drafted from local knowledge of construction and material used, because these have a major impact on the statistical attenuation.
- **Percentage of floor space.** Statistical by nature. It applies mainly to large venues where the users can be equally distributed throughout. In such venues, the percentage of floor

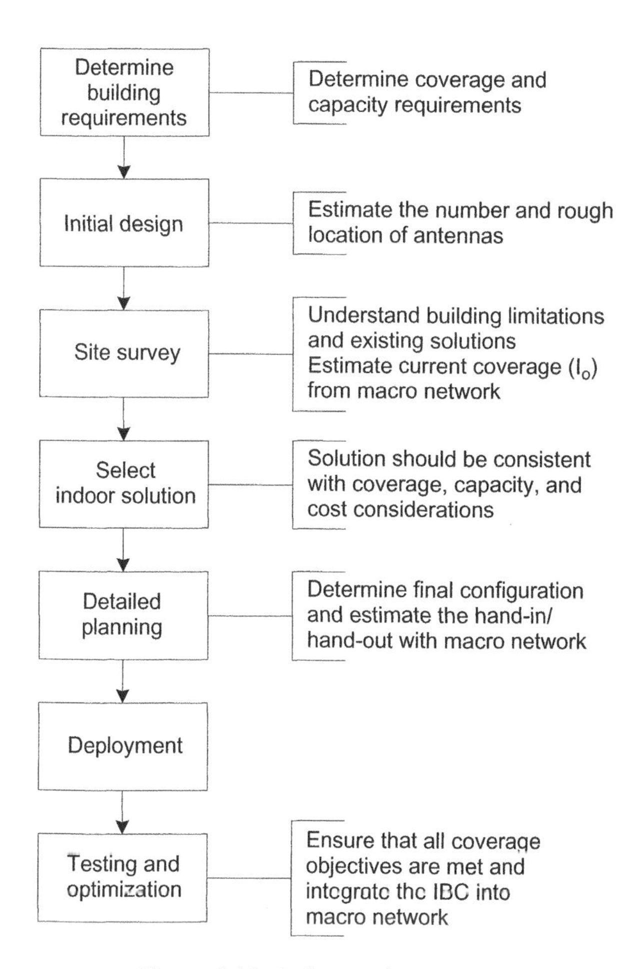

Figure 8.16 Indoor project lifecycle

Table 8.12 Types of coverage requirements

Coverage requirement	Typical value	Comments
Statistical building penetration	15 to 20 dB, first wall 30 to 50 dB, entire ground floor	Most appropriate for coverage provided by an outdoor macronetwork
Percentage of floor space	90 to 95%	Most appropriate for large structures where users are evenly distributed, for example, stadiums, exhibition halls, airport terminals
Specific floor plan coverage	None	Specific to the type of building: for example, commercial, industrial, residential, offices

space covered can be assimilated to the cell area probability in the macronetwork design.
- **Specific floor plan coverage.** Has different implications if indoor coverage extends to private buildings such as large shopping malls, supermarkets, and department stores. Because of high expected end user demand, coverage in these privately owned buildings may be quite sought after. However, installing equipment inside them may require the permission of their owners, and maintaining equipment (such as active devices) could be complicated by the arrangements needed for access.

It is important to accurately classify the type of indoor coverage required for a given indoor area. Is coverage mandatory, needed only for limited service, or not required?

- **Mandatory full coverage areas.** Areas of dense UE traffic: foyers; entrances to cinemas, supermarkets, and restaurants; waiting areas; escalators.
- **Limited coverage areas.** Areas of medium importance that require coverage only for a limited service type, namely, voice and low-to-medium data rate for PS data; for example, car parking structures and elevators where people do not stay for long.
- **No required coverage areas.** Areas such as loading docks, restrooms, and inside performance halls. Within this category, there may be areas where coverage is not required and areas where coverage is not desired:
  - Coverage Not Required. To reduce infrastructure and planning costs, dedicated equipment must not be installed in these areas. Radio signals spilling in may provide some limited coverage.
  - Coverage Not Desired. Trying to block service may be more difficult because of the conflict with other coverage requirements. As an example, providing coverage in the foyer of a movie theater will conflict with the need to block service within the theater itself.

Estimating end user traffic demand can be difficult. Extending coverage indoors can dramatically change users' behavior and it may not be possible to accurately predict the new generated traffic. When there is no indoor service, users resort to other communication methods. However, when indoor coverage is introduced they start using the UE service. This adds to the traffic that can be offloaded from the macrocells. In that respect, while indoor systems can increase the UE coverage penetration and capacity, they can also increase revenues for mobile operators. If macrocells provide only limited coverage, a new indoor system is likely to capture additional traffic from users, or, in a competitive landscape, lead to churning when users switch to the operator with better indoor coverage.

User applications also contribute to variations in system traffic. With 3G multimedia services, the system is not limited to voice service only, but can also provide PS data service at a high bit rate that ultimately supersedes the voice capacity requirement. Apart from capacity concerns, different applications can have different traffic characteristics, which also affect radio resource management (the allocation and reallocation of radio resources).

To summarize, Table 8.13 lists the main factors that affect the capacity requirement. Their impact is represented as a scaling factor, or as a ratio of comparison with the macronetwork traffic.

Table 8.13 Capacity requirement impacts

Capacity factor	Scaling factor	Comment
Existing voice traffic on macronetwork	Low	Scaling is mainly due to users gaining reliable service
Existing PS data traffic on macronetwork	Medium to high	Scaling is not only due to gaining reliable service but also due to the type of location where the service is provided: areas where users can stay for extended periods have a large impact
PS data traffic due to new services	High	Scaling is affected by both the services and their quality For scalability, WLAN traffic is a better indicator than cellular traffic

8.4.1.2 Initial Design

After defining the coverage and capacity requirements, the first step in indoor network planning is to make an initial assessment about which coverage solution to use. To determine the radio solution components, the macrocell site locations must be analyzed to see whether it is possible to find a dominant macrocell, either directly or as the donor for a repeater radio system. Two conditions largely determine whether the building can be served from the macronetwork: the location of the serving site and the type of coverage defined. Working from the Link Budget defined in Chapter 2, along with a typical statistical RF propagation model, Table 8.14 defines maximum BPLs that can be overcome from distant sites. These distances or BPLs will change significantly depending on local conditions.

To drive a repeater from an existing macrosite, the main requirement is a LOS view to a single donor, which would result in both signal strength and quality of the donor signal (SNR and E_c/N_o), as discussed in Section 8.3.1.1.

If no dominant serving macrocell is available, a standalone dedicated cell (macro, micro, or pico) is the only radio system option for feeding the subsequent antenna system. Even if a dominant serving cell exists, before an indoor repeater or dedicated cell is chosen, the traffic expected to be generated within the building, and loading of the donor site are first evaluated. A repeater is well suited for lightly loaded situations, allowing

Table 8.14 Macronetwork coverage of building based on distance and BPL

Distance between building and cell [m]	Maximum BPL [dB]
<50	40
50 to 500	30
>500	20

traffic multiplexing onto the donor macrocell. Alternatively, a dedicated cell–ultimately a standalone Node B–provides additional capacity to the network instead of sharing the macrocell capacity. If the indoor traffic cannot be estimated accurately, we could start with a repeater as a temporary solution, and then upgrade to a dedicated cell when the indoor traffic grows or becomes better determined. This approach does not prohibit the selection of an antenna system, and upgrading from a repeater to a picocell does not require changes to the antenna system.

However, starting with a repeater is not always possible, depending on the power requirements of the antenna system. The first step in selecting an antenna system is to determine which option is suitable for the entire area to be covered. To assess the area, we must obtain and analyze construction drawings and facility layouts before making a site visit. All indoor regions should be classified into different coverage categories with different priorities. The categories should follow the coverage requirements documentation: high importance for the areas to be covered for all services, medium for the areas to be covered for basic services, and low for the areas where coverage is not required. After all the areas have been categorized, we can estimate the antenna type and the number of antenna nodes required, based on the size of the indoor areas that the macrocells will not cover. We can also estimate the required input power on the antennas.

On the basis of the estimates, we can use the decision tree in Figure 8.17 to determine a preliminary indoor solution. The following sections discuss the decision process in more detail.

8.4.1.3 Site Survey

After the preliminary system design is complete, the next step is to survey the site. Site surveys allow us to measure the level of coverage that the macronetwork provides, verify the building details, and understand the surroundings and existing facilities in the target buildings.

Preparations for these site surveys should include a best server coverage plot showing all the surrounding serving cells in the neighborhood. This plot can be completed with RSCP and E_c/N_o predictions and with the results of the drive test performed around the building to verify the measurements against the predictions. This will identify or confirm whether a dominant serving cell exists, and at the same time assess the potential interactions between macrocells and indoor cells. On a copy of all the floor plans and technical drawings, all the coverage areas must be classified so that they can be marked or corrected as needed during the site survey. The coverage objectives must be reviewed so that on-the-spot coverage classification can be done if necessary. If a BDA or Repeater is likely to be the radio system option, a Pilot scanner or test UE must be available for on-site measurements to verify the quality of the available donor PSC.

During the site surveys, as much useful planning information as possible is collected. Pilot scanners or test UE devices are used to collect signal strength and quality measurements from the surrounding macrocells. This also provides a measure of I_o (or RSSI) for use in the Link Budget and MAPL calculation.

If a new DAS is designed, potential antenna types and positions are identified. Approximate antenna coverage range can be based on the preliminary Link Budget design. Obstructions are noted and appropriate antenna locations are marked on the floor plan.

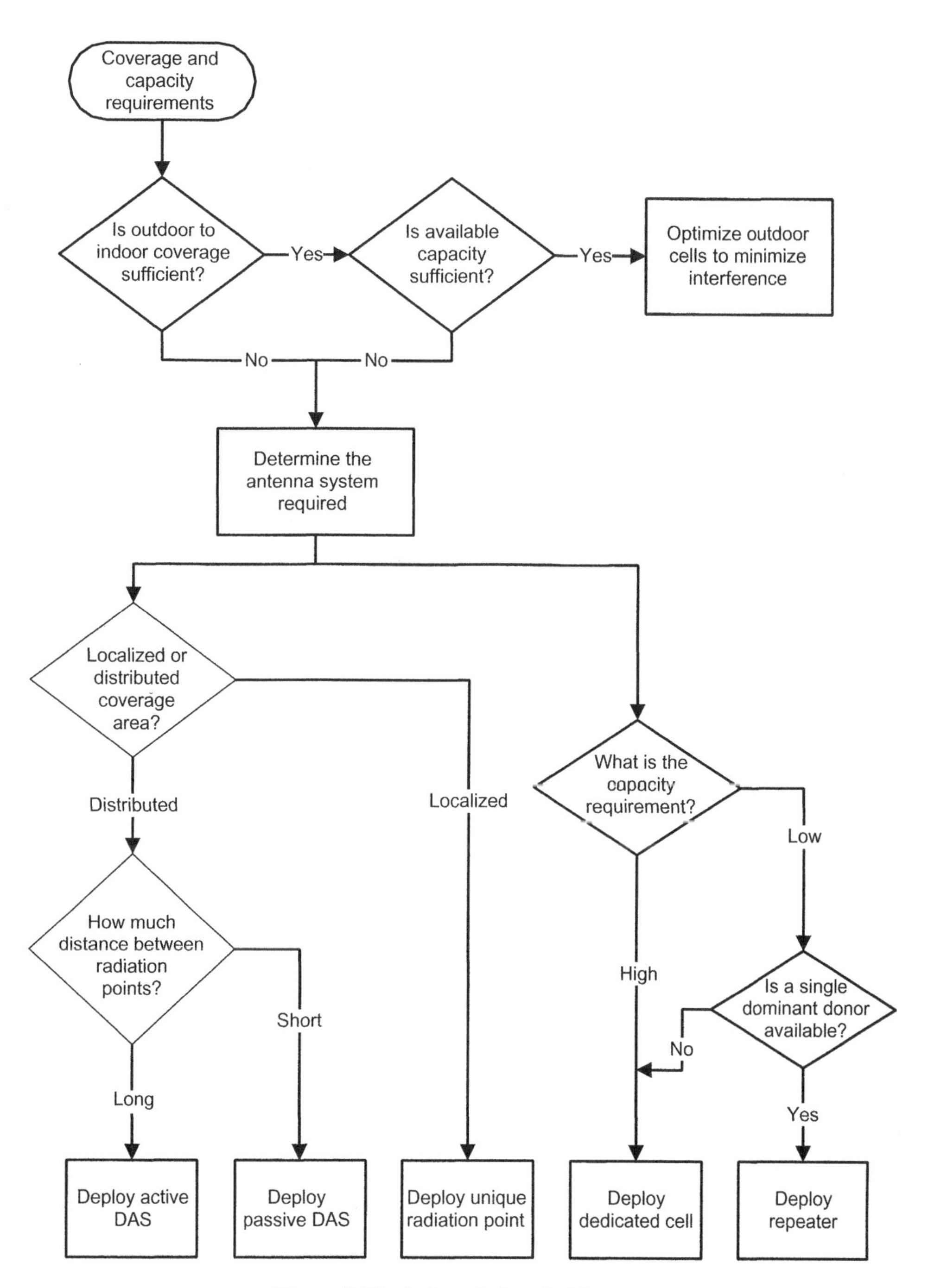

Figure 8.17 Indoor design decision tree

If an existing DAS is likely to be reused, all existing antennas are identified. It is verified whether they are broadband antennas and are consistent with the design provided by the infrastructure owner. Couplers and splitters are difficult to verify, and the same applies to antennas if they are hidden above false ceilings.

After pre-design site surveys have been carried out, the indoor system solution can be finalized. Preliminary studies are important and can narrow down the options, but site visits provide a much more accurate foundation for determining the most suitable indoor system solution.

8.4.1.4 Selecting the Indoor Solution

Is an indoor solution required? First, analyze whether the macrocell coverage is sufficient. Even if the macronetwork design review or a macronetwork drive test indicates sufficient coverage, an additional site survey must confirm that all targeted indoor areas are covered and, more importantly, that a dominant server can be detected in all areas. If this latter requirement is not met, then an indoor system may be required along with adjusting the macronetwork.

Second, estimate the capacity requirement for a specific traffic model and the number of users. Because this process begins with the macronetwork, use the typical macronetwork capacity as described in Chapter 3.

If the first two steps of the analysis indicate that an indoor solution is needed, determine the proper radio and antenna system. The choice depends on the status of any preexisting systems:

- **Existing antenna systems are available.** Perform additional site surveys and analysis to ensure that the number of indoor antennas and their separation is sufficient for the 3G technology operating in the 2.1 GHz frequency band. Indoor systems designed for 2G technologies may be compatible with only lower frequency bands, namely, 900 and 1800 MHz. The design verification should account for considerable path loss differences, especially when passive systems were deployed to cater to GSM900. When active systems are available, the main focus should be the compatibility of existing equipment with the higher frequency band. In all cases, verify that the resulting antenna distribution is compatible with the design target.
- **No existing equipment, or the existing equipment is not compatible.** Plan a new antenna system.
  - If the indoor area is small, consider a passive DAS. Radio signals can come from a repeater or a standalone picocell, depending on the expected indoor traffic.
  - If the indoor area is large, consider an active DAS. Radio signals come from a picocell due to the sheer size of area to be covered and expected traffic (say >200 indoor antennas, long feeder runs, big area, considerable separation of indoor areas, and so on).
  - Use leaky coaxial cable in subway tunnels, underground areas, and walkways, or wherever the users and the coax are in close proximity. The low-profile nature of this solution makes leaky coax attractive. It can be used in areas where a DAS antenna is impractical or unsuitable, such as in subway tunnels where a low-profile antenna must avoid physical interference with passing trains. The design of the

radiating coaxial cable provides uniform coverage throughout the tunnel, which is an effective solution for overcoming radio dead zones. Leaky coax can be used in conjunction with BDAs or repeater systems to increase a system's in-building (or in-tunnel) coverage.

Assuming that a DAS is required, a Link Budget can be used to determine the MAPL that would include both antenna system gains/losses and propagation losses. The antenna system gains/losses directly relate to the hardware used, while the propagation losses are a function of the selected propagation model and any applicable margins. To address the first point further, all the common indoor network components (multicarrier combiner, couplers, splitters, antennas) to be used in the analysis are identified and listed. Positions of the antennas and precise feeder lengths are not needed at this stage. The main purpose is to identify the suitable radio and antenna system solutions.

8.4.1.5 Detailed Antenna System Design

During the detailed design, the main task is to determine the exact antenna type and placement to ensure sufficient coverage to meet the requirements.

During this process, the physical properties of the antenna must be kept in mind, since they may affect the antenna system design. As discussed in Section 8.3.1.4, many indoor antennas are available with patterns suitable for different situations. These antennas can be ceiling or wall mounted. Omnidirectional antennas have the lowest gain (about 2 dBi) but radiate in 360 degrees of azimuth. The omnidirectional antenna pattern permits it to collect signals from all directions, which partially compensates for the imperfect signal reflections and scatterings found in the indoor environment. Directional antennas have a single beam focused in one direction, and the gain in the bore site can reach 7 dBi, depending on the antenna horizontal beamwidth. Such an antenna can be placed at one building corner, providing good coverage to a long narrow passage area. Bidirectional antennas, sometimes called railway antennas, have two focused beams on each side but a reduced directional gain of 5 dBi in the bore sight due to the two back-to-back radio beams. They can serve thoroughfares and corridors, or provide contiguous coverage between two long passage areas. Suitable antenna types and orientation can be determined either on the basis of experience (manually), or by using a network planning tool.

Antenna placement is a function of the three-dimensional building geometry. High-rise buildings with vertically stacked floors may require one or more antennas on each floor because floor and wall penetration are significant and also because other-cell (outdoor) interference may be high. We can use the typical value discussed in Section 8.3.2, but for more accurate planning, the actual construction must be verified and used as a guideline. The required cable length increases with the number of floors in the building. The number of power splitters or directional couplers that tap the signals out of the main cable run increases as well. Consequently, some high office towers need optical fiber cables and repeaters to carry signals for long distances from the radio system. In contrast, malls or stadiums may require fewer antennas for coverage, because they have fewer internal obstructions and more LOS type propagation environments. However, when multiple cells must be used, usually for capacity reasons, it is much more difficult to control the cell overlap and handover in these environments. Directional antennas and clever positioning of antennas can help increase the isolation between different antenna beams.

Antenna placement within a building should not only ensure proper coverage but also handle transitions between indoor and outdoor Node Bs. The two cases of indoor coverage are treated separately, using the same or different frequencies relative to the outdoor cells. When different carriers are used indoors and outdoors, whether the transition zones are adequate is verified: that is, large enough to allow the UE to perform Compressed Mode measurements and to support inter-frequency handovers. When the same carrier(s) are used indoors and outdoors, isolation between the indoor/outdoor cells is ensured to minimize the amount of handover and interference between the systems.

Antenna placement must also ensure EMR safety for the general public moving in close proximity to radiating elements. EMR regulations limit the maximum radio signal strength that can be emitted, and specify the minimum distance that must be maintained between a radiating antenna element and the public. These restrictions guarantee that RF exposure is below hazard limits defined by the applicable regulations. Among other things, EMR regulations define the maximum input power at the antenna, which in turn defines the distance that can be sustained between the Node B and the mobile subscriber.

In summary, an indoor network planner must consider all these issues and plan the equipment loss such that the combined total loss is within the desired range. Unlike macronetwork planning, for which an RF planning tool is always used, indoor planning can be done either manually or using an indoor network planning tool. The difference is mainly due to the need to plan only for coverage for the indoor system, rather than coverage and capacity (although certain types of indoor coverage must be ensured for various traffic densities). Capacity can be provisioned relatively easily in indoor systems, by upgrading from repeaters to a dedicated Node B or by adding multiple sectors/carriers.

8.4.1.6 Manual Design

For a manual design, the starting point is the calculated MAPL, derived from the Link Budget. The MAPL includes two losses: the loss of the DAS and the propagation loss based on the range. The main issue is that both losses interact; thus the design becomes an iterative process, as illustrated in Figure 8.18.

When estimating the antenna system attenuation, every element of the DAS is considered. This includes not only the hardware used but also the likely cable routing, because this directly affects the cable length and thus the attenuation. This simple process may yield different values for each branch of the DAS, which would lead to a different MAPL for each branch. Figure 8.19 shows an example of DAS antenna locations in a building. Figure 8.20 represents a schematic for a possible DAS to cover this building.

The attenuation calculation for each branch can easily be calculated on the basis of the insertion loss of each splitter or combiner, and the cable loss of each connecting cable run, as presented in Table 8.15.

From the per-node antenna system loss and a Link Budget, the MAPL for each node can be determined. Once this number is calculated, an indoor RF model is selected. During manual planning, according to the propagation models presented in Section 8.3.4, it is easier to use a non-site-specific model although it is less accurate. If directional antennas are used, the limitations of a manual design are quickly noticed. In that case, the maximum propagation path loss can be estimated easily. A maximum range can also be estimated, but this range has limited value for verifying the fulfillment of coverage requirements, unless an antenna pattern is applied.

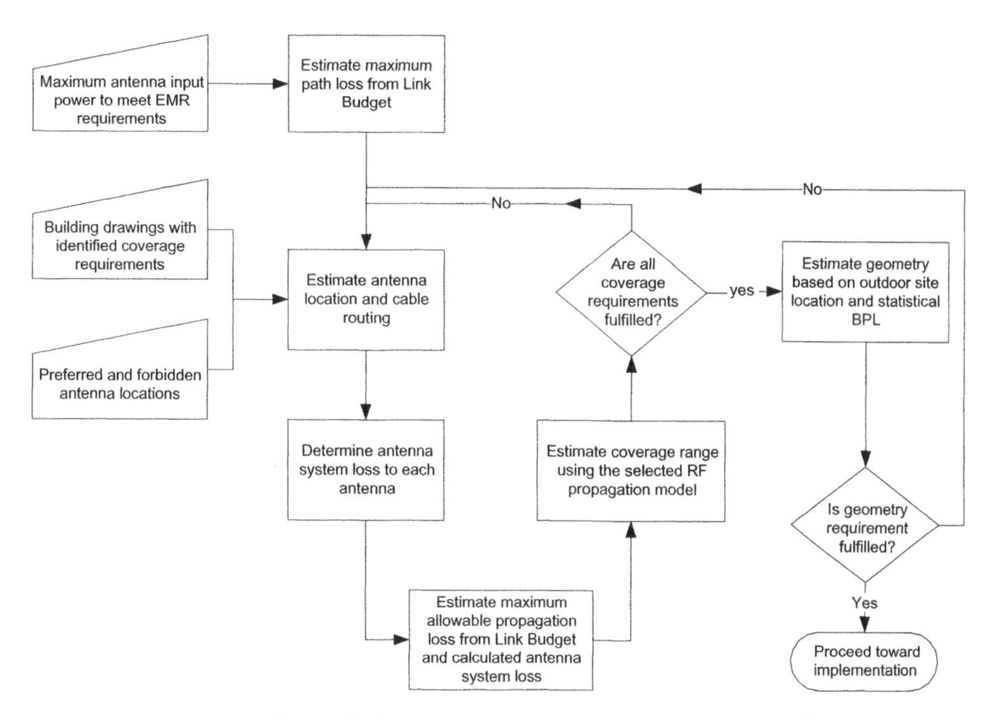

Figure 8.18 Manual indoor planning process

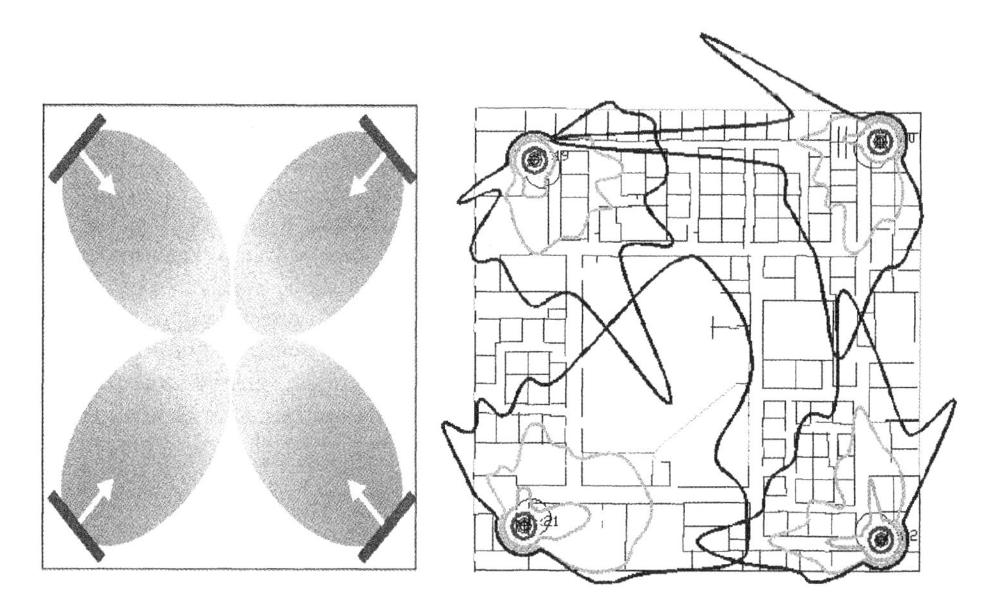

Figure 8.19 Example of DAS antenna locations in an office building

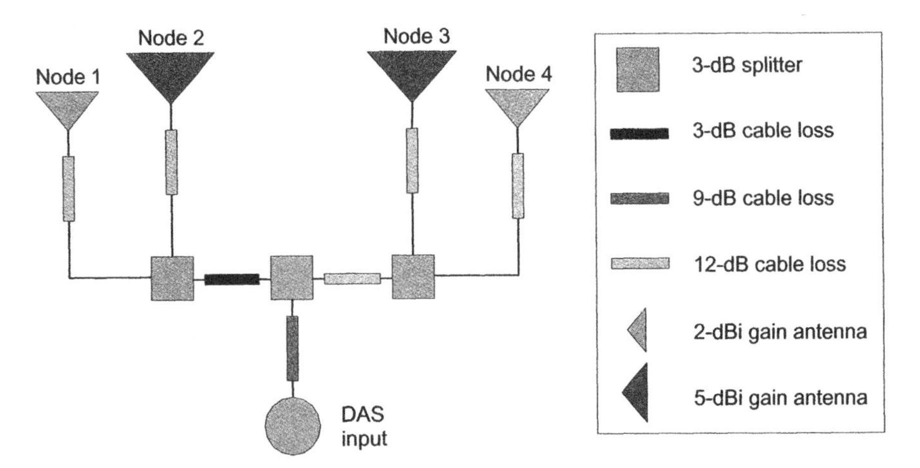

Figure 8.20 Example of schematic DAS layout

Table 8.15 Per-node antenna system loss

	Gain/loss [dB]	Node 1	Node 2	Node 3	Node 4
Splitter	3	2	2	2	2
Cable loss	3	1	1	2	2
Cable loss	9	1	1		
Cable loss	12	1	1	1	1
Antenna	2	1			1
Antenna	5		1	1	
Per-node antenna system loss		28	25	19	22

Nevertheless, during this process it can be decided whether to deploy an active or passive antenna system by comparing the available propagation path loss and the antenna density. Using the comparison of non-site-specific models from Section 8.3.4, the available propagation path loss should be at least 60 dB to ensure that fading margins are available. Beyond that point, the relationship between path loss and distance is almost linear and corresponds to a path loss exponent of 3, which allows the maximum site-to-site distance in meters to be estimated as approximately three times the attenuation in dB.

8.4.1.7 Tool-assisted Design

Network planning tools can automate the iterative process required for manual planning. Several tools have emerged in the last few years. Some of them can simultaneously plan indoor and outdoor networks, or at least include a few outdoor cells to estimate the interference or handover between the systems.

When selecting an indoor network planning tool, ask the following important questions. Does it have appropriate RF propagation models? How are the required inputs entered in the tool? The relationship between required inputs and the RF model is weak for statistical models, but becomes increasingly important–requiring more effort–for ray launching or (even more so) ray tracing models.

Most indoor network planning tools (for picocells) use propagation models that can be categorized as follows:

- **Empirical models.** These consider only the single spatial relationship between one transmitter and one receiver with no building data. Empirical models in indoor network planning tools correspond to the models summarized in Section 8.3.4.5.
- **Semiempirical models.** These are consistent with the Motley–Keenan model defined in Section 8.3.4.5. In addition to this simple distance-partition modeling, an indoor planning tool can include diffraction modeling that would simulate both vertical and horizontal plane effects. Vertical plane diffraction, usually considered for an outdoor or macronetwork planning tool, is of limited use indoors, because it simulates an entire floor all at once. On the other hand, horizontal plane diffraction is useful indoors to estimate the effect of walls and openings (i.e., doors and windows).
- **Deterministic models.** These consider the radio propagation environment (wave-guiding effects of corridors, tunnels, etc.) and physical characteristics of partitions (permittivity, conductivity, and reflections) to accurately estimate the signal received at the receiver. These models can be further divided into 2-D or 3-D ray tracing models, and the simplified dominant paths model. Ray tracing models can resolve the various possible paths from the transmitter to the receiver for individual path loss predictions. Final prediction at each point is the combined signal strength summing up all the multipath components reaching the specified location. Predictions can also account for diffraction effects around corners, which is particularly significant for outdoor microcell environments. Because of the computer calculations required, deterministic models take more time to process than empirical models do. The relatively high accuracy of deterministic models should nevertheless be balanced against the required accuracy of the necessary input (both the building's outline and its material characteristics) and the relative slowness of the predictions. A dominant paths model is a faster alternative to a true 3-D ray tracing. It basically selects only the most relevant paths from all those determined in the 3-D ray tracing calculations, and then determines the propagation attenuation, wall penetration, reflection, and diffraction losses.

The level of effort needed to calculate an accurate building definition, combined with the lengthy prediction time–especially for the ray tracing model–has so far discouraged the widespread use of such planning tools.

8.4.2 Indoor System Deployment and Postdeployment Optimization

Once the design is complete, installation takes place. For aesthetic reasons, the components of an indoor system are often deliberately hidden. As a result, after installation, it is difficult to verify or test them fully. Craftsmanship during the installation is therefore very important because troubleshooting faulty components is difficult.

Perform a full RF survey and call sample testing to make sure the design meets all coverage requirements. Use RF scanners and test mobile handsets to collect measurement points for further analysis. In addition to any performance indicators considered for the macrolayer, pay special attention to the following areas:

- **CPICH performance (E_c/N_o and RSCP).** As in outdoor systems, use CPICH RSCP to estimate the path loss because CPICH is transmitted at constant power (not power-controlled). From the path loss, and in particular the minimum measured path loss, the loss of the antenna system can be estimated and compared to the designed value. E_c/N_o, together with RSCP and RSSI, can be used to estimate the geometry as in Equation 3.19. By itself, as in any WCDMA system, RSSI is of limited use because the quality (E_c/N_o) of the received signal is more critical than the strength.
- **UE transmit power.** Confirm that the UE dynamic range is not exceeded. In particular, focus on the lower transmit power (below −50 dBm) because this is typically the limit of UE transmit power. If the UE transmit power is often at or below the lower limit, this could result in ineffective power control; specifically, the power control should maintain the transmit power within the dynamic range. When power control drives the UE outside the dynamic range, the UE becomes an Uplink interferer.
- **Uplink interference (SIB 7).** As a complement of UE transmit power, this value would indicate that UEs are not properly power-controlled and therefore generate Uplink interference. Uplink interference can also be caused, in an indoor system, by intermodulation components produced by a poorly functioning neutral host system, or simply by other system leakage (similar to adjacent channel interference).

Indoor systems have less tuning flexibility than macrocells, because most equipment cannot be relocated or reoriented. However, indoor systems often have the advantage of being quite isolated compared to the macrocell environment. As a result, they have fewer constraints and fewer of the complex interactions that accompany macrocell tuning, where a change in one cell affects many surrounding cells. Changes for an indoor system include adding/changing attenuators, couplers, splitters, and antennas. Major antenna location changes are not common because this requires negotiation with several concerned parties.

In some rare cases, intermodulation between different systems sharing the same DAS may arise, requiring the installation of additional band-limited filters and equipment changes. This can be serious and should be anticipated before deployment. It can be very tricky to troubleshoot such a problem in a live system.

8.4.3 How Indoor Parameter Settings Differ from Outdoor Systems

Enhancing indoor coverage with repeaters or indoor Node Bs means that UEs move across the boundaries of indoor and outdoor coverage. To maintain quality of service both indoors and outdoors at the coverage borders, the boundary transitions for Idle and Connected Modes must be examined.

Because of Link Budget differences, macrocell and indoor cell coverage quality can be different outdoors and indoors. In Idle Mode, the UE monitors only the serving cell and neighbor cells, so the primary concern is to ensure that cell reselection is smooth and successful. This includes reacquiring the new indoor/outdoor cell without going to an Out-of-Service (OOS) state and performing a full search. Such a condition results in long periods during which the UE is not reachable, and should be avoided if possible. It is common practice to plan the indoor coverage to overlap with the macronetwork around the entrances. This ensures that sufficient time is available for cell reselection when a user enters the building.

In Connected Mode, on the other hand, the handover area should be minimized to conserve resources. In addition, in Connected Mode, the path loss to the serving cell may change abruptly when transitioning from outdoors to indoors. Closed loop power control adjusts the UE transmit power, potentially resulting in abrupt transmit power changes. A well-designed indoor system limits these power differences to minimize transitions and eliminate UL blocking or degradation of link quality.

In practice, the settings for either reselection or handover between the indoor and macronetworks greatly depend on the selected carrier and the flexibility of setting parameters in the network. Carrier selection can be divided into two categories: same carrier or different carrier.

When the same carrier is selected, the reselection between both systems is primarily influenced by the offset (*Qoffset*, as described in Chapter 4). For handover, if parameters cannot be set at a cell level, the main parameter available is the Cell Individual Offset (CIO) because all other parameters would be shared. In either case, the offset can delay or speed up the transitions. Either one can have drastic effects. Speeding up the transition, by setting a positive offset, increases the risk of "ping-pong effect" for reselection but, more detrimentally, it increases the resource utilization in Connected Mode because the Active Set would likely increase. Delaying the transition, by setting a negative offset, affects the quality.

For reselection, delaying the transition increases the risk of the serving cell's quality falling below Qqualmin. Similarly, delaying intra-frequency handover would degrade the geometry at the border, negatively affecting the cell capacity. In either case, to find the correct trade-off, the serving cell quality versus the reselection rate for Qhyst and Active Set size versus total Downlink power for handover are weighed. When handover parameters can be set on a per-cell basis, the reporting range can be reduced and the Time-to-Trigger can be increased, compared to macrocells. These settings are only possible on the indoor cell; thus they affect the handover from indoors to outdoors. For handover from outdoors to indoors, different measurement controls for different cells are not possible. Any intelligence in the handover decisions must be implemented at the RNC level.

For same-carrier selection, a special case occurs when multiple RF carriers are deployed on the macronetwork. In this case, deploying all the RF carriers indoors as well would simplify mobility management. However, this drives up the cost and may not increase capacity as much as deploying a singular RF carrier on the indoor system. Deploying only one carrier indoors means using two different mobility management strategies for the two different carriers. Because most traffic is expected to come from indoor locations, this leads to unbalanced loading between carriers. This would have to be addressed via inter-frequency transitions. Ultimately, the transition from outdoors to indoors is dictated by coverage, while the transition from indoors to outdoors is dictated by load. The coverage-based transition in this same-carrier case would be equivalent to the need for coverage-based transitions in the different-carrier case.

Coverage-based transitions are required in both directions when different carriers are used indoors and outdoors. The main advantage of this deployment is the limited other-cell interference for both indoor and outdoor layers, at the expense of having to set up intercarrier transitions. Unlike the same-carrier case, transition can be delayed (through negative offsets or a timer) without a major impact on geometry and thus signal quality. The difference is that other-cell interference, on different carriers, would not be a major factor. In this condition, the only issue is ensuring that sufficient time is available to complete the transition before the signal quality degrades beyond Qrxlevmin in Idle Mode, or to a point where high BLER or dropped calls occur in Connected Mode.

In both the same-carrier or different-carrier cases, one task of mobility management is setting appropriate Neighbor Lists. If only a few buildings have indoor systems, the limitation of 31 neighbors per cell is unlikely to be an issue if the macronetwork is well optimized. In the macrolayer Neighbor Lists, with up to 16 neighbors being common, as many as 15 indoor neighbors would be possible. On the other hand, in dense urban areas, where controlling cell overlap is difficult and large numbers of buildings or subway stations must be covered, the 31-neighbor limitation may be reached, especially in the same-RF-carrier case. If a separate RF carrier is used indoors, 32 inter-frequency neighbors can be defined for each additional carrier, making limitations unlikely.

References

[1] Kishino Y. 3G indoor strategy. DoCoMo Engineering presentation; *IIR'05*; Jul. 2005; Las Vegas.
[2] Blackard KL, Feuerstein MJ, Rappaport TS, Seidel SY, Xia HH. Path loss and delay spread models as functions of antenna height for microcellular system design. *1992 IEEE Vehicular Technology Conference*; May 1992; Denver, CO.
[3] ATOLL Global RF Planning Solution. User Manual 2.2.0.
[4] Hugl K, Uykan Z. HSDPA system performance of optical fiber distributed antenna systems in an office environment. *PIMRC 2005*; Sep. 2005; Berlin.
[5] Johnson C, Morris P, Fraley D. Indoor WCDMA solutions co-channel with the macrocell layer. *Fifth IEE International Conference on 3G Mobile Communication Technologies (3G 2004)*; Oct. 18–20; 2004, London, UK, pp 469–473.
[6] Perez-Romero J, Sallent O, Agusti R. Impact of indoor traffic on W-CDMA capacity. *15th IEEE International Symposium on Personal, Indoor and Mobile radio Communications, PIMRC 2004*; Sep. 5–8 2004, Barcelona, Spain, v 4, pp 2861–2865.
[7] Xia HH, Herrera A, Kim S, Rico F. A CDMA-distributed antenna system for in-building personal communications services. *IEEE J. Sel, Areas Commun*; May 1996, Boluder, , v 14, n 4, pp 644–650.
[8] Vargas Jr. R, Victor EB, Baker KR. Polarization diversity for indoor cellular and PCS CDMA reception. *Proceedings of the IEEE 47th Vehicular Technology Conference*; May 4–7 1997, Phoenix, Az, USA, v 2, pp 1014–1018.
[9] Delogne P. *Leaky Feeders and Subsurface Radio Communications*. Peregrinus; 1982.
[10] Motley AJ, Palmer DA. Directed radio coverage within buildings. *Radio Spectrum Conservation Techniques Conference*; Sep. 6–8 1983, Boulder, CO, USA, pp 56–60.
[11] 25.101. UE Radio transmission and reception (FDD). 3GPP. 2004.
[12] Parsons JD. *The Mobile Radio Propagation Channel*. New York: John Wiley & Sons; 1992.
[13] *WinProp Documentation: Propagation Models Background Information, Version 5.43*. AWE Communications Gmbh; Oct. 5 2002.
[14] Turkmani AMD, Parsons JD, Lewis DG. Measurement of building penetration loss on radio signals at 441, 900 and 1400 MHz. *J. IERE*; 1988, vol. 58, no. 6, pp S169–S174.
[15] de Toledo AF, Turkmani AMD. Propagation into and within buildings at 900, 1800 and 2300 MHz. *IEEE 42th Vehicular Technology Conference*; May 10–13 1992; v 2, pp 633–636.
[16] Jakes WC, editor. *Microwave Mobile Communications*. New York: John Wiley & Sons; 1974.
[17] Suzuki H. A statistical model for urban radio propagation. *IEEE Trans. Commun.*; Jul. 1977; v 25, pp 673–680.
[18] Hashemi H. The indoor radio propagation channel. *Proc. IEEE*; Jul. 1993; vol. 81, no. 7, pp 943–968.
[19] Aulin T. Characteristics of a digital mobile radio channel. *IEEE Trans. Veh. Tech.*; May 1981; v 30.
[20] Rec ITU-R. Propagation data and prediction methods for the planning of indoor radio communication systems and radio local area networks in the frequency range 900 MHz to 100 GHz. 2001; pp 1238–2.

[21] Pahlavan K, Ganesh R. Statistical characterization of a partitioned indoor radio channel. *Proceedings of the IEEE International Conference on Communications ICC'92*; Jun. 14–17 1992; Chicago, IL. pp 1252–1256.

[22] Zaghloul H, Morrison G, Tholl D, Davis RJ, Kazeminejad S. Frequency response measurements of the indoor channel. *Proceedings from the ANTEM'90 Conference*; Aug. 1990; Winnipeg, Manitoba. pp 267–272.

[23] Murray RR, Arnold HW, and Cox DC. 815 MHz radio attenuation measured within a commercial building. *IEEE Antennas Propagation Internationl Symposium*; Jun. 8–13 1986; Philadelphia, PA. vol. 1, pp 209–212.

[24] Akerber D. Properties of a TDMA pico cellular office communication system. *Proceedings from IEEE Globecom'88*; Dec. 1988, Hollywood, FL. pp 1343–1349.

[25] Devasirvatham DMJ, Banerjee C, Murray RR, and Rappaport DA. Four frequency radiowave propagation measurements of the indoor environment in a large metropolitan commercial building. *Proceedings from IEEE GLOBECOM'91*, Phoenix, Az, USA, pp 1282–1286.

[26] Patsiokas SJ, Johnson BK, and Dailing JL. Propagation of radio signals inside buildings at 150, 450 and 850 MHz. *Proceedings from the IEEE Vehicular Technical Conference*; 1986, Montreal, Canada, pp 66–71.

[27] Patsiokas SJ, Johnson BK, Dailing JL. The effects of buildings on the propagation of radio frequency signals. *Proceedings of the IEEE International Conference on Communications*; 1987. pp 63–69.

[28] Motley AJ., Keenan JM Radio coverage in buildings. *Br. Telecom J*; Jan. 1990; v 8, n 1, pp 19–24.

[29] Pratt T, Bostian C, Allnutt J. Chapter 4. *Satellite Communications*. 2nd ed. New York: John Wiley & Sons; 2003.

[30] Aguirre S, Loew LH, Yeh L. Radio propagation into buildings at 912, 1920, and 5990 MHz using microcells. *Universal Personal Communications, Third Annual International Conference*; Sep. 27-Oct 1; 1994, San Diego, CA, USA, pp 129–134.

[31] Rappaport TS, Seidel SY, Singh R. 900 MHz multipath propagation measurements for U.S. digital cellular radio telephone. *IEEE Trans. VT*; May 1990; v VT-39, n 2, pp 132–139.

[32] Rappaport TS, Seidel SY. Multipath propagation models for in-building communications. *Proceedings of the IEE Fifth International Conference on Mobile Radio and Personal Communications*; Dec. 1989; Warick. pp 69–74.

[33] Motley AJ, Keenan JMP. Personal communication radio coverage in buildings at 900 MHz and 1700 MHz. *Electron. Lett*; 1988; v 24, n 12, pp 763–764.

[34] Seidel SY, Rappaport TS, Feuerstein MJ, Blackard KL, Grindstaff L. The impact of surrounding buildings on propagation for wireless in-building personal communications system design. *IEEE 42nd Vehicular Technology Conference*; May 10–13 1992; Denver, Co, USA, v 2, pp 814–818.

Index

Printed and bound by CPI Group (UK) Ltd, Croydon, CR0 4YY

16/04/2025

14658555-0004